Manganese Catalysis in Organic Synthesis

Manganese Catalysis in Organic Synthesis

Edited by
Jean-Baptiste Sortais

WILEY-VCH

Editor

Prof. Jean-Baptiste Sortais
University of Toulouse III
CNRS (LCC) - UPR 8241
205 route de Narbonne
31077 Toulouse, cedex 4
France

Cover

Cover Image:
© RHJPhtotoandilustration/Shutterstock

Library of Congress Card No.:
applied for

British Library Cataloguing-in-Publication Data
A catalogue record for this book is available from the British Library.

Bibliographic information published by the Deutsche Nationalbibliothek
The Deutsche Nationalbibliothek lists this publication in the Deutsche Nationalbibliografie; detailed bibliographic data are available on the Internet at <http://dnb.d-nb.de>.

Print ISBN: 978-3-527-34730-8
ePDF ISBN: 978-3-527-82612-4
ePub ISBN: 978-3-527-82611-7
oBook ISBN: 978-3-527-82613-1

Typesetting Straive, Chennai, India
Printing
CPI Group (UK) Ltd, Croydon CR04YY

C100756-110821

Contents

Preface

Homogeneous catalysis using transition metal complexes has been in continuous development since the late 1960s. These great achievements have conferred to organometallic catalysis a central role in organic synthesis, as a powerful tool not only to promote new reactions but also to develop highly chemo-, regio-, and enantioselective transformations, resulting in the award of three Nobel Prizes in the field of homogeneous catalysis at the beginning of this century in less than 10 years (2001, 2005, and 2010). Most of the historic advances were realized with noble metals, laying the groundwork for basic reactions of organometallic chemistry and mechanisms in catalysis. However, over the last 20 years, the evolution of economic, geopolitical, and environmental constraints, as well as scientific curiosity, has paved the way for the use of non-noble transition metals in homogeneous catalysis. The chemistry of iron, the most abundant transition metal, was one of the first to be revisited in the light of sustainable development, and fantastic advances have been made in a short period of time. Nickel, cobalt, and copper have also experienced a resurgence of interest. On the opposite, while manganese compounds are generally known as stoichiometric oxidation reagents ($KMnO_4$ or MnO_2), their use in homogeneous catalysis has been less investigated compared with other late 3d transition metals, albeit manganese has several advantages. Manganese is indeed the third most abundant transition metal after iron and titanium, with an abundance of about 1000 ppm in the Earth's crust. Unlike cobalt, this metal is not considered a critical resource by the European Union, i.e. there is no supply risk, nor too much price volatility, which is a recurrent issue with precious metals. It is also a heavy metal with low toxicity, as evidenced by the oral permitted daily exposure (PDE) of 250 ppm according to the European Medicines Agency. This is an undeniable advantage when considering late state transformations in drug synthesis without extensive purification to remove metal residues. Finally, manganese has the widest range of oxidation states of the metals in the first row of block d, from −III to +VII, a large playground for developing efficient catalysts when combined with the properly designed ligand. All these appealing characteristics make manganese an ideal candidate to build up complementary catalysts compared with the other 3d metals.

The aim of this book is to give the reader a general overview of what is possible to achieve in homogeneous catalysis with manganese-based catalysts nowadays. It will address various aspects: from organometallic precursors to applications in organic synthesis. The areas covered, in terms of transformations, are spanning from reduction to oxidation; from C—C, C—N, and C—X bond formations; and from cross-coupling reactions to C—H bond activations, including photo- and electrochemical transformations. My personal feeling is that the state of the art regarding manganese can be compared to some extent with the case of iron a decade ago, when the book titled *Iron Catalysis in Organic Chemistry: Reactions and Applications* edited by Dr. Bernd Plietker was published by Wiley in 2008. I hope that reading the present book will be highly stimulating for specialists and non-specialists in the field.

As editor, I would like to warmly thank all contributors to this book, for their participation and their enthusiasm in this project and, in some cases, for their patience, as a global pandemic attempted to slow the project down. I also acknowledge the support of the staff from Wiley VCH, namely, Anne Brenfuehrer, Elke Maase, and Katherine Wong.

Toulouse, France
15th December 2020

1

Organometallic Manganese Compounds in Organic Synthesis

Alina A. Grineva[1,2], Noël Lugan[1], and Dmitry A. Valyaev[1]

[1] *LCC-CNRS, Université de Toulouse, CNRS, 31077 Toulouse cedex 4, France*
[2] *Kurnakov Institute of General and Inorganic Chemistry, Russian Academy of Sciences, 31 Leninsky Pr., Moscow, 119991 Russia*

1.1 Introduction

Organometallic manganese chemistry emerged in 1937 with the *in situ* generation of the first Mn(II) species, PhMnI[1], followed by the preparation of two other emblematic Mn(0) and Mn(I) derivatives, $Mn_2(CO)_{10}$ [2] and $CpMn(CO)_3$ [3] in 1949 and 1954, respectively. Despite a long history, this research field remained for many years a purely fundamental area with only sporadic applications in organic synthesis [4, 5] and homogeneous catalysis [6]. The situation started to change rapidly in the late 1970s, and Mn(II) σ-complexes soon became valuable tools for organic synthesis acting as soft and chemoselective nucleophilic reagents. Recent excellent reviews by Gérard Cahiez, one of the pioneers and key players in the field, are available on this topic [7], and therefore it will not be covered in our contribution. Even if the chemistry of Mn(I) complexes directed to organic synthesis is much less developed, some interesting results have been obtained during the last 30 years, but they have never been systematically reviewed. The main goal of this chapter is to provide an overview of the multiple facets of manganese organometallic chemistry, from the information on various metallic precursors and basic reactivity patterns to the design of Mn-containing chiral ligands and application of Mn(I)-mediated processes in organic synthesis.

1.2 Basic Manganese Precursors Relevant to Organometallic Chemistry

Unlike more electropositive alkaline and alkaline earth metals, commercial manganese powder undergoes the insertion across C—halogen bond to form organometallic RMnX species uniquely for highly activated allylhalides and α-halogenated esters under Barbier conditions [8]. Still, the reduction of Mn(II)

Manganese Catalysis in Organic Synthesis, First Edition. Edited by Jean-Baptiste Sortais.

salts MnX_2 or manganates $MnX_2 \times 2LiX$ (X = Cl, Br, I) with metallic Mg [9], Li/naphthalene [10] or Li/2-phenylpyridine [11] systems, and potassium graphite KC_8 [12] was shown to generate *in situ* a very reactive form of metallic manganese, capable of activating a variety of alkenyl-, aryl-, and heteroarylhalide substrates under mild conditions.

Commercially available anhydrous manganese salts MnX_2 represent the simplest manganese-containing starting materials. While $MnCl_2$ and $MnBr_2$ are very stable, MnI_2 is light sensitive, and its commercial form is often considered as too impure for given synthetic applications [7]. Fortunately, this compound can be easily prepared on demand upon reaction of Mn powder with iodine in ether [13]. The solubility of MnX_2 salts in ethereal solvents is markedly different and strongly decreases in $MnI_2 > MnBr_2 \gg MnCl_2$ order. Yet, when necessary, $MnCl_2$ can be easily transformed by refluxing in tetrahydrofurane (THF) into the pink solvated complex $MnCl_2 \times 2THF$ [14], which can be used for the synthesis of Mn(II) complexes, even in toluene [15].

Reactions of MnX_2 with organolithium RLi or Grignard RMgX reagents remain the most convenient route to organometallic Mn(II) complexes. In particular, mixed RMnX, homoleptic R_2Mn, and even manganate $[R_3Mn]^-$ or $[R_4Mn]^{2-}$ species can be obtained, depending on the reagents stoichiometry [7]. In general, their thermal stability follows the trend $[R_4Mn]^{2-} \approx [R_3Mn]^- > RMnX \gg R_2Mn$, and the less stable dialkylmanganese compounds typically decompose by a β-elimination process. However, some stable homoleptic Mn(II) σ-complexes such as $Mn_3(Mes)_6$ [16], (Mes – 2,4,6-trimethylphenyl), or $Mn_4(CH_2tBu)_8$ [17] can be conveniently produced on a multi-gram scale. In contrast to ubiquitous ferrocene, manganocene Cp_2Mn [18] (Cp = η^5-C_5H_5 herein and throughout the chapter) is not stable under ambient conditions due to the strongly ionic character of the Mn–Cp bond and thus has found quite a limited use in synthetic chemistry [19]. We can also mention the highly sensitive complex $[Mn_2(N(SiMe_3)_2)_4]$, yet accessible on a large scale [14], which has been applied recently to the preparation of some Mn(II) amidinate complexes [20] and well-defined clusters [21] using the ability of amide ligands to serve as internal base.

Manganese carbonyl $Mn_2(CO)_{10}$ is the foremost accessible Mn(0) compound, which can be easily transformed into Mn(I) halide complexes $Mn(CO)_5X$ (X = Cl, Br, I) upon the oxidation with the corresponding free halogen [22]. To date, commercially available air-stable $Mn_2(CO)_{10}$ and $Mn(CO)_5Br$ remain the most popular manganese precursors, which could be either directly used in catalysis [6, 23] or exploited for the synthesis of various Mn(I) precatalysts [24]. The chemistry of $Mn(CO)_5Cl$ and $Mn(CO)_5I$ is much less developed because of more difficult preparations and lower stability of these compounds. Half-sandwich Mn(I) complex $CpMn(CO)_3$, also known as cymantrene, and its methylated analogue $Cp'Mn(CO)_3$ (Cp′ = η^5-C_5H_4Me), belongs to a few examples of quite sophisticated transition metal organometallic compounds produced industrially at a scale of hundred tons per year as anti-knock gasoline additives. Although their use in such application is currently declining in many countries, technical grade $CpMn(CO)_3$ or $Cp'Mn(CO)_3$ are actually available at a cost 20–30 USD per kilogram from chosen suppliers, analytically pure samples suitable for laboratory use being eventually prepared upon recrystallization from hexane [25] or vacuum distillation, respectively. The availability and rich reactivity of these organometallic complexes discussed in Sections 1.4 and 1.5 make them attractive candidates for application in organic synthesis.

1.3 Overview of the Synthetic Chemistry for Main Classes of Mn(I) Complexes

Besides thermal CO ligand(s) substitution, the most important reactivity pattern of $Mn_2(CO)_{10}$ (**1**) deals with different types of activation of the metal–metal bond (Scheme 1.1). In particular, under UV (350 nm) or visible light (430 nm) irradiation, a smooth generation of the metal-centered radical **2** is achieved; this transformation has recently found numerous applications in photo-induced polymerization processes [26]. In addition to the oxidative Mn–Mn bond cleavage in **1**, induced by halogens already mentioned in Section 1.2, the reduction of this compound with sodium amalgam [27], Na/K alloy [28], or commercial trialkylborohydrides $M[BR_3H]$ [29] (M = Li, K) leads to the formation of anionic metallocarbonylate species **3**. Reactions of the latter with various electrophiles constitute a powerful approach to the synthesis of Mn(I) σ-complexes $Mn(CO)_5E$ ($\mathbf{4^E}$). Upon protonation of **3** with aqueous H_3PO_4, the hydride complex $\mathbf{4^H}$ can be isolated as a volatile liquid [30]. However, due to its extreme air sensitivity and toxicity, $\mathbf{4^H}$ is preferably generated and used *in situ* [31]. Substitution of one or two CO groups in $\mathbf{4^H}$ by P-donor ligands strongly improves the stability of the resulting hydrides $Mn(CO)_{5-n}(L)_nH$ [31, 32]. The alkylation of **3** with MeI proceeds easily to form air-stable $\mathbf{4^{Me}}$, which has been recently applied for expedient preparation of 16-electron Mn(I) complexes of bifunctional pincer ligands [28]. Similarly, the acylation affords isolable σ-acyl derivatives $\mathbf{4^{COAr}}$, which upon heating undergo selective CO deinsertion, providing a viable route to Mn(I) aryl complexes $\mathbf{4^{Ar}}$ [33].

Carbonyl ligand substitution plays a central role in the chemistry of $Mn(CO)_5Br$ (**5**) in the context of the synthesis of various Mn(I) precatalysts (Scheme 1.2) [24]. While the first CO ligand can be easily replaced at room temperature by a donor ligand L to form *cis*-$Mn(CO)_4(L)Br$ (**6**) [34], heating at 50–60 °C is often required to remove the second CO group. For some monodentate ligands, besides the kinetically controlled formation of *fac*-**7** species, formation of more thermodynamically stable *mer,trans*-**7** products can eventually be observed at higher temperatures [35]. Chelating ligands necessarily lead to *fac*-isomers of complexes **8**. The thermal reaction of **5**

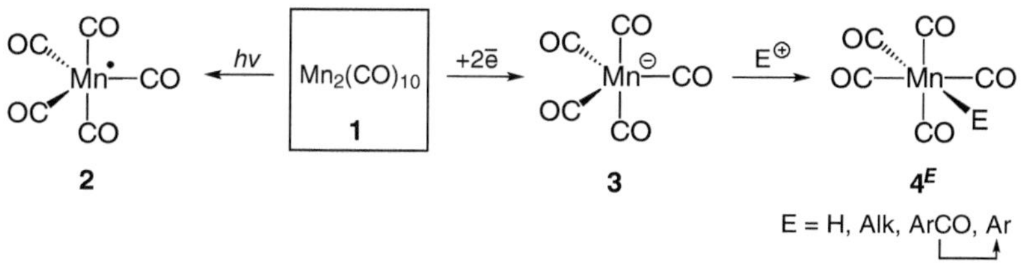

Scheme 1.1 Photo-induced and reductive cleavage of Mn—Mn bond in $Mn_2(CO)_{10}$ and the associated reactivity with electrophiles.

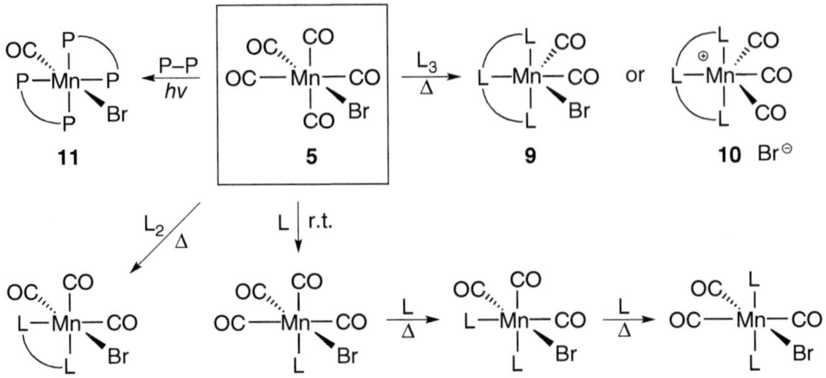

11 5 9 10 Br⊖

fac-8 cis-6 fac-7 mer,trans-7

Scheme 1.2 Substitution of CO groups in $Mn(CO)_5Br$ with various donor ligands. Source: Garbe et al. [24a]; Mukherjee et al. [24b].

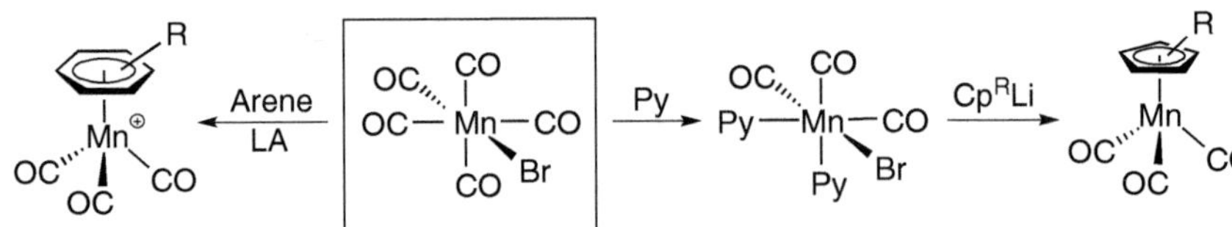

12 5 13 14

Scheme 1.3 Synthesis of half-sandwich Mn(I) complexes from $Mn(CO)_5Br$

with L_3 pincer-type systems can afford either the neutral dicarbonyl $Mn(CO)_2(L_3)Br$ (**9**) or the cationic tricarbonyl $[Mn(CO)_3(L_3)]Br$ (**10**) complexes upon substitution of three CO or two CO and Br^- ligand combinations, respectively [24]. However, the formation of monocarbonyl Mn(I) species *trans*-$Mn(P–P)_2(CO)Br$ (**11**) from **5** and chelating diphosphines involving the substitution of four CO ligands can be achieved only under UV irradiation [36].

Complex **5** is also a useful precursor for the synthesis of half-sandwich Mn(I) complexes (Scheme 1.3). Cationic π-arene complexes $(\eta^6\text{-}RC_6H_5)Mn(CO)_3^+$ (**12**) are easily obtained upon heating **5** with the corresponding aromatic compound in the presence of Lewis acids as bromide scavengers [37]. While for simple arenes, cheap $AlCl_3$ can be typically applied for the generation of "$[(CO)_5Mn^+]$" intermediate, the use of silver salts is essential for a better functional group tolerance [38]. Alternatively, the $[Mn(CO)_5]^+$ synthon can be prepared by the oxidation of $Mn_2(CO)_{10}$ with strong acids under heating in $(CF_3CO)_2O$ [37]. Direct reaction of **5** with cyclopentadienide salts led mostly to its reduction to $Mn_2(CO)_{10}$, but the use of its easily available bis-pyridine-substituted derivative **13** [39] allowed the efficient preparation of substituted cymantrene derivatives **14**, for example, $Cp^*Mn(CO)_3$ ($Cp^* = \eta^5\text{-}C_5Me_5$) [40].

Cymantrene **15** can undergo reversible one-electron oxidation at +0.92 V vs. Fc/Fc^+ in a CH_2Cl_2 medium, but the resulting radical cation **16** is much less stable than ferrocenium and therefore has been isolated only recently, using non-nucleophilic $B(C_6F_5)_4^-$ anion (Scheme 1.4) [41]. As expected, the incorporation

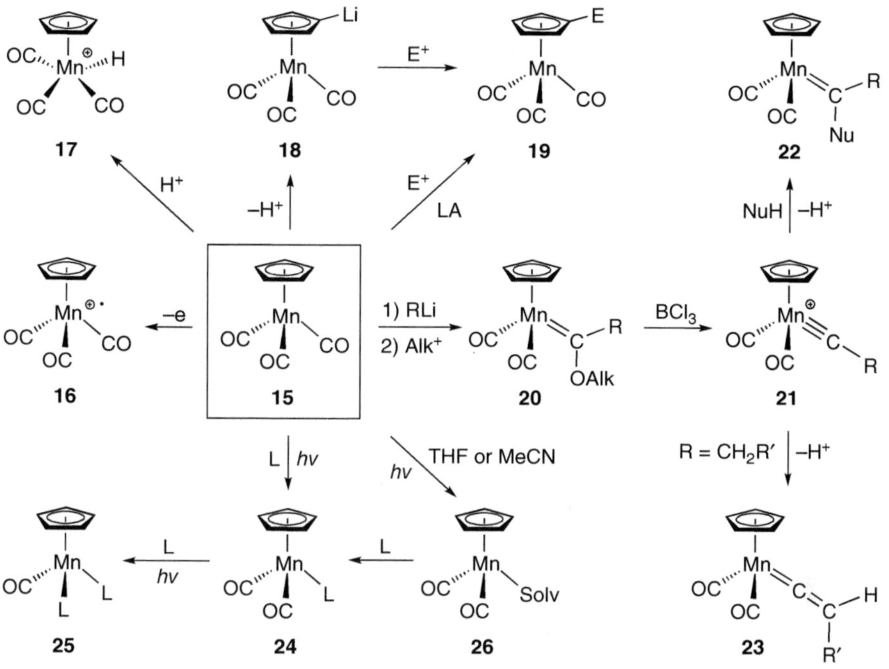

Scheme 1.4 General overview of reactivity for cymantrene $CpMn(CO)_3$. Source: Based on Laws et al. [41].

of more donating ligands into the cymantrene moiety decreased the oxidation potentials [42], which can reach −0.3 V vs. Fc/Fc^+ in CH_2Cl_2 for $Cp(CO)_2Mn(NHC)$ derivatives [43], and increased the stability of the resulting 17-electron radical cations.

Reversible protonation of **15** to form the cationic hydride **17** can be achieved only at −100 °C in HSO_3F/CH_2Cl_2 mixture [44]. For more basic analogues such as CpMn(CO)(P—P) bearing chelating diphosphine ligands, even the strength of CF_3COOH (added in excess) is sufficient for the spectroscopic detection of the corresponding hydride species at room temperature [45]. In line with this trend, hydride complex $[CpMn(PMe_3)_3H]BF_4$ obtained by protonation of the corresponding triphosphine precursor could effectively be isolated [46]. Deprotonation of the Cp ligand in **15** upon treatment with alkyllithium reagents and subsequent reaction of the lithiated derivative **18** with various electrophiles represent the main strategy for the modification of the Cp ring in cymantrene [47]. The selectivity in the reaction strongly depends on the choice of base, solvent, and reaction temperature, providing under optimized conditions (*n*BuLi, THF, −80 °C, 1.5–2 hours [48a] or *t*BuLi, THF, −80 °C to −50 °C, 30 minutes [48b]), the Cp-substituted derivatives **19** in 80–90% yield. An alternative approach for the **15** → **19** transformation consists in the use of Friedel–Crafts electrophilic substitution processes. While this method is synthetically valuable for preparation of cymantrenyl (Cym) ketones [49], it generally suffers from a low regioselectivity [50].

Reactions of **15** with organolithium reagents *at room temperature* in ether lead to the selective nucleophilic attack of a carbonyl ligand to form acyl derivatives, the alkylation of which affords Fischer carbenes **20** in good yield [51]. The alkoxy group in the latter can be smoothly removed by the treatment with excess of BCl_3 at low temperature to form cationic carbyne complexes **21** [52]. While complexes $[Cp^{(\prime)}(CO)_2Mn{\equiv}C{-}R]BCl_4$ typically start decomposing above −20 °C and are extremely sensitive to air and moisture, replacing the BCl_4^- counteranion by non-nucleophilic BPh_4^- considerably improves the stability of the resulting salts, allowing, for instance, the isolation of $[Cp^{(\prime)}(CO)_2Mn{\equiv}C{-}Ar]BPh_4$ on *ca.* 10 g scale as a solid, perfectly stable under ambient conditions [53]. Complexes $[\mathbf{21}]^+$ are highly electrophilic species reacting in solution with a wide variety of nucleophiles to form the corresponding Fischer carbenes **22** (see Section 1.5.3 for some examples). Deprotonation of alkyl-substituted derivatives **21** with Et_3N leads to the quantitative *in situ* formation of the manganese vinylidene complexes **23** [52b].

In contrast to $Mn(CO)_5Br$, the substitution of CO ligand in $CpMn(CO)_3$ for two-electron donors requires a photochemical activation ($\lambda < 400$ nm). Again, the replacement of the first CO ligand to get monosubstituted derivatives **24** proceeds easily in aliphatic or aromatic solvents, whereas the formation of the disubstituted species **25** is much less common and can be achieved only with a few phosphines or arsines. The direct substitution of *all* CO ligands in **15** is known uniquely for the molecules having significant π-acidity such as PF_3 [54] or isocyanides [55], whereas carbonyl-free half-sandwich Mn(I) complexes bearing donor phosphorous ligands can be alternatively prepared from Cp_2Mn or $CpMn(\eta^6\text{-}C_7H_8)$ by thermal substitution of the π-ligand [56]. Whether the resulting dicarbonyl products **24** are unstable under irradiation conditions or the reaction proceeds with low selectivity between **24** and **25**, an alternative approach consisting in the preparation of solvate complexes **26** can be employed. In most cases, thermally unstable $Cp(CO)_2Mn(THF)$ is generated *in situ* under UV irradiation at low temperature to provide the best conversion of the starting material, and then the easily dissociating THF ligand in **26** is replaced upon warming to room temperature by the given incoming ligand L. Alternatively, it is possible to use the acetonitrile complex $Cp(CO)_2Mn(NCMe)$, which is isolable on 10 g scale and air-stable in the solid state [57]. While the MeCN ligand substitution can be performed simply under heating, in some cases this reaction may advantageously be carried out instantaneously at room temperature in the presence of a small amount of the oxidant ($Cp_2^{Ac}FeBF_4$) [57] via a redox catalytic process [42b].

Thanks to a persistent interest of many researchers to fundamental organometallic chemistry since 1960s, now a complete toolbox for "on demand" synthesis of various Mn(I) complexes is available. However, it should be pointed out that this chemistry is sometimes quite delicate and often requires a rigorous purification of the target Mn(I) products in order to obtain NMR data of good quality. Indeed, trace amounts of paramagnetic Mn(II) decomposition products lead to a significant line broadening (sometimes with singlet signal half-width up to 200–300 Hz) in ^1H spectra, thus precluding reliable NMR characterization.

1.4 Planar Chiral Ligands Based on Cymantrene Scaffold

Though the overwhelming majority of bidentate ligands having planar chirality elements are based on ferrocene [58], some interesting results were also obtained from cymantrene scaffold. In 1998, looking for efficient chiral ligands for Pd-catalyzed allylic substitution with cyclic substrates, Helmchen et al. designed the phosphine–oxazoline ligands **28** (Scheme 1.5). Diastereoselective C–H lithiation of the oxazoline-substituted cymantrene **27**, available in two steps from CymCOCl and (*S*)-*tert*-leucenol in *ca.* 70% overall yield, followed by quenching the organolithium intermediate with chlorophosphines afforded the mixture of epimeric P-chirogenic phosphines **28** and **28′**, which were separated by crystallization and column chromatography [59].

The configuration of the phosphorous atom in these ligands emerged as a crucial factor for the chirality induction in Pd-catalyzed allylic substitution of cyclic substrates. While the use of **28b′** led to only 30% enantiomeric excess (ee), comparable to a similar system showing a benzene spacer between phosphine and oxazoline moieties [60], the catalytic system based on **28b** provided excellent results with 98% ee (Scheme 1.6). Further increase of the steric hindrance (**28a**, 44% ee) or variation of the electronic properties of the aryl substituents (**28c**, 87% ee; **28d**, 92% ee) resulted in a lower asymmetric induction. The outstanding efficiency of the ligand **28b** was rationalized by the stabilization of its specific conformation in the cationic complexes **29** (Scheme 1.6), in which the bulky $[Mn(CO)_3]$ fragment forces the 2-biphenylyl group to point toward the π-allyl ligand. The ligand **28b** was then successfully used in the allylic substitution with non-stabilized enolates [61] and for the preparation of building blocks for natural product synthesis [62].

More recently, Kamikawa et al. reported a series of chiral phosphine–alkene ligands **36** (Scheme 1.7) [63]. Starting from optically active acetal **30**, easily available from CymCHO and (*S*)-1,2,4-butantriol [64], the aldehyde (*S*)-**32** was obtained in more than 99.9% enantiomeric purity by a sequence of metallation, bromination, and deprotection steps. The subsequent Wittig olefination, photochemical CO

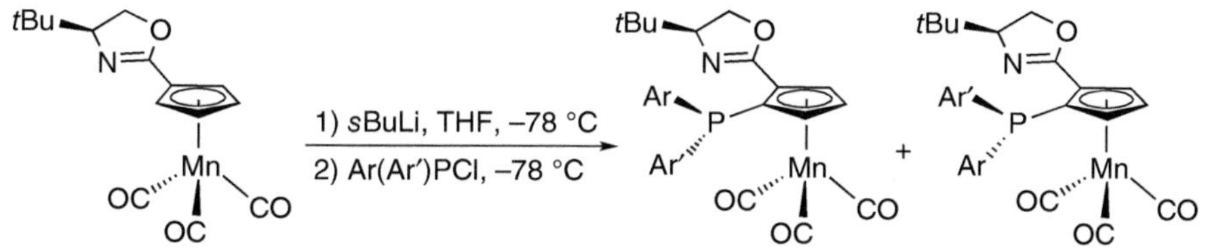

27 **28a–d′** **28a–d**

	28a–d′	28a–d
a: Ar = Ar′ = 2-C_6H_4Ph	–	28%
b: Ar = Ph; Ar′ = 2-C_6H_4Ph	31%	26%
c: Ar = Ph; Ar′ = 3,5-$C_6H_3Me_2$	n.r.	32%
d: Ar = Ph; Ar′ = 3,5-$C_6H_3(CF_3)_2$	n.r.	31%

Scheme 1.5 Synthesis of planar chiral phosphine-oxazoline ligands **28** based on cymantrene scaffold.

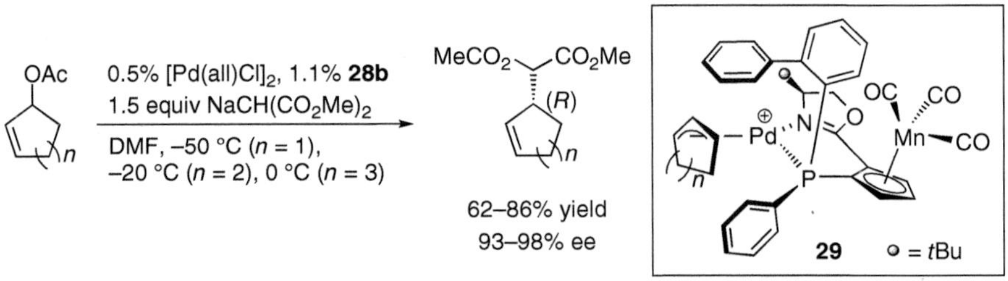

Scheme 1.6 Asymmetric allylic substitution catalyzed by Pd(II) complex bearing cymantrene-based chiral phosphine-oxazoline ligand **28b**.

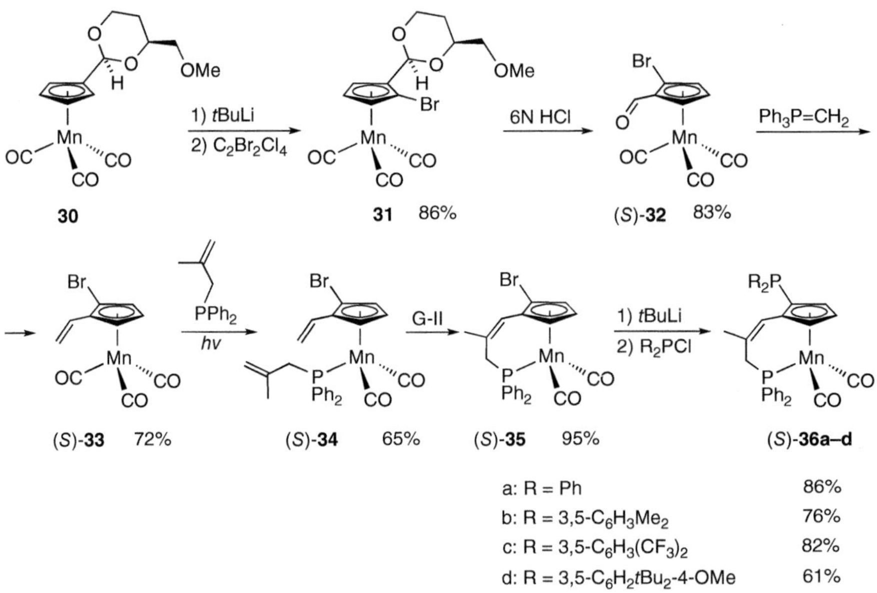

Scheme 1.7 Synthesis of planar chiral phosphine-alkene ligands **36** based on cymantrene scaffold. Source: Based on Kamikawa et al. [63].

substitution for (methallyl)PPh_2, and ring closing metathesis afforded the key intermediate (*S*)-**35**, which was finally transformed into the target ligands (*S*)-**36** using Grubbs-II (G-II) catalyst bearing different substituents at the phosphorous atom. The corresponding enantiomers (*R*)-**36** were prepared either in the same way starting from (*R*)-**30**, or through the separation of racemic **35** by chiral preparative HPLC.

Phosphine–alkene ligands **36** have appeared extremely efficient in a variety of Rh-catalyzed arylation processes (Scheme 1.8) [63]. In particular, the $[Rh(C_2H_4)Cl]_2$/**36b** rhodium system catalyzes the 1,4-arylation of both cyclic and acyclic enones with arylboronic acids in very high yields and typically more than 99% ee, whereas the use of less bulky **36a** or very crowded **36d** ligands provided only modest enantioselectivity (30–60% ee). In contrast, 1,2-arylation of arylimines using phenylboroxine can be more efficiently performed using ligands **36c** and **36d**. Finally, in the highly challenging 1,2-arylation of aromatic aldehydes, the bulkiest ligand **36d** was the most efficient, providing the best induction level reported to date.

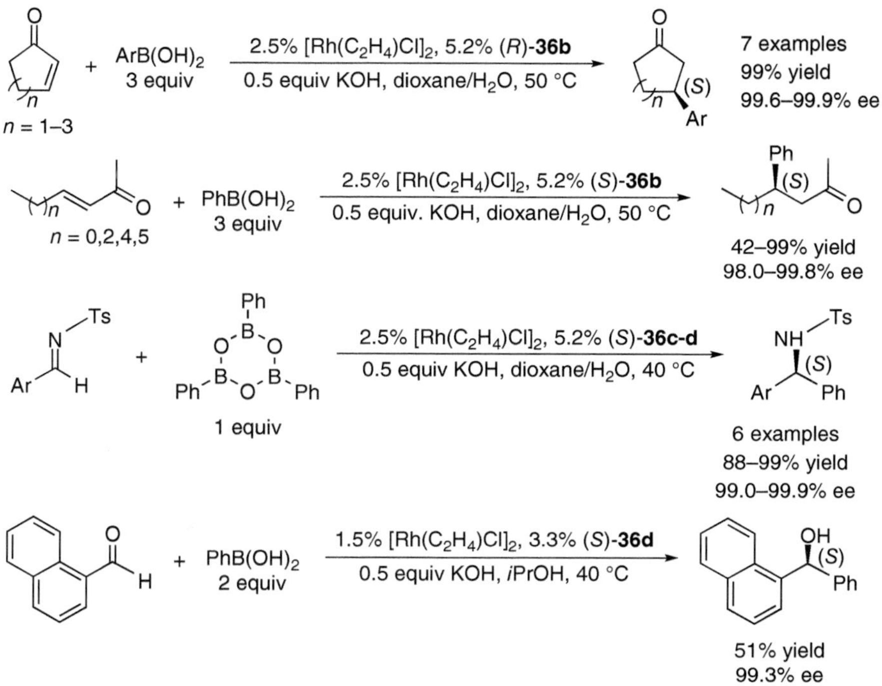

Scheme 1.8 Asymmetric arylation of enones, imines and aldehydes catalyzed by Rh(I) complexes bearing cymantrene-based chiral phosphine-alkene ligands **36**. Source: Based on Kamikawa et al. [63].

Later on, the same group published an alternative strategy for the preparation of enantiopure alkene-bromide derivatives **35** using kinetic resolution of the racemic complex **34** with optically active Schrock metathesis catalysts (Scheme 1.9) [65]. The catalyst (*S*)-**Mo1** was highly discriminant, cyclizing mainly (*S*)-**34** at the expense of the other enantiomer (selectivity factor $k_{rel} = 127$). The cyclization of the recovered (*R*)-**34** (98% ee) using antipodal (*R*)-**Mo1** catalyst afforded the analogous complex (*R*)-**35** (44% yield from *rac*-**34**, >99.9% ee), which was used for the preparation of planar chiral diene ligand (*R*)-**36** similarly as it was described earlier. Contrary to the previous family of ligands, (*R*)-**36** was much less efficient in Rh-catalyzed 1,4-arylation of cyclohexenone under the same conditions (Scheme 1.8), both in terms of activity (23% yield) and enantioselectivity (52% ee).

Chiral molybdenum alkylidene complex (*R*)-**Mo2** was used for the desymmetrization of C_s-symmetric phosphacyclopentadienyl manganese complexes **38** to give quantitatively the corresponding cyclized products (*R*)-**39** with excellent optical purity (Scheme 1.10) [66]. The preliminary tests for these alkene-phospha Cp ligands in Pd-catalyzed allylic substitution demonstrated the potential utility of these new class of ligands in asymmetric catalysis.

Although ferrocene and cymantrene virtually display the same synthetic chemistry at the Cp ligand, the design of cymantrene-based chiral ligands and their application in homogeneous catalysis still are in their infancy. Yet, cymantrene-derived ligands can have a significant potential of their own due to an easy modulation of the steric bulk of the metal moiety by CO substitution and the possibility to create additional chirality at the metal atom.

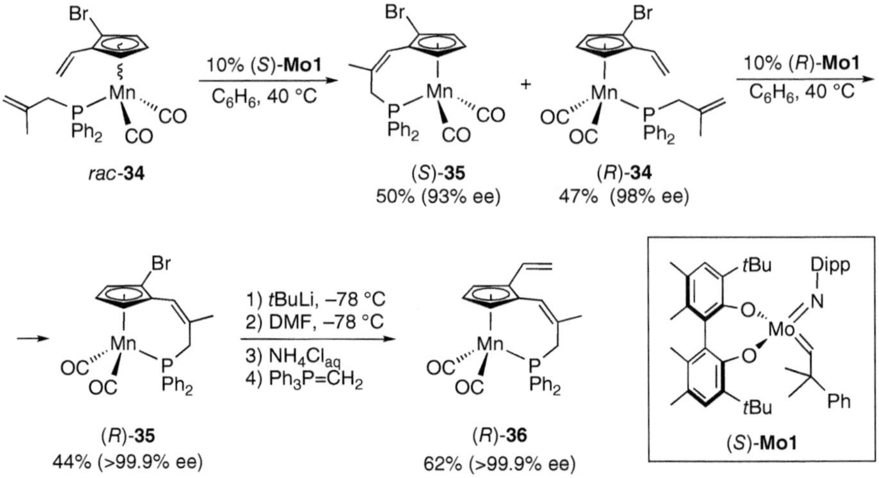

Scheme 1.9 Synthesis of planar chiral diene ligands 36 using kinetic resolution with Mo-based ofefin metathesis catalyst as a key step Source: Based on Ogasawara et al. [65].

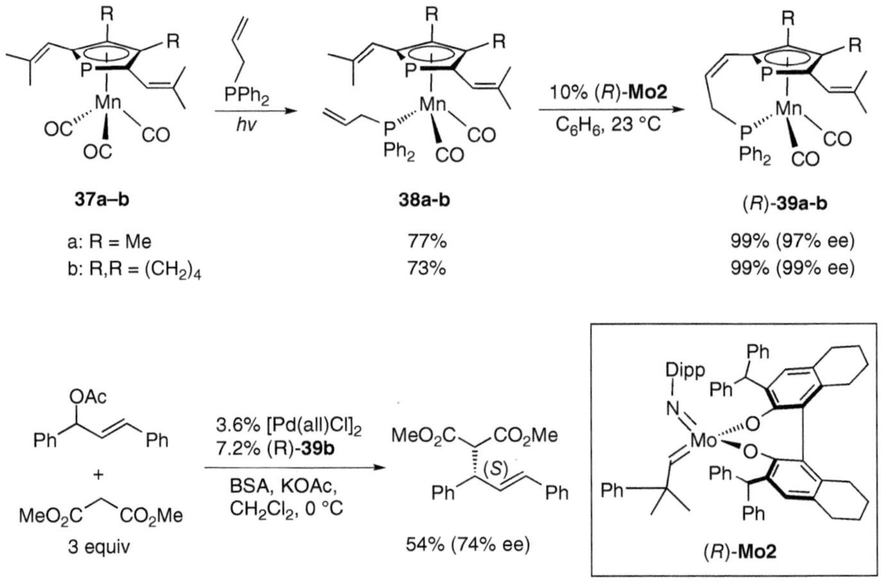

Scheme 1.10 Synthesis of planar chiral phosphacyclopentadienyl-alkene ligands **39** and their evaluation in Pd(II)-catalyzed asymmetric allylic substitution.

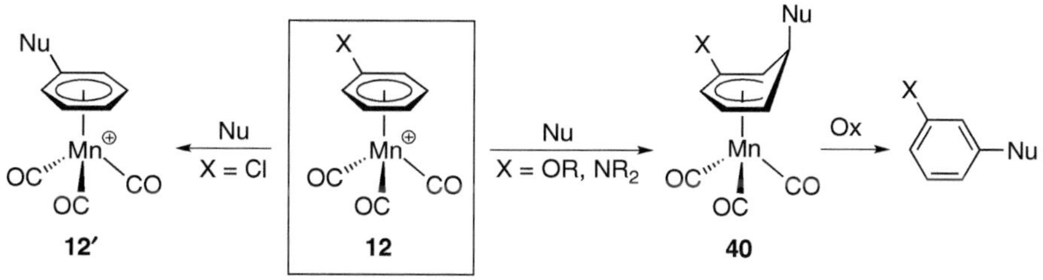

Scheme 1.11 Reactions of cationic Mn(I) π-arene complexes with nucleophiles.

1.5 Mn(I)-Mediated Transformations in Organic Synthesis

1.5.1 Ring-Centered Reactivity in Half-Sandwich Mn(I) π-Complexes

In addition to the S_NAr substitution (Scheme 1.11, left), nucleophilic *C*-attack on π-coordinated arene ligands in complexes **12** to form selectively *exo*-η^5-cyclohexadienyl products **40** is of primary importance for these compounds from a practical point of view (Scheme 1.11, right) [67]. Noteworthy, due to their overall positive charge, Mn(I) complexes **12** are much more electrophilic than the more studied (π-arene)$Cr(CO)_3$ species, which lead to an acceleration of S_NAr processes [68] and allow the use of weaker nucleophiles (Grignard reagents, enolates, etc.) in the preparation of **40** [69]. While in the latter process, the directing effect of a simple carbon or halogen substituents is rather small, highly selective *meta*-addition that is a characteristic of phenol and aniline-based π-ligands [69], provided interesting applications in organic synthesis [70, 71]. The resulting functionalized arenes can be easily displaced from neutral complexes **40** by oxidation either with Jones reagent [69], 2,3-dichloro-5,6-dicyanobenzoquinone (DDQ) [70], or *N*-bromosuccinimide (NBS) [71] under mild conditions, the latter procedure being compatible with some sensitive functional groups.

The representative application of this methodology to the synthesis of deoxyristomycinic acid derivative **45**, a structural motif presented in Ristocetin antibiotics family, is illustrated in Scheme 1.12 [71a, b]. The reaction of complex **41** with phenoxide **42** in the presence of silver salt as chloride scavenger afforded in good yield the diarylether derivative **43**, which was treated with Schöllkopf's nucleophile to form the η^5-cyclohexadienyl product **44**. Separation of the diastereomers, followed by the oxidative decomplexation and deprotection of Boc group, led to the enantiomerically pure compound **45**. This approach was later used for the preparation of even more elaborated polysubstituted aromatic structures containing sensitive functional groups [71c, e].

Treatment of the π-arene complex **46** (R = R′ = H) with the Evans's enolate **47** afforded complex **48** in a 9 : 1 dr, the major diastereomer being isolated in 75% yield (Scheme 1.13) [70]. The cleavage of the chiral auxiliary in the latter, followed by demetallation and Bn group removal, afforded 2-phenylpropionic acid **50** in more

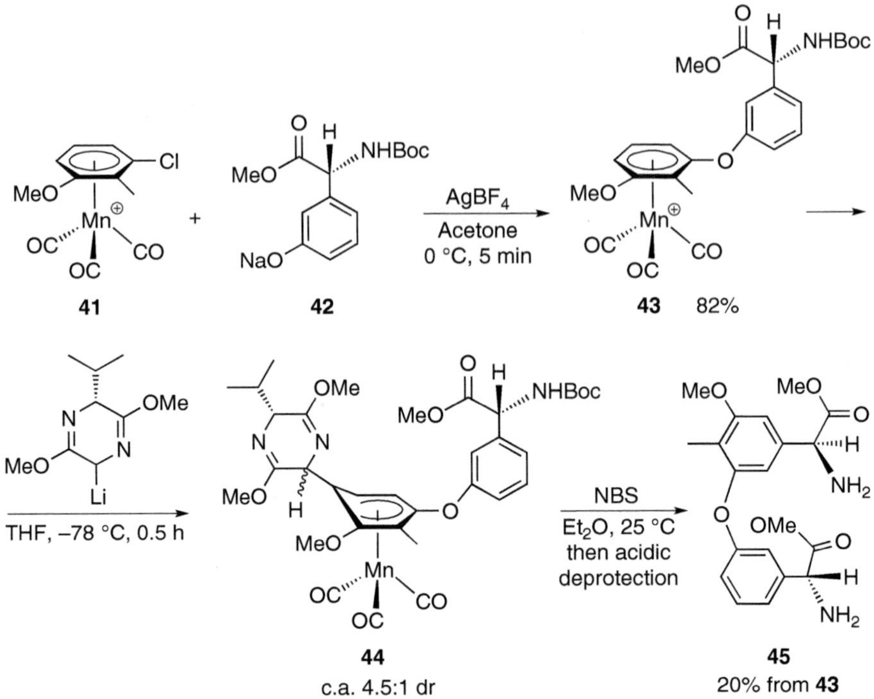

Scheme 1.12 Synthesis of deoxyristomycinic acid derivative **45** from Mn(I) π-arene complex **41**. Source: Based on Pearson et al. [71].

than 95% ee. Lower diastereoselectivity level observed for η^5-cyclohexadienyl complexes **48**, derived from substituted π-arene precursors, actually did not affect the overall synthesis of analogous acids **50**, isolated in similar yield and optical purity, as the destruction of the second chiral center occurred during the oxidative decomplexation of **49**.

Besides oxidative decomplexation to release free arenes, Mn(I) η^5-cyclohexadienyl complexes can be selectively transformed back into π-arene derivatives, typically using *exo*-hydride abstraction by the Ph_3C^+ cation [68, 72]. Noticeably, Rose and coworkers applied the combination of nucleophilic addition and rearomatization to design an elegant route to enantiomerically pure manganese π-arene complexes (Scheme 1.14) [73]. The reaction of complexes **51** based on 1,3-substituted arenes with the enolate obtained from natural *D*-camphor led to the formation of four diastereomers **52** resulting from the presence of central and planar chirality elements. Controlled epimerization allowed an enrichment of the mixture in the two *endo*-isomers (*R*,2p*R*)-**52** and (*R*,2p*S*)-**52** (for the stereochemistry definition of these molecules, see Ref. [74]), which were isolated in pure form with >98% de after column chromatography and crystallization. The treatment of the latter species with a mixture of $AgBF_4$ and Me_3SiCl, or simple protonation with $HBF_4 \times OMe_2$, induced a smooth elimination of the enolate to form the desired optically active π-arene products **53** with a perfect retention of planar chirality. This "round-trip" approach

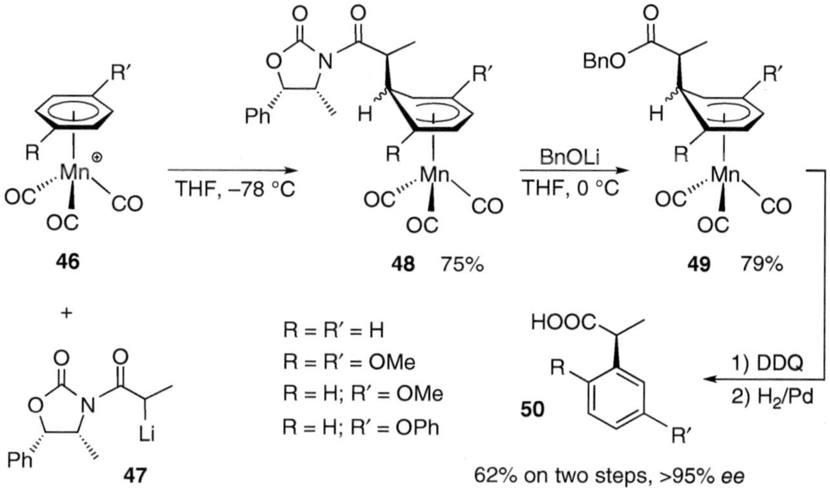

Scheme 1.13 Asymmetric synthesis of 2-arylpropionic acids using cationic Mn(I) π-arene complexes **46**. Source: Based on Miles et al. [70].

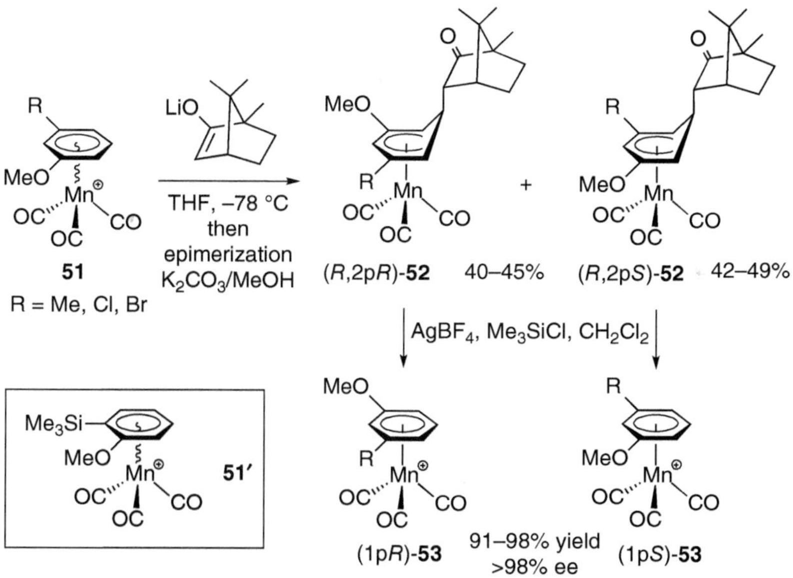

Scheme 1.14 Synthesis of enantiomerically pure planar chiral Mn(I) π-arene complexes.

was quite general and could be applied for the resolution of the 1,2-disubstituted π-arene complex **51′** as well (Scheme 1.14).

Besides the nucleophilic attack on cationic π-arene precursors described earlier, direct η^5-cyclohexadienyl ligand functionalization appeared as a viable complementary strategy to access more structurally elaborated Mn(I) species. The first approach dealt with a C–H bond deprotonation in complexes **40** with *n*BuLi/TMEDA at −78 °C and quenching the resulting lithiated species with

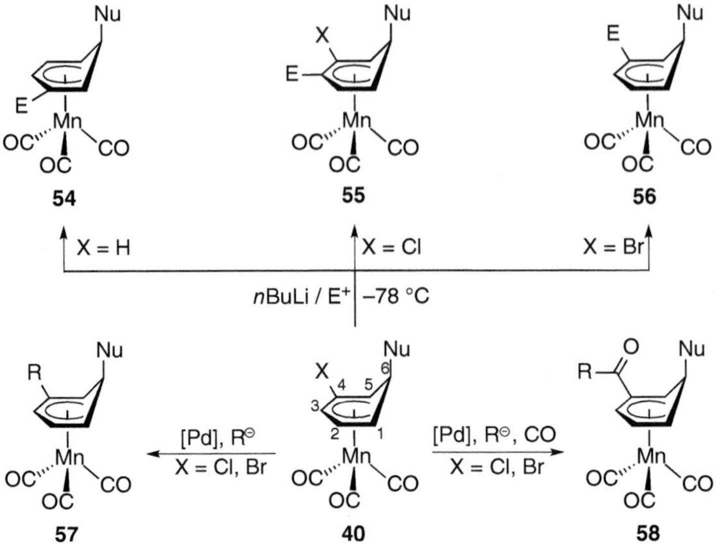

Scheme 1.15 Various strategies for the functionalization of η^5-cyclohexadienyl manganese complexes.

various electrophiles to form the substituted η^5-cyclohexadienyl products **54–56** in moderate to high yield (Scheme 1.15) [75]. Direct use of acyl chloride as electrophile led to a significant degradation of the organolithium intermediates, but the yield of the targeted products could be strongly improved by applying transmetallation with $Li_2[MnCl_4]$, followed by the acylation of the organomanganese derivatives, catalyzed with $Fe(acac)_3$ [75e]. While for non-substituted derivatives, the deprotonation occurred mostly at the C2 carbon atom, a perfect regioselectivity for the *ortho*-position relative to the chloride, ether, or amine directing groups were observed with a net preference for the C3 carbon to form complexes **55**. Similar initial treatment of the bromo-substituted **40** with *n*BuLi led to the selective metal/halogen exchange to form *in fine* complexes **56** [77]. Interestingly, for the η^5–1,2,4,5-tetramethyl-6-phenylcyclohexadienyl substituted version of complex **40**, deprotonation took place exclusively at the benzylic (C2)C–H bond, leaving intact the usual aromatic C3 position [76].

The second powerful strategy, largely exploited for the functionalization of halogen-containing η^5-cyclohexadienyl complexes, was based on Pd-catalyzed cross-coupling processes including Suzuki-Miyaura [78], Sonogashira [72b, 72d, 79b], Stille [72a–c, 80], and Negishi [72b] reactions as well as the efficient installation of O-, S-, N-, and P-nucleophiles to form a variety of products **57** [72b]. Importantly, the majority of these processes could be performed with a simple $Pd_2(dba)_3/AsPh_3$ catalytic system and worked well under CO atmosphere to afford the corresponding carbonylated products **58** in good yield [72b, 72d, 80c].

The formation of 1,3-cyclohexadienes with good stereoselectivity upon nucleophilic addition onto the Mn(I) η^5-cyclohexadienyl complexes **40** represents the most important synthetic application of these compounds (Scheme 1.16). The reaction of tricarbonyl species **40** with strong nucleophiles led to the formation of

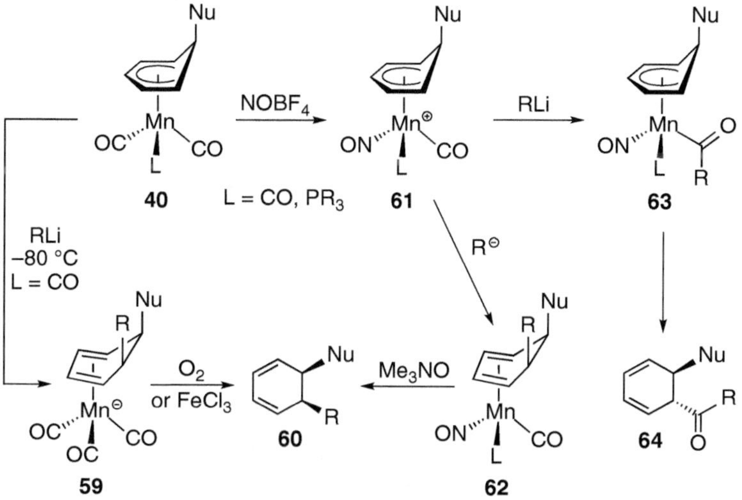

Scheme 1.16 Synthesis of 1,3-cyclohexadienes using the reactions of Mn(I) η^5-cyclohexadienyl complexes with nucleophiles.

anionic η^4-diene intermediates **59** that release the corresponding dienes **60** upon oxidation with $FeCl_3$ [73] or upon simple exposure to air [81]. The substitution of one CO ligand in **40** for NO led to the cationic species **61**, capable of readily accommodating a variety of weaker nucleophiles to yield the neutral η^4-diene products **62** in moderate to good yield [82]. The latter reactions were more efficient for phosphine-containing complexes **61** due to the suppression of side electron transfer processes, resulting in 10–20% higher yield [82b]. Neutral diene complexes **62** have been more stable than their anionic analogues **59** and typically required Me_3NO for the demetallation step. Interestingly, the treatment of complexes **61** with MeLi or PhLi led *in fine* to the selective formation of dienes **64** incorporating an acyl group in *trans*-arrangement relative to Nu. This unexpected selectivity was rationalized by the occurrence of the initial nucleophilic attack at the carbonyl ligand, followed by reductive elimination in the resulting σ-acyl intermediates **63** [83].

A representative application of η^5-cyclohexadienyl complexes to the enantioselective synthesis of 2-cyclohexenones is shown on Scheme 1.17 [73]. The reaction of complexes (2p*R*)-**65** obtained from optically pure π-arene precursors (1p*R*)-**53** (see Scheme 1.14) with a variety of C-nucleophiles proceeded with complete stereoselectivity to afford the target enones (*R*)-**67** in excellent yield after the demetallation of the anionic diene intermediates **66** with $FeCl_3$ and acid hydrolysis. Interestingly, the stereoselective reduction of a selected cationic η^5-cyclohexadienyl Mn(I) complex was used by Miles et al. as one of the key steps for the formal total synthesis of (+)-Juvabione [84].

It is important to mention that η^5-cyclopentadienyl ligands in cymantrene derivatives can be also efficiently transformed into the corresponding conjugated dienes upon UV irradiation in a 2 : 1 Et_2O/MeOH mixture acting as a proton source [85]. This reaction was applied by Top, Jaouen et al. as a key step for the preparation of substituted cyclopentadiene derivatives (Scheme 1.18) [85a, b]. As

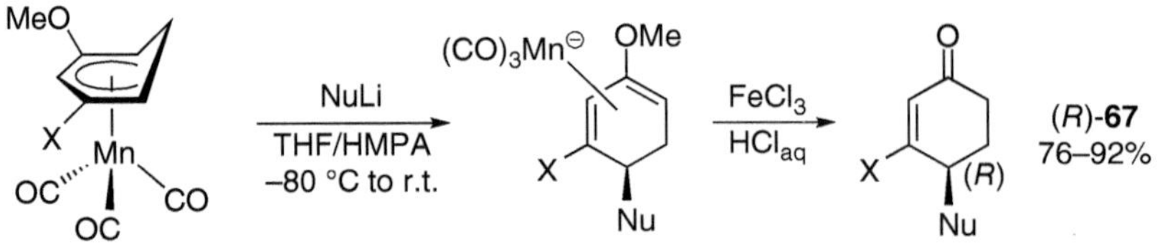

(2p*R*)-**65**

66

X = H, Br, Cl
Nu = $CHPh_2$, CPh_3, CMe_2CN, CMe_2CO_2Et

Scheme 1.17 Enantioselective synthesis of 2-cyclohexenones from optically pure Mn(I) η5-cyclohexadienyl complexes (2p*R*)-**65**. Source: Eloi et al. [73].

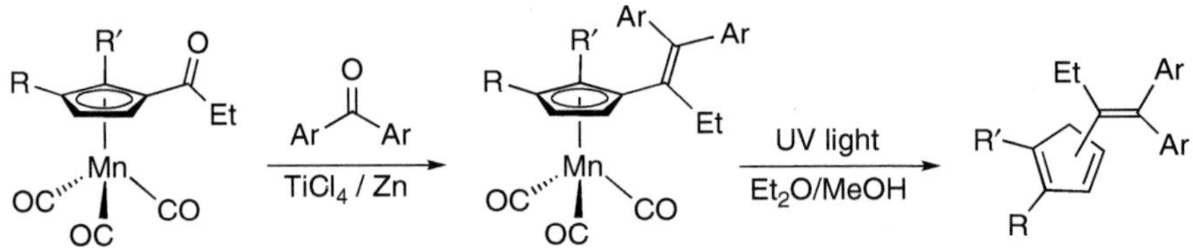

68 Ar = *p*-C_6H_4OMe **69** 48–83% **70** 80–91%

R = R′ = H
R = H, R′ = Me
R = Me, R′ = H

Scheme 1.18 Synthesis of substituted cyclopentadienes by McMurry reaction of cymantrenyl ketones followed by photochemical demetallation. Source: Jaouen and coworkers [85].

an example, the cymantrenyl ketones **68**, easily obtained upon Friedel–Crafts acylation, were first engaged into the McMurry reaction with substituted benzophenone to produce **69**, from which the target alkenyl-substituted cyclopentadienes **70** were obtained as a mixture of regioisomers upon such a demetallation. This procedure was further applied to the synthesis of rather elaborated molecules, including ethynylestradiol-substituted cyclopentadiene [85a, b].

1.5.2 Preparation of Allenes Using [Cp′(CO)$_2$Mn] Auxiliary

In 1979, Franck-Neumann et al. reported the isomerization of the Mn(I) η^2-alkyne complexes **71** bearing electron-withdrawing substituents into the corresponding η^2-allene products **72** in the presence of basic alumina (Scheme 1.19) [86a]. The manganese moiety in complexes **72** was always coordinated to more electron-poor double bond. While under these conditions, a mixture of *exo*- and *endo*-isomers, differing by the relative positions of the metal fragment and the larger allene substituent R or R′, was typically produced; the use of stoichiometric amounts of stronger base such as DBU or *t*BuOK both accelerated the reaction rate and afforded preferentially the *exo*-**72** derivative [86b]. These species were generally air-stable and could be purified by chromatography on silica, while the corresponding allenes **73** were readily released upon oxidation with Fe(III) or Ce(IV) salts (Scheme 1.19) [86].

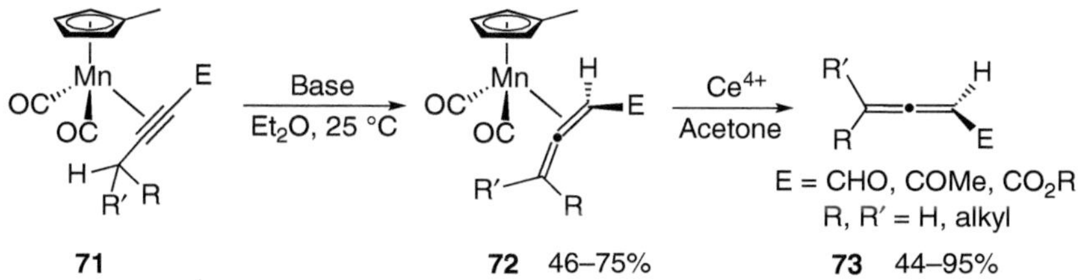

Scheme 1.19 Synthesis of allenes by base-catalyzed isomerization of Mn(I) π-alkyne complexes and oxidative dematallation. Source: Based on Franck-Neumann et al. [86].

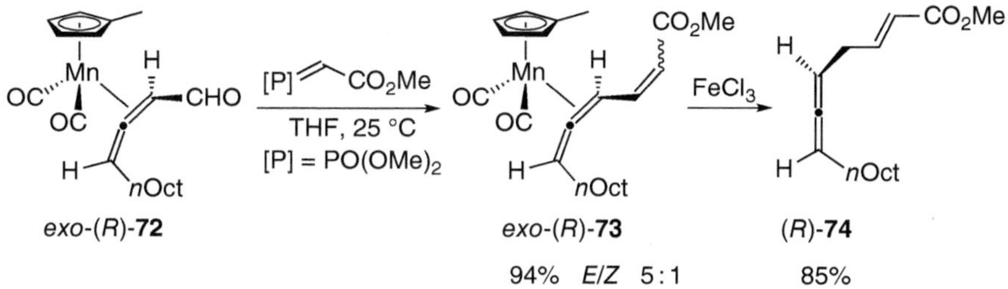

Scheme 1.20 Asymmetric synthesis of bee pheromone using Mn(I) η^2-allene complex *exo-(R)*-**72**. Source: Based on Franck-Neumann et al. [86].

Allenylcarbaldehyde complexes *exo*-**72** (E = CHO, R = H, R′ = *n*Hex, *n*Oct) can be resolved using (*S*)-5-(α-phenylethyl)semioxamazide chiral auxiliary to afford both optically pure products in *ca.* 35% yield each [86c]. These optically pure species were used for the efficient asymmetric synthesis of bee pheromone (Scheme 1.20) [86c]. Horner–Evans olefination of *exo*-(*R*)-**72** (R = *n*Oct) led to the clean formation of the corresponding alkene derivatives *exo*-(*R*)-**73** as a mixture of *E*/*Z* isomers, from which the desired *E*-compound could be obtained in 79% yield providing the target molecule upon oxidative decomplexation. Importantly, the olefination of free allenyl aldehyde led to lower yield (41%) and significant racemization.

Later, Lepore and coworkers showed that the enantioenriched Mn(I) complexes *exo*-(*R*)-**72** can be prepared under biphasic conditions using a specially designed chiral phase-transfer catalyst **75** based on cinchonidinium scaffold (Scheme 1.21) [87]. Though the ee values were not very high, this method represents the first catalytic route to protected allenyl aldehydes from readily available alkynyl aldehydes. While the optical purity of *exo*-(*R*)-**72** was not improved by crystallization due to their oily aggregation state, the replacement of Cp′ for Cp, which typically facilitates crystallization, could be a possible solution for this problem in light of further applications in organic synthesis.

The reaction of *exo*-**72** with Grignard reagents led to a 1,2-addition onto the carbonyl group of the coordinated allenyl aldehyde with complete diastereoselectivity, to form the Mn(I) complexes *exo*-**76**, from which the corresponding allenols **77** are easily displaced upon oxidation with $PhI(OAc)_2$ (Scheme 1.22) [86c, 88]. Organolithium and organozinc nucleophiles gave similar results; no ee erosion was detected for optically pure substrates *exo*-**72** [86c]. This approach was also applied for disubstituted allene precursors, providing reasonable dr ratios in **77** even for small alkyl groups (5 : 1 for Me/Et). The presence of the coordinated $[Cp'(CO)_2Mn]$

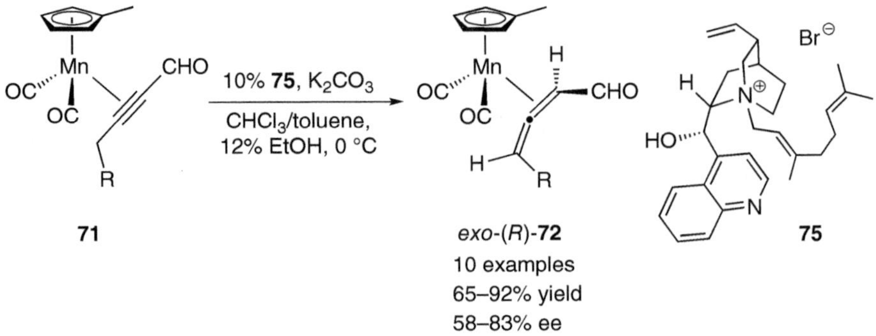

Scheme 1.21 Synthesis of enantio enriched Mn(I) η^2-allene complexes *exo-(R)*-**72** using cinchonidinium-based phase-transfer catalyst **75**. Source: Based on Roy et al. [87].

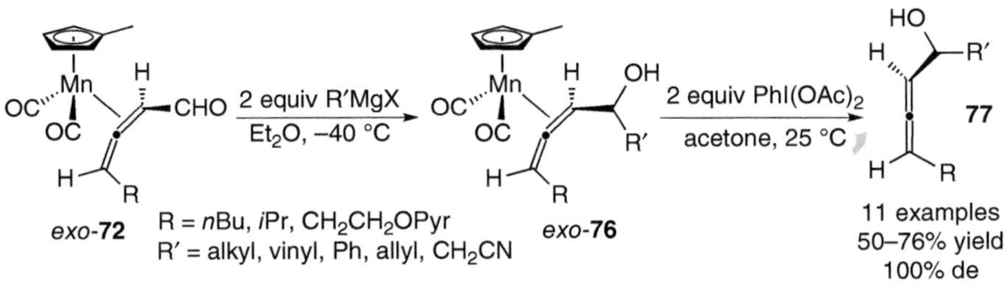

Scheme 1.22 Synthesis of allenols using the reaction of Mn(I) η^2-allene complexes with nucleophiles. Franck-Neumann et al. [86]; Roy et al. [88].

fragment plays a crucial role in this process, controlling both the discrimination between the remote allene substituents in the γ-position and the preferred side of the nucleophilic attack. Noteworthy, for free allenals, a reasonably high level of diastereoselectivity was achieved only with substrates containing the very bulky Ph_2tBuSi group. The sequence of enantioselective preparation of complex *exo*-(*R*)-**72** (R = *n*Bu) and its diastereoselective reaction with $NCCH_2MgX$ was used as a basis of one of the most concise asymmetric total syntheses of Hagen's gland lactone [87].

The reactions of anionic cumulenolate derivatives, obtained *in situ* upon treatment of the π-alkyne complexes **71** by an excess of strong base with arylaldehydes, followed by oxidative demetallation, were found to proceed exclusively at the α-position, to afford aldol products **79** in moderate to excellent yields (Scheme 1.23) [89]. While the diastereoselectivity for ester substrates was rather low (dr 1.5–2 : 1), alkynyl ketones ensured much better results (dr 9–14 : 1). Importantly, the aldol-type processes in the case of free alkynyl ketones and esters produced the addition products uniquely at the γ-position and with rather poor yields.

1.5.3 Synthetic Applications of Mn(I) Fischer Carbenes

Stoichiometric reactions of group 6 Fischer carbenes have found numerous applications in organic synthesis [90]. Despite the related manganese complexes have attracted much less attention, in some cases the reactivity of these species was remarkably different from their chromium and tungsten analogues leading to

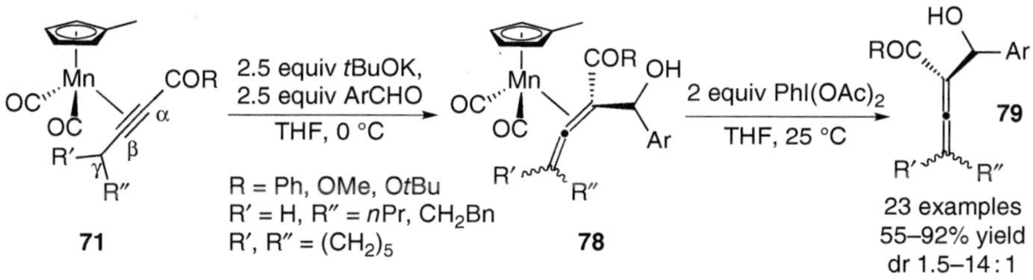

Scheme 1.23 Synthesis of functionalized allenes by aldol-type reaction of Mn(I) π-alkyne complexes followed by oxidative dematallation. Source: Based on Bhowmick et al. [89].

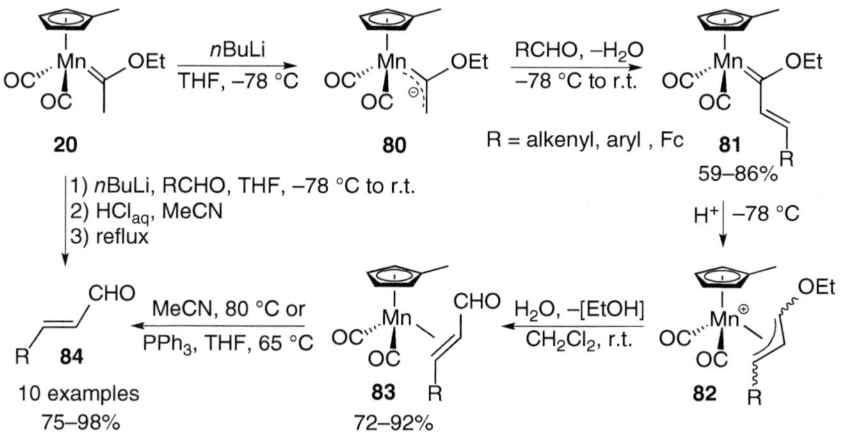

Scheme 1.24 Vinylogation of aldehydes using Mn(I) alkoxycarbene complexes. Source: Yi et al. [91]; Mongin et al. [91].

some valuable results. The less electrophilic character of the $[Cp(CO)_2Mn]$ moiety, compared with $[(CO)_5M]$ (M = Cr, W) resulted in a much higher nucleophilicity of the carbene enolates derivatives. As a representative example of such behavior, condensation of the carbene enolate **80** with a variety of aldehydes readily afforded alkenylcarbenes **81** (Scheme 1.24) [91]. Protonation of the latter species at low temperature, followed by treatment with water, afforded complexes **83** via the intermediacy of the cationic π-allyl intermediates **82**. The α,β-unsaturated aldehyde ligands in **83** were easily released upon thermal substitution for PPh_3 or MeCN. The overall process, which could eventually be performed in "one pot" turned as an efficient aldehyde vinylogation procedure with synthetically useful yields.

The outcome of the Michael addition of the enolate **80** to α,β-unsaturated ketones was dramatically dependent on the hydrolysis temperature (Scheme 1.25) [92]. The acidic quenching of the conjugated addition product **85** led to the formation of carbene complexes **86** as a mixture of *cis* and *trans*-isomers, eventually separated by column chromatography. In contrast, room temperature evolution of **85**, followed by protonation, gave the cyclohexenone complexes **87** as *cis* and *trans*-isomers through a sequence of intramolecular nucleophilic attack of the remote enolate to the carbene atom, with concomitant elimination of EtO^- group, protonation, and carbene-to-alkene isomerization. Noteworthy, for chromium Fischer carbenes, the

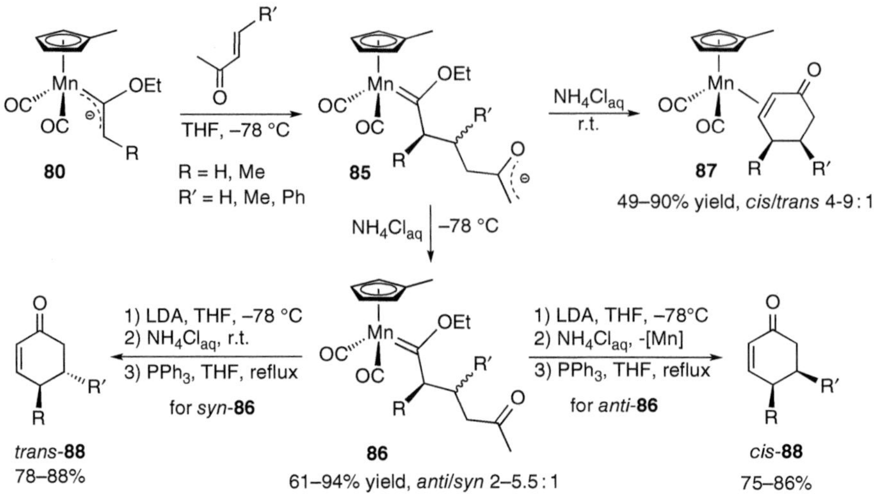

Scheme 1.25 Stereoselective synthesis of 4,5-disubstituted cyclohexenones from enolates of Mn(I) alkoxycarbene complexes. Source: Based on Mongin et al. [92].

remote enolates similar to **85** typically undergo the proton transfer from α-carbene position due to its higher C–H acidity, to form stable carbene enolate isomer, thus precluding the further cyclization. Diastereoselective high yield formation of *cis*- and *trans*-4,5-disubstituted cyclohexenones **87** was achieved from pure stereoisomers of **86** upon a sequence of low temperature deprotonation, room temperature protonation, and finally demetallation upon heating with PPh_3 in refluxing THF (Scheme 1.25).

The reaction of the enolates **85** with benzaldehyde led to the diastereoselective formation of six-membered cyclic Fischer carbenes **88** in good yield (Scheme 1.26) [93]. Further methylation of the α-position relative to the carbene atom occurred in totally stereoselective manner, providing complex **88′** showing four contiguous stereogenic carbon centers. The corresponding lactones were conveniently released upon decomposition of the complexes under air atmosphere. This reaction represents an illustrative example of highly stereoselective assembly of polysubstituted cyclic molecules from simple Fischer carbenes, enone, and aldehyde, proceeding as tandem Michael addition/aldol condensation/transesterification process.

The reactivity of manganese Fischer carbenes with alkynes has been scarcely explored. Thermal reaction of complexes **20** with enynes leads to the formation of bicyclic vinylcyclopropanes **90** (Scheme 1.27, left equation) [94]. The only example of Dötz benzannulation for manganese complexes bearing acceptor $Cp_2Ti(Cl)O$ carbene substituent was achieved under UV irradiation to form, after oxidation in acidic medium, quinones **91** in moderate yield (Scheme 1.27, right equation) [95]. Notably, regular alkoxycarbene **20** (R = Me) was completely unreactive showing that the increased electrophilicity of carbene ligand was a crucial factor enabling this reactivity.

The Mn(I) alkynylcarbene complexes **92**, easily available from carbyne precursors **21** and alkynyllithium derivatives [53a, 96, 98a], can undergo coupling of the remote

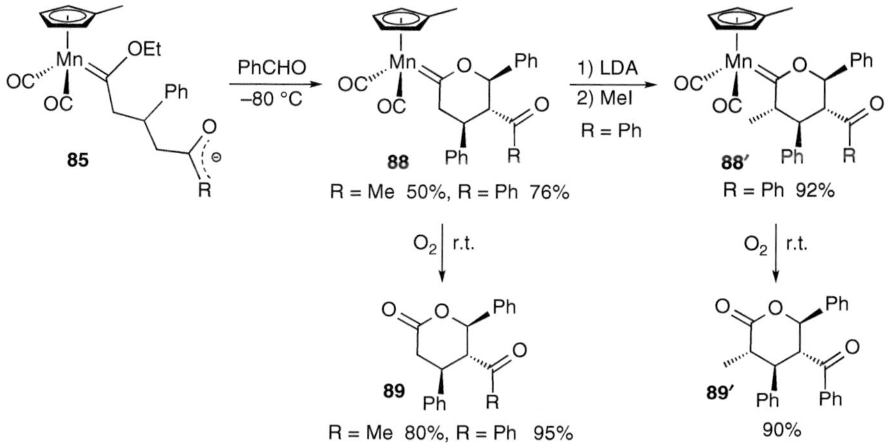

Scheme 1.26 Stereoselective synthesis of polysubstituted lactones using enolates of Mn(I) alkoxycarbene complexes. Source: Based on Mongin et al. [93].

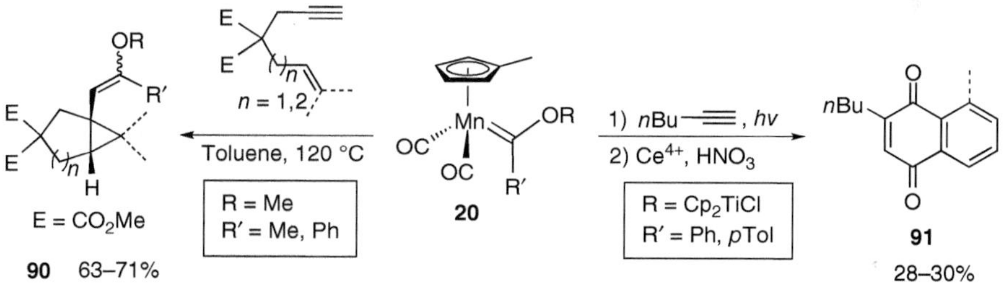

Scheme 1.27 Cyclization reactions of Mn(I) alkoxycarbene complexes with terminal alkynes and enynes. Source: Based on Hoye et al. [94].

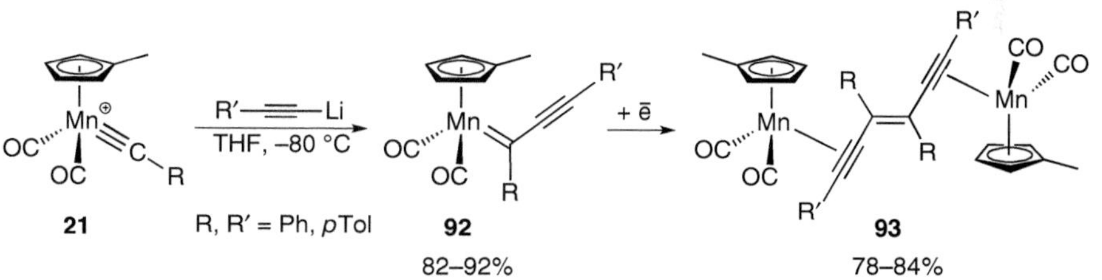

Scheme 1.28 Synthesis and reductive dimerization of Mn(I) alkynylcarbene complexes.

alkynyl carbon atoms, to give binuclear ene-diyne products **93** (Scheme 1.28). While the thermal reaction was shown to proceed with low *E*/*Z* selectivity and was accompanied by partial R/R′ scrambling [96], under electrocatalytic reduction conditions, the *E*-isomers of **93** were formed exclusively in excellent yield [97].

The intramolecular version of this dimerization process constitutes a highly efficient route to cyclic ene-diynes **96**, starting from readily available terminal dialkynes (Scheme 1.29) [98]. Under optimized conditions, binuclear alkynylcarbenes could be generated *in situ* by copper-mediated alkyne addition to trifluoroacetylcarbenes

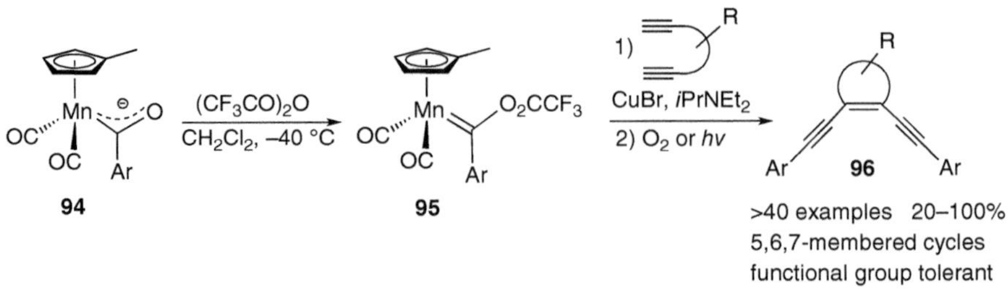

Scheme 1.29 Synthesis of cyclic endiynes by the dimerization of Mn(I) alkynylcarbene complexes. Source: Casey et al. [98].

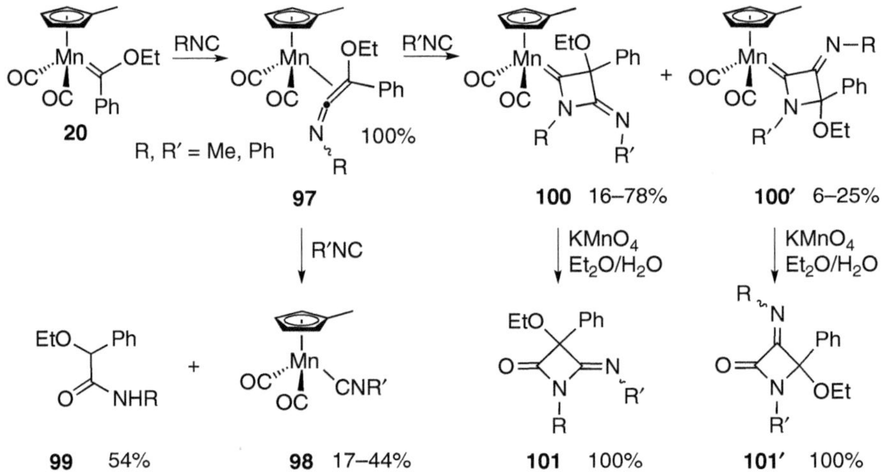

Scheme 1.30 Reactions of Mn(I) alkoxycarbene complexes with isonitriles. Source: Based on Aumann et al. [99].

95 acting as carbyne surrogates upon the partial dissociation of CF_3COO^- anion. The demetallation of the resulting products was achieved either by Cu(II)-catalyzed O_2 oxidation [98a] or by visible light irradiation in CH_2Cl_2/THF solution [98b]. This approach showed an ample scope, tolerated numerous functional groups, and could even be applied for the preparation of non-symmetric ene-diynes **96** in 18–40% yields, starting from an equimolar mixture of σ-acyl complexes **94** [98b].

The insertion of isocyanides across Mn=C bond in complexes **20** afforded quantitatively the η^2-keteneimine products **97** (Scheme 1.30) [99]. The thermal reaction of the latter with a second equivalent of isocyanide led to three competitive processes of keteneimine liberation (selective for R = cyclohexyl (Cy) to form the amide **99** after hydrolysis), and the formation of two isomeric four-membered cyclic aminocarbenes **100** and **100′** via [3+1] and [2+2] cycloadditions, respectively. The oxidation of these complexes with $KMnO_4$ under biphasic conditions efficiently produced the corresponding 4-imino- (**100**) and 3-imino-2-azetidinones (**100′**), respectively.

The reactions of **97** with a variety of substrates having polar C=X bonds efficiently produced the five-membered cyclic aminocarbenes **102** (Scheme 1.31) [100]. While the diastereoselectivity in reactions with aldehydes and ketones was modest, these

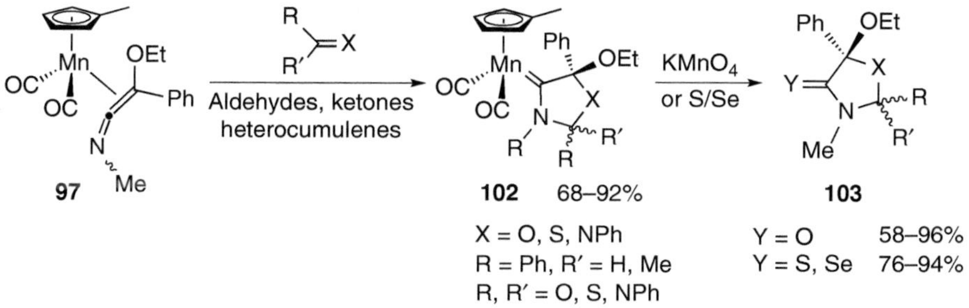

Scheme 1.31 [3+2] cycloaddition of Mn(I) η^2-keteneimine complexes with the substrates containing polar double bonds. Source: Based on Aumann et al. [100].

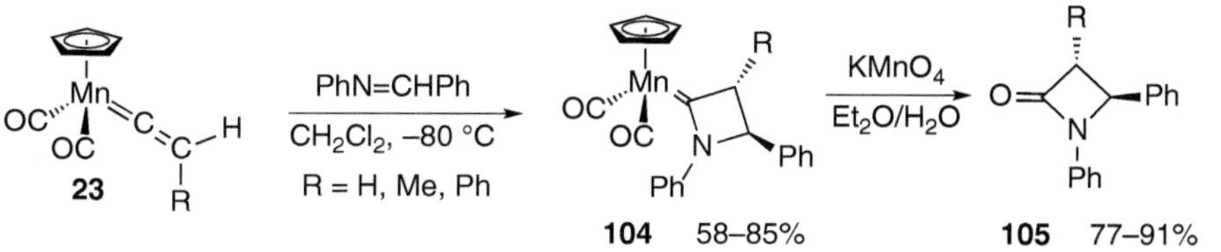

Scheme 1.32 Stereoselectively synthesis of β-lactams from Mn(I) vinylidene complexes. Source: Based on Terry et al. [52].

reactions were especially efficient for different heterocumulenes. Besides oxidation, the cleavage of Mn=C bond could be readily achieved by the treatment with elemental sulfur or selenium.

The reaction of the vinylidene complexes **23** with imines was found to proceed as a formal [2+2] cycloaddition to form stereoselectively aminocarbene complexes **104** (Scheme 1.32) [52b]. The latter products could be readily transformed to the corresponding β-lactams **105** in good yield.

1.5.4 Carbonyl-Containing Manganese σ-Complexes in Organic Synthesis

Synthetic application of Mn(I) complexes $(CO)_5MnR$ was extensively developed by group of DeShong in the late 1980s (Scheme 1.33) [101]. Beyond the classic migratory CO insertion leading to σ-acyl complexes being trapped with nucleophiles (Reppe carbonylation) to form the corresponding carbonyl derivatives **106**, highly selective sequential insertion of CO and unsaturated C—C bonds can be achieved at room temperature to form cyclometallated species **107** and **108**. Though the most reactive $(CO)_5MnMe$ can readily insert several terminal alkynes (R′ = Ph, CO_2Et) [101d], in all other cases these transformations were performed only using solution high-pressure techniques, allowing a better stabilization of the $(CO)_4$(substrate)MnCOR intermediates. While the substrate scope was restricted to strained cyclic systems like norbornene and disubstituted olefins bearing electron-withdrawing groups, both terminal and internal alkynes could be engaged, albeit with modest regioselectivity in the latter case. Photolysis of complexes **107** in MeCN afforded ketones **109** in good yield [101b]. Protonation of complexes **108**

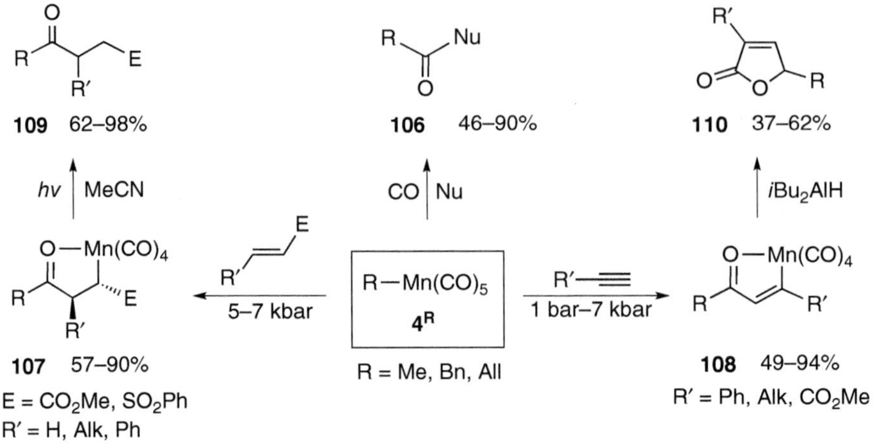

Scheme 1.33 Synthesis of various carbonyl derivatives from Mn(I) σ-complexes $(CO)_5MnR$ under high-pressure conditions. Source: DeShong et al. [101].

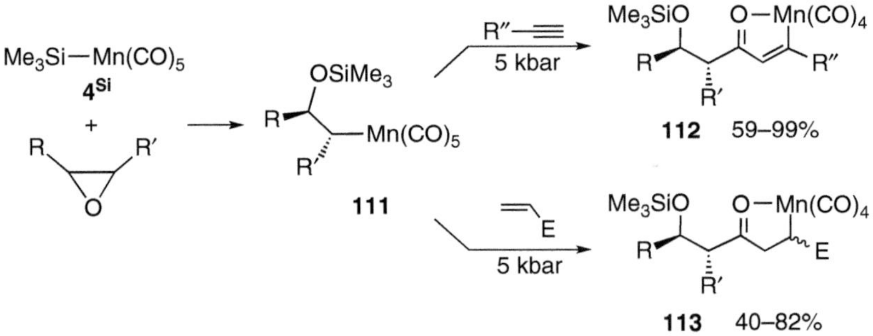

Scheme 1.34 Three-component reaction between Mn(I) complex $(CO)_5MnSiMe_3$, epoxides and organic substrates containing unsaturated C-C bonds. Source: Based on DeShong et al. [102].

induced the Mn—C bond cleavage, liberating the corresponding enones, whereas their reduction with $i\text{Bu}_2\text{AlH}$ gave butenolide products **110** in moderate yield [101c].

Trimethylsilyl-substituted manganese σ-complex $\mathbf{4^{Si}}$ could be applied even to a three-component synthesis of metallacyclic products **112–113** from epoxides to unsaturated hydrocarbons (Scheme 1.34) [102]. Thermodynamically favorable epoxide insertion into the Mn—Si bond to form σ-alkyl species **111**, having *anti*-configuration, seems to be the key feature allowing the excellent level of selectivity observed in this reaction.

The most important practical application of this methodology dealt with the synthesis of glucoside derivatives (Scheme 1.35) [103]. The representative insertion reactions of the β-anomer $\mathbf{4^{GL}}$, available in a nearly quantitative yield from α-glucopyranosyl bromide and $(CO)_5MnK$, always proceeded with a complete retention of the configuration for the reacting carbon center. Reppe carbonylation of complex $\mathbf{4^{GL}}$ in the presence of various O-, N-, and S-nucleophiles led to a highly efficient formation of the corresponding carbonyl derivatives $\mathbf{106^{GL}}$, suitable for

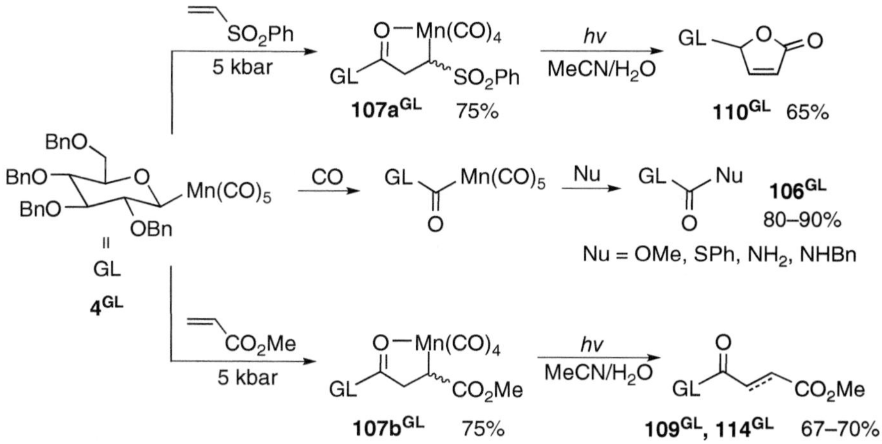

Scheme 1.35 Application of Mn(I) σ-complexes for synthesis of glucoside derivatives. Source: DeShong et al. [103].

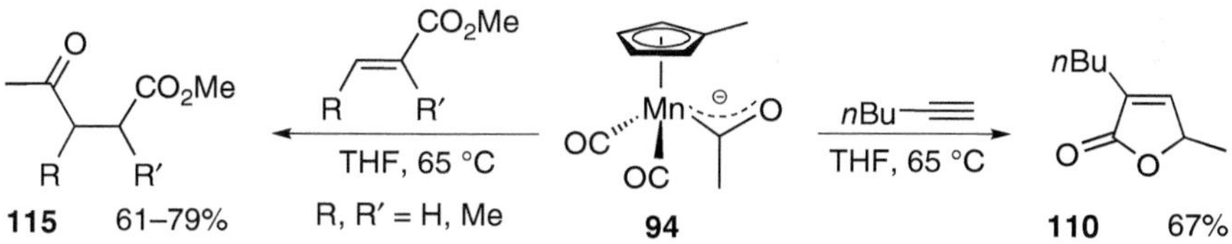

Scheme 1.36 Carbonylation of alkenes and terminal alkyne with anionic Mn(I) σ-acyl complex under atmospheric pressure. Source: Based on Hoye et al. [94].

further derivatization [103d]. UV irradiation of sulfonylated compound **107a**^{GL} in aqueous acetonitrile afforded butenolide **110**^{GL} in good yield. Interestingly, the outcome of the photochemical demetallation in the case of acrylate insertion product **107b**^{GL} can be easily switched by the presence of oxygen, selectively affording either the enone **114**^{GL} or its saturated analogue **109**^{GL} for air-saturated or degassed MeCN/H_2O solutions, respectively [103d].

Similar transformations can also be performed in one step under atmospheric pressure for the anionic half-sandwich σ-acyl manganese complex **94**, to afford the butenolide **110** and the γ-ketoester **115** derivatives (Scheme 1.36) [94]. Despite the availability of cymantrene-derived σ-acyl complexes and a convenient reaction setup, this approach was not extended to more elaborated substrates.

1.5.5 Mn-Mediated Synthesis of Organophosphorous Compounds

The manganese alkynylcarbene complex **92** reacts at low temperature with lithium phosphides to afford after protonation *syn/anti* mixtures of the η^2-allene complexes **115**, resulting from a nucleophilic attack of the phosphide on the remote γ-carbon atom (Scheme 1.37) [104]. Photochemical decomposition of the resulting Mn(I) species under UV irradiation in CH_2Cl_2 solution readily released the corresponding phosphinoallenes **116** in good yield.

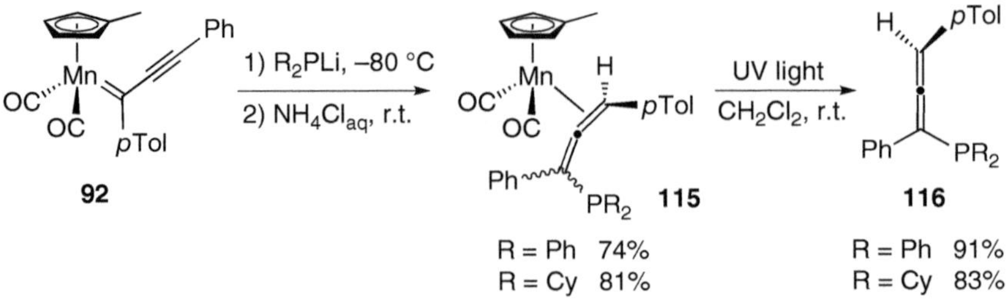

Scheme 1.37 Synthesis of allenylphosphines from Mn(I) alkynylcarbene complexes. Source: Based on Sentets et al. [104].

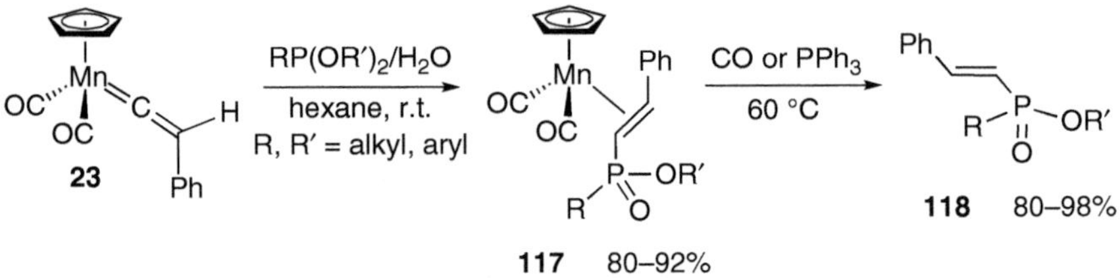

Scheme 1.38 Synthesis of styrylphosphonates from Mn(I) vinylidene complexes and phosphites. Source: Antonova et al. [105].

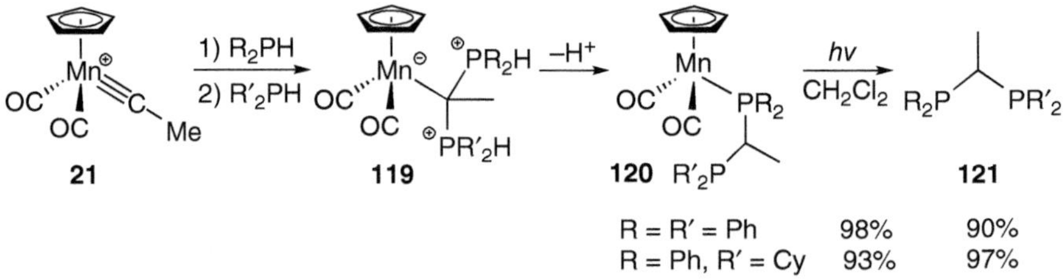

Scheme 1.39 Synthesis of diphosphinomethane derivatives by the coupling of Mn(I) carbyne complex with secondary phosphines. Source: Based on Valyaev et al. [107].

The reaction of vinylidene complex **23** with phosphites readily afforded η^2-styrylphosphonate complexes **117**, to yield free alkenes upon substitution with CO or P-ligands (Scheme 1.38) [105]. The key step of this reaction includes the hydrolysis of zwitterionic phosphite adducts $Cp(CO)_2Mn^{-}{-}C(RP^{+}(OR')_2 = C(H)Ph$ with elimination of R′OH [106].

Some of us have recently shown that the cationic carbyne complex **21** is able to accommodate two molecules of secondary phosphines, the same or different, to afford zwitterionic intermediates **119**, which are prone to undergo a 1,2-proton shift from more acidic phosphonium site to the adjacent carbon atom, to form *in fine* the Mn(I) complexes **120** bearing κ^1-coordinated bridge-substituted diphosphinomethane ligands (Scheme 1.39) [107]. The resulting organophosphorous products **121** can be quantitatively obtained upon visible light irradiation of **120** in CH_2Cl_2 solution, provided the pendant phosphine moiety is blocked from potential coordination by complexation with borane or protonation with strong acid such as $HBF_4 \times OEt_2$ [107].

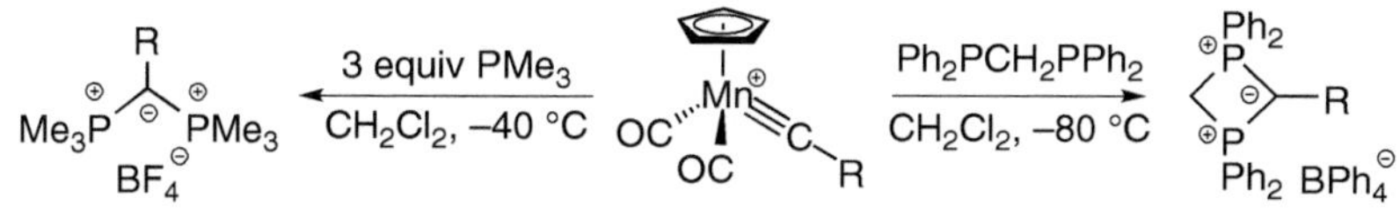

122

R = Ph, *p*Tol, SiPh$_3$

55–70%

21

123

R = Ph, *p*Tol

81–83%

Scheme 1.40 Synthesis of phosphorus semi-ylides by the reaction of Mn(I) carbyne complex with two tertiary phosphines. Source: Valyaev et al. [107].

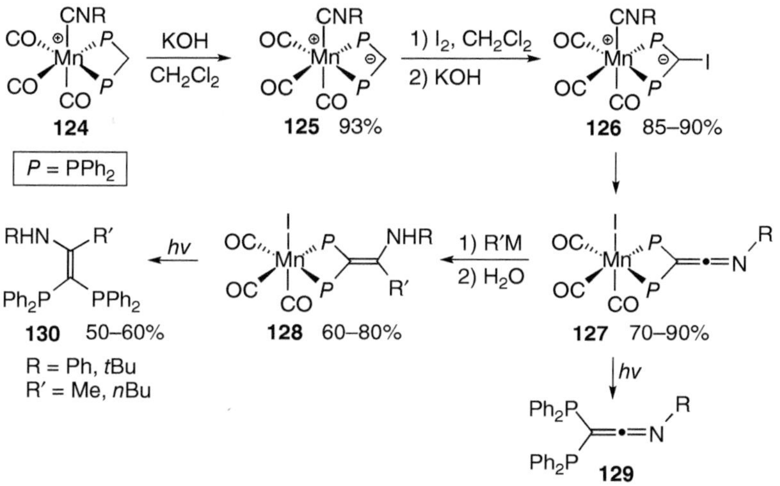

124

125 93%

126 85–90%

130 50–60%

R = Ph, *t*Bu

R′ = Me, *n*Bu

128 60–80%

127 70–90%

129

R = Ph, *t*Bu 33–37%

Scheme 1.41 Bridge modification of dppm ligand in manganese coordination sphere. Source: Ruiz et al. [109].

A similar reaction of **21** with tertiary phosphines led to the formation of the semi-ylides **122–123** in good yield (Scheme 1.40) [107, 108]. While for the reactions with PMe$_3$ the semi-ylide **112** was released upon reaction with the excess of phosphine to form Cp(CO)$_2$Mn(PMe$_3$) as a by-product [108a], the intermediate cyclic four-membered derivatives similar to **119** obtained from bis[diphenylphosphino]methane (dppm) underwent spontaneous loss of the [Cp(CO)$_2$Mn] moiety upon simple warming to room temperature affording **123** [107, 108b].

An interesting modification of the bridge in dppm ligand in manganese coordination sphere was reported by Ruiz and coworkers (Scheme 1.41) [109]. The readily available zwitterionic complexes **126**, bearing iodo-substituted κ^2-diphosphinomethanide ligand, undergo a metal-to-carbon migration of the coordinated isocyanide ligand to form neutral species **127** exhibiting an aza-allene dppm scaffold. The cumulene fragment in the latter complexes can be further modified by adding organolithium or Grignard reagents, followed by hydrolysis to

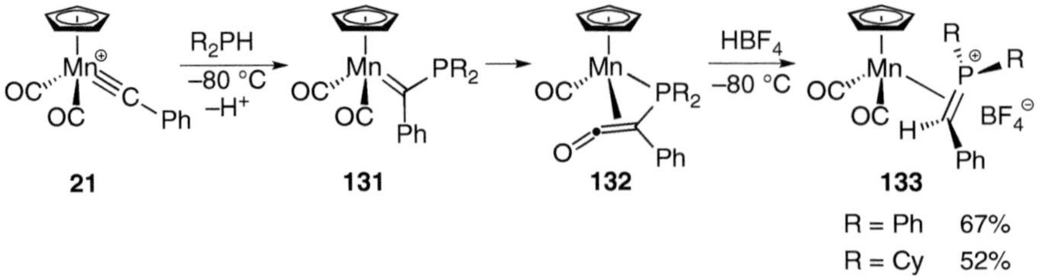

Scheme 1.42 Synthesis of Mn(I) methylenephosphonium complexes from carbyne precursor and secondary phosphine. Source: Based on Valyaev et al. [110].

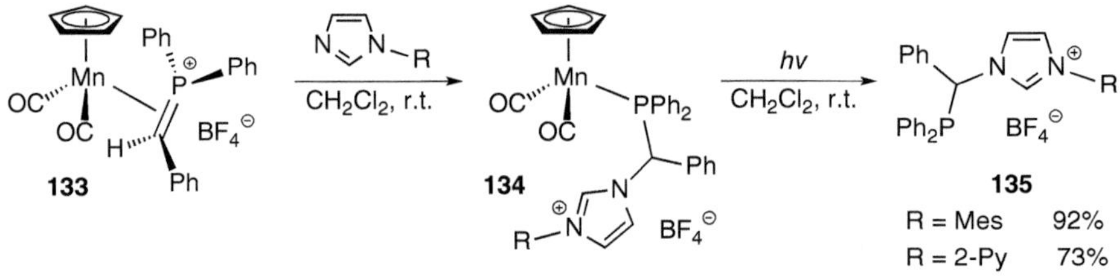

Scheme 1.43 Synthesis of phosphine-imidazolium salts from Mn(I) methylenephosphonium complexes. Source: Valyaev et al. [110].

complexes **128**. Photochemical decomposition of manganese products **127** or **128** affords the corresponding diphosphines **129–130** in moderate yields.

Other important organophosphorous synthons, the cationic half-sandwich Mn(I) complexes **133**, bearing η^2-methylenephosphonium ligands, have been prepared recently by some of us via a sequence of intramolecular migratory CO insertion in Fischer-type phosphinocarbenes **131**, followed by protonation-induced CO-deinsertion in the resulting η^3-phosphinoketene products **132** (Scheme 1.42) [110a]. The overall transformation sequence of **21** into **133** can be performed in one pot on a 20 mmol scale, the final synthons **133** being stable in the solid state under ambient conditions.

In contrast to iminium salts, the chemistry of their phosphorous analogues, methylenephosphonium species, has been barely developed, likely because of their relative instability. Yet, their reactivity was characterized by their propensity to react with nucleophiles at the phosphorous atom, giving the corresponding phosphonium salts [111]. Beyond stabilizing methylenephosphonium cations, the metal moiety in complexes **133** totally inverted the regioselectivity of the eventual nucleophilic addition, thus opening new synthetic perspectives. As a representative example, reaction of **133** with easily available substituted imidazoles instantaneously afforded the phosphine complexes **134**, featuring a pendant imidazolium moiety (Scheme 1.43) [110]. Direct irradiation of crude **134** with visible light affords the corresponding phosphine–imidazolium salts **135**, precursors of bi- and tridentate (for R = 2-Py) N-heterocyclic carbene (NHC)-phosphine ligands, in good yield.

The reaction of **133** with 1 equiv of imidazole afforded an equimolar mixture of binuclear meso- and racemic phosphine complexes **136** (Scheme 1.44) [110a]. These air-stable complexes can be readily separated by column chromatography, to provide (after photochemical demetallation) the imidazolium salts **137**,

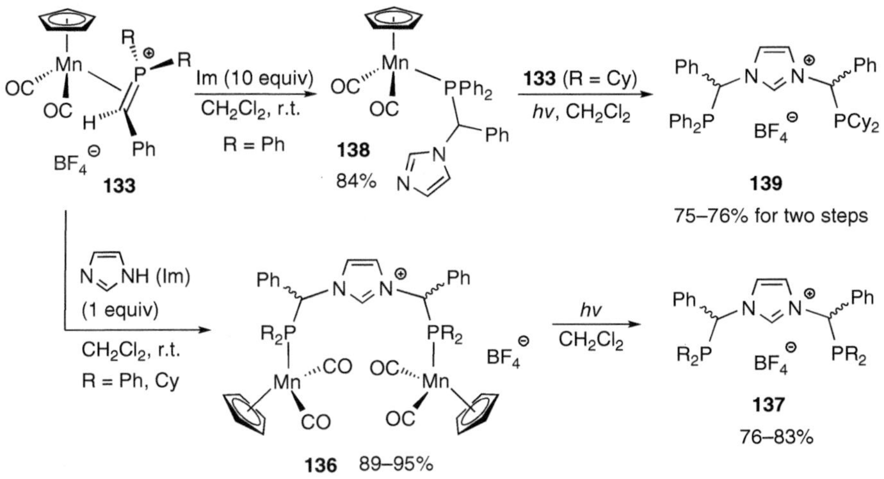

Scheme 1.44 Synthesis of precursors of chiral pincer-type phosphine-NHC-phosphine ligands from Mn(I) methylenephosphonium complexes. Source: Based on Valyaev et al. [110a].

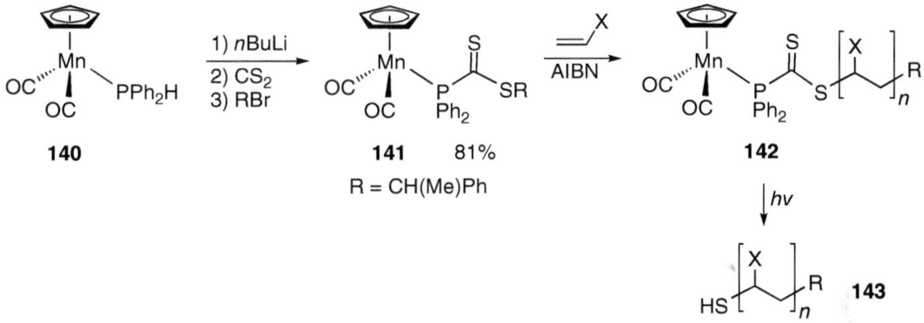

Scheme 1.45 Synthesis of Mn-based agent for RAFT polymerization and its photochemical decomposition. Source: Based on Kulai et al. [57].

precursors of pincer-type phosphine-NHC-phosphine ligands, in good yields. Notably, non-symmetric version of these pre-ligands, **139**, can be readily prepared upon reaction of **133** with an excess of imidazole, followed by the treatment of the resulting complex **138** with a different methylenephosphonium complex, separation of diastereoisomers, and release of the corresponding organophosphorous products **139**.

Novel phosphorous-containing agent **141** for Reversible Addition Fragmentation chain Transfer (RAFT) polymerization for styrene and acrylate monomers was recently prepared from cymantrene-derived diphenylphosphine complex **140** using a classic approach (Scheme 1.45) [57]. Interestingly, unlike all previously reported examples in this section, visible light irradiation of the resulting polymers **142** proceeded with complete destruction of the Ph_2PCS_2 fragment, allowing the efficient preparation of thiol-terminated polymers **143** under mild conditions.

Very recently, Bullock and coworkers showed that triple hydrogen atom abstraction from the ammonia manganese complex **144** with stable

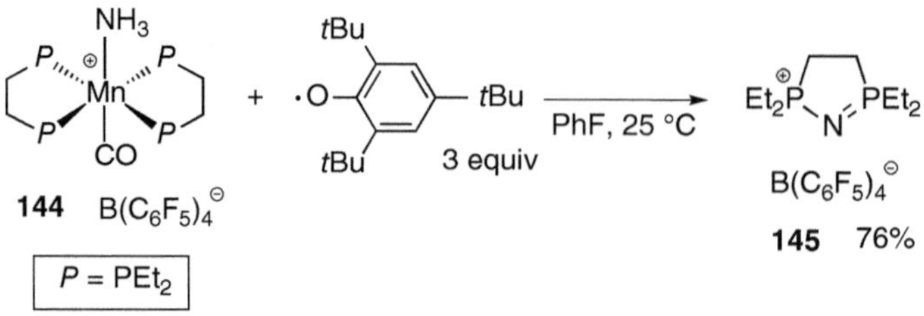

Scheme 1.46 Synthesis of cyclic phosphazenium salt by hydrogen atoms abstraction in Mn(I) ammonia complex. Source: Based on Cook et al. [112].

2,4,6-tri-*tert*-butylphenoxyl radical leads to the formation of the rare cyclic phosphazenium salt **145** in good yield (Scheme 1.46) [112].

1.5.6 Backbone Modification of N-heterocyclic Carbenes in Mn(I) Coordination Sphere

We have recently shown that deprotonation of the half-sandwich manganese complex **146**, readily available from cymantrene and free carbene on 10 g scale [43a], with *n*BuLi at room temperature selectively proceeds at the backbone position of the NHC ligand to form bimetallic complex **147**, bearing an anionic imidazol-2,4-diylidene ligand (Scheme 1.47) [25]. In contrast to other complexes of this type, the [Cp(CO)$_2$Mn] transition metal fragment did not migrate from the "normal" to the "abnormal" carbene position, thus permitting to exploit the highly nucleophilic character of the latter. In particular, this approach was used for the stepwise electrophilic fluorination with *N*-fluorobenzenesulfonimide (NFSI) to form complexes **148** and **149**, bearing unprecedented IMes^{F} and IMes^{F2} ligands, respectively (Scheme 1.47) [113]. Very importantly, simple treatment of complexes **148** and **149** with triflic acid in CH_2Cl_2 solution led to the quantitative release of the corresponding imidazolium salts **150** and **151**. This reaction, which likely involves protonation at the manganese center, followed by elimination of the imidazolium for the resulting cationic hydride intermediates [Cp(CO)$_2$(NHC)MnH]$^+$, appears to be quite general for Cp(CO)$_2$Mn(NHC) complexes.

Synthetic potential of the anionic imidazol-2,4-diylidene manganese complexes **146** can be further illustrated by their oxidative coupling, mediated by $CuCl_2$, leading to bimetallic complexes **152** exhibiting new Janus-type bis(carbene) ligands, in which two imidazol-2-ylidene moieties are connected by a simple C—C bond (Scheme 1.48) [43a]. As in the previous case, the corresponding bis(imidazolium) salts **153** are readily released from **152** upon protonation.

1.6 Summary and Conclusions

The information systematically collected in this chapter illustrates a significant potential of various classes of organometallic Mn(I) complexes in organic synthesis. In this chapter, we deliberately selected reactions leading to effective formation of organic compounds, whereas many other relevant transformations existing in

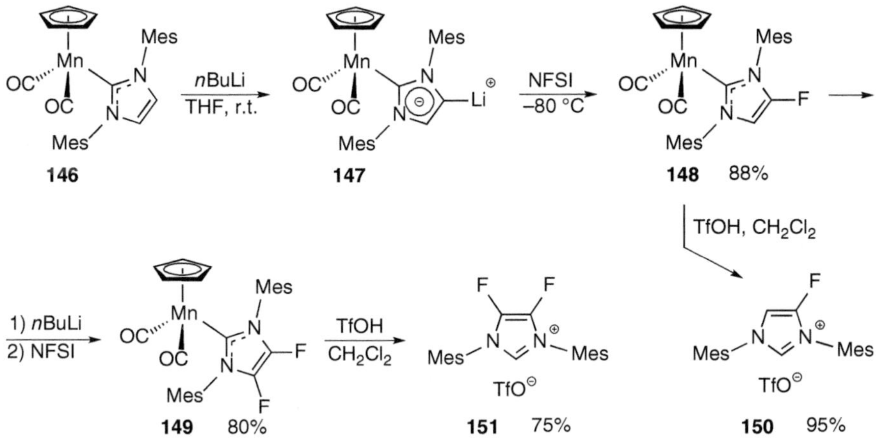

Scheme 1.47 Synthesis of backbone fluorinated imidazolium salts in Mn(I) coordination sphere. Source: Based on Grineva et al. [113].

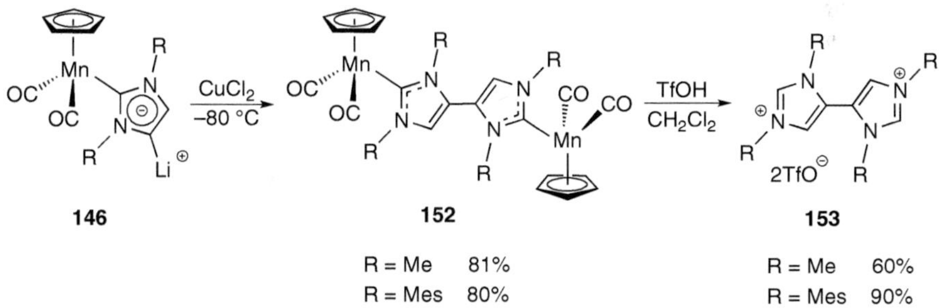

Scheme 1.48 Synthesis of bis-imidazolium salts by oxidative coupling of abnormal carbenes in Mn(I) NHC complex **146**. Source: Based on Grineva et al. [113].

the literature were not covered. The role of the most common Mn(I) fragments ($Cp(CO)_2Mn$, $(CO)_5Mn$, $(CO)_3Mn$) in these processes can vary from simple action as a protecting group for allene, phosphine, and NHC moieties to the direct influence on the ligand-centered reactivity in the case of η^5-cyclohexadienyl, carbene, and methylenephosphonium complexes. It is particularly important to point out that numerous demetallation protocols including oxidation, ligand substitution, protonation, and photochemical destruction have been already developed for different types of Mn(I) complexes, thus ensuring this crucial step for metal-mediated synthesis. While catalysis undoubtedly remains the privileged way for the design of highly selective and atom economical processes, stoichiometric reactivity of complexes based on cheap and non-toxic 3d metals may sometimes provide a valuable alternative. Considering a tremendous progress in homogeneous catalysis with organometallic manganese complexes over the last few years, and existing solid background in synthetic chemistry of Mn(I) species, we expect that Mn-mediated organic synthesis will attract continuously growing attention of the research community in the near future.

References

1 Gilman, H. and Bailie, G.C. (1937). *J. Org. Chem.* 2: 84–94.

2 Hurd, D.T., Sentell, J.W., and Norton, F.J. (1949). *J. Am. Chem. Soc.* 71: 1899–1899.

3 Fischer, E.O. and Jira, R. (1954). *Z. Naturforsch.* 9b: 618–619.

4 Tsutsui, M. and Zeiss, H. (1961). *J. Am. Chem. Soc.* 83: 825–827.

5 Corey, E.J. and Posner, G.H. (1970). *Tetrahedron Lett.* 11: 315–318.

6 Valyaev, D.A., Lavigne, G., and Lugan, N. (2016). *Coord. Chem. Rev.* 308: 191–235.

7 (a) Cahiez, G., Duplais, C., and Buendia, J. (2009). *Chem. Rev.* 109: 1434–1476. (b) Cahiez, G. and Gager, O. (2011). Preparation and reactivity of organomanganese compounds. *Chemistry of Organomanganese Compounds* (eds. Z. Rappoport, I. Marek, J.F. Liebman and S. Patai), 305–418. Germany: Wiley-Blackwell.

8 Cahiez, G. and Chavant, P.-Y. (1989). *Tetrahedron Lett.* 30: 7373–7376.

9 Tang, J., Shinokubo, H., and Oshima, K. (1998). *Synlett.* 1998: 1075–1076.

10 Rieke, R.D., Kim, S.-H., and Wu, X. (1997). *J. Org. Chem.* 62: 6921–6927.

11 Cahiez, G., Martin, A., and Delacroix, T. (1999). *Tetrahedron Lett.* 40: 6407–6410.

12 Fürstner, A. and Brunner, H. (1996). *Tetrahedron Lett.* 37: 7009–7012.

13 Friour, G., Cahiez, G., and Normant, J.F. (1984). *Synthesis.* 1984: 37–40.

14 Horvath, B., Möseler, R., and Horvath, E.G. (1979). *Z. Anorg. Allg. Chem.* 460: 165–177.

15 (a) Reardon, D., Aharonian, G., Gambarotta, S., and Yap, G.P.A. (2002). *Organometallics* 21: 786–788. (b) Mukhopadhyay, T.K., Flores, M., Groy, T.L., and Trovitch, R.J. (2014). *J. Am. Chem. Soc.* 136: 882–885.

16 Solari, E., Musso, F., Gallo, E. et al. (1995). *Organometallics* 14: 2265–2276.

17 Price, J.S., Chadha, P., and Emslie, D.J.H. (2016). *Organometallics* 35: 168–180.

18 Köhler, F.H. (1997). Sandwich complexes with two cyclic ligands C_nH_n. *Synthetic Methods of Organometallic and Inorganic Chemistry*, vol. 8 (ed. W.A. Herrmann), 7–9. Stuttgart: George Thieme.

19 Layfield, R.A. (2008). *Chem. Soc. Rev.* 37: 1098–1107.

20 Kelly, C.M., McDonald, R., Sydora, O.L. et al. (2017). *Angew. Chem. Int. Ed.* 56: 15901–15904.

21 (a) Chakraborty, U., Reyes-Rodriguez, E., Demeshko, S. et al. (2018). *Angew. Chem. Int. Ed.* 57: 4970–4975. (b) Chakraborty, U., Demeshko, S., Meyer, F., and Jacobi von Wangelin, A. (2019). *Angew. Chem. Int. Ed.* 58: 3466–3470.

22 Reimer, K.J. and Shaver, A. (1990). *Inorg. Synth.* 28: 154–159.

23 Hu, Y., Zhou, B., and Wang, C. (2018). *Acc. Chem. Res.* 51: 816–827.

24 (a) Garbe, M., Junge, K., and Beller, M. (2017). *Eur. J. Org. Chem.* 2017: 4344–4362. (b) Mukherjee, A. and Milstein, D. (2018). *ACS Catal.* 8: 11435–11469.

25 Valyaev, D.A., Uvarova, M.A., Grineva, A.A. et al. (2016). *Dalton Trans.* 45: 11953–11957.

26 Ciftci, M., Tasdelen, M.A., and Yagci, Y. (2016). *Polym. Int.* 65: 1001–1014.

27 Gismondi, T.E. and Rausch, M.D. (1985). *J. Organomet. Chem.* 284: 59–71.

28 Kadassery, K.J. and Lacy, D.C. (2019). *Dalton Trans.* 48: 4467–4470.

29 Gladysz, J.A., Williams, G.M., Tam, W. et al. (1979). *Inorg. Chem.* 18: 553–558.

30 King, R.B. and Stone, F.G.H. (1963). *Inorg. Synth.* 7: 196–201.

31 Gladysz, J.A., Tam, W., Williams, G.M. et al. (1979). *Inorg. Chem.* 18: 1163–1165.

32 Booth, B.L. and Haszeldine, R.N. (1966). *J. Chem. Soc.*: 157–160.

33 Motz, P.L., Sheeran, D.J., and Orchin, M. (1990). *J. Chem. Soc. A* 383: 201–212.

34 (a) Angelici, R.J. and Basolo, F. (1962). *J. Am. Chem. Soc.* 84: 2495–2499. (b) Martin, T.A., Ellul, C.E., Mahon, M.F. et al. (2011). *Organometallics* 30: 2200–2211.

35 (a) Angelici, R.J., Basolo, F., and Poë, A.J. (1963). *J. Am. Chem. Soc.* 85: 2215–2219. (b) Reimann, R.H. and Singleton, E. (1973). *J. Chem. Soc., Dalton Trans.*: 841–846.

36 (a) Reimann, R.H. and Singleton, E. (1972). *J. Organomet. Chem.* 38: 113–119. (b) Hulley, E.B., Helm, M.L., and Bullock, R.M. (2014). *Chem. Sci.* 5: 4729–4741.

37 Jackson, J.D., Villa, S.J., Bacon, D.S. et al. (1994). *Organometallics* 13: 3972–3980. and references therein.

38 Sun, S., Yeung, L.K., Sweigart, D.A. et al. (1995). *Organometallics* 14: 2613–2615.

39 Pons, M. and Herberich, G.E. (2014). *Inorg. Synth.* 36: 152–153.

40 Kirk, P.M. and Castellani, M.P. (2014). *Inorg. Synth.* 36: 63–65.

41 Laws, D.R., Chong, D., Nash, K. et al. (2008). *J. Am. Chem. Soc.* 130: 9859–9870.

42 (a) Connelly, N.G. and Kitchen, M.D. (1977). *J. Chem. Soc. Dalton Trans.*: 931–937. (b) Hershberger, J.W., Klingler, R.J., and Kochi, J.K. (1983). *J. Am. Chem. Soc.* 105: 61–73. (c) Huang, Y., Carpenter, G.B., Sweigart, D.A. et al. (1995). *Organometallics* 14: 1423–1428.

43 (a) Grineva, A.A., Valyaev, D.A., César, V. et al. (2018). *Angew. Chem. Int. Ed.* 57: 7986–7991. (b) Wu, K., Conger, M.A., Waterman, R. et al. (2019). *Polyhedron* 157: 442–448.

44 Ginzburg, A.G., Petrovskii, P.V., Setkina, V.N., and Kursanov, D.N. (1985). *Russ. Chem. Bull.* 34: 176–179.

45 Ginzburg, A.G., Fedorov, L.A., Petrovskii, P.V. et al. (1974). *J. Organomet. Chem.* 73: 77–84.

46 Green, M.L.H., Joyner, D.S., and Wallis, J.M. (1987). *J. Chem. Soc., Dalton Trans.*: 2823–2830.

47 Ginzburg, A.G. (1993). *Russ. Chem. Rev.* 62: 1025–1045.

48 (a) Valyaev, D.A., Wei, D., Elangovan, S. et al. (2016). *Organometallics* 35: 4090–4098. (b) Assim, K., Jeschke, J., Jakob, A. et al. (2016). *Thin Solid Films* 619: 265–272.

49 (a) Fischer, E.O. and Pleszke, K. (1958). *Chem. Ber.* 91: 2719–2726. (b) Gogan, N.J. and Chu, C.-K. (1975). *J. Organomet. Chem.* 93: 363–376.

50 (a) Egger, H. and Nikiforov, A. (1968). *Monat. Chem.* 99: 2296–2310. (b) Egger, H. and Nikiforov, A. (1968). *Monat. Chem.* 99: 2311–2323.

51 Geoffroy, G.L. (1997). Transition metal carbene and ketenyl complexes. *Synthetic Methods of Organometallic and Inorganic Chemistry*, vol. 7 (ed. W.A. Herrmann), 173–179. Stuttgart: George Thieme.

52 (a) Fischer, E.O., Besl, G., Meineke, E.W., and Kreissl, F.R. (1997). Transition metal carbyne complexes. *Synthetic Methods of Organometallic and Inorganic Chemistry*, vol. 7 (ed. W.A. Herrmann), 208–210. Stuttgart: George Thieme. (b) Terry, M.R., Mercando, L.A., Kelley, C. et al. (1994). *Organometallics* 13: 843–865.

53 (a) Ortin, Y., Lugan, N., and Mathieu, R. (2005). *Dalton Trans.*: 1620–1636. (b) Valyaev, D.A., Lugan, N., Lavigne, G., and Ustynyuk, N.A. (2008). *Organometallics* 27: 5180–5183.

54 Kruck, T. and Krause, V. (1972). *Z. Naturforsch. B* 27: 302–304.

55 Treichel, P.M. and Mueh, H.J. (1977). *Inorg. Chim. Acta* 22: 265–268.

56 (a) Werner, H. and Juthani, B. (1977). *J. Organomet. Chem.* 129: C39–C42. (b) Unseld, D., Krivykh, V.V., Heinze, K. et al. (1999). *Organometallics* 18: 1525–1541.

57 Kulai, I., Karpus, A., Soroka, L. et al. (2019). *Polym. Chem.* 10: 267–277.

58 (a) Atkinson, R.C.J., Gibson, V.C., and Long, N.J. (2004). *Chem. Soc. Rev.* 33: 313–328. (b) Gómez Arrayás, R., Adrio, J., and Carretero, J.C. (2006). *Angew. Chem. Int. Ed.* 45: 7674–7715.

59 Kudis, S. and Helmchen, G. (1998). *Angew. Chem. Int. Ed.* 37: 3047–3050.

60 Sennhenn, P., Gabler, B., and Helmchen, G. (1994). *Tetrahedron Lett.* 35: 8595–8598.

61 Weiß, T.D., Helmchen, G., and Kazmaier, U. (2002). *Chem. Commun.*: 1270–1271.

62 (a) Bergner, E.J. and Helmchen, G. (2000). *Eur. J. Org. Chem.* 2000: 419–423. (b) Schleich, S. and Helmchen, G. (1999). *Eur. J. Org. Chem.* 1999: 2515–2521.

63 Kamikawa, K., Tseng, Y.-Y., Jian, J.-H. et al. (2017). *J. Am. Chem. Soc.* 139: 1545–1553.

64 Ferber, B., Top, S., and Jaouen, G. (2004). *J. Organomet. Chem.* 689: 4872–4876.

65 Ogasawara, M., Tseng, Y.-Y., Liu, Q. et al. (2017). *Organometallics* 36: 1430–1435.

66 Ogasawara, M., Tseng, Y.-Y., Uryu, M. et al. (2017). *Organometallics* 36: 4061–4069.

67 Pape, A.P., Kaliappan, K.P., and Kündig, E.P. (2000). *Chem. Rev.* 100: 2917–2940.

68 Pauson, P.L. and Segal, J.A. (1975). *J. Chem. Soc., Dalton Trans.*: 1677–1682.

69 Chung, Y.K., Williard, P.G., and Sweigart, D.A. (1982). *Organometallics* 1: 1053–1056.

70 Miles, W.H., Smiley, P.M., and Brinkman, H.R. (1989). *J. Chem. Soc., Chem. Commun.*: 1897–1899.

71 (a) Pearson, A.J., Bruhn, P.R., Gouzoules, F., and Lee, S.-H. (1989). *J. Chem. Soc., Chem. Commun.* 1897–1899: 659–661. (b) Pearson, A.J., Lee, S.-H., and Gouzoules, F. (1990). *J. Chem. Soc., Perkin Trans*: 2251–2254. (c) Pearson, A.J.

and Bruhn, P.R. (1991). *J. Org. Chem.* 56: 7092–7097. (d) Pearson, A.J. and Shin, H. (1992). *Tetrahedron* 48: 7527–7538. (e) Pearson, A.J. and Shin, H. (1994). *J. Org. Chem.* 59: 2314–2323. (f) Lee, S.-H. and Nam, S.-W. (1998). *Bull. Korean Chem. Soc.* 19: 613–615.

72 (a) Prim, D., Auffrant, A., Rose-Munch, F. et al. (2001). *Organometallics* 20: 1901–1903. (b) Auffrant, A., Prim, D., Rose-Munch, F. et al. (2003). *Organometallics* 22: 1898–1913. (c) Jacques, B., Tranchier, J.-P., Rose-Munch, F. et al. (2004). *Organometallics* 23: 184–193. (d)Schouteeten, S., Tranchier, J.-P., Rose-Munch, F. et al. (2004). *Organometallics* 23: 4308–4316. (e) Cetiner, D., Tranchier, J.-F., Rose-Munch, F. et al. (2008). *Organometallics* 27: 784–788.

73 (a) Eloi, A., Rose-Munch, F., and Rose, E. (2009). *J. Am. Chem. Soc.* 131: 14178–14179. (b) Eloi, A., Rose-Munch, F., Rose, E. et al. (2010). *Organometallics* 29: 3876–3386.

74 (a) Rose-Munch, F., Rose, E., and Eloi, A. (2011). Cationic(η^6-arene)- and neutral (η^5-cyclohexadienyl)-tricarbonylmanganese complexes: Synthesis and reactivity. *Chemistry of Organomanganese Compounds* (eds. Z. Rappoport, I. Marek, J.F. Liebman and S. Patai), 489–558. Germany: Wiley-Blackwell. (b) Rose-Munch, F. and Rose, E. (2011). *Org. Biomol. Chem.* 9: 4725–4735.

75 (a) Jacques, B., Chavarot, M., Rose-Munch, F., and Rose, E. (2006). *Angew. Chem. Int. Ed.* 45: 3481–3484. (b) Jacques, B., Chanaewa, A., Chavarot-Kerlidou, M. et al. (2008). *Organometallics* 27: 626–636. (c) Jacques, B., Eloi, A., Chavarot-Kerlidou, M. et al. (2008). *Organometallics* 27: 2505–2517. (d) Cetiner, D., Jacques, B., Payet, E. et al. (2009). *Dalton Trans.*: 27–29. (e) Eloi, A., Rose-Munch, F., Rose, E. et al. (2009). *Organometallics* 28: 925–928.

76 Offner, J.D., Rose-Munch, F., Rose, E. et al. (2010). *Organometallics* 29: 4643–4646.

77 (a) Eloi, A., Rose-Munch, F., Rose, E., and Lennartz, P. (2009). *Organometallics* 28: 5757–5764. (b) Eloi, A., Poizat, M., Hautecoeur, A. et al. (2011). *Organometallics* 30: 5564–5567.

78 Rose-Munch, F., Marti, A., Cetiner, D. et al. (2011). *Dalton Trans.* 40: 1567–1575.

79 Rose-Munch, F., Li, M., Rose, E. et al. (2012). *Organometallics* 31: 92–104.

80 (a) Auffrant, A., Prim, D., Rose-Munch, F. et al. (2001). *Organometallics* 20: 3214–3216. (b) Cetiner, D., Norel, L., Tranchier, J.-P. et al. (2010). *Organometallics* 29: 1778–1788. (c) Eloi, A., Rose-Munch, F., Jonathan, D. et al. (2006). *Organometallics* 25: 4554–4559.

81 (a) Brookhart, M. and Lukacs, A. (1984). *J. Am. Chem. Soc.* 106: 4161–4166. (b) Roell, B.C. Jr., McDaniel, K.F., Vaughan, W.S., and Macy, T.S. (1993). *Organometallics* 12: 224–228.

82 (a) Pike, R.D., Ryan, W.J., Carpenter, G.B., and Sweigart, D.A. (1989). *J. Am. Chem. Soc.* 111: 8535–8537. (b) Lee, T.-Y., Kang, Y.K., Chung, Y.K. et al. (1993). *Inorg. Chim. Acta* 214: 125–134.

83 (a) Pike, R.D., Ryan, W.J., Lennhoff, N.S. et al. (1990). *J. Am. Chem. Soc.* 112: 4798–4804. (b) Pike, R.D. and Sweigart, D.A. (1990). *Synlett.* 1990: 565–571.

84 Miles, W.H. and Brinkman, H.R. (1992). *Tetrahedron Lett.* 33: 589–592.

85 (a) Jaouen, G., Top, S., Vessières, A., and Alberto, R. (2000). *J. Organomet. Chem.* 600: 23–36. (b) Top, S., Kaloun, E.B., Toppi, S. et al. (2001). *Organometallics* 20: 4554–4561. (c) Telegina, L.N., Ezernitskaya, M.G., Godovikov, I.A. et al. (2009). *Eur. J. Inorg. Chem.* 2009: 3636–3643. (d) Kee, J.W., Tan, Y.Y., Swennenhuis, B.H.G. et al. (2011). *Organometallics* 30: 2154–2159.

86 (a) Franck-Neumann, M. and Brion, F. (1979). *Angew. Chem., Int. Ed. Engl.* 91: 688–689. (b) Franck-Neumann, M., Neff, D., Nouali, H. et al. (1994). *Synlett.* 1994: 657–659. (c) Franck-Neumann, M., Martina, D., and Neff, D. (1998). *Tetrahedron Asymm.* 9: 697–708. (d) Lepore, S.D., Khoram, A., Bromfield, D.C. et al. (2005). *J. Org. Chem.* 70: 7443–7446.

87 Roy, A., Bhat, B.A., and Lepore, S.D. (2016). *Org. Lett.* 18: 1230–1233.

88 Roy, A., Bhat, B.A., and Lepore, S.D. (2015). *Org. Lett.* 17: 900–903.

89 Bhowmick, M. and Lepore, S.D. (2010). *Org. Lett.* 12: 5078–5080.

90 (a) Zaragoza Dörwald, F. (1999). *Metal Carbenes in Organic Synthesis.* Wiley–VCH: Weinheim. (b) Dötz, K.H. and Stendel, J. Jr., (2009). *Chem. Rev.* 109: 3227–3274.

91 (a) Yi, C.S., Geoffroy, G.L., White, C.A., and Rheingold, A.R. (1993). *J. Am. Chem. Soc.* 115: 3806–3807. (b) Mongin, C., Ortin, Y., Lugan, N., and Mathieu, R. (1999). *Eur. J. Inorg. Chem.* 1999: 739–742.

92 Mongin, C., Lugan, N., and Mathieu, R. (1997). *Organometallics* 16: 3873–3875.

93 Mongin, C., Gruet, K., Lugan, N., and Mathieu, R. (2000). *Tetrahedron Lett.* 41: 7341–7345.

94 Hoye, T.R. and Rehberg, G.M. (1990). *Organometallics* 9: 3015–3019.

95 Balzer, B.L., Cazanoue, M., Sabat, M., and Finn, M.G. (1992). *Organometallics* 11: 1759–1761.

96 Casey, C.P., Dzwiniel, T.L., Kraft, S. et al. (2003). *Inorg. Chim. Acta* 345: 320–326.

97 Ortin, Y., Sournia-Saquet, A., Lugan, N., and Mathieu, R. (2003). *Chem. Commun.*: 1060–1061.

98 (a) Casey, C.P., Dzwiniel, T.L., Kraft, S., and Guzei, I.A. (2003). *Organometallics* 22: 3915–3920. (b) Casey, C.P. and Dzwiniel, T.L. (2003). *Organometallics* 22: 5285–5290.

99 Aumann, R. and Heinen, H. (1988). *Chem. Ber.* 121: 1085–1091.

100 Aumann, R. and Heinen, H. (1989). *Chem. Ber.* 122: 77–82.

101 (a) DeShong, P. and Slough, G.A. (1984). *Organometallics* 3: 636–638. (b) DeShong, P., Slough, G.A., and Rheingold, A.L. (1987). *Tetrahedron Lett.* 28: 2229–2232. (c) DeShong, P., Sidler, D.R., and Slough, G.A. (1987). *Tetrahedron Lett.* 28: 2233–2236. (d) DeShong, P., Sidler, D.R., Rybczynski, P.J. et al. (1988). *J. Am. Chem. Soc.* 110: 2575–2585.

102 DeShong, P. and Sidler, D.R. (1988). *J. Org. Chem.* 53: 4892–4894.

103 (a) DeShong, P., Slough, G.A., Elango, V., and Trainor, G.L. (1985). *J. Am. Chem. Soc.* 107: 7788–7790. (b) DeShong, P., Slough, G.A., and Elango, V. (1987). *Carbohydr. Res.* 171: 342–345. (c) DeShong, P., Soli, E.D., Slough, G.A. et al. (2000). *J. Organomet. Chem.* 593–594: 49–62.

104 Sentets, S., Serres, R., Ortin, Y. et al. (2008). *Organometallics* 27: 2078–2091.

105 (a) Antonova, A.B., Kovalenko, S.V., Cherkasov, R.A. et al. (1987). *Z. Obsh. Khim.* 57: 1030–1042. (b) Antonova, A. B., Gulbis, G. R., Ioganson, A. A., Kovalenko, S. V., Korniyets, E. D. Diaryl(alkyl)styrylphosphonates. Patent RU1077896 (1993).

106 Utegenov, K.I., Krivykh, V.V., Chudin, O.S. et al. (2018). *J. Organomet. Chem.* 867: 113–124.

107 Valyaev, D.A., Bastin, S., Utegenov, K.I. et al. (2014). *Chem. Eur. J.* 20: 2175–2178.

108 (a) Uedelhoven, W., Eberl, K., Siebert, W., and Kreissl, F.R. (1982). *J. Organomet. Chem.* 236: 301–307. (b) Valyaev, D.A., Utegenov, K.I., Krivykh, V.V. et al. (2018). *J. Organomet. Chem.* 867: 353–358.

109 (a) Ruiz, J., Riera, V., Vivanco, M. et al. (1998). *Organometallics* 17: 3835–3837. (b) Mosquera, M.E.G., Ruiz, J., García, G., and Marquínez, F. (2006). *Chem. Eur. J.* 12: 7706–7716.

110 (a) Valyaev, D.A., Filippov, O.A., Lugan, N. et al. (2015). *Angew. Chem. Int. Ed.* 54: 6315–6319. (b) Valyaev, D.A., Willot, J., Mangin, L.P. et al. (2017). *Dalton Trans.* 46: 10193–10196.

111 (a) Grützmacher, H. and Marchand, C.M. (1997). *Coord. Chem. Rev.* 163: 287–344. (b) Guerret, O. and Bertrand, G. (1997). *Acc. Chem. Res.* 30: 486–493.

112 Cook, B.J., Johnson, S.I., Chambers, G.M. et al. (2019). *Chem. Commun.* 55: 14058–14061.

113 Grineva, A.A., Filippov, O.A., Nefedov, S.E. et al. (2019). *Organometallics* 38: 2330–2337.

2

Manganese-Catalyzed Hydrogenation and Hydrogen Transfer Reactions

Jean-Baptiste Sortais[1,2], *Ruqaya Buhaibeh*[1], *and Yves Canac*[1]

[1] *LCC-CNRS, Université de Toulouse, CNRS, UPS, 205 route de Narbonne, 31077 Toulouse Cedex 4, France*
[2] *Institut Universitaire de France, 1 rue Descartes, F-75231 Paris Cedex 05, France*

2.1 Introduction

The apparent simplicity of the hydrogenation reaction, which consists of the formal addition of elemental hydrogen to an unsaturated bond, does not reflect the essential role of this transformation in synthetic chemistry. Hydrogenation with molecular dihydrogen constitutes indeed a clean and economical reaction, which makes it one of the most used catalytic processes, thus presenting a great interest for pharmaceutical and industrial sectors [1]. Following the pioneering work of Döbereiner [2], Sabatier brought in 1912 the first significant contribution in the field, demonstrating that elemental hydrogen could add to an unsaturated bond in the presence of nickel as a catalyst [3]. This breakthrough was followed by many others in the field of heterogeneous hydrogenation catalysis until the preparation of the first homogeneous rhodium hydrogenation catalyst by Wilkinson and coworkers [4]. Later on, another major advance in this area was achieved by the development of asymmetric versions based on phosphine ligands due to the successive contributions of Knowles [5] and Noyori [6], which allowed them to obtain the Nobel Prize a few years later, like Sabatier in his time [7]. All these advances in hydrogenation catalysis have been made possible by the exceptional properties of noble transition metals belonging to groups 8–10 such as ruthenium, palladium, or rhodium metals [1]. However, the scarcity of these metals has recently led to the emergence of alternatives based on more abundant metals, such as those of the first row of the periodic table. Inspired by concepts known from its well-established ruthenium congeners, iron was the first representative of the first row to give significant results comparable to those of noble metals in the field of hydrogenation [8]. Aware that abundant metals can compete to some extent with noble metals [9], other representatives such as manganese have then emerged, knowing that the latter constitute the third most abundant transition metal after iron and titanium in the Earth's crust and therefore its natural abundance makes it a serious candidate for applications in catalysis [10].

With this objective in mind, the groups of Beller and Milstein described simultaneously in 2016 the first applications of manganese in hydrogenation-type reactions

Manganese Catalysis in Organic Synthesis, First Edition. Edited by Jean-Baptiste Sortais.

of polar bonds, respectively, for the hydrogenation [11] and for the dehydrogenative condensation of alcohols and amines into imines [12]. In both cases, the Mn(I) center was stabilized by a rigid and strongly chelating bis(phosphine) amine pincer ligand that allowed in the case of hydrogenation to reduce a large range of unsaturated substrates. Following these promising results, many other related Mn systems exhibiting a pincer-type structure with different N- and P-based donor extremities were elaborated for the hydrogenation of organic substrates containing multiple polar bonds, such as aldehydes, ketones, nitriles, esters, amides, imines, carbonates, or CO_2 [13]. Bidentate systems combining P-, N-, and C-donor ends were more recently developed, leading to spectacular advances in the field. Other Mn-containing systems have been more punctually envisaged as molecular complexes based on tetradentate ligands and/or ligands with different donor atoms (O, Si), as well as nanoclusters [14]. As an alternative to atom–economic molecular hydrogen, alcohols were also shown to act as a valuable source of hydrogen, thus rendering transfer hydrogenation a process complementary to classical hydrogenation. Whatever the hydrogenation process, the catalytic efficiency of these metal complexes was largely rationalized with reference to the concept of metal–ligand cooperation, which consists in taking advantage of a strong synergy created between the ligand and the metal to promote H_2 heterolysis followed by H-atom transfer onto unsaturated bonds [15]. Most of catalytic Mn systems were thus developed on the basis of two functional architectures verifying the pre-established concept, either through the involvement of an amide/amine interplay (the so-called NH effect) or via the aromatization/dearomatization process of a pyridine moiety. A third mode of metal–ligand cooperation involving a bidentate *N*-heterocyclic carbene (NHC)–phosphine ligand through a formal λ^5–λ^3 phosphorous valence change was more recently described, leading to an efficient catalyst for ketone hydrogenation.

In this chapter, a selection of the most significant advances in hydrogenation-type reactions catalyzed by well-defined manganese complexes will be reviewed. Such metal complexes will be classified according to the ligand structure and in particular the denticity and the nature of donor ends; the main purpose of this presentation is to provide the reader with a general overview complementary to those generally proposed [10c–j, 13].

2.2 Pincer-Type Manganese Complexes

2.2.1 PNP Ligands

The pioneering reports of Beller and coworkers in the field of Mn-catalyzed hydrogenation have led to the tremendous development of tridentate ligands of PNP-type, these chelating ligands being indeed part of the first catalytic systems to lead to significant results in this area [11]. Beyond the nature of P-substituents, the developed PNP pincer ligands differ also by the hybridization state of the central N-atom. The N-atom can be either of sp^3-type in the form of an amine moiety or of sp^2-type via a pyridine or a triazine nucleus (Scheme 2.1).

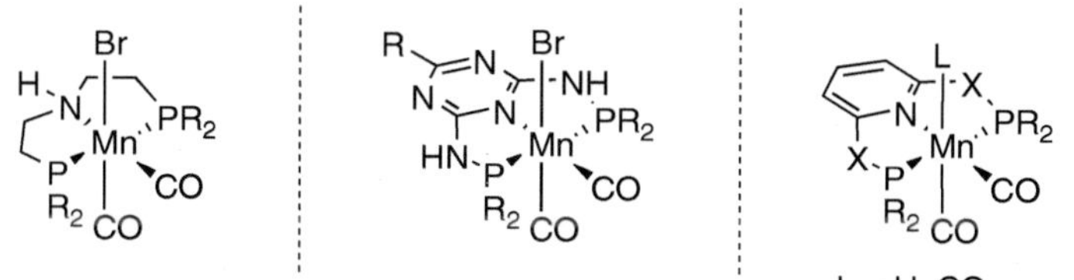

Scheme 2.1 General representation of Mn(I) pincer-type complexes featuring a PNP ligand.

2.2.1.1 PN-sp^{3}P Ligand

Thanks to their expertise in hydrogenation-type reactions using Fe and Ru-based pincer complexes [16], Beller et al. became interested in the preparation of well-defined Mn analogues. The first category of Mn pincer complexes envisaged was based on a PN-sp^{3}P structure consisting of a central amine substituted with two lateral phosphines [17]. Two air-stable Mn(I) complexes were thus conveniently prepared upon the reaction of $Mn(CO)_5Br$ complex with corresponding PNP ligands differing by the nature of P-substituents (*i*Pr vs. Cy) [11]. These complexes were tested for the hydrogenation of unsaturated substrates such as aldehydes, ketones, and nitriles (Scheme 2.2, left). For nitriles, the catalytic system (cat. 3 mol%, *t*BuONa 10 mol%, 50 bar H_2, 120 °C, 24 hours) was demonstrated to be effective for several representatives, including substituted aromatic, benzylic, and aliphatic nitriles and dinitriles. Electron-donating and electron-withdrawing groups (halogen, Me, MeO, NH_2, CF_3) as well as heterocycles were generally also well tolerated. For ketones, the reduction occurred in the presence of 1 mol% of catalyst and 3 mol% of *t*BuONa at 100 °C in toluene with 30 bar of H_2. Under these conditions, other reducible groups such as C=C bonds, lactams, and esters were not affected by the reaction. As an added value, natural products as illustrated with citronellal, perillaldehyde, and 5-hydroxymethylfurfural could be converted to corresponding primary alcohols in satisfactory yield. On the mechanism, an outer-sphere mechanism was proposed involving successive amido and hydride Mn species. The amido complex first formed after the reaction of the precatalyst with a base would react with dihydrogen to afford a hydride complex, which would be able by a simultaneous transfer of the hydride from the Mn center (Mn–H) and the proton from the nitrogen (N–H) to reduce the unsaturated bond.

The more challenging hydrogenation of esters was achieved with less hindered precatalysts based on the cationic ethyl-substituted PNP backbone that exhibits an unusual cis-coordination of the P-atoms (Scheme 2.2, right) [18]. No activity was indeed observed with more crowded PNP Mn precatalysts featuring P-*i*Pr and P-Cy donor ends. This catalytic system (Cat. 2 mol%, *t*BuOK 10 mol%, 110 °C, dioxane, 24 hours, 30 bar H_2) enabled the transformation of aromatic and aliphatic ester derivatives, as well as diesters and lactones, in good yields. The corresponding neutral complex showed similar activity under the optimized reaction conditions.

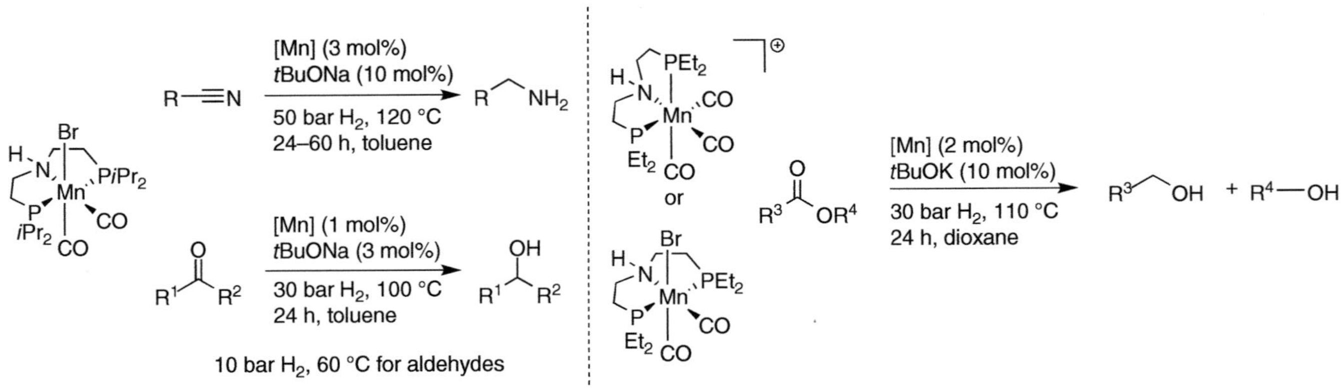

Scheme 2.2 Mn(I) pincer complexes bearing a PN-sp^{3}P ligand for the hydrogenation of nitriles, aldehydes or ketones (left), and esters (right).

Supported by DFT calculations, the efficiency of these catalysts was rationalized through an outer-sphere mechanism involving amido and hydride Mn complexes.

Air-stable PNP Mn(I) pincer complexes active for aldehyde, ketone, and nitrile reductions catalyzed also the sequential one-pot CO_2 hydrogenation to CH_3OH [19]. The transformation involves first the N-formylation of an amine in the presence of CO_2 and H_2 followed by the formamide reduction to methanol and amine products (Scheme 2.3). Thanks to the PN-sp^{3}P precatalyst bearing two strongly donating P*i*Pr$_2$ moieties, CH_3OH was produced in good yield from benzylamine or morpholine substrates with a maximum TON of 36 comparable to the values reported earlier with Co-based catalysts [20].

A Mn(I) complex of such type of PN-sp^{3}P ligand was reported by Leitner and coworkers for the catalytic hydrogenation of cyclic carbonates to diols and methanol [21]. This catalytic system afforded high yields for diols and methanol products under relatively mild conditions (120 °C, 30–60 bar H_2) with respective TONs up to 620 and 400. In the same family of ligands, we can mention the Mn(I) complex bearing an amino-bis(phosphinite) ligand (of PONOP-type), which was shown to be effective for the hydrogenation of ketones [22].

Mn(I) pincer complexes bearing the chiral PNP ligand bis(2-((2*R*,5*R*)-2,5-dimethylphospholanoethyl))amine allowed the enantioselective hydrogenation of ketones (Scheme 2.4, left) [23]. Compared with related pincer complexes based on Ru, Re, and Fe metal centers [23], the Mn representative showed better performance not only for aromatic substrates but also for aliphatic substrates affording enantioselectivities up to 99% ee. For instance, the catalytic conditions required for the hydrogenation of cyclic aliphatic ketones were the following: 1 mol% of catalyst, 5 mol% of *t*BuOK, 30 bar H_2, four hours, 40 °C, *t*-amyl alcohol. Based on DFT calculations, the asymmetric induction was rationalized considering the approach of the prochiral ketone perpendicular to the *cis*-N–H and Mn–H groups of the hydride complex with either the *re* or *si* enantioface.

Other variants of P-stereogenic PN(H)P tridentate ligands were reported by Mezzetti and coworker for the Mn-catalyzed asymmetric hydrogenation of ketones (Scheme 2.4, right) [24]. Derived from a Mn(I)/Fe(II) comparison, kinetic and DFT studies concluded that a bifunctional mechanism for H^+/H^- transfer occurs here.

2.2.1.2 PN-sp^{2}P Ligand

Modification of the central nitrogen donor has made it possible to develop effective hydrogenation catalysts. A revealing example was provided by Kempe and coworkers who described the preparation of highly active PN$_5$P Mn(I) complexes based on a triazine core for the hydrogenation of carbonyl derivatives, precatalysts that involve generally milder conditions than systems containing an aliphatic PN-sp^{3}P skeleton [25]. Related $Mn(CO)_2Br$ and $MnCl_2$ complexes were easily prepared and tested in the hydrogenation of ketones and aldehydes (Scheme 2.5). While no activity was observed with the Mn(II) representative, a broad scope of substrates with high functional group tolerance could be reduced by the PN$_5$P $Mn(CO)_2Br$ complex. The system was able to reduce acetophenone under the following conditions: Cat. 0.1 mol%, 80 °C, toluene, four hours, 20 bar H_2. Aryl–alkyl, diaryl, dialkyl, and

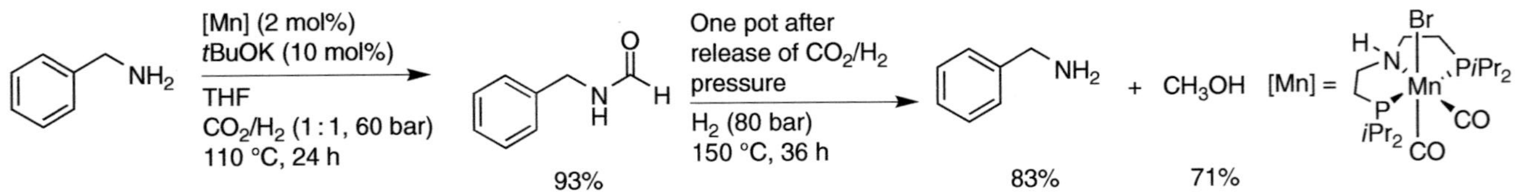

Scheme 2.3 Mn(I) pincer complex featuring a PN-sp^{3}P ligand for the hydrogenation of CO_2 to CH_3OH.

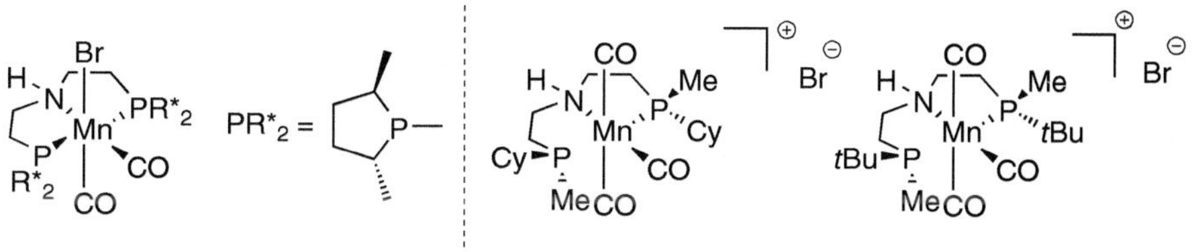

Scheme 2.4 Mn(I) complexes of chiral PN-sp^{3}P ligands involved in enantioselective hydrogenation.

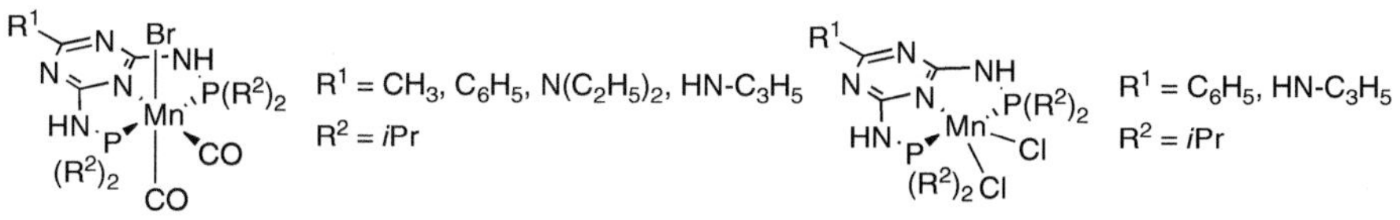

Scheme 2.5 PN$_5$P Mn(I) complexes for the hydrogenation of carbonyl derivatives.

cycloalkyl ketones, as well as aldehydes, were also converted into corresponding alcohols. Thanks to the possibility of using milder conditions, improved selectivity was obtained in particular for unsaturated ketones.

Despite lower activity compared with previous systems, Mn(I) complexes exhibiting a PN$_3$P core also catalyzed the hydrogenation of ketones [26]. The system was based on an air-stable cationic Mn precatalyst bearing a tridentate 2,6-(diaminopyridinyl)diphosphine ligand [27]. Under 50 bar of H_2 at 130 °C, various ketones were reduced to the corresponding alcohols with moderate to good yield (Scheme 2.6, left). Experimental studies have shown that the active species is likely a Mn–H complex formed *in situ* in the presence of a base and H_2, through a N–H deprotonation–dearomatization process [28].

A well-defined hydride Mn(I) complex of the same PN$_3$P ligand was found to be one of the most active catalyst for the hydrogenation of aldehydes (Scheme 2.6, right) [29]. The reaction that takes place at room temperature under base-free conditions with catalyst loadings between 0.1 and 0.05 mol% and 50 bar of H_2 allowed the selective hydrogenation of aldehydes in the presence of ketones and other reducible groups such as C=C bonds, esters, or nitriles. The same catalyst also enables the hydrogenation of CO_2 to HCOOH affording TONs up to 10 000 and quantitative yields in the presence of DBU as a base at 80 °C and 80 bar of H_2 [29]. Noteworthy in the presence of a Lewis acid (LiOTf) and catalyst loading about 0.002 mol%, TONs up to 30 000 could be reached, which stand as the highest activities reported for CO_2 hydrogenation using base metal catalysts (Scheme 2.6, right).

More recently, Milstein-type-Mn PNP catalysts built around a central pyridine were considered for the challenging deoxygenative hydrogenation of amide to amines (Scheme 2.7) [30]. The first experiments focused on the feasibility of the reaction using *N*-phenylbenzamide as a model substrate. After optimization, C—O bond cleavage was observed affording in 89% yield the corresponding secondary amine using the following catalytic conditions: Cat. 5 mol%, *t*BuOK 6 mol%, $B(C_6F_5)_3$ 1.5 equiv, 150 °C, *m*-xylene, 72 hours. The transformation was shown

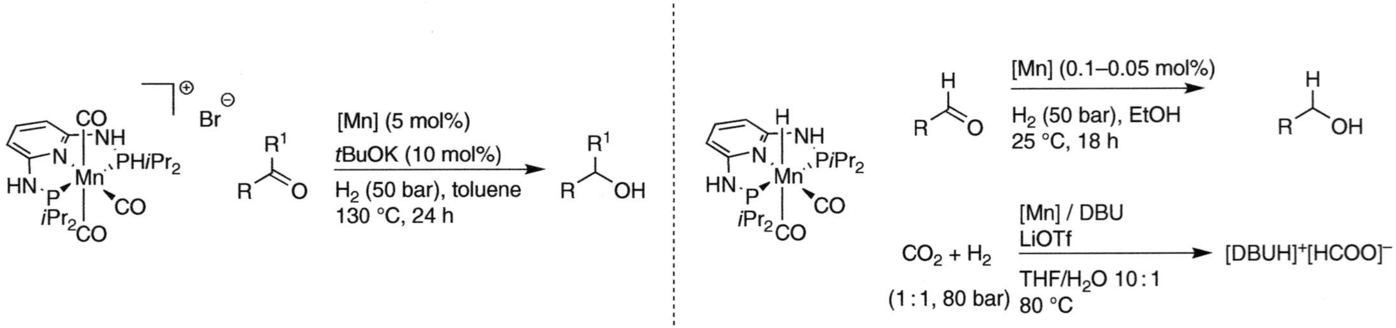

Scheme 2.6 Mn(I) complexes based on PN_3P ligands for the hydrogenation of ketones (left) and aldehydes or CO_2 (right).

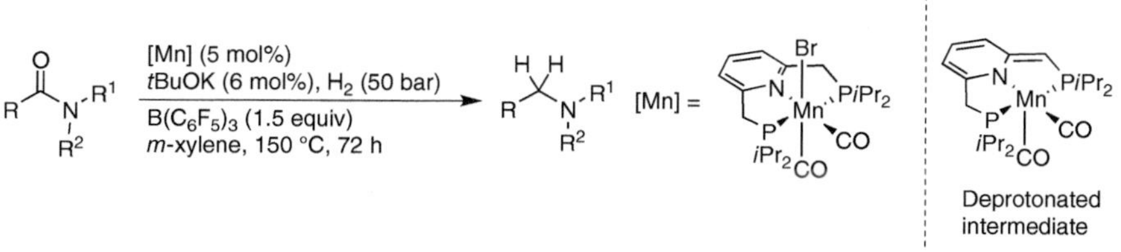

Scheme 2.7 Mn(I) catalyst featuring a PN-sp^{2}P ligand for the hydrogenation of amides into amines.

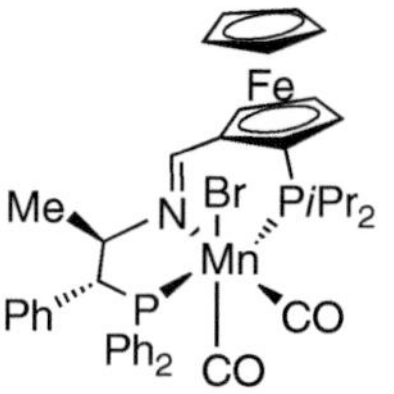

Scheme 2.8 Mn(I) complex of a chiral PN-sp^{2}P ligand for the enantioselective transfer hydrogenation.

to be highly selective, and only traces of C—N bond cleavage products, namely, aniline and benzyl alcohol, could be detected. This system enabled the efficient and selective reduction of a broad scope of amides, such as benzamides bearing substituents of different electronic nature on the phenyl group (halogen, Me, MeO), *N*-benzyl-, cyclohexyl-, hexyl benzamides, and aliphatic *N*-phenyl amides. The hydrogenation of tertiary amides was also possible, although with a lower yield. According to NMR and mechanistic studies, the authors suggested that an imine could be an intermediate in the hydrogenation of amides and that Lewis acid may accelerate the hydrogenation of the imine. On this basis, a catalytic cycle involving metal–ligand cooperation through the dearomatization of the pyridine nucleus was proposed (Scheme 2.7, right).

The enantioselective transfer hydrogenation of ketones was achieved by a Mn(I) complex containing an unsymmetrical chiral PN-sp^{2}P pincer ligand based on a planar chiral ferrocene and a central chiral aliphatic unit (Scheme 2.8) [31]. For a large range of ketones, the reactions proceeded with conversion up to 96% and enantiomeric excess values up to 86% under mild conditions at room temperature in the presence of *t*BuOK as a base. On the basis of DFT calculations, hydrogen bonding and steric factors were proposed to explain the origin of the (S) selectivity observed.

2.2.2 PNN Ligands

Following the emergence of PNP-type pincer ligands, considerable work has been realized to adjust the ligand structure, in particular by modifying the nature of the coordinating ends in order to study the impact on catalytic properties. In a logical approach, tridentate ligands featuring a PNN skeleton have thus appeared and coordinated with manganese following similar procedures to those developed in the

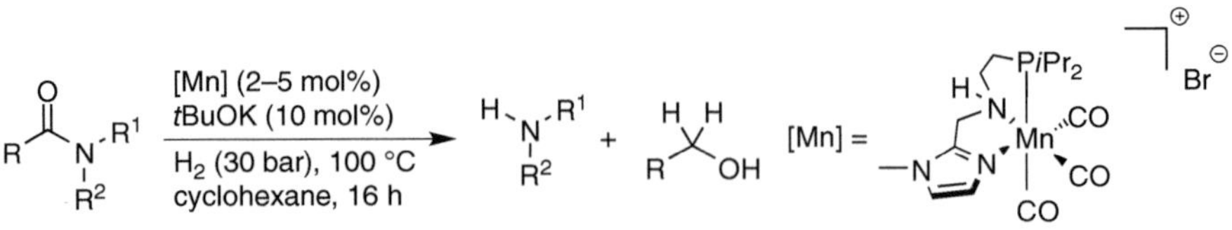

Scheme 2.9 Mn(I) complex of a PN-sp^{3}N ligand for the hydrogenation of amides.

PNP series. PNN Mn pincer complexes can be also classified in two main categories depending on the hybridization state of the central N-atom, the latter being either of sp^3-type or of sp^2-type. It should be noted that the advent of these ligands has allowed significant progress in hydrogenation Mn catalysis compared with their PNP counterparts.

2.2.2.1 PN-sp^{3}N Ligand

PNN ligands allowed significant advances in the reduction of amides. Indeed, after the successful applications of Ru complexes of such pincer ligand in this area of catalysis [32], Mn analogues were naturally envisaged [33]. For this specific purpose, neutral and cationic Mn(I) complexes containing an imidazolylaminophosphine ligand were prepared in good yield by addition of $Mn(CO)_5Br$ to the corresponding imidazolylaminophosphine precursor (Scheme 2.9). The cationic complex PNN $Mn(CO)_3{}^+Br^-$ exhibits a distorted octahedral geometry with the PNN moiety adopting a facial coordination mode. Compared with its neutral counterparts, this tripodal-type Mn complex afforded high activity and selectivity in the reduction of secondary and tertiary amides into corresponding alcohols and amines under relatively mild conditions. The selectivity obtained in this case is different to that observed with Mn(I) catalysts based on a PNP ligand [30]. The catalytic system proceeds generally using 2 mol% of catalyst and 5–10 mol% of *t*BuOK in cyclohexane at 100–120 °C. Several reducible functional groups were tolerated under these conditions such as aryl halides. While the efficiency of the catalyst was demonstrated through the reduction of primary amides and formamides, its selectivity was illustrated by the reduction of amides in the presence of other reducible groups like carbamates or ureas. From mechanistic considerations, experiments in the absence of base confirmed the role of amido complexes as active catalyst species suggesting the existence of an outer-sphere mechanism.

The imidazole nucleus can be replaced by a pyridine to form PNN Mn(I) complexes that are able to reduce CO_2-derived carbonates to alcohols [34]. Considering the interest of converting industrially produced cyclic organic carbonates to value-added alcohols, Rueping, El-Sepelgy et al. prepared Mn(I) pincer complexes featuring a tridentate pyridine amine phosphine ligand by reacting the corresponding air-stable ligands [35] with $Mn(CO)_5Br$ complex (Scheme 2.10) [34]. As a model reaction, the complex featuring a central NH donor with 0.25 mol% catalyst loading showed excellent reactivity for the reduction of ethylene carbonate affording ethylene glycol in quantitative yield and methanol in 92% yield in 1,4-dioxane at 140 °C under 50 bar of H_2. The role of the NH function was confirmed by the lower activity

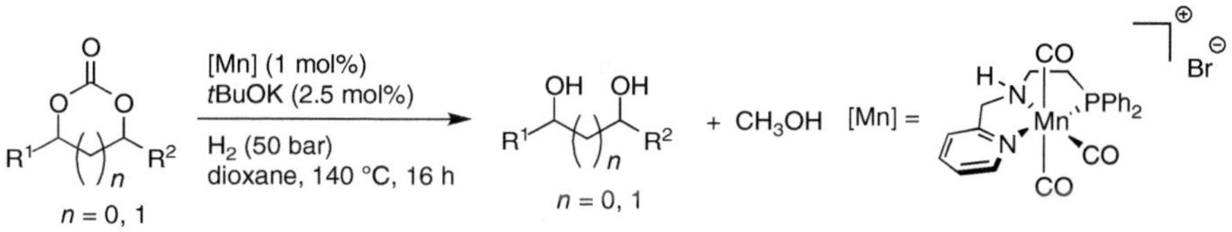

Scheme 2.10 Mn(I) complex of a PN-sp³N ligand for the hydrogenation of CO_2-derived carbonates.

obtained with the N–Me analogue complex (1 mol%), where only 14% of ethylene glycol was formed and no methanol detected. Under optimized conditions, a range of cyclic organic carbonates including five- and six-membered representatives with different alkyl and aryl substituents, as well as polycarbonates, were successfully converted to corresponding diols and methanol products with yields between 80% and 99%. Based on deuterium labeling experiments and DFT calculations, a metal–ligand cooperative catalysis mechanism was evidenced involving the heterocyclic cleavage of three H_2 molecules by metal–ligand cooperation through three successive catalytic cycles. The present system opens promising perspectives for the recycling of wastes.

Asymmetric Mn(I) complexes of such PN-sp³N ligand were considered for enantioselective ketone and ester hydrogenation (Scheme 2.11, top) [36]. From the enantiopure PNN ligand (S_C, R_P) based on a ferrocenyl backbone, cationic Mn(I) complexes exhibiting planar chirality were prepared stepwise by reaction with $Mn(CO)_5Br$. The bromo-representative was found to be active for the hydrogenation of ketones with high level of enantioselectivity (up to 97% ee) superior in some cases than those obtained with non-abundant metal catalysts, as in the case of methyl ketones bearing aromatic rings with o-substituents. A range of functionalized ketones could be thus hydrogenated using the following conditions: cat. 1 mol%, *t*BuOK 10 mol%, 50 bar H_2, 50 °C, ethanol, 16 hours. In addition to ketones, this system exhibited high activity for the hydrogenation of esters down to 0.1 mol% catalyst loading. Using 1 mol% of catalyst and 10 mol% of *t*BuOK at 75 °C in *i*PrOH, a variety of esters was reduced with good yield and functional group tolerance, including free amino *p*-phenyl substituted esters. The same catalytic system was successfully applied for the reduction of α-chiral esters (Scheme 2.11, bottom)

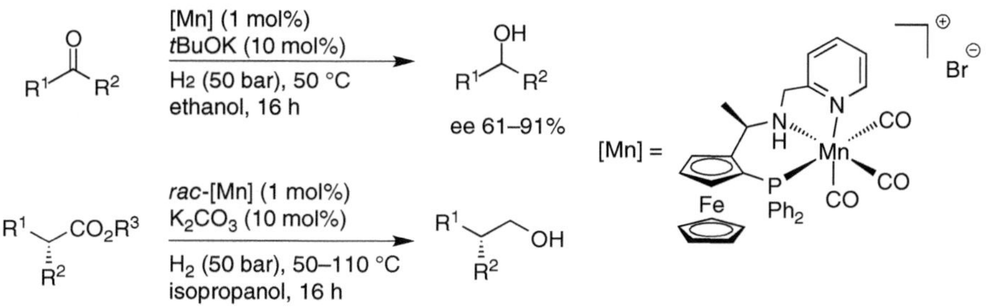

Scheme 2.11 Mn(I) complex of a chiral PN-sp³N ligand for the enantioselective reduction of ketones and α-chiral esters.

[37]. This transformation accomplished with TONs up to 1000 in the presence of potassium carbonate and alcoholic solvents occurred mainly with no loss of enantiomeric purity.

The asymmetric transfer hydrogenation of acetophenone in isopropanol could be achieved with chiral Mn(I) complexes based on tridentate P–NH–NH- and tetradentate P–NH–NH–P-type ligands [38]. Despite moderate enantioselectivity, the so-called NH effect described by Noyori and coworkers [39] was herein reported to be a requirement to obtain catalytic activity.

2.2.2.2 PN-sp^{2}N Ligand

The combination of P- and N-donor atoms also amounts to modifying the hardness vs. softness of the coordinating ends, which should make it possible to develop heteroleptic ligands with well-defined electronic properties. In this idea, Milstein and coworkers described the preparation of a PNN-type Mn pincer complex based on a central pyridine and a lateral amine [40]. This complex directly obtained by the reaction of a PNN(H) ligand and $Mn(CO)_5Br$ allowed the hydrogenation of a wide range of esters under rather mild conditions (Cat. 1 mol%, KH 2 mol%, 20 bar H_2, toluene, 100 °C) (Scheme 2.12, left) [40a]. Under these conditions, aliphatic and aromatic esters as well as esters bearing terminal alkenes, CF_3, and CN groups were converted to corresponding alcohols in excellent yields. To get more insight into the mechanism and identify active species, the Mn complex was reacted with a base knowing that the deprotonation can take place either at the benzylic position leading to the dearomatization of the pyridine or at the N–H position. A solid analysis of the deprotonated complex indicated rather the formation of an amido complex able of activating H_2 to form *cis*- and *trans*-dihydride isomers, which slowly isomerize in solution. The same Mn complex was also considered for the hydrogenation of organic carbonates to methanol and alcohols [40b]. A large variety of carbonates were thus converted under moderate catalytic conditions (Cat. 2 mol%, KH 4 mol%, 30–50 bar H_2, 110 °C). Considering the mechanism, the authors proposed the formation of formate and aldehyde intermediates through metal–ligand cooperation between the Mn center and the N–H coordinating moiety of the ligand.

Introducing a chiral phospholane resulted in the formation of asymmetric PNN pincer Mn(I) complexes combining the features of Milstein's achiral motif PNN [40] and those of a stereogenic phosphine as in the Beller's system (Scheme 2.12, middle) [23a]. Upon addition of $Mn(CO)_5Br$ complex, corresponding Mn(I) complexes were

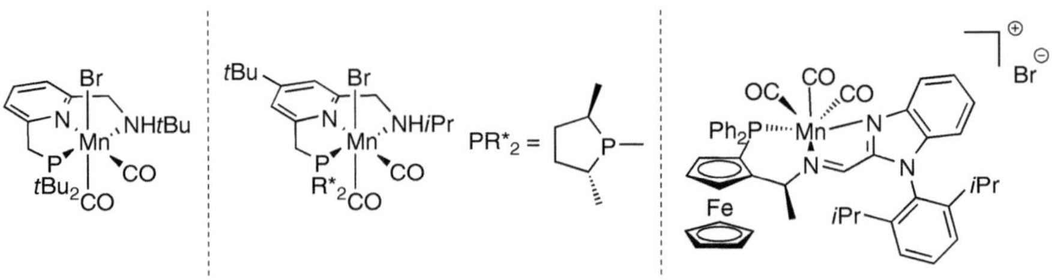

Scheme 2.12 Mn(I) complexes based on PN-sp^{2}N ligand in the chiral and achiral series.

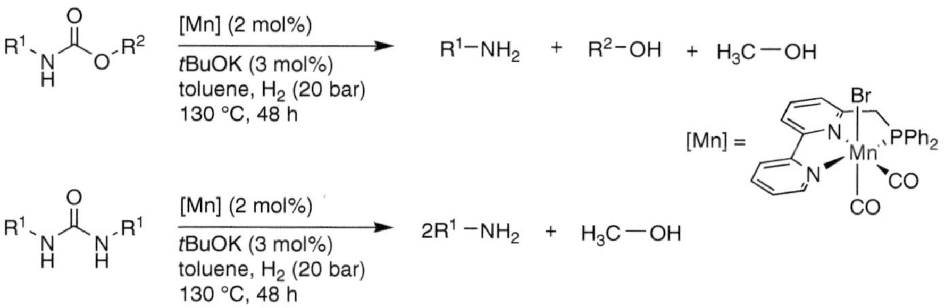

Scheme 2.13 Mn(I) complex based on a PNsp^{2}N ligand for the hydrogenation of carbamates and ureas.

obtained as a mixture of two isomers due to the syn- and anti-orientations of the flexible NH donor relative to the metal–halogen bond [41]. Under 1 mol% catalyst loading, the (*S,S*)-complex afforded full conversion in the hydrogenation of acetophenone with 67% ee. Addition of alcohols such as *i*PrOH or $(CF_3)_2CHOH$ resulted in the improvement of both activities and selectivities. The best result was obtained with the (*R,R*)-complex where the catalyst loading was diminished to 0.01 mol% while maintaining quasi-full conversion with 86% ee and a TON of 9800. With this system, a wide range of aryl–alkyl and heteroaromatic ketones could be converted to corresponding chiral secondary alcohols in 85–97% ee. Benzofused cyclic- and diaryl ketones could be also transformed with high enantioselectivity. Experimental studies confirmed the critical role of the NH moiety in agreement with the existence of an outer-sphere mechanism.

Mn(I) catalysts bearing imidazole-based chiral PN-sp^{2}N tridentate ligands have been prepared allowing the asymmetric hydrogenation of unsymmetrical benzophenones (Scheme 2.12, right) [42]. This system characterized by a high activity (up to 13 000 TON) and enantioselectivity (up to >99% ee) tolerates a large range of substrates and presents a good tolerance of the functional group. The imine group was invoked to play a key role in the cleavage of H_2 and the activation of the substrate.

The hydrogenation of challenging carbamate and urea substrates was recently reported, thanks to the efficiency of the Mn(I) pincer complex of a PN-sp^{2}N pincer ligand (Scheme 2.13) [43]. In both cases, under slight H_2 pressure (20 bar), methanol formation was observed in addition to the amine and alcohol products. This transformation, which represents a sustainable method for the conversion of CO_2 to methanol, implies a cooperative metal–ligand mode via the dearomatization of the pyridine nucleus.

2.2.3 NNN Ligands

Transfer hydrogenation is a complementary process to classical hydrogenation, and by varying the nature of donor extremities, the group of Beller discovered that Mn(I) complexes featuring a NNN pincer ligand act as very efficient catalysts for the transfer hydrogenation of ketones (Scheme 2.14) [44]. These Mn complexes

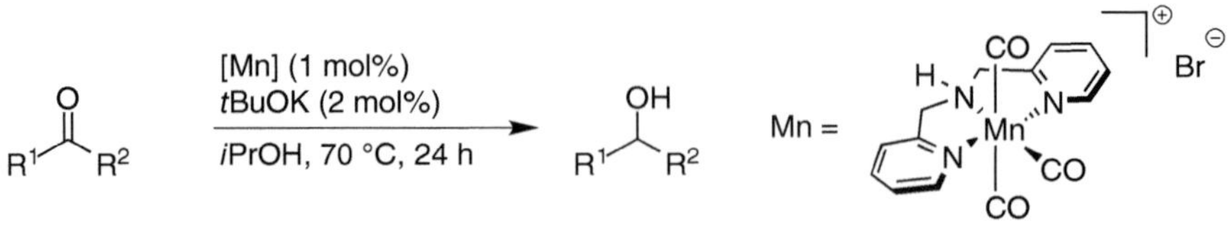

Scheme 2.14 Mn(I) complex of a NN-sp^{3}N pincer ligand for the transfer hydrogenation of ketones.

were easily obtained from commercially available tridentate amine (dipicolylamine) and $Mn(CO)_5Br$ complex. Under optimized conditions (Cat. 1 mol%, *t*BuOK 2 mol%, 70 °C, *i*PrOH), a broad range of hetero(aromatic) and aliphatic ketones could be reduced with good to excellent yields, especially substrates containing nitrile and ester groups. Cyclic ketones, such as cyclohexanone or 1-ethyl-4-piperidone were also transformed into corresponding alcohols in 90% and 85% yield, respectively. Chemoselectivity was illustrated through the selective reduction of dihydro-β-ionone in 97% yield. The only limitation concerns aldehydes, which were found to be unreactive in the developed catalytic conditions. Mechanistic studies indicate that the reaction occurs through a mono-hydride species but that cooperation with the NH moiety is not essential since the N–Me analogue led to similar catalytic results [45]. Even if the mechanism of this process remains unclear to date, an outer-sphere mechanism can be excluded and the catalytic results obtained raise the question of whether the aromatization/dearomatization process through the activation of the benzylic arm that takes place in related PNP pincer complexes applies here. Noteworthy, the same type of NNN ligand was also reported to be efficient for the transfer dehydrogenation of secondary alcohols [45].

2.3 Non-pincer-Type Manganese Complexes

The tridentate coordination mode is not a prerequisite for carrying out hydrogenation-type reactions with manganese complexes. By combining N- and P-donor atoms, different bidentate Mn(I) systems were thus successfully developed as was previously done in the pincer series. The singularity of the bidentate series concerns the presence of the carbon element as coordinating atom. Carbon was introduced here into its sp^2-hybridization state via an NHC.

2.3.1 NN Ligands

Complexes based on NN ligand are the most developed systems in the bidentate series. Three families can be distinguished depending on the hybridization state of the N-donor atoms: the N-sp^3,N-sp^3; N-sp^2,N-sp^2; and the N-sp^3,N-sp^2 families.

2.3.1.1 N-sp^3,N-sp^3 Ligand

In order to develop simple, sustainable, and practical catalytic systems, Sortais and coworkers reported that nitrogen-based bidentate Mn(I) catalysts featuring a chelating diamine ligand were effective for the transfer hydrogenation of

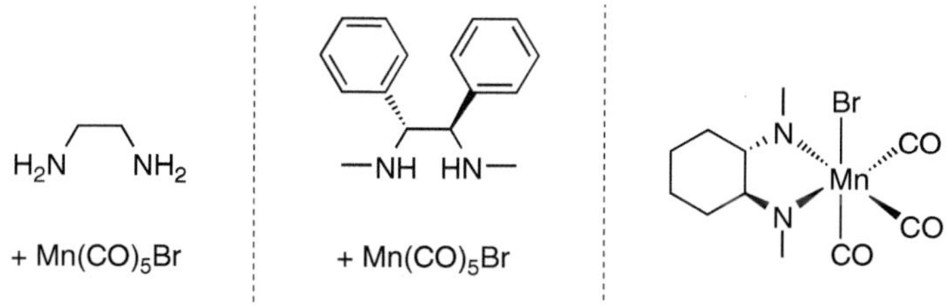

Scheme 2.15 N-sp^3,N-sp^3 ligands involved in hydrogenation-type reactions.

ketones (Scheme 2.15, left) [46]. A series of simple diamine ligands in the presence of $Mn(CO)_5Br$ as a metal precursor (0.5 mol%), *t*BuOK as a base (1 mol%), and 2-propanol as a reductant was screened for the reduction of acetophenone. Using ethylene diamine as a ligand, the transfer hydrogenation reaction was shown to proceed in high yield at 80 °C in three hours. The generality of the reduction with the *in situ* prepared Mn(I) ethylene diamine system was then demonstrated through the effective conversion of various para-substituted acetophenone substrates into corresponding alcohols. Encouraged by these results, chiral diamines were also engaged as illustrated with the formation of sterically hindered alcohols with enantiomeric excess up to 90% in the presence of the chiral (1*R*,2*R*)-*N*,*N*′-dimethyl-1,2-diphenylethane-1,2-diamine (Scheme 2.15, middle) [46]. This approach demonstrated for the first time that simple bidentate phosphine free ligands can be associated with manganese to promote (enantio)selective reactions in the field of hydrogenation.

Based on this preliminary report, Pidko and coworkers described the preparation and the catalytic performance of a series of Mn(I) complexes bearing simple chiral diamine ligands in the asymmetric transfer hydrogenation of acetophenones to the corresponding alcohols with 75–87% ee (Scheme 2.15, right) [47]. From these complexes, experimental and DFT calculations, as well as kinetic studies, were performed evidencing the complexity of the different intermediates responsible for the catalytic activity in terms of (enantio)selectivity.

2.3.1.2 N-sp^2,N-sp^2 Ligand

Inspired by nature with the role of Fe hydrogenases based on an *ortho*-OH-substituted pyridine backbone, related Mn(I) catalysts were developed for the hydrogenation of CO_2 (Scheme 2.16). A series of air-stable neutral and cationic Mn(I) complexes bearing readily available substituted 2,2′-bipyridyl ligands were thus prepared and fully characterized [48]. In the series, 6,6′-dihydroxy-2,2′-bipyridyl-based Mn catalysts were shown to present high activity for CO_2 hydrogenation to formate in the presence of DBU base at relatively low temperature, leading to TON up to 6250. The hydrogenation of CO_2 to formamide was also achieved in 72% yield with such complex in the presence of diethylamine as a base with a TON of 588 after 24 hours at 80 °C.

In the continuity of this work, the same type of Mn complex based on a bidentate bipyridyl-type ligand was used for the transfer hydrogenation of ketones, aldehydes, and imines [49]. In addition to reduce a broad scope of ketones and aldehydes

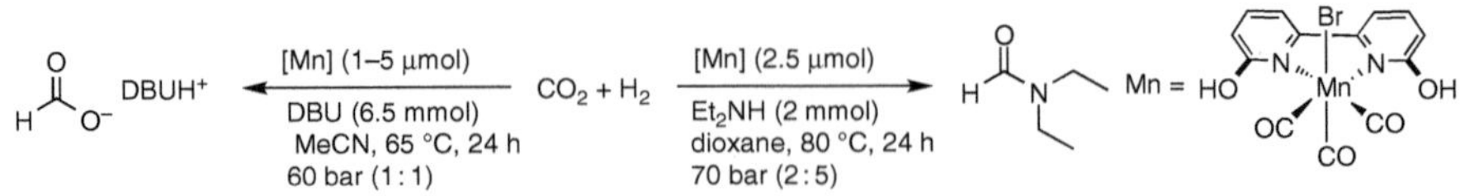

Scheme 2.16 Mn(I) complex based on a N-sp²,N-sp² ligand for the hydrogenation of CO_2.

with good functional group tolerance, the specificity of the present system lies in the reduction of aromatic N-heterocycles into corresponding saturated cyclic amines. For instance, acridine, quinoxaline, and 1,5-naphthydrine could be reduced with 0.5 mol% cat. loading, 1.5 mol% of *t*BuOK at 80 °C in 84%, 94%, and 92% yield, respectively. Among different substituted bipyridine ligands, the representative bearing two hydroxyl groups appeared to be the best catalyst, surpassing its analogue based on two methoxy groups. Labeling experiments with the help of DFT calculations concluded on the active role of the hydroxyl groups in the hydrogen transfer steps through the formation of mono-hydride species.

Catalytic activity in the transfer hydrogenation of ketones and aldehydes was also observed with 2-hydroxy-substituted NN ligands when a pyridine nucleus is replaced by a quinoline or naphthyridine moiety. The best representative allowed the formation of a series of secondary alcohols in satisfactory yields according to the following catalytic conditions: Cat. 0.5 mol%, *t*BuOK 20 mol%, 85 °C, 24 hours [50]. Noteworthy, the preparation of 1,2-disubstituted benzimidazoles and quinolines via acceptorless dehydrogenation condensations was shown to be possible with the latter type of catalysts.

2.3.1.3 N-sp²,N-sp³ Ligand

In the quest for simple and inexpensive catalytic systems, mixed N-sp²,N-sp³ ligands were also envisaged as exemplified with the transfer hydrogenation of carbonyl derivatives catalyzed by Mn(I) complexes bearing 2-(aminomethyl) pyridine ligands (Scheme 2.17, left) [51]. The Mn(I) complexes of interest were prepared stepwise by reacting aminomethyl pyridine precursors with the $Mn(CO)_5Br$ complex. A large variety of ketones and aldehydes, including α,β-unsaturated aldehydes could be reduced under mild conditions, that is, at room temperature with low catalyst loading down to 0.1 mol% and 2-propanol as hydrogen source while having a high tolerance toward functional groups. TONs and turnover frequencies (TOFs) up to 2000 and 3600 h^{-1}, respectively, were observed under these catalytic conditions. In the understanding of the catalytic cycle, a dimeric Mn(I) complex resulting from the deprotonation of the NH moiety was isolated as an intermediate species consistent with the generally accepted mechanism of ligand-assisted hydrogen transfer reactions [52].

The well-defined Mn(I) complex of the bidentate 2-(aminomethyl) pyridine ligand was also used for the transfer hydrogenation of imines [53]. Using 2-propanol as a reductant, in the presence of *t*BuOK (4 mol%) and the precatalyst (2 mol%), a large variety of aldimines (30 examples) was reduced in three hours at 80 °C with good to excellent yield. Here again, the catalytic conditions were reported to be tolerant with

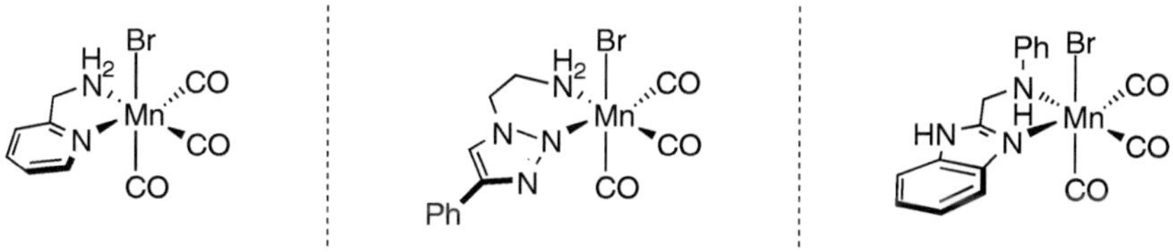

Scheme 2.17 Mn(I) complexes bearing N-sp^2,N-sp^3 ligands for hydrogenation-type reactions.

a broad range of functional groups (halogen, methoxy, cyano, ester, amido, acetal, vinyl, and alkynyl).

On the basis of encouraging results obtained with simple NN ligands, alternative N-donor units were then envisaged. In this logic, several Mn(I)$(CO)_3$Br complexes based on bidentate amino triazole ligands with different substitution pattern were prepared and investigated in the transfer hydrogenation of ketones (Scheme 2.17, middle) [54]. Under mild conditions, at 80 °C for 20 hours with 3 mol% of catalyst loading using either *t*BuOK or NaOH as a base, a wide scope of ketones with large functional group tolerance was reduced to the corresponding alcohols in good to excellent yield. The critical role of the amine function was demonstrated by comparing the catalytic activity of the amino triazole complex with a related complex based on an imino triazole backbone. The superiority of the former evidences the existence of an outer-sphere hydrogen transfer with the NH_2 coordinating group acting as a proton source.

Catalytic performances were also observed in the hydrogenation of carbonyl derivatives and imines with Mn(I) complexes featuring cooperative amino benzimidazole ligands (Scheme 2.17, right) [55]. It was established that both benzimidazole and NH amine donors contribute to the high catalytic activity observed as clearly demonstrated with the higher reactivity of the *N*-((1*H*-benzimidazol-2-yl)methyl) aniline-based Mn complex. From this observation, a concerted outer-sphere mechanism was proposed supported by DFT calculations. Concerning its catalytic activity, the present system (Cat. 0.2 mol%, *i*PrONa 20 mol%, 90 °C, two hours) allowed the efficient reduction of a wide range of ketones, aldehydes, and challenging imines, as well as unsaturated ketones with good chemoselectivity.

2.3.2 PP Ligands

Bis-phosphine ligands were first considered by Kirchner and coworkers for the manganese-catalyzed hydrogenation of nitriles and ketones [56]. The catalytic active Mn(I) complex was obtained in 71% yield by treating the chelating bis-phosphine $n\mathrm{Pr_2PCH_2CH_2P}n\mathrm{Pr_2}$ with the $Mn(CO)_5Br$ complex (Scheme 2.18, right). This catalytic system allowed the effective hydrogenation of a range of (hetero)aromatic and aliphatic nitriles into primary amines. Electron-donating (Me, MeO, H_2N), electron-withdrawing groups (Cl, Br), and heterocycles were well tolerated in the selected conditions. Chemoselectivity was observed as illustrated with the reduction of the nitrile function in the presence of an ester group or a C=C bond.

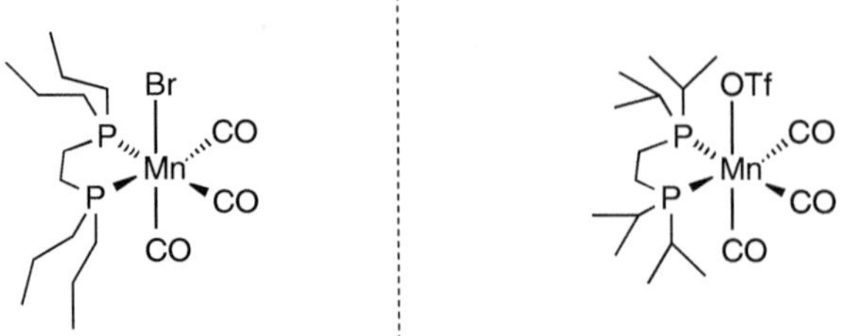

Scheme 2.18 Mn(I) complexes bearing bidentate PP ligands for hydrogenation reactions.

Also, 4-((trimethylsilyl)ethynyl) benzonitrile was reduced into the corresponding amine in 67% yield without affecting the CC triple bond. Linear aliphatic nitriles and decanedinitrile could be also converted in high yields. The best conditions developed for the hydrogenation of nitriles were the following: Cat. 2 mol%, *t*BuOK 20 mol%, 100 °C, 18 hours. Aromatic and aliphatic ketones were also reduced into the corresponding alcohols in excellent yields (71–96%). The hydrogenation reaction occurred in this case at 50 °C with a catalyst loading of 1 mol% in the presence of 5 mol% of base. On the mechanism, the authors proposed the role of a hydride complex as an active species, arguing that metal–ligand cooperation cannot take place and that an outer-sphere process without involvement of the ligand can probably happen.

Almost simultaneously, Garcia and coworker reported an analogous non-pincer Mn(I) complex for the same catalytic purpose (Scheme 2.18, right) [57]. The cationic complex *fac*-[$(CO)_3Mn(iPr_2P(CH_2)_2PiPr_2)(OTf)$] based on an electron-rich bisphosphine ligand as a catalyst precursor (3 mol%) exhibited catalytic activity toward the hydrogenation of (hetero)aromatic and aliphatic nitriles in the presence of the *t*BuOK base (10 mol%) affording primary amines in good yield (83–98%) under rather mild conditions (7–35 bar H_2, 90 °C, 15–30 minutes, 2-BuOH). The efficient hydrogenation of dinitriles is worth noting, such as terephthalonitrile and adiponitrile into corresponding diamines. As in the case of Kirchner's report, unsaturated Mn hydride complexes have been discussed to participate in the determining catalytic steps.

The same authors observed that secondary alcohols such as 2-butanol can substitute for molecular hydrogen in the hydrogenation of nitriles [58]. For such an application, they used the *fac*-[$(CO)_3Mn(iPr_2P(CH_2)_2PiPr_2)(Br)$] complex more easily synthesizable than the cationic Mn–OTf analogue. The Mn–Br complex (3 mol%) exhibited catalytic activity in the presence of *t*BuOK (10 mol%) for the transfer hydrogenation of benzonitrile leading to a mixture of benzylamine and *N*-*sec*-butylidenebenzylamine. After acidic treatment of the mixture, benzylamine hydrochloride can be formed in 96% yield. According to this strategy, a series of amine hydrochlorides was readily prepared (39–92% isolated yields). Aromatic nitriles bearing electron-donating groups (Me, MeO) were more easily converted than those containing electron-withdrawing groups (CF_3). Like in the conventional hydrogenation reaction, unsaturated Mn hydride species were proposed to be involved in the catalytic cycle.

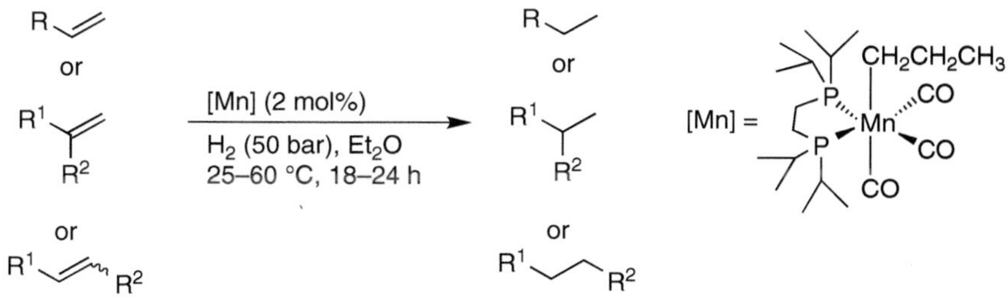

Scheme 2.19 Mn(I) complex of a bidentate PP ligand for the hydrogenation of alkenes.

Manganese-based catalysts capable of hydrogenating inactivated C=C bonds are still rare. On the basis of this finding, Kirchner and coworkers took advantage of the readily available *fac*-[$(CO)_3Mn(i\text{Pr}_2P(CH_2)_2Pi\text{Pr}_2)(CH_2CH_2CH_3)$] complex to perform the Mn-catalyzed hydrogenation of alkenes to alkanes (Scheme 2.19) [59]. From a mechanistic point of view, the catalytic transformation is initiated by migratory insertion of a CO ligand into the Mn—$CH_2CH_2CH_3$ bond to form an acyl intermediate that affords the catalytically active Mn(I) hydride complex after hydrogenolysis. Except for 4-vinylpyridine (39% of conversion), this strategy allowed the hydrogenation of mono- and disubstituted alkenes into corresponding alkanes in very satisfactory yields (70–99%). Additionally, the developed conditions (Cat. 2 mol%, 50 bar H_2, 25–60 °C, 18–24 hours) are compatible with a high functional group tolerance. The only difference is relative to the temperature, that is, the hydrogenation of monosubstituted alkenes and 1,1-disubstituted alkenes take place at 25 °C, while 1,2-disubstituted representatives require a higher temperature of about 60 °C. According to DFT calculations and experimental studies, an inner shell mechanism was proposed with catalytic steps involving the successive protonation of the internal C=C bond in the Mn hydride complex followed by H_2 coordination, the hydride insertion into the Mn—C bond, and finally the release of the alkane product.

2.3.3 NP Ligands

2.3.3.1 N-sp^3,P-sp^3 Ligand

Non-pincer PN Mn(I) complexes based on simple and easily accessible bidentate aminophosphine ligands were first reported by Pidko and coworkers (Scheme 2.20, left) [60]. They were easily prepared by reacting the $Mn(CO)_5Br$ complex with corresponding PN precursors, and depending on the ratio metal/ligand used, mono- or diligated PN Mn(I) complexes were selectively obtained. In the cationic octahedral bis(PN) $Mn(CO)_2$ complex, the two N-donor groups are located in cis-position, while the two P-donor groups bind in trans-position to each other. Both types of Mn(I) complexes were found to be active for ester hydrogenation. However, by using methyl benzoate as a model substrate, the mono-ligated representative exhibited higher activity compared with diligated complexes. Aliphatic and aromatic esters were effectively reduced to very low catalytic levels down to 0.2 mol% in the presence of *t*BuOK base. Sterically demanding aliphatic esters were almost quantitatively hydrogenated, while functional group tolerance (Cl,

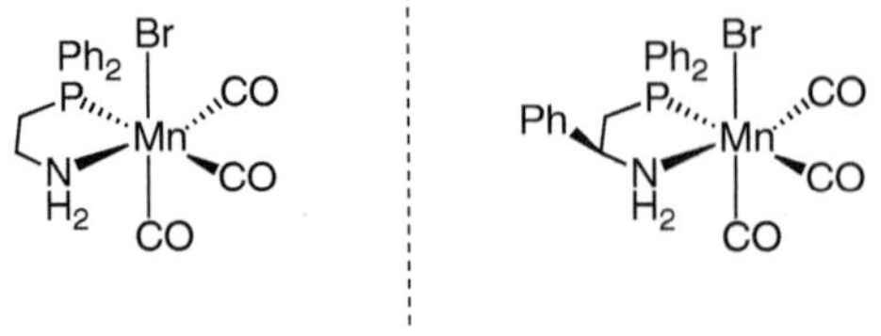

Scheme 2.20 Mn(I) complexes based on bidentate NP ligands in hydrogenation reactions.

MeO) was observed with para-substituted aromatic substrates. Chemoselectivity of unsaturated esters was only reached for substrates where the C=C bond is located far from the ester group, such as in fatty acid methyl esters. The exact role of the base was experimentally studied and rationalized by DFT calculations, suggesting possible inhibition through the formation of stable Mn–alkoxide species acting as resting states [61].

PN ligands based on amino acids were also considered for hydrogenation-type reactions. Mn(I) complexes bearing a β-amino phosphine ligand derived from phenylglycine were reported to act as efficient catalysts for the transfer hydrogenation of ketones and chalcones (Scheme 2.20, right) [62]. The main advantage lies in the high accessibility of β-amino phosphines that can be easily produced from cheap amino acid feedstocks. With such bidentate PN Mn(I) complex, transfer hydrogenation of acetophenone in 2-propanol was found to be quantitative at 60 °C with 0.5 mol% catalyst loading and 1 mol% of base. The catalytic conditions tolerated a broad scope of para- and ortho-substituted acetophenones. Good functional tolerance was generally achieved under mild conditions excepted for phenolic acetophenones, which appeared to be non-reactive. The challenging *p*-nitroacetophenone was reduced in 71% yield after refluxing three days. Thanks to the unique properties of β-amino phosphine ligands, chalcone derivatives were selectivity reduced to the saturated ketones, further reduction to the corresponding alcohols being observed only under drastic conditions. The isomerization of allylic alcohols was also described with β-aminophosphine-supported Mn(I) catalysts.

2.3.3.2 N-sp^2,P-sp^3 Ligand

Mn(I) complexes featuring bidentate 2-aminopyridinyl phosphine ligands were reported to be highly active in the hydrogenation of carbonyl derivatives (Scheme 2.21) [63]. Different Mn(I) complexes were readily prepared in excellent yield upon treatment of $Mn(CO)_5Br$ with corresponding PN ligands derived from the reaction between a chlorophosphine R_2PCl (R = Ph, *i*Pr) and 2-aminopyridine, 2-picoline, or 2-methylquinoline. All complexes exhibit a typical octahedral geometry for the Mn center, the PN ligand being coordinated in a κ^2P,N-fashion with the three CO ligands adopting a facial arrangement. The catalytic activity of these heteroleptic complexes was evaluated for the hydrogenation of a wide range of carbonyl derivatives, such as para-substituted aromatic ketones, aliphatic ketones, unsaturated ketones, and para-substituted benzaldehydes. The reaction proceeds generally with low catalyst loading (0.5 mol%) under mild conditions (50 °C) with yields up to 96%. High functional group tolerance was also observed compared

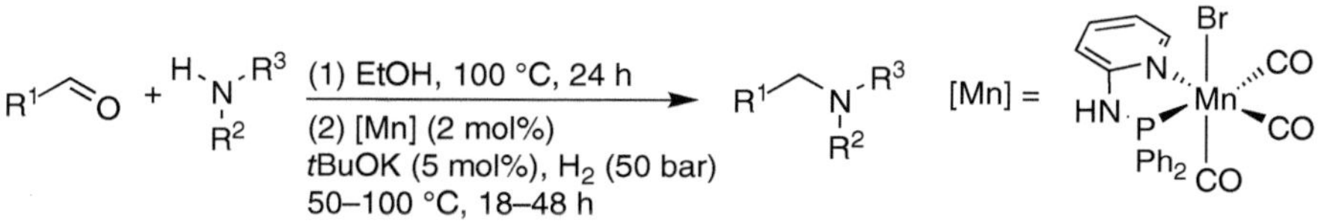

Scheme 2.21 Mn(I) complex featuring a PN ligand for the reductive amination of aldehydes.

to the most active catalytic systems. Based on these results, the authors point out that bidentate complexes can compete with pincer systems, knowing that these complexes bearing 2-aminopyridinyl phosphine are more effective than related PNP-type pincers [26], reducing the catalyst loading by a factor of 10, and lowering the temperature from 130 to 50 °C with the same activity and chemoselectivity. Noteworthy such Mn pyridinyl phosphines complexes were also used for the alkylation of amines via reductive amination of aldehydes using molecular dihydrogen as a reductant [64]. After the initial condensation step, the reduction of imines formed *in situ* was achieved under mild conditions with 2 mol% cat. loading, 5 mol% of *t*BuOK at 50 °C in ethanol under 50 bar of hydrogen. High yields were obtained for a large range of aldehydes and amines (Scheme 2.21).

2.3.4 CP Ligand

Considering the unique steric and electronic properties of NHC ligands and the recent developments in the field of NHC–Mn(I) catalysis, bidentate ligands incorporating electron-rich NHC donors were naturally envisaged for hydrogenation reactions (Scheme 2.22). In this direction, the Mn(I) NHC-phosphine complex *fac*-[MnBr(CO)$_3$(κ^{2}P,C-Ph$_2$PCH$_2$NHC)] was readily prepared in 86% yield from the corresponding phosphine imidazolium salt through the sequential addition of KHMDS base and Mn(CO)$_5$Br complex (Scheme 2.22, left) [65]. The facial arrangement of the three CO co-ligands was confirmed by a solid-state analysis. This complex could be selectively deprotonated at the carbon position located between the two donor moieties to afford an original 18-e NHC-phosphinomethanide complex (Scheme 2.22, middle), which behaves as a reservoir for an unconventional 16-e NHC-phosphonium ylide complex able to activate dihydrogen. Thanks to this new mode of metal–ligand cooperation based on the formal interplay between λ^5- and λ^3-P species, this Mn(I) complex acts as one of the most efficient system for the hydrogenation of ketones. For the conversion of acetophenone into

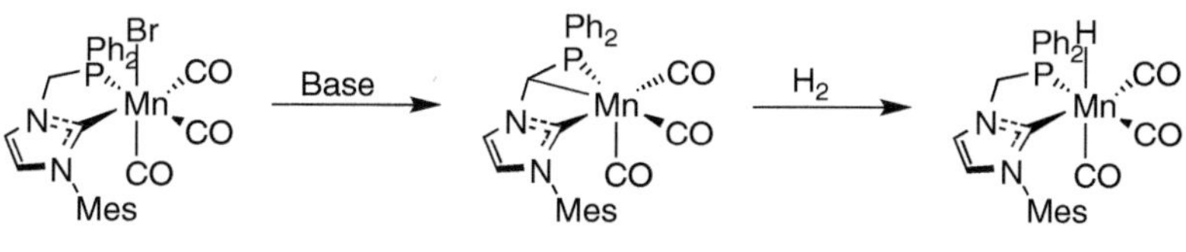

Scheme 2.22 Mn(I) complex based on a bidentate CP ligand for the hydrogenation of ketones.

1-phenylethanol, a maximum TON of 6200 was indeed achieved using 0.01 mol% of catalyst in *t*AmOH at 100 °C and with *t*BuOK as a base. A large variety of aryl(alkyl)ketones could be readily reduced including sterically hindered representatives inaccessible with Mn catalysts based on bidentate phosphine aminopyridine ligand. The reaction was reported to be tolerant to aryl groups substituted with halogen atoms and Me, MeO, and CF_3 substituents. Aliphatic 2-decanone as well as heterocyclic substrates bearing coordinating groups were also reduced but with lower efficiency. Concerning the mechanism involved, the authors noted that the Mn(I) hydride complex formed by adding dihydrogen to the deprotonated intermediate (Scheme 2.22, right) required the presence of a base to catalyze the hydrogenation reaction. Anyway, the presence of the R_2PCH_2 motif in many transition metal complexes opens new perspectives in the field of non-innocent ligands for applications in homogeneous catalysis [66].

2.3.5 CN Ligand

After the association of NHC and phosphine donor moieties, Pidko and coworkers reported the preparation of Mn(I) complexes bearing strongly donating bidentate NHC-amine ligands (Scheme 2.23) [67]. The pre-ligands were prepared by reacting 1-mesityl-1*H*-imidazole with 2-bromoethylamine hydrobromide salts. The Mn(I) complexes were then obtained in satisfactory yield through the reaction of *in situ* generated free NHCs with Mn(CO)Br. While the complex featuring the N(Me)H donor was found to be highly active for the transfer hydrogenation of ketones with *i*PrOH, the $N(Me)_2$ analogue appeared to be inactive under the same conditions. The Mn(I) NHC–N(Me)H complex exhibited unprecedented catalytic activity affording quasi-quantitative reduction of acetophenone with very low metal concentration down to 75 ppm (0.0075 mol%). A wide range of aromatic ketones could be reduced, except for very hindered representatives, such as 2,4,6-trimethylacetophenone. Aliphatic cyclic and linear ketones were also converted to corresponding alcohols but generally with lower efficiency. Chemoselectivity was observed for the reduction of unsaturated ketones, as illustrated in the case of benzylideneacetone with the formation of the corresponding unsaturated alcohol in 82% yield. Functional group tolerance was demonstrated with the efficient reduction of para-substituted acetophenones bearing halide atoms, phenyl, and esters groups. Lower conversions were however observed for substrates containing hydroxyl, cyano, and nitro substituents. Mechanistic studies evidenced that metal–ligand cooperativity operates through a bifunctional protonation/deprotonation mechanism, thanks to the

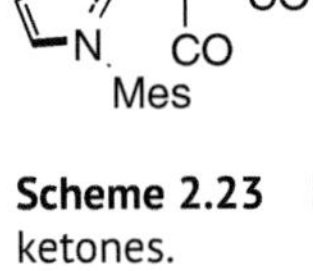

Scheme 2.23 Mn(I) complex based on a bidentate CN ligand for the hydrogenation of ketones.

presence of the N(Me)H donor group. The authors concluded that the present catalytic Mn system can compete with more conventional catalysts based on Ru or Ir noble metals.

2.4 Other Manganese Complexes

Bidentate and pincer-type manganese complexes based on C, N, P-donor atoms are not the only possible systems for catalyzed hydrogenation-type reactions. Mono- and tetradentate manganese complexes built from different atoms such as oxygen and silicon exist.

2.4.1 Monodentate Ligand

N-Heterocyclic silylenes (NHSi) are not only exotic species but can act as strong donor ligands toward transition metals as behave their famous NHC analogues. Driess and coworkers thus reported the preparation of monodentate NHSi, bidentate bis(NHSi) ferrocene, and pincer-type bis(NHSi) pyridine Mn(II) complexes from the reaction between $MnCl_2$ and related ligands (Scheme 2.24) [68]. Whatever the denticity of the ligand, all complexes exhibit a distorted tetrahedral environment for the Mn(II) center. In the pincer series, the central pyridine moiety is reluctant to coordinate the metal because of the stronger σ-donor character of the two NHSi extremities. All NHSi Mn(II) complexes were successfully engaged in the transfer semi hydrogenation of alkynes using ammonia–borane as a hydrogen source. From 1-phenyl-substituted alkynes, using 1 mol% of catalyst, all systems led to high conversions and *E*-selectivity, the pincer-type Mn complex being the most efficient catalyst showing excellent *E*-stereoselectivity up to 98%. Despite a certain tolerance to functional groups, the presence of CN, NH_2, NO_2, and OH substituents at the phenyl group of 1-phenyl substituted alkynes was found to inhibit the reaction, probably due to a coordination of the active species. On the mechanism, Mn hydride active species were proposed to be involved in the catalytic cycle supported by IR measurements that evidence the presence of the corresponding Mn–H frequency at $1900\ cm^{-1}$. These results offer new perspectives in Mn-catalyzed hydrogenation reactions by the use of other types of ligands.

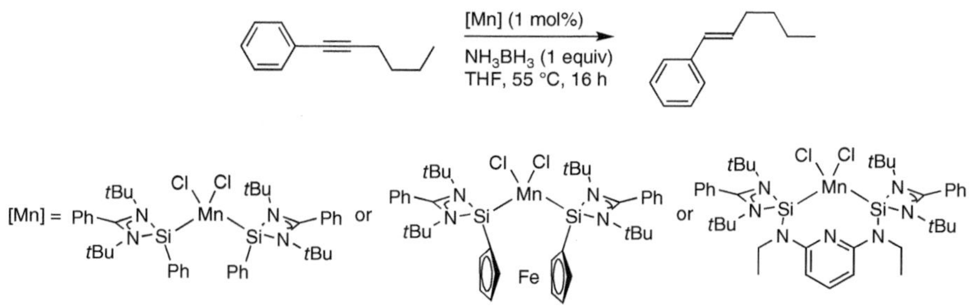

Scheme 2.24 Mn(II) complexes featuring Si-based ligands for the semi hydrogenation of alkynes.

2.4.2 Tetradentate Ligand

In the presence of $Mn(CO)_5Br$, the ONNO-type (*S*,*S*)-Jacobsen ligand was reported by Beller and coworkers to catalyze the asymmetric transfer hydrogenation of acetophenone with 36% ee (Scheme 2.25, left) [69]. Following this observation, other tetradentate ligands combining N- and O-donor atoms, especially those based on inexpensive C_2-symmetric bisoxalamides, were evaluated for such application [69]. The ligands were easily prepared by amidation of dimethyl oxalate with different amino alcohols. The *in situ* generated Mn(I) precatalyst derived from *N*,*N*′-bis(2-hydroxyethyl)oxamide ligand exhibited high activity for the reduction of aliphatic ketones with ee up to 93% (Scheme 2.25, middle). Under optimized conditions ($[MnBr(CO)_5]$ 6 mol%, ligand 2 mol%, *t*BuOK 20 mol%, *i*PrOH, 80 °C, 20 hours), a variety of cyclic aliphatic prochiral ketones could be reduced with satisfactory yields and selectivities opening the door to a new class of ligands.

In this direction, a bis(carbonyl) Mn(I) complex containing a chiral $(NH)_2P_2$ macrocyclic ligand was recently reported to be the most effective catalyst for the asymmetric transfer hydrogenation of ketones (Scheme 2.25, right) [70]. Under optimized conditions (Cat. 2 mol%, *t*BuONa 10 mol%, 50 °C, three hours), acetophenone was reduced to (*R*)-1-phenylethanol in 97% yield and 94% ee in the presence of 2-propanol as a hydrogen source. As well, a broad scope of aryl–alkyl ketones was converted to secondary alcohols with yields up to 99% and excellent enantioselectivity (90–99%). High functional group tolerance was observed as demonstrated with the effective reduction of phenyl ketones bearing Me, MeO, halogen, F_3C, CF_3O, Me_2N, MeO_2S groups, and CC triple bonds. Polyaromatic, heteroaromatic, and sterically hindered aryl–alkyl ketones were also substrates well tolerated in terms of activity and enantioselectivity. In the latter case, the present system outperforms all asymmetric transfer hydrogenation catalysts described so far, thus evidencing the critical role of the macrocyclic structure. Based on DFT calculations, attractive CH–π interactions between the ketone and the catalyst would be mainly at the origin of the high enantioselectivity observed.

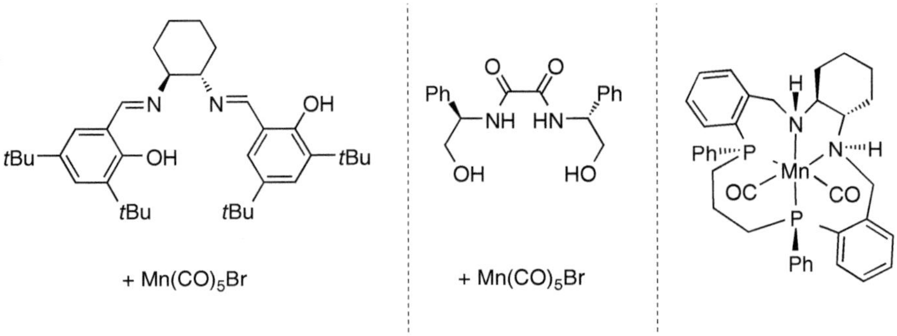

Scheme 2.25 Polydentate systems combining N- and O-donor atoms for the hydrogenation of ketones.

2.5 Conclusions

In a very short period of time, the use of base metals and especially that of manganese for catalytic purposes has grown considerably not only due to its availability but also to its unique properties. Hydrogenation-type reactions represent one of the most extensively developed fields of manganese catalysis, and the recent accomplishments confirm that manganese can provide a valuable alternative to noble metals. The design of a large diversity of chelating ligands, mainly in the bi- and tridentate series combining different coordinating extremities based on N-, P-, and C-donor atoms, largely explains this huge and rapid development. In addition to electronic (donation vs. acceptation) and steric properties (rigidity vs. flexibility) intrinsic to each ligand, the metal–ligand cooperativity has undoubtedly played a determinant role in the successful applications of manganese complexes in hydrogenation catalysis. The metal here is not the only actor but it is its role to be associated with that of the ligand, which is at the origin of the catalytic transformation. In order to maximize the catalytic potential of such metal complexes in terms of activity and selectivity, a better understanding of the structure–activity relationship will be however needed and in particular how the metal reactivity is influenced by the ligand structure. If the access to simple and cheap catalytic systems is clearly of great interest for industrial purposes, further improvement of these systems will require the development of alternative ligands. It is very likely that carbon-based ligands will soon become very popular in this field of manganese catalysis. The selective reduction of less polar bonds is one of the many challenges that could be overcome with this latter type of ligand.

References

1 De Vries, J.G. and Elsevier, C.J. (2007). *The Handbook of Homogeneous Hydrogenation*. Weinheim: Wiley-VCH.
2 Kauffman, G.B. (1999). *Platinum Met. Rev.* 43: 122–128.
3 *Nobel Lectures, Chemistry 1901–1921*. Amsterdam: Elsevier Publishing Company, (1966).
4 Osborn, J.A., Jardine, F.H., Young, J.F., and Wilkinson, G. (1966, https://doi.org/10.1039/J19660001711). *J. Chem. Soc. A*: 1711–1732.
5 Knowles, W.S. (2002). *Angew. Chem. Int. Ed.* 41: 1998–2007.
6 Noyori, R. (2002). *Angew. Chem. Int. Ed.* 41: 2008–2022.
7 Sedgwick, D.M. and Hammond, G.B. (2018). *J. Fluorine Chem.* 207: 45–58.
8 Wei, D. and Darcel, C. (2019). *Chem. Rev.* 119: 2550–2610.
9 Bullock, R.M. (2013). *Science* 342: 1054–1055.
10 (a) Khusnutdinov, R.I., Bayguzina, A.R., and Dzhemilev, U.M. (2012). *Russ. J. Org. Chem.* 48: 309–348. (b) Valyaev, D.A., Lavigne, G., and Lugan, N. (2016). *Coord. Chem. Rev.* 308: 191–235. (c) Trovitch, R.J. (2017). *Acc. Chem. Res.* 50: 2842–2852. (d) Kallmeier, F. and Kempe, R. (2018). *Angew. Chem. Int. Ed.* 57: 46–60. (e) Filonenko, G.A., van Putten, R., Hensen, E.J.M. et al. (2018). *Chem.*

Soc. Rev. 47: 1459–1483. (f) Gorgas, N. and Kirchner, K. (2018). *Acc. Chem. Res.* 51: 1558–1569. (g) Mukherjee, A. and Milstein, D. (2018). *ACS Catal.* 8: 11435–11469. (h) Alig, L., Fritz, M., and Schneider, S. (2019). *Chem. Rev.* 119: 2681–2751. (i) Vogiatzis, K.D., Polynski, M.V., Kirkland, J.K. et al. (2019). *Chem. Rev.* 119: 2453–2523. (j) Zell, T. and Langer, R. (2018). *ChemCatChem* 10: 1930–1940.

11 Elangovan, S., Topf, C., Fischer, S. et al. (2016). *J. Am. Chem. Soc.* 138: 8809–8814.

12 Mukherjee, A., Nerush, A., Leitus, G. et al. (2016). *J. Am. Chem. Soc.* 138: 4298–4301.

13 (a) Garbe, M., Junge, K., and Beller, M. (2017). *Eur. J. Org. Chem.*: 4344–4362. (b) Maji, B. and Barman, M.K. (2017). *Synthesis* 49: 3377–3393. (c) Maser, L., Vondung, L., and Langer, R. (2018). *Polyhedron* 143: 28–42. (d) Gorgas, N. and Kirchner, K. (2018). *Pincer Compounds: Chemistry and Applications*. Elsevier. https://doi.org/10.1016/B978-0-12-812931-9.00002-5, 19–45. (e) Kulkarni, N.V. and Jones, W.D. (2018). *Pincer Compounds: Chemistry and Applications*. Elsevier. https://doi.org/10.1016/B978-0-12-812931-9.00024-4, 491–518. (f) van der Vlugt, J.I. (2018). *Pincer Compounds: Chemistry and Applications*. Elsevier. https://doi.org/10.1016/B978-0-12-812931-9.00028-1, 599–621. (g) Valdés, H., García-Eleno, M.A., Canseco-Gonzalez, D. et al. (2018). *ChemCatChem* 10: 3136–3172. (h) Wei, Z. and Jiao, H. (2019). *Advances in Inorganic Chemistry*, vol. 73 (eds. R. van Eldik and R. Puchta), 323–384. Academic Press.

14 (a) Chakraborty, U., Reyes-Rodriguez, E., Demeshko, S. et al. (2018). *Angew. Chem. Int. Ed.* 57: 4970–4975. (b) Chakraborty, U., Demeshko, S., Meyer, F. et al. (2019). *Angew. Chem. Int. Ed.* 58: 3466–3470.

15 (a) Grützmacher, H. (2008). *Angew. Chem. Int. Ed.* 47: 1814–1818. (b) van der Vlugt, J.I. (2012). *Eur. J. Inorg. Chem.* 2012: 363–375. (c) Khusnutdinova, J.R. and Milstein, D. (2015). *Angew. Chem. Int. Ed.* 54: 12236–12273.

16 (a) Elisabetta, A., Peter, S., Christoph, C. et al. (2013). *Angew. Chem. Int. Ed.* 52: 14162–14166. (b) Nielsen, M., Alberico, E., Baumann, W. et al. (2013). *Nature* 495: 85–89.

17 Kuriyama, W., Matsumoto, T., Ogata, O. et al. (2012). *Org. Process Res. Dev.* 16: 166–171.

18 Elangovan, S., Garbe, M., Jiao, H. et al. (2016). *Angew. Chem. Int. Ed.* 55: 15364–15368.

19 Kar, S., Goeppert, A., Kothandaraman, J. et al. (2017). *ACS Catal.* 7: 6347–6351.

20 Schneidewind, J., Adam, R., Baumann, W. et al. (2017). *Angew. Chem. Int. Ed.* 56: 1890–1893.

21 Kaithal, A., Hölscher, M., and Leitner, W. (2018). *Angew. Chem. Int. Ed.* 57: 13449–13453.

22 Li, H., Wei, D., Bruneau-Voisine, A. et al. (2018). *Organometallics* 37: 1271–1279.

23 (a) Garbe, M., Junge, K., Walker, S. et al. (2017). *Angew. Chem. Int. Ed.* 56: 11237–11241. (b) Garbe, M., Wei, Z., Tannert, B. et al. (2019). *Adv. Synth. Catal.* 361: 1913–1920.

24 Passera, A. and Mezzetti, A. (2019). *Adv. Synth. Catal.* 361: 4691–4706.

25 Kallmeier, F., Irrgang, T., Dietel, T. et al. (2016). *Angew. Chem. Int. Ed.* 55: 11806–11809.

26 Bruneau-Voisine, A., Wang, D., Roisnel, T. et al. (2017). *Catal. Commun.* 92: 1–4.

27 Schirmer, W., Flörke, U., and Haupt, H.J. (1987). *Z. Anorg. Allg. Chem.* 545: 83–97.

28 Bruneau-Voisine, A., Wang, D., Dorcet, V. et al. (2017). *J. Catal.* 347: 57–62.

29 (a) Bertini, F., Glatz, M., Gorgas, N. et al. (2017). *Chem. Sci.* 8: 5024–5029. (b) Glatz, M., Stöger, B., Himmelbauer, D. et al. (2018). *ACS Catal.* 8: 4009–4016. (c) Mastalir, M., Glatz, M., Gorgas, N. et al. (2016). *Chem. Eur. J.* 22: 12316–12320.

30 Zou, Y.-Q., Chakraborty, S., Nerush, A. et al. (2018). *ACS Catal.* 8: 8014–8019.

31 Zirakzadeh, A., de Aguiar, S.R.M.M., Stöger, B. et al. (2017). *ChemCatChem* 9: 1744–1748.

32 (a) Cabrero-Antonino, J.R., Alberico, E., Drexler, H.-J. et al. (2016). *ACS Catal.* 6: 47–54. (b) Adam, R., Alberico, E., Baumann, W. et al. (2016). *Chem. Eur. J.* 22: 4991–5002.

33 Papa, V., Cabrero-Antonino, J.R., Alberico, E. et al. (2017). *Chem. Sci.* 8: 3576–3585.

34 Zubar, V., Lebedev, Y., Azofra, L.M. et al. (2018). *Angew. Chem. Int. Ed.* 57: 13439–13443.

35 Spasyuk, D., Smith, S., and Gusev, D.G. (2012). *Angew. Chem. Int. Ed.* 51: 2772–2775.

36 Widegren, M.B., Harkness, G.J., Slawin, A.M.Z. et al. (2017). *Angew. Chem. Int. Ed.* 56: 5825–5828.

37 Widegren, M.B. and Clarke, M.L. (2018). *Org. Lett.* 20: 2654–2658.

38 Demmans, K.Z., Olson, M.E., and Morris, R.H. (2018). *Organometallics* 37: 4608–4618.

39 Doucet, H., Ohkuma, T., Murata, K. et al. (1998). *Angew. Chem. Int. Ed.* 37: 1703–1707.

40 (a) Espinosa-Jalapa, N.A., Nerush, A., Shimon, L.J.W. et al. (2017). *Chem. Eur. J.* 23: 5934–5938. (b) Kumar, A., Janes, T., Espinosa-Jalapa, N.A. et al. (2018). *Angew. Chem. Int. Ed.* 57: 12076–12080.

41 Ding, K., Zhang, L., Tang, Y. et al. (2019). *Angew. Chem. Int. Ed.* 58: 4973–4977.

42 Ling, F., Hou, H., Chen, J. et al. (2019). *Org. Lett.* 21: 3937–3941.

43 Das, U.K., Kumar, A., Ben-David, Y. et al. (2019). *J. Am. Chem. Soc.* 141: 12962–12966.

44 Perez, M., Elangovan, S., Spannenberg, A. et al. (2017). *ChemSusChem* 10: 83–86.

45 Budweg, S., Junge, K., and Beller, M. (2019). *Chem. Commun.* 55: 14143–14146.

46 Wang, D., Bruneau-Voisine, A., and Sortais, J.-B. (2018). *Catal. Commun.* 105: 31–36.

47 van Putten, R., Filonenko, G.A., Gonzalez de Castro, A. et al. (2019). *Organometallics* 38: 3187–3196.

48 Dubey, A., Nencini, L., Fayzullin, R.R. et al. (2017). *ACS Catal.* 7: 3864–3868.

49 Dubey, A., Rahaman, S.M.W., Fayzullin, R.R. et al. (2019). *ChemCatChem* 11: 3844–3852.

50 Zhang, C., Hu, B., Chen, D. et al. (2019). *Organometallics* 38: 3218–3226.

51 Bruneau-Voisine, A., Wang, D., Dorcet, V. et al. (2017). *Org. Lett.* 19: 3656–3659.
52 Zhao, B., Han, Z., and Ding, K. (2013). *Angew. Chem. Int. Ed.* 52: 4744–4788.
53 Wei, D., Bruneau-Voisine, A., Dubois, M. et al. (2019). *ChemCatChem* 11: 5256–5259.
54 Martínez-Ferraté, O., Werlé, C., Franciò, G. et al. (2018). *ChemCatChem* 10: 4514–4518.
55 Ganguli, K., Shee, S., Panja, D. et al. (2019). *Dalton Trans.* 48: 7358–7366.
56 Weber, S., Stöger, B., and Kirchner, K. (2018). *Org. Lett.* 20: 7212–7215.
57 Garduño, J.A. and García, J.J. (2019). *ACS Catal.* 9: 392–401.
58 Garduño, J.A., Flores-Alamo, M., and García, J.J. (2019). *ChemCatChem* 11: 5330–5338.
59 Weber, S., Stöger, B., Veiros, L.F. et al. (2019). *ACS Catal.* 9: 9715–9720.
60 van Putten, R., Uslamin, E.A., Garbe, M. et al. (2017). *Angew. Chem. Int. Ed.* 56: 7531–7534.
61 Liu, C., van Putten, R., Kulyaev, P.O. et al. (2018). *J. Catal.* 363: 136–143.
62 Vigneswaran, V., MacMillan, S.N., and Lacy, D.C. (2019). *Organometallics* 38: 4387–4391.
63 Wei, D., Bruneau-Voisine, A., Chauvin, T. et al. (2018). *Adv. Synth. Catal.* 360: 676–681.
64 Wei, D., Bruneau-Voisine, A., Valyaev, D.A. et al. (2018). *Chem. Commun.* 54: 4302–4305.
65 Buhaibeh, R., Filippov, O.A., Bruneau-Voisine, A. et al. (2019). *Angew. Chem. Int. Ed.* 58: 6727–6731.
66 Kireev, N.V., Filippov, O.A., Gulyaeva, E.S. et al. (2020). *Chem. Commun.* https://doi.org/10.1039/C9CC07713H.
67 van Putten, R., Benschop, J., de Munck, V.J. et al. (2019). *ChemCatChem* 11: 5232–5235.
68 Zhou, Y.P., Mo, Z., Luecke, M.P. et al. (2018). *Chem. Eur. J.* 24: 4780–4784.
69 Schneekönig, J., Junge, K., and Beller, M. (2019). *Synlett* 30: 503–507.
70 Passera, A. and Mezzetti, A. (2020). *Angew. Chem. Int. Ed.* 59: 187–191.

3

Manganese-Catalyzed Hydrogen-Borrowing and Dehydrogenative Coupling Reactions

Stefan Weber and Karl Kirchner

Vienna University of Technology, Institute of Applied Synthetic Chemistry, Getreidemarkt 9, A-1060 Vienna, Austria

3.1 Introduction

The development of sustainable, efficient, and selective synthesis is one of the fundamental research goals in chemistry. In this respect, catalysis is a key technology, since a large amount of chemical and pharmaceutical products are made by catalysts. However, most catalytic reactions today rely on precious metals such as ruthenium, rhodium, iridium, or palladium. The limited availability of these precious metals, their high price, and their toxicity will diminish their attractivity in the long run, and more environmentally friendly and also economical alternatives have to be found, which are in line with green chemistry guidelines. Accordingly, in order to develop greener, safer, and more cost-effective chemical processes, it is highly important to replace precious metal catalysts by Earth-abundant non-precious metal catalysts [1]. Moreover, these metals possess high biocompatibility and catalytic potential being an essential part of numerous enzymes that are involved in a wide variety of biological processes. Among these metals, manganese comprises a promising candidate since it belongs to one of the most abundant metals in the Earth's crust, is inexpensive, and exhibits a rather low environmental impact [2].

Manganese has long been neglected for applications in hydrogenation and dehydrogenation reactions, and as late as 2016 it became a subject of intense investigations from many different academic and industrial research groups. As demonstrated by a large number of recent publications, well-defined manganese carbonyl complexes bearing in many cases strongly bound PNP pincer ligands or related tridentate and bidentate ligands proved to be highly effective and versatile systems for hydrogenations and dehydrogenations [3]. Common structural motifs are depicted in Chart 3.1. The reason that manganese pincer complexes received tremendous attention in recent years might be attributed to the wide structural diversity arising from such ligand scaffolds that can be synthesized in a simple

Manganese Catalysis in Organic Synthesis, First Edition. Edited by Jean-Baptiste Sortais.

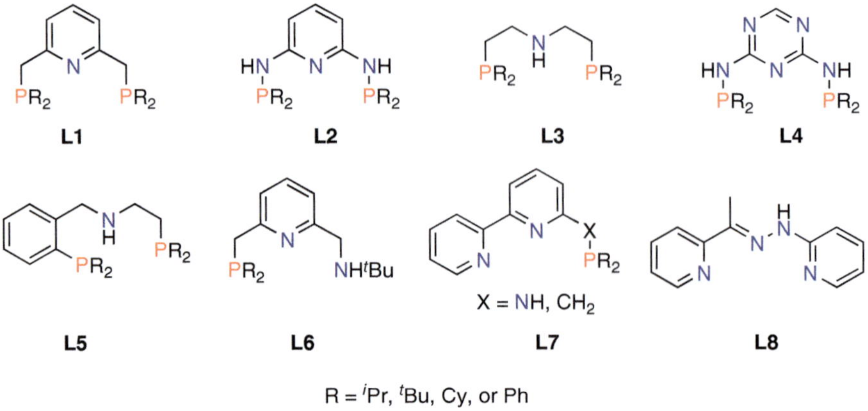

Chart 3.1 Frequently used ligands that are capable of metal–ligand bifunctionality.

and modular fashion. Consequently, the synthesis of isolated and structurally well-defined complexes combined with a fundamental understanding of reaction mechanisms appears to be a primary objective for the rational development of novel catalytic systems. In this context the acceptorless alcohol dehydrogenation (AAD), often in conjunction with hydrogen-borrowing (HB) reactions, is becoming an important area of research [4, 5].

In this chapter, we describe well-defined Mn(I)-based catalysts featuring PNP, PNN, and NNN pincer ligands, among others. We highlight characteristic features that influence the activity and selectivity of these systems. In many cases, metal–ligand cooperation (metal–ligand bifunctionality) plays an important role [6, 7]. Pincer complexes with aliphatic ligand backbones possess a secondary amine functionality adjacent to the metal center that is involved through an interplay between the amine/amide forms in H—H and H—heteroatom bond activation and formation steps. Lutidine-, 2,6-diaminopyridine-, and triazine-derived PNP pincers, on the other hand, are able to undergo reversible deprotonation of a methylene and NH linker, respectively, which is accompanied by dearomatization.

3.2 Acceptorless Dehydrogenative Coupling (ADC)

An elegant way of utilizing alcohols as sustainable carbon source for selective mono-alkylation of primary amines is the oxidation of alcohol moieties to carbonyl groups such as aldehydes and ketones producing hydrogen gas as sole by-product. The formed carbonyl motive forms an imine with the primary amine, thereby liberating water. The general reaction sequence is depicted in Scheme 3.1.

Within this context, a broad variety of different well-defined manganese complexes may be employed for the acceptorless dehydrogenative coupling (ADC) of alcohols and amines. As seen in Chart 3.2, PNP pincer ligands are frequently used as tridentate ligands. Furthermore, most manganese complexes contain two strongly

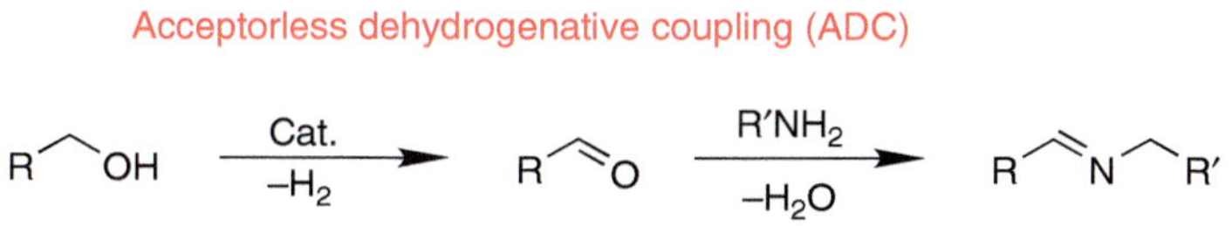

Scheme 3.1 General reaction pattern of acceptorless dehydrogenative coupling (ADC).

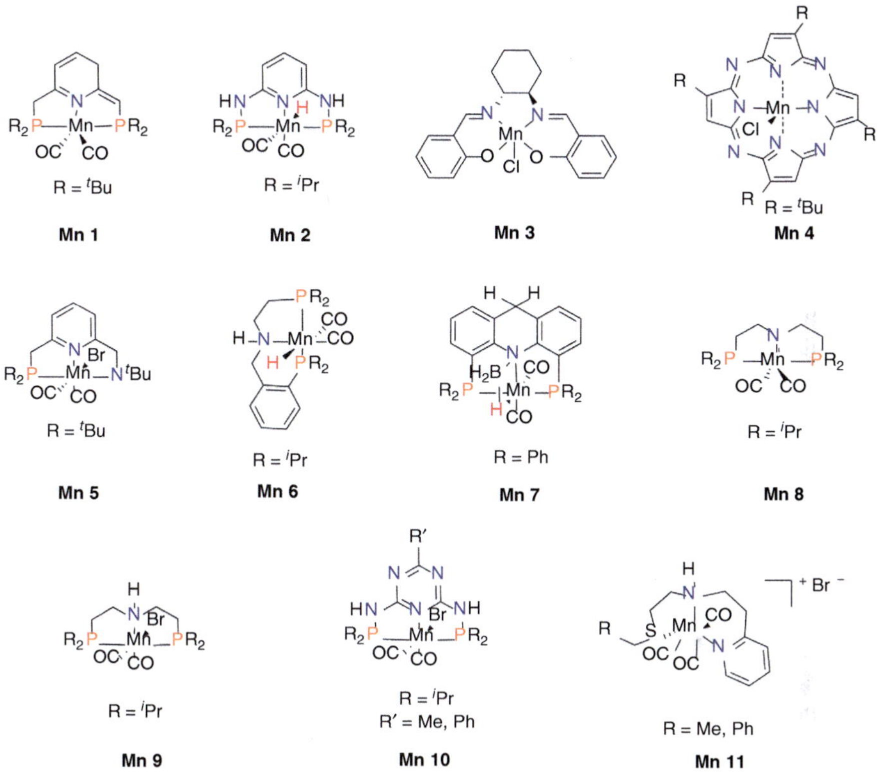

Chart 3.2 Frequently used well-defined manganese complexes for ADC.

bonded carbonyl ligands in order to stabilize the low oxidation state of the Mn(I) center. Regarding the anionic ligand in these coordination compounds, halides and hydrides are frequently employed. Apart from that, a deprotonated pincer ligand may also serve as anionic ligand, yielding a coordinatively unsaturated compound.

3.2.1 Synthesis of Aldimines, Cyclic Imides, and Amides

Milstein and coworkers were the first one to utilize the elegant ADC of primary alcohols with amines, yielding aldimines [8]. During this reaction, water and hydrogen gas were formed as the only by-product (Scheme 3.2). Interestingly, this reaction operated without the addition of a base. This protocol allows the usage of benzylic

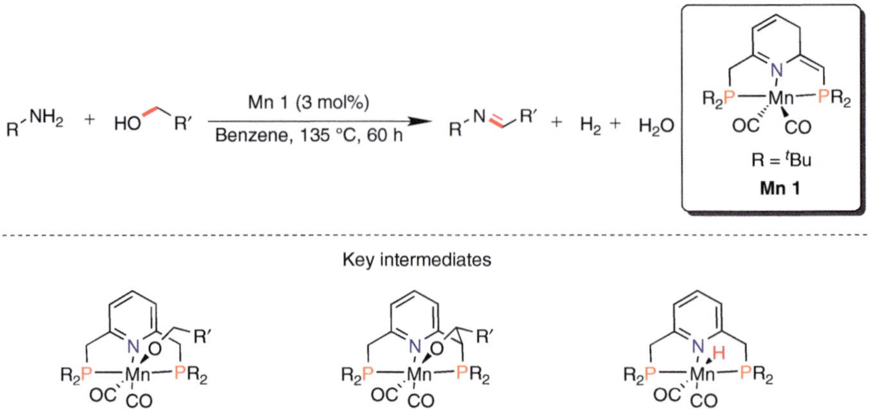

Scheme 3.2 Synthesis of aldimines catalyzed by **Mn 1**.

and aliphatic alcohols as well as aromatic and aliphatic amines as coupling partner.

Detailed mechanistic studies revealed the formation of two different alkoxide-containing complexes and the presence of a hydride species.

Shortly after that, our group reported the divergent coupling of alcohols with amines [9]. Within this context, iron and manganese based upon the same PNP ligand containing *NH* linkers (L2) were investigated. Remarkably, striking differences in product formation were observed. The iron-based system catalyzed the formation of secondary amines, whereas the manganese compound **Mn 2** yielded aldimines in a selective fashion (Scheme 3.3).

The reason for this interesting behavior is attributed to the fact that the deprotonated Fe(II) species contains a hydride ligand, which is able to insert into the C—N bond, yielding an amido ligand and consecutively leading to the formation of amine as product. In case of the manganese-centered complex, only carbonyl ligands stabilized the deprotonated species and insertion can therefore not occur.

Very recently, the group of Madsen reported the dehydrogenative coupling of alcohols with amines using a salen-based manganese(III) complex **Mn 3** [10]. This procedure relied on a catalytic amount of Ca_3N_2. Shortly after that, a porphyrin-supported manganese(III) complex **Mn 4** was reported, whereas the catalyst-loading could be decreased [11]. Furthermore, this procedure operated without base addition. Unfortunately, harsh reaction conditions were required (reflux in mesitylene) as depicted in Scheme 3.4.

A highly atom-efficient way of synthesizing cyclic imides was reported by the group of Milstein in 2017 [12]. The treatment of amines with an equimolar amount of 1,4-butanediol under ACD condition using **Mn 5** resulted in the formation of cyclic imides as seen in Scheme 3.5. The used PNN systems (e.g., L6) differ from the typical PNP, as one phosphine donor was replaced by a secondary amine group.

This allowed the operation via metal–ligand bifunctionality, under which the amine was deprotonated. Within this concept, no dearomatization of the pyridine ring occurred. Apart from that, the authors described a hemilabile behavior of the

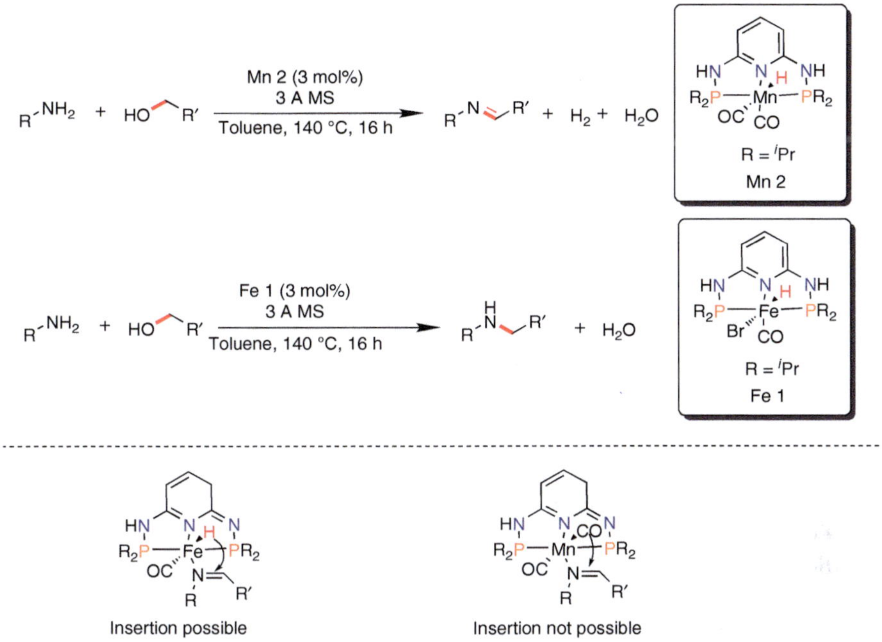

Scheme 3.3 Synthesis of aldimines and secondary amines catalyzed by **Mn 2** and **Fe 1**.

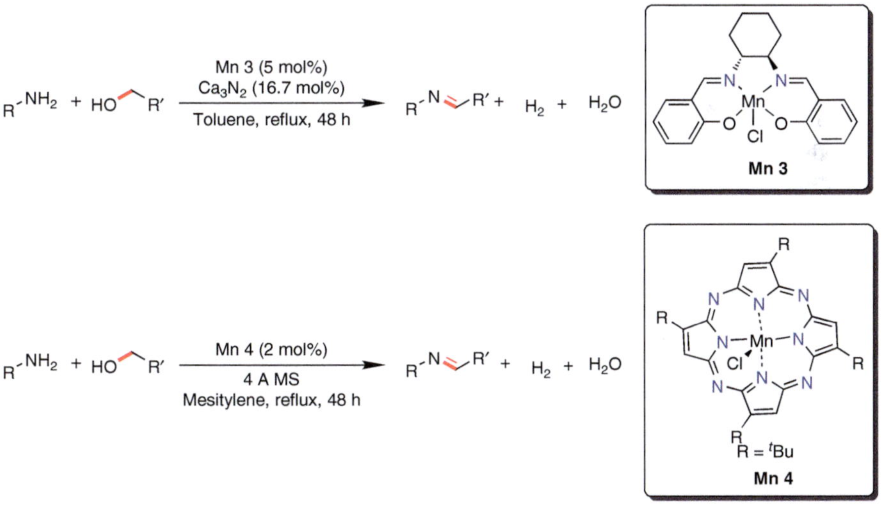

Scheme 3.4 Synthesis of aldimines catalyzed by Mn(III) complexes.

amino ligand during catalysis. Since a bromide ligand is used in this context, an external base had to be added in order to initiate the deprotonation of the complex.

The catalytic reaction probably proceeded via an oxidation of one alcohol group. In this step, the hemilabile ligand dissociated from the metal center and allowed

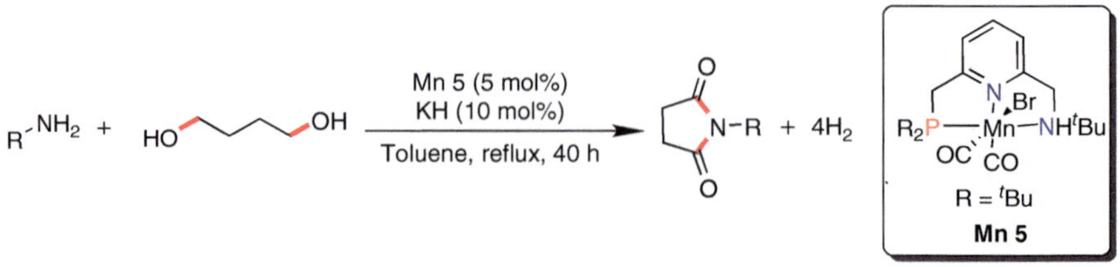

Scheme 3.5 Synthesis of cyclic imides via ADC catalyzed by **Mn 5**.

β-hydride elimination of the coordinated substrate (Scheme 3.7). This led to the formation of a hydride complex under recoordination of the dissociated amine and aldehyde release.

Condensation to a lactone or alternatively direct amide formation with an amine were found to be the next step during the reaction. Further oxidation of the alcohol moiety and ring closure with the amide motive resulted in the formation of a cyclic hemiacetal intermediate. Finally, oxidation of the hemiacetal resulted in product formation.

The catalytic dehydrogenation of methanol to formaldehyde is more challenging in comparison to substituted alcohols, due to a higher enthalpy of the oxidation. Apart from that, formaldehyde is unstable toward decomposing to CO and hydrogen gas. Nevertheless, Milstein and coworkers reported the manganese-catalyzed selective N-formylations of amines using methanol as formylation agent in an additive-free way in 2017 [13] (Scheme 3.6a).

Furthermore, the used PNP system (L5) **Mn 6** showed interesting metal–ligand bifunctionality modes. When the catalyst was heated, the amido ligand was protonated by the neighboring benzylic carbon, resulting in the formation of a tetradentate coordination mode of the former PNP pincer system. This complex was found to catalyze the formylation of amines, although showing lower reactivity.

According to DFT calculations, this intermediate is not likely to be formed during the catalytic cycle. Instead, a typical metal–ligand cooperativity under loss of hydrogen, forming a fivefold coordinated manganese complex was found to be energetically favored.

In the following step, methanol was deprotonated, forming a methoxide ligand. Another interesting feature of this ligand system is the possibility of the hemilabile amino moiety to dissociate from the metal center. This allowed the formation of a C–H agnostic intermediate, followed by β-hydride elimination to release formaldehyde (Scheme 3.7).

The amine reacted with the aldehyde, forming a hemiacetal, which was consecutively oxidized to yield formamides as desired products.

Very recently, Milstein and coworkers, reported the direct synthesis of secondary amides via ADC of two primary alcohols with ammonia using **Mn 7** [14]. It is noteworthy to say that a stoichiometric amount of base was used for this transformation. If a catalytic amount of KH was used, mainly aldimine formation was observed (Scheme 3.6b).

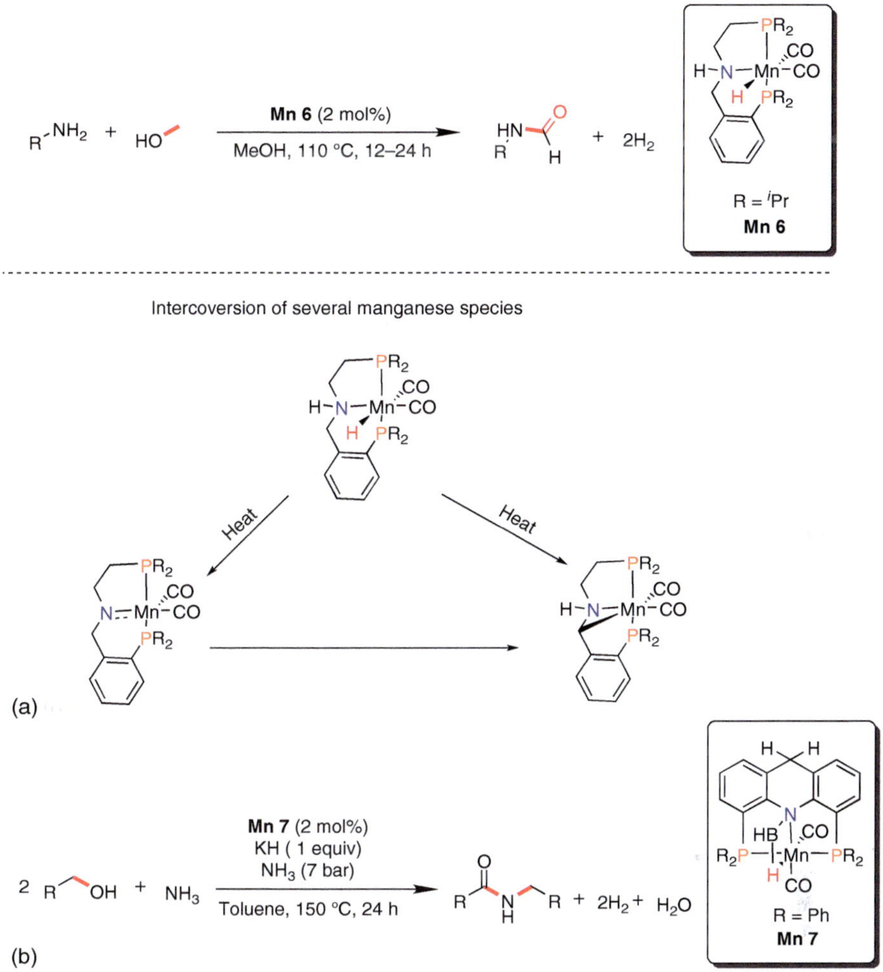

Scheme 3.6 Synthesis of amides via ADC catalyzed by **Mn 6** and **Mn 7**. (a) N-formylation of primary pmines via ADC (Milstein and coworkers [12]). (b) Direct amide synthesis of primary alcohols and ammonia via ADC (Milstein and coworkers [14]).

3.2.2 Synthesis of Esters and Functionalizations of Nitriles and Alkanes

In 2017, the group of Gauvin reported the additive-free dehydrogenative homo-coupling of primary alcohols to yield esters [15]. For this purpose, a coordinatively unsaturated dicarbonyl complex bearing an aliphatic PNP ligand, **Mn 8,** was used (Scheme 3.8). This procedure allowed the ADC of benzylic as well as aliphatic alcohols.

Detailed mechanistic investigation concluded that the release of hydrogen gas forming **Mn 8** during catalysis from an amino hydride species is likely to be rate-determining.

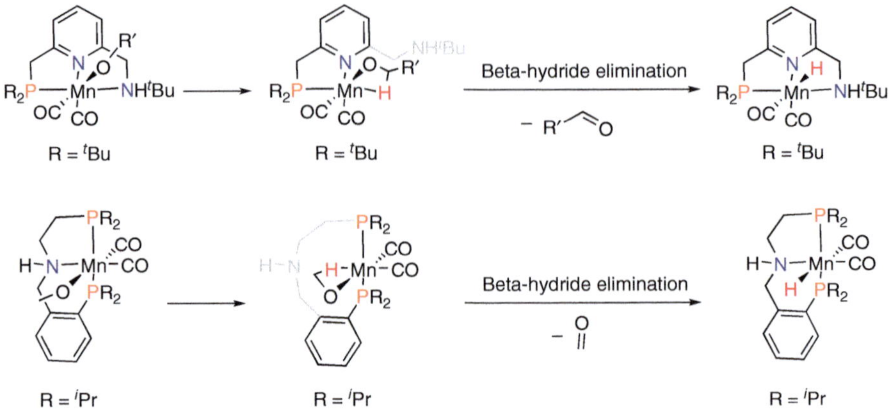

Scheme 3.7 Hemilabile behavior of **Mn 5** and **Mn 6** in ADC reactions.

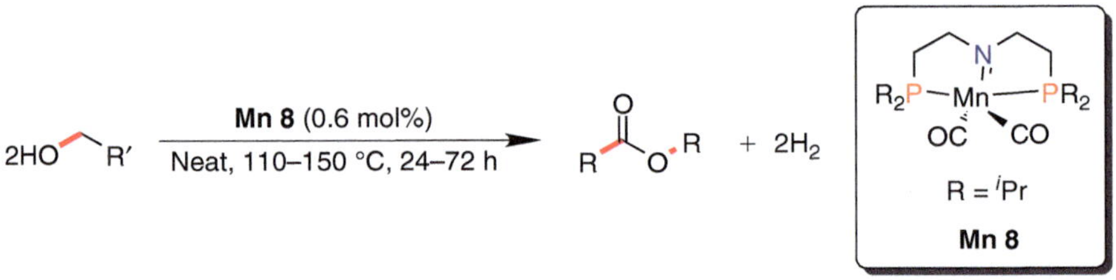

Scheme 3.8 Manganese-catalyzed synthesis of esters via ADC.

The group of Liu reported the dual-deoxygenative coupling of alcohols employing **Mn 9** in 2018 [16]. This protocol revealed high reactivity for the homo-coupling of 2-aryl-ethanols, yielding diphenylpropene derivatives. Under these reaction conditions, sodium formate was formed as by-product, leading to the loss of one carbon-fragment during the coupling process. The proposed reaction pathway is depicted in Scheme 3.9.

Furthermore, this procedure could be used for cross-coupling of 2-aryl-ethanols with primary alcohols, using an excess of the primary alcohol as coupling partner. Consecutive hydrogenation by a heterogeneous nickel catalyst yielded fully saturated systems.

A different type of olefination reaction was introduced by Milstein and coworkers in 2017 using **Mn 6** [17]. Within this context, the authors described the selective α-olefination of nitriles, using alcohols as surrogates (Scheme 3.10).

Detailed DFT calculation unveiled that the manganese complex was likely to serve as a template during the carbon–carbon bond forming step, therefore speeding up the condensation reaction [18].

The groups of Kempe [19] and Maji [20] independently reported the α-olefination of alkyl-substituted N-heteroarenes with alcohols. The general reaction pattern and proposed mechanism using **Mn 10** can be seen in Scheme 3.11. Investigation of the reaction mechanisms revealed that the role of the manganese-based catalyst laid beyond the AAD of the primary alcohol. In presence of the manganese complex,

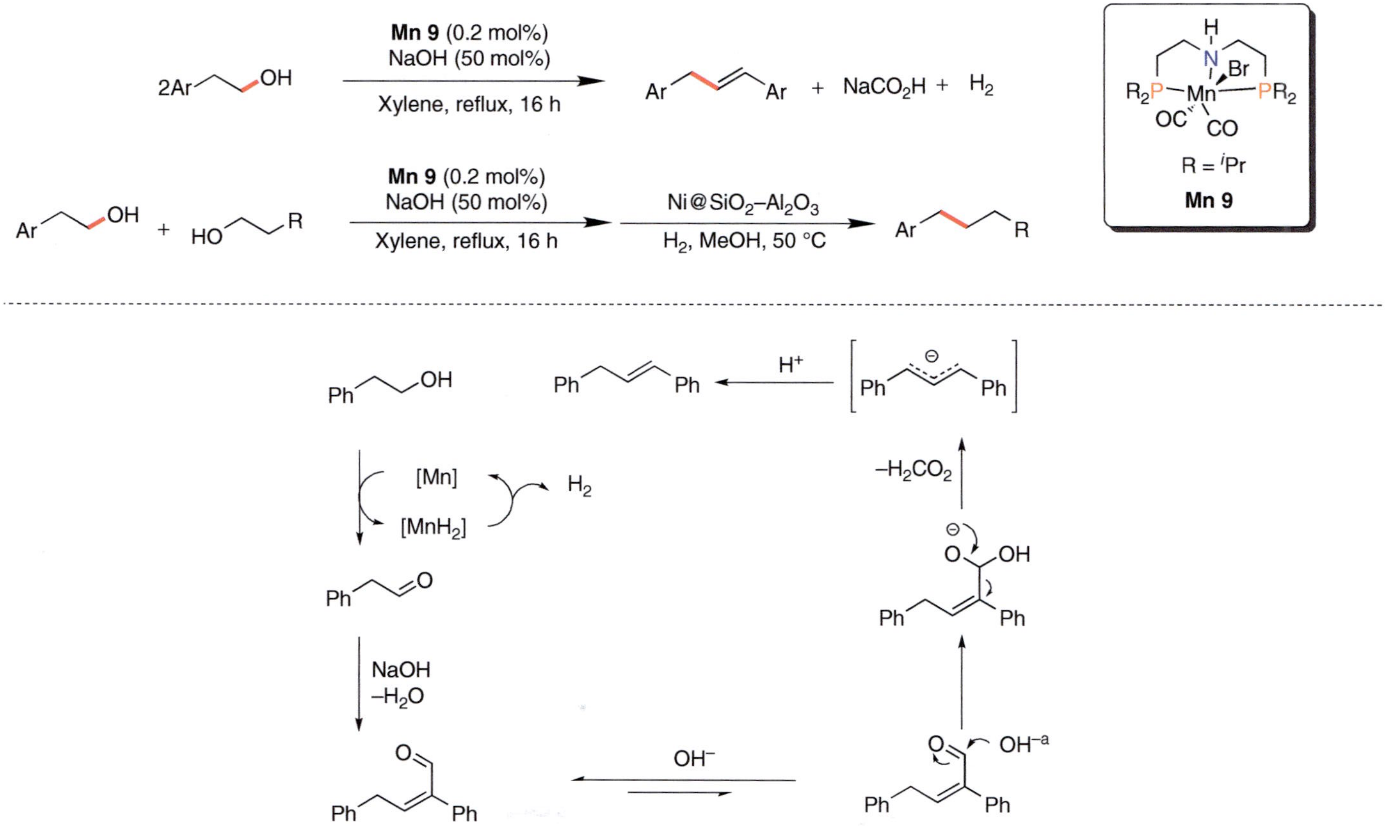

Scheme 3.9 Dual-oxidative coupling of alcohols catalyzed by **Mn 9**.

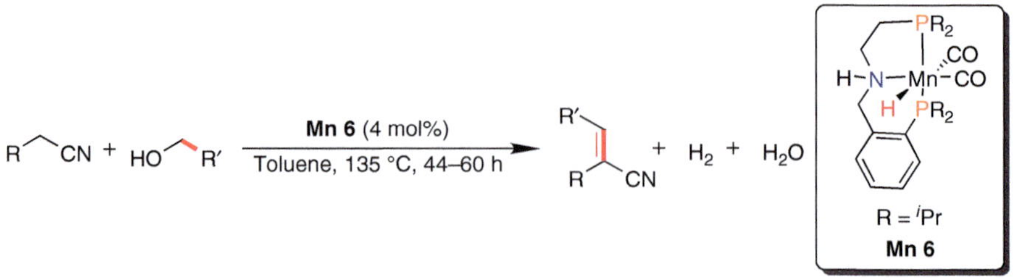

Scheme 3.10 α-Olefination of nitriles via ADC using **Mn 6**.

the condensation of aldehyde with enamine was accelerated. This indicated that the manganese complex may serve as platform for the C—C bond formation.

3.2.3 Synthesis and Derivatization of Heterocycles

Our group was the first one to report the sustainable synthesis of pyrimidines and quinolines via ADC catalyzed by a well-defined manganese PNP ligand containing *NH* linkers, **Mn 2** [21]. This protocol showed high functional group tolerance, resulting in the formation of a broad variety of different substituted quinolines and pyrimidines (Scheme 3.12a).

Kempe and coworker reported a four-component reaction, yielding pyrimidines employing L4-type ligands [22]. High selectivity within this reaction could be achieved via consecutive addition of primary alcohols (Scheme 3.12b).

Furthermore, an elegant way of synthesizing substituted pyrroles via the oxidative coupling of amino alcohols with secondary alcohols was reported by the group of Kempe (Scheme 3.12c) [23].

An overview for the synthesis of various heterocycles via multicomponent reactions is given in Scheme 3.12.

In 2017, our group reported the selective aminomethylation of different aromatic compounds, using methanol (Scheme 3.13) [24]. Within this context, **Mn 2** was employed as catalyst. In contradiction to Mn, the iron-centered complex **Fe 1** was found to catalyze the methylation of naphthol under the same reaction conditions.

Milstein and coworkers reported the synthesis of pyrazines mediated by **Mn 7** using an oxidation protocol of β-amino alcohols followed by condensation and final oxidation of the partially saturated heterocycle [25].

Furthermore, the synthesis of quinoxalines using 1,2-diaminobenzene and different 1,2-diols was reported within this context as seen in Scheme 3.14.

Srimani and coworkers reported the synthesis of benzimidazoles (Scheme 3.15, **Mn 11**), quinoxalines, pyrazines, benzothiazoles, and quinolines (Scheme 3.16, **Mn 11′**), catalyzed by a cationic manganese tricarbonyl complex [26, 27]. In contradiction to other Mn(I) complexes, containing a tridentate ligand, the reported NNS ligand coordinated in a facial fashion.

Recently Madsen and coworkers reported the synthesis of quinolines via ADC using a porphyrin-supported Mn(III) complex **Mn 4** (Scheme 3.17) [11]. This protocol required an over-stochiometric amount of potassium hydroxide and ^{t}BuOK as

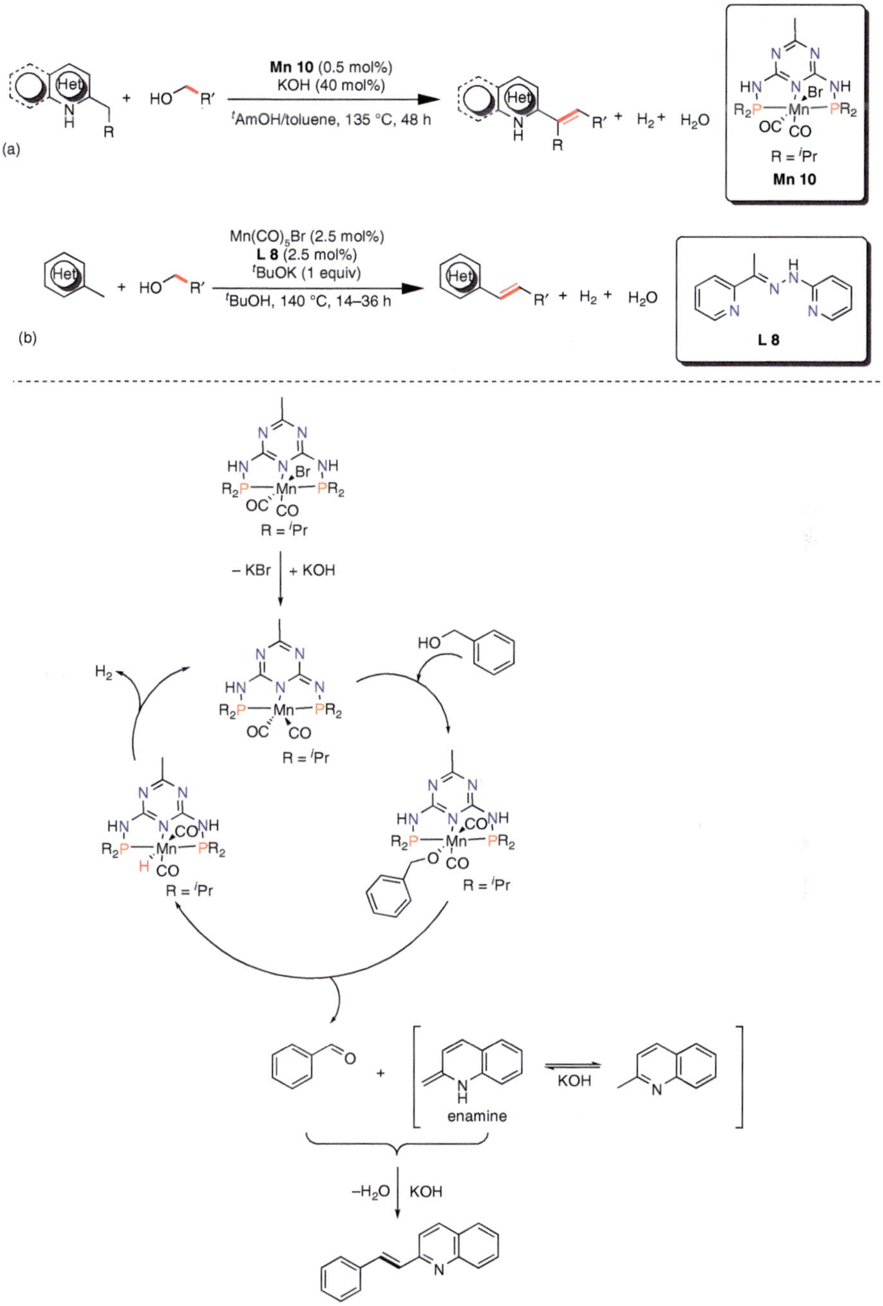

Scheme 3.11 Direct olefination of heteroarenes catalyzed by **Mn 10**. (a) Olefination of heteroaromatics via ADC (Kempe and coworkers [19]). (b) Olefination of methyl-substituted heteroaromatics via ADC (Maji and coworkers [20]).

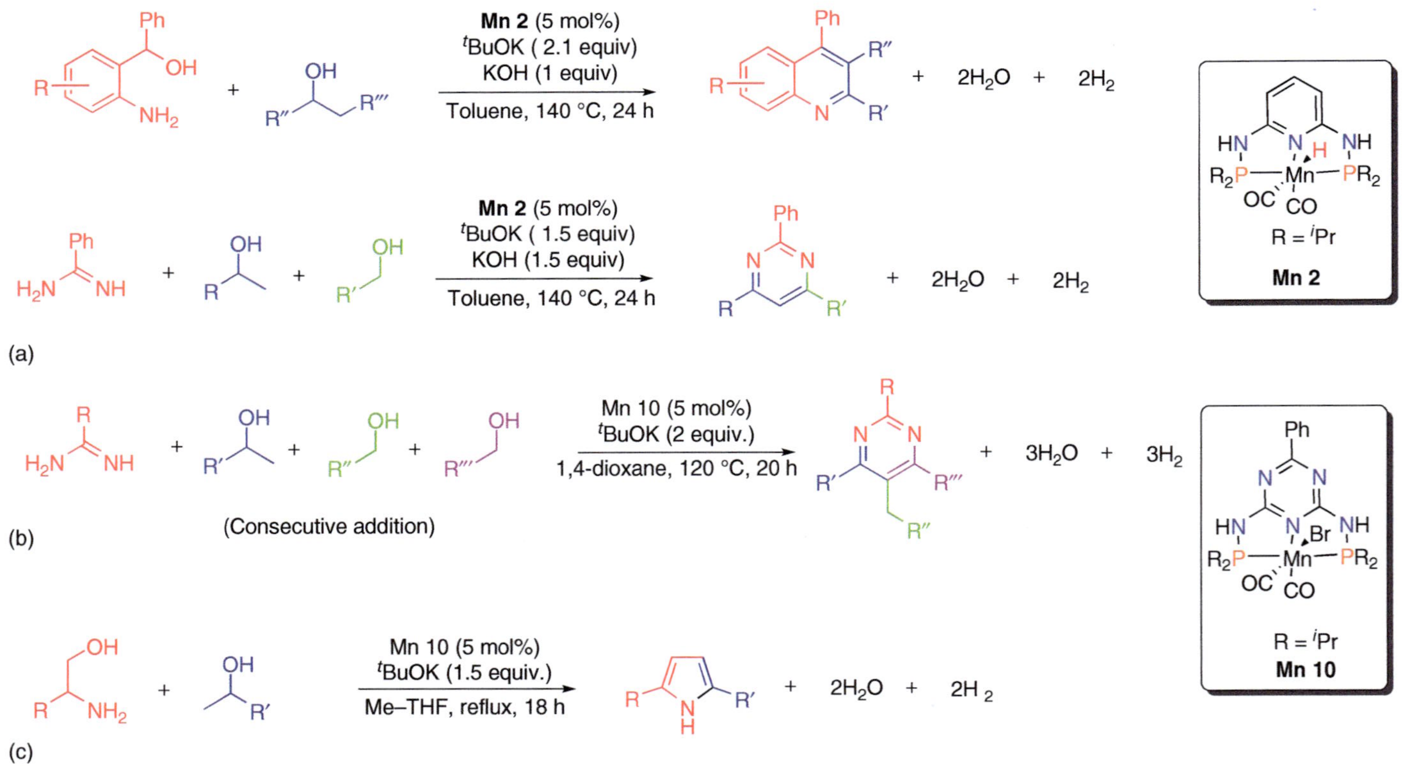

Scheme 3.12 Multicomponent synthesis of quinolines, pyrimidines, and pyrroles catalyzed by **Mn 2** and **Mn 10**. (a) Synthesis of quinolines and pyrimidine via ADC (Kirchner and coworkers [21]). (b) Synthesis of pyrimidine via ADC (Kempe and coworkers [22]). (c) Synthesis of pyrroles via ADC (Kempe and coworkers [23]).

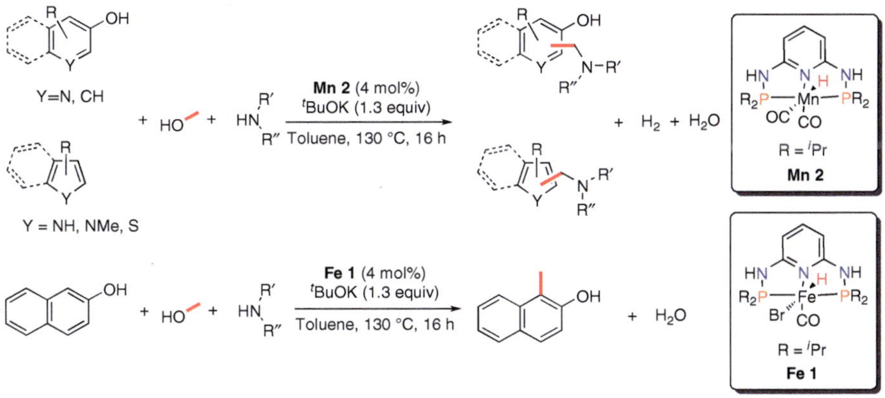

Scheme 3.13 Aminomethylation and methylation of aromatic compounds catalyzed by **Mn 2** and **Fe 1**.

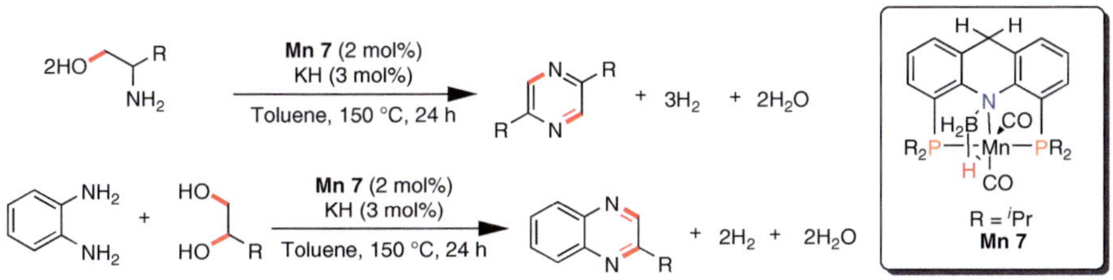

Scheme 3.14 Synthesis of pyrazines and quinoxalines catalyzed by **Mn 7**.

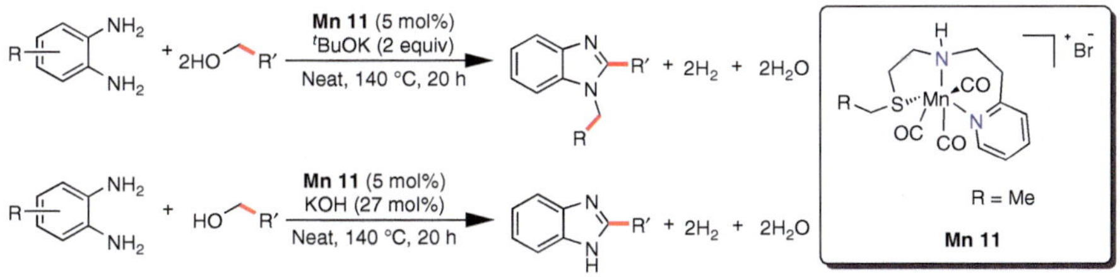

Scheme 3.15 Synthesis of benzimidazoles catalyzed by **Mn 11**.

well as a catalytic amount of pyridine and the addition of molecular sieve as water scavenger. Furthermore, the reactions had to be carried out in refluxing mesitylene and with comparingly long reaction time.

3.3 Hydrogen-Borrowing Reactions

A similar approach for the usage of alcohols as feeding stock in organic synthesis is the HB or hydrogen autotransfer (HAT). Within this concept, alcohols are oxidized to carbonyl groups. The general reaction pattern of HB reactions is depicted in Scheme 3.18. In contradiction to ADC, the formed hydrogen is stored around the

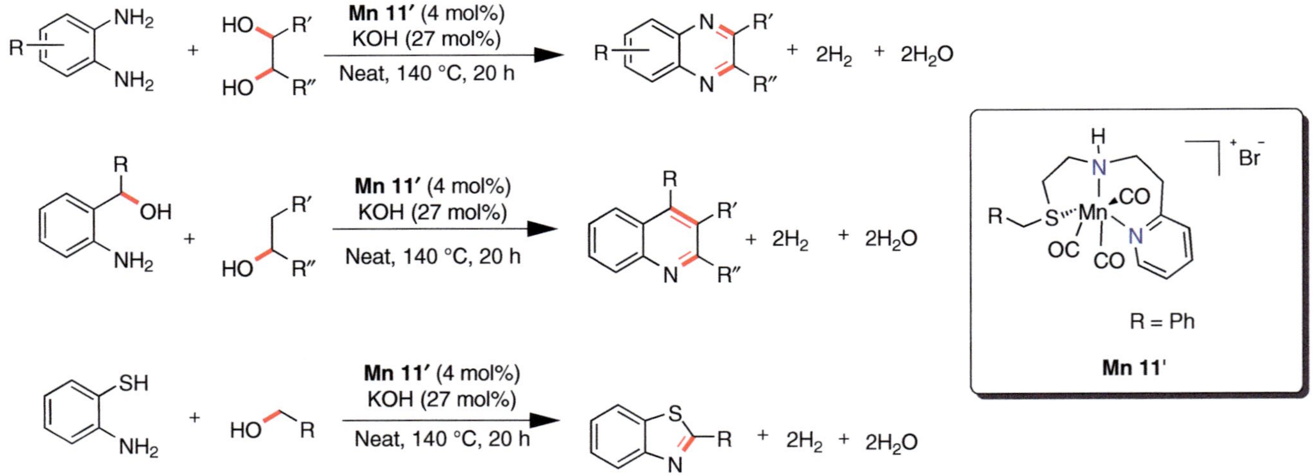

Scheme 3.16 Synthesis of various heterocycles catalyzed by **Mn 11′**.

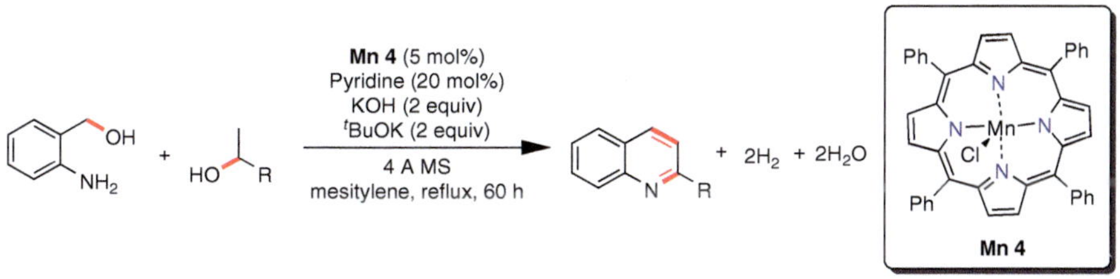

Scheme 3.17 Synthesis of quinolines via ADC catalyzed by **Mn 4**.

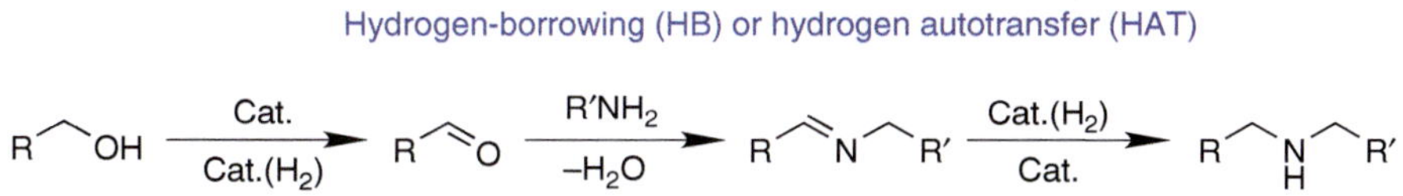

Scheme 3.18 General reaction scheme of hydrogen-borrowing (HB) or hydrogen autotransfer (HAT).

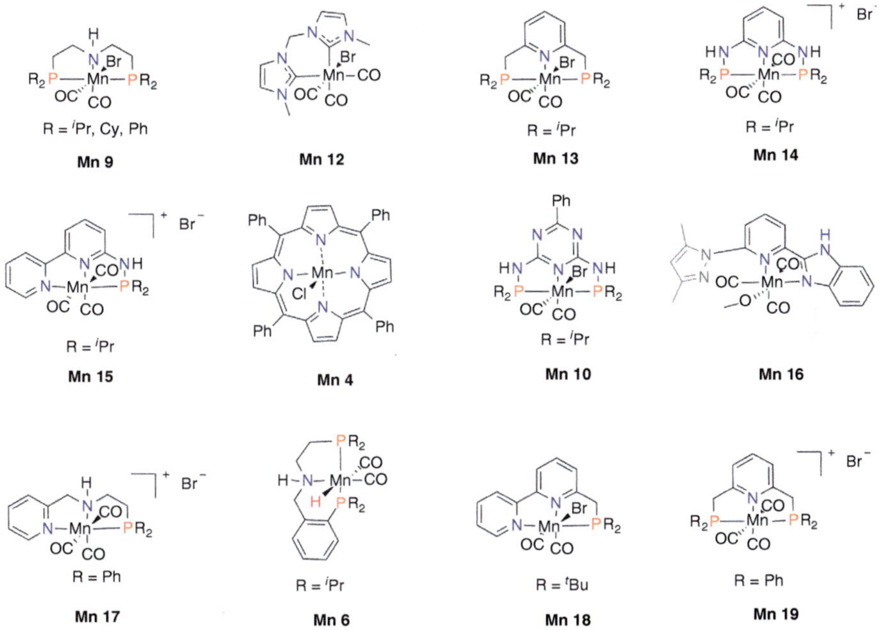

Chart 3.3 Frequently used well-defined manganese complexes for HB.

metal center and can be used for the reduction of an unsaturated compound, e.g. imines or conjugated carbon–carbon double bonds.

A broad variety of different well-defined manganese complexes can be used for HB reactions, whereas pincer-based complexes play a major role in these transformations as seen in Chart 3.3.

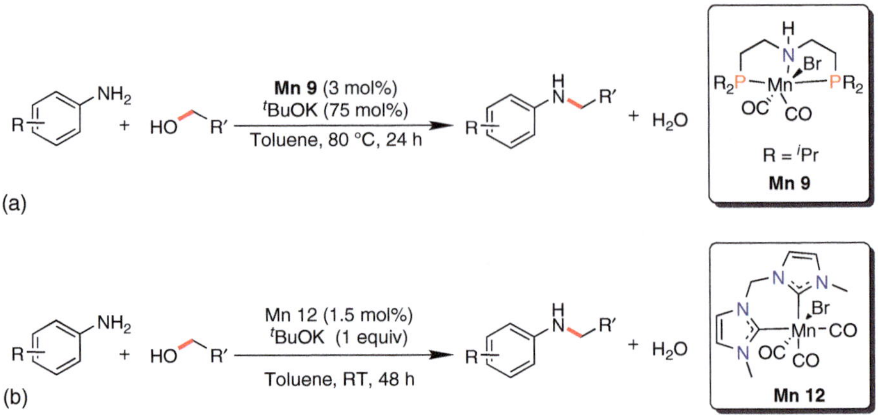

Scheme 3.19 Alkylation of amines catalyzed by **Mn 9** and **Mn 12** via HB. (a) Alkylation of amines via HB reaction (Beller and coworkers [28]). (b) Alkylation of amines via HB reaction (Ke and coworkers [29]).

3.3.1 Alkylation of Amines

Beller and coworkers were the first ones to report a manganese-catalyzed selective mono-alkylation of aromatic amines with alcohols via HB in 2016 as seen in Scheme 3.19a [28]. Within this context aliphatic PNP pincer ligands showed the best reactivity. This reaction proceeded at comparingly mild conditions and 3 mol% of **Mn 9**. In addition to that, the investigated PNP Mn(II) complex revealed almost no reactivity.

Within this seminal work, a broad variety of (hetero)aromatic amines could be alkylated with benzylic alcohols as well as with (hetero)aromatic and aliphatic alcohols. This protocol showed high functional group tolerance, even toward phenols, acetals, terminal alkenes, and several heterocycles such as pyridine-, thiophene-, furane-, and pyrozole-based substrates. Furthermore, methanol could be used for selective mono-methylation of aromatic amines underlining the wide applicability of this procedure under mild reaction conditions.

Recently, Ke and coworkers investigated the use of bis-NHC and bidentate NC pyridine-based NHC systems [29]. The NC-based complexes showed neglectable reactivity toward the alkylation of anilines with primary alcohols, whereas the bis-NHC system was found to be highly active for the alkylation of amines via a HB protocol with stoichiometric amount of base. Remarkably, this transformation took place at room temperature, using **Mn 12**. A broad variety of aromatic amines could be alkylated with a catalyst loading as low as 1.5 mol% as seen in Scheme 3.19b. In order to use methanol as C_1 source for alkylation, 100 °C had to be applied.

It should be noted that this type of ligand set cannot operate with the usual metal–ligand bifunctionality pattern. No deprotonation/deprotonation may occur at the ligand under these reaction conditions.

DFT calculations were carried out to get insight in the reaction mechanism, where two different pathways were compared. An inner-sphere mechanism, including a

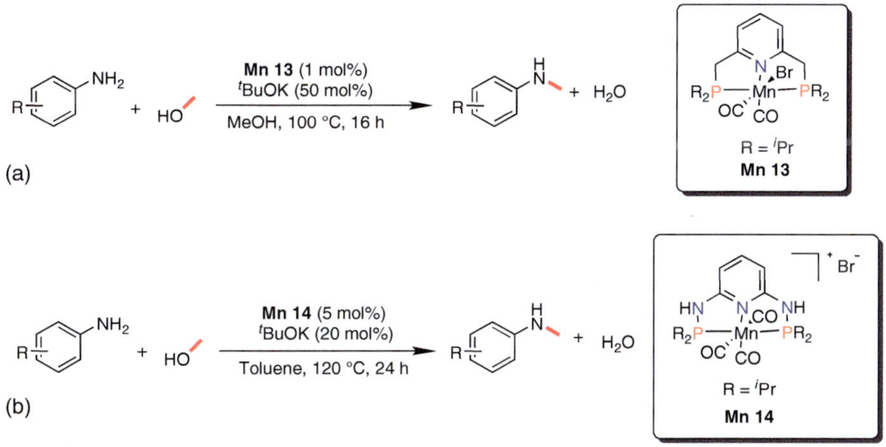

Scheme 3.20 Selective methylation of anilines via HB catalyzed by **Mn 13** and **Mn 14**. (a) Selective methylation of amines via HB reaction (Beller and coworkers [30]). (b) Selective methylation of amines via HB reaction (Sortais and coworkers [31]).

loss of a CO ligand, followed by the oxidation of the primary alcohol to an aldehyde, imine formation, and hydrogen transfer had a significant higher energy barrier then an outer-sphere pathway. Infrared analysis during the reaction gave rise to a new set of CO stretches, indicating the formation of a tricarbonyl complex, probably bearing an alkoxide ligand. These measurements support the presence of an outer-sphere-type mechanism.

Since a reliable procedure for selective mono-methylation of amines is highly desirable, the group of Beller reported a protocol using a pyridine-based PNP ligand, containing CH_2 linkers [30]. The use of **Mn 13** allowed to operate with only 1 mol% of catalyst in methanol as solvent and reagent. The general reaction pattern is depicted in Scheme 3.20a. A broad variety of aniline derivatives could be mono-methylated under these conditions. It is noteworthy to say that no alkylation of ketones, esters, or amides was observed under these reaction conditions, yielding a highly chemoselective protocol for the methylation of aromatic amines.

Sortais and coworkers introduced a protocol employing a cationic pyridine-based PNP with *NH* linkers, **Mn 14**, for the mono-alkylation of aniline derivatives with methanol [31]. This protocol operated with a catalyst loading of 5 mol% and a comparingly low amount of ^{t}BuOK (20 mol%) as seen in Scheme 3.20b. This procedure tolerated a broad variety of functional groups, including halides, ethers, esters, amides, ketones, nitriles, and nitro groups. No alkylation of ketones, amides, or esters was reported and reducible groups such as ketones ester, amides, nitriles, or nitro groups stayed unaltered.

Balaraman and coworkers investigated the role of several phosphine-free chelate ligands in combination with $Mn(CO)_5Br$ as manganese precursor for the alkylation of amines employing alcohols as carbon sources [32]. The *in situ* generated catalyst system (5 mol% $Mn(CO)_5$ Br and 5 mol% **L9** or **L10**) operated at 140 °C in toluene and an over-stoichiometric amount of ^{t}BuOK (Scheme 3.21a).

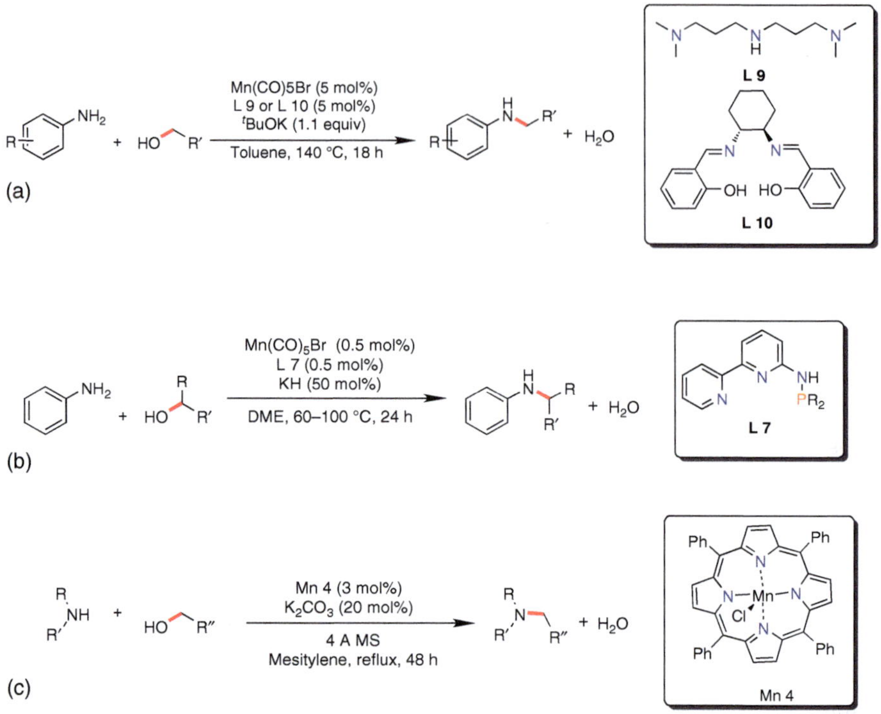

Scheme 3.21 Alkylation of amines via HB reactions catalyzed by *in situ* generated catalysts and **Mn 4**. (a) Alkylation of amines via HB reaction (Balaraman and coworkers [32]). (b) Alkylation of aniline with secondary alcohols via HB reaction (Hultzsch and coworkers [33]). (c) Alkylation of secondary amines via HB reaction (Madsen and coworkers [10]).

Among the investigated ligands, a nitrogen-based *NNN* tridentate ligand and interestingly a salen-based diamine ligand with a possible fourfold coordination mode performed best. The coordination mode of the investigated ligands stayed unclear within this context. Sulfur- and oxygen-containing tridentate ligands showed lower reactivity toward the alkylation of aromatic amines. It is noteworthy to say that dppb was as well employed within these investigations, giving moderate yields, although a typical metal–ligand bifunctionality mode cannot proceed with this type of ligand.

An overview of several alkylation reactions of anilines via HB reaction is given in Scheme 3.21.

An *in situ* generated cationic complex (**Mn 15**), bearing a bipyridine-based PNN ligand (**L7**) with an *NH* linker, was used by Hultzsch and coworkers for the alkylation of amines with alcohols (Scheme 3.21b) [33]. It should be noted that only 0.5 mol% of catalyst was required to give good to excellent yields.

In contradiction to other systems, this protocol can also be applied for the alkylation of benzylic and even aliphatic amines. Furthermore, secondary alcohol can be used as alkylation agents, which is the only example for the use of these kinds of alcohols as alkyl surrogates.

Madsen and coworkers used a porphyrin-based Mn(III), **Mn 4**, for alkylation of secondary amines with primary alcohols, yielding tertiary alcohols (Scheme 3.21c) [11]. This is the only literature-known manganese-catalyzed HB reaction, for the synthesis of tertiary amines so far.

Kempe and coworkers introduced an interesting base-switchable reaction path for the alkylation of amines via ADC or HB using **Mn 10** (Scheme 3.22) [34]. Interestingly, when using ^{t}BuOK or KHMDS, amines were generated with a typical selectivity >98%. Under similar conditions, the use of ^{t}BuONa yielded imines in excellent selectivity. Furthermore, iron-, cobalt-, and iridium-based systems were investigated, which could not compete with the remarkable selectivity of the manganese-centered systems.

Mechanistic studies of the system indicated the formation of sodium and potassium manganate species. Under the same reaction conditions, the potassium manganate complex showed a reaction rate that was about 40 times higher for the reduction of the *in situ* formed imine to the corresponding amine than the sodium manganate. This might be the reason for this highly interesting base-switchable reactivity pattern.

A broad variety of alcohol/amine combinations were examined for this transformation. In case of imine formation, even aliphatic amines could be used, but only benzylic alcohols were suitable as alkyl surrogates. The formation of secondary amines was restricted to anilines, which is in line with the most literature-known examples of these transformations catalyzed by manganese complexes.

3.3.2 Alkylation of Alcohols and Ketones

An overview of β-alkylation of secondary alcohols with primary alcohols is given in Scheme 3.23.

Yu and coworkers introduced an NN(N) ligand that is coordinated in a bidentate fashion for the β-alkylation of secondary alcohols with primary alcohols [35]. Apart from that, the impact of the anionic ligand was investigated. Substitution of the halide ligand through a methoxide ligand, yielding **Mn 16** resulted in higher reactivity for the given reaction. This protocol operated at 110 °C with 30 mol% ^{t}BuOK and 2.1 mol% catalyst (Scheme 3.23a). A variety of primary benzylic and aliphatic alcohols could be used as alkylation agent, tolerating halides, ethers, and pyridine moieties. In case of cyclopentanol, symmetric dialkylation could be achieved, using benzylic alcohols as alkyl donors.

Remarkably, complex substrates such as cholesterol could selectively be alkylated at only one the β-positions under slightly more forcing conditions (4.2 mol% catalyst, 140 °C, 3 equiv of alcohol) underlining the potential use of this procedure for the fast and selective modification of nature-derived compounds.

El-Sepelgy et al. investigated the role of PNN and PNP ligands on the β-alkylation of alcohols [36]. The pyridine-based PNN system with phenyl groups on the phosphorous donor (**Mn 17**) turned out to be the most suitable for this transformation with a catalyst loading of 1–3 mol% (Scheme 3.23b). Furthermore, coupling of two

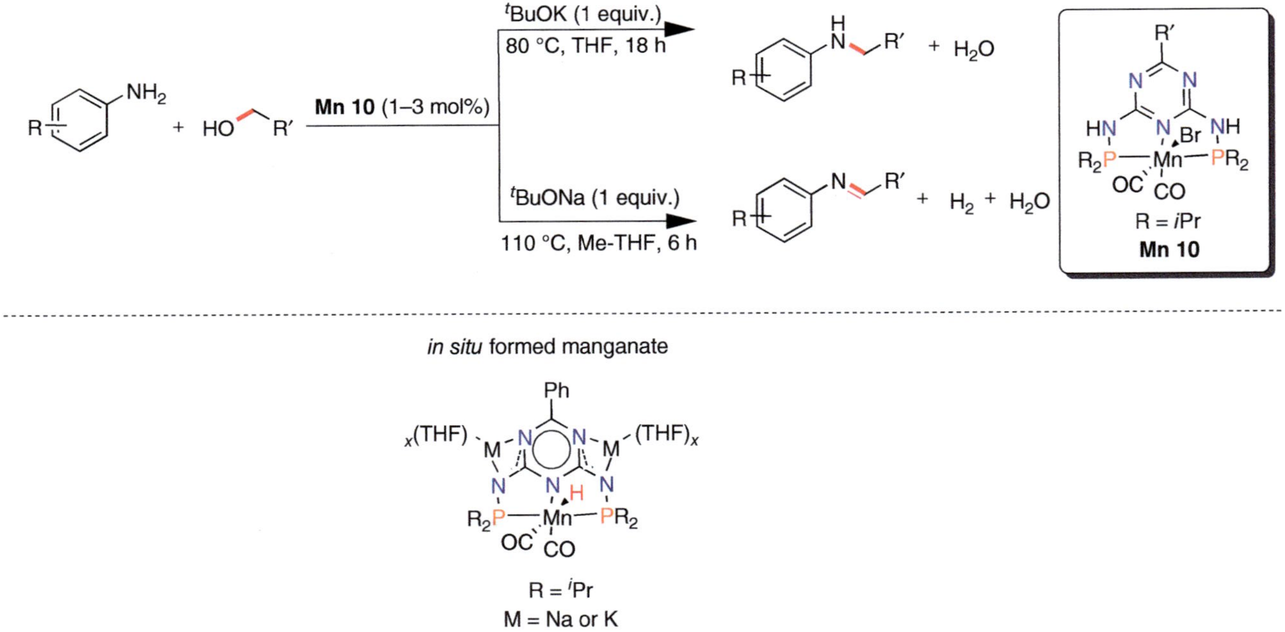

Scheme 3.22 Base-switchable alkylation of amines catalyzed by **Mn 10** yielding amines via HB or imines via ADC.

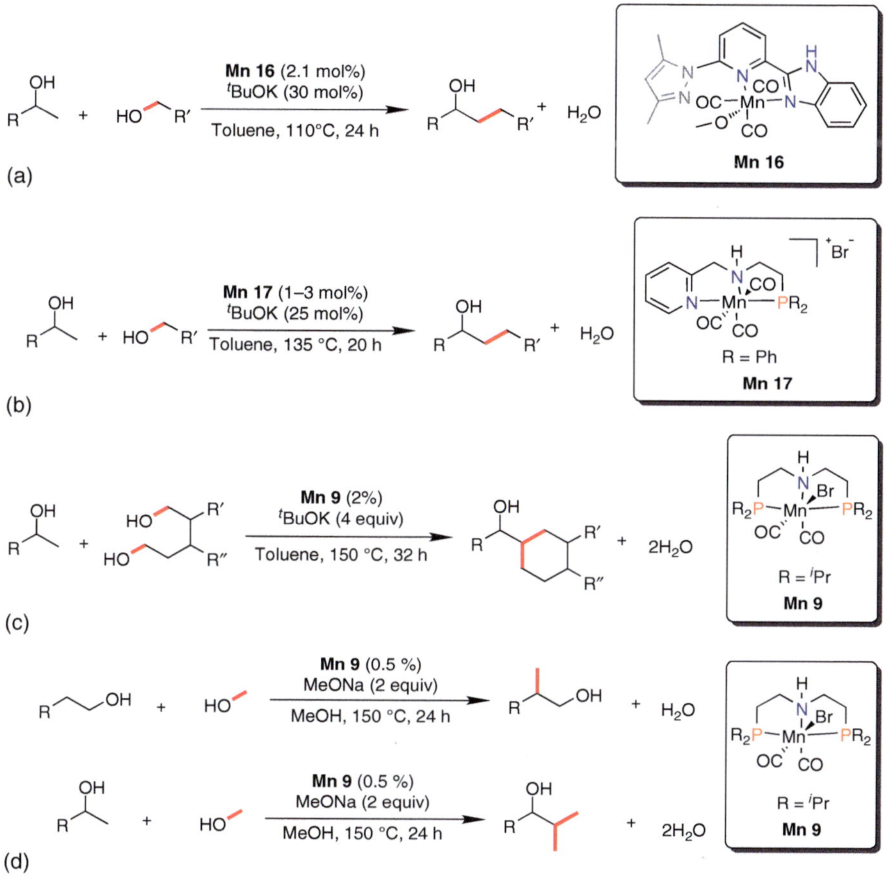

Scheme 3.23 Selective β-alkylation of alcohols catalyzed by well-defined Mn(I) catalysts via HB. (a) Alkylation of secondary alcohols via HB reaction (Yu and coworkers [35]). (b) Alkylation of secondary alcohols via HB reaction (El-Sepelgy and coworkers [36]). (c) Synthesis of cyclohexanes from secondary alcohols and diols via HB reaction (Leitner and coworkers [37]). (d) Methylation of alcohols using methanol as C_1 source via HB reaction (Leitner and coworkers [37]).

aliphatic alcohols was possible, which shows a unique reactivity in comparison to other manganese-containing systems.

In 2019, the group of Leitner reported on the direct synthesis of cyclohexanes from primary diols and secondary alcohols [37]. This transformation involved a cascade of dehydrogenation, condensation, and hydrogenation via HAT reactions catalyzed by **Mn 9** (Scheme 3.23c). Unfortunately, a large excess of diols (typically 4 equiv) were needed in order to suppress side reactions, such as lactone formation.

Very recently, Leitner and coworkers introduced a procedure for β-methylation of primary and secondary alcohols using methanol as alkylation agent using **Mn 9** (Scheme 3.23d) [38]. In case of primary alcohols, selective mono-methylation was described using 2 equiv of base. This procedure included 2-arylethanols as well as aliphatic alcohols. In case of aliphatic diols, methylation on both β-positions could be achieved. Interestingly, the use of 1-phenylethanol-based substrates gave di-methylated compounds if 4 equiv of base were used.

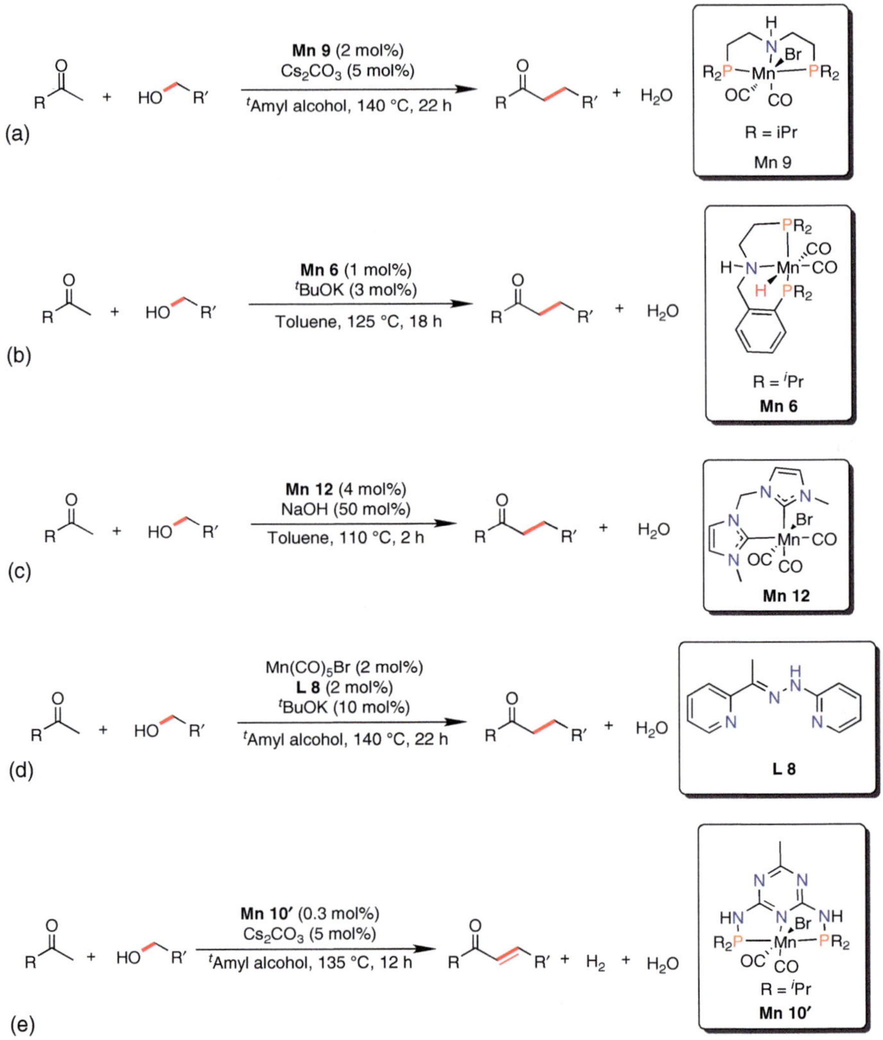

Scheme 3.24 Selective α-alkylation and alkenylation via HB or ADC using Mn(I) catalysts. (a) Alkylation of ketones via HB reaction (Beller and coworkers [39]). (b) Alkylation of ketones via HB reaction (Milstein and coworkers [40]). (c) Alkylation of ketones via HB reaction (Ke and coworkers [41]). (d) Alkylation of ketones via HB reaction (Maji and coworkers [42]). (e) Alkenylation of ketones via ADC reaction (Gunanathan and coworkers [43]).

Beller and coworkers were the first to report α-alkylation of ketones with primary alcohols [39]. Investigation of the alkyl groups on the phosphorous donors revealed that ^{i}Pr groups were the best candidates for this kind of transformation (**Mn 9**). The reactions were carried out at 140 °C in ^{t}amyl alcohol with only 5 mol% Cs_2CO_3 and 2 mol% catalyst as seen in Scheme 3.24a.

Within this context, aromatic and heteroaromatic ketones could successfully been coupled with benzylic and (hetero)aromatic alcohols. This procedure could be extended for the selective α-alkylation of 2-oxindoles without the introduction of a nitrogen protection group. This type of transformation resulted in a shorter protocol in comparison to classic organic chemistry. Apart from that, hormones, such as testosterone and an estrone derivative, could selectively been alkylated in only one α-position, making this procedure interesting for fast and simple derivatization of several hormones in clinical trials.

In 2018, Milstein and coworkers reported the α alkylation of ketones using an aliphatic PNP system (**Mn 6**), with an aliphatic amine donor [40]. The use of pyridine-based PNP or PNN ligands resulted in a slightly decreased reactivity.

This protocol used an equimolar amount of ketone and alcohol substrate, resulting in a highly atom-efficient transformation. Apart from that, the reaction proceeded at a catalyst loading of only 1 mol% and remarkable low amount of base (3 mol% ^{t}BuOK) (Scheme 3.24b). This protocol showed a broad reaction scope, tolerating halides, ethers, alkenes, amines, pyridines, thiophenes, and even the strongly coordinating nitrile group.

It should be noted that secondary benzylic alcohols could be alkylated with primary alcohols, followed by oxidation of the secondary alcohol, yielding α-alkylated ketones. This displays an elegant route toward the synthesis of α-substituted ketones using alcohols as feeding stock.

Ke and coworkers reported a rapid HB protocol under comparingly mild conditions for the alkylation of ketones in 2019 using **Mn 12** (Scheme 3.24c) [41].

Maji and coworkers investigated the role of NN and NNN bi- and tridentate ligands for the α-alkylation of ketones with primary alcohols [42]. Within this context, a hydrazone-based NNN ligand was found to be the most suitable ligand for this transformation (Scheme 3.24d). For two hours prior the catalytic reaction, 2 mol% of **L8** and $Mn(CO)_5$ Br were reacted at 80 °C in ^{t}amyl alcohol.

This procedure shows a suitable functional group tolerance, whereas halides, ethers, and heteroaromates were well tolerated. Ester and nitrile functionalities were found not to be compatible within this procedure.

In 2019, Gunanathan and coworkers implemented an ADC of ketones with primary alcohols, yielding α,β-unsaturated ketones [43]. Remarkably, this protocol operated at a very low catalyst loading (0.3 mol% **Mn 10′**) and the addition of a small amount of Cs_2CO_3 as base, as seen in Scheme 3.24e.

Sortais and coworkers were the first ones to report a reliable protocol for the α-methylation of ketones with methanol using 3 mol% of a cationic PNP complex with *NH* linkers (**Mn 14**, Scheme 3.25) [44]. In the presence of an alkyl group on the α-carbon, selective mono-methylation was observed. Surprisingly, in the case of unsubstituted α-carbons, di-methylation took place, which is quite a remarkable finding.

3.3.3 Alkylation of Amides and Esters

The group of Milstein reported the selective α-alkylation of aliphatic amides and esters, depicted in Scheme 3.26a [40]. These reactions took place under more forcing conditions: 5 mol% **Mn 6** and 1.5 equiv of ^{t}BuOK were needed to yield moderate to

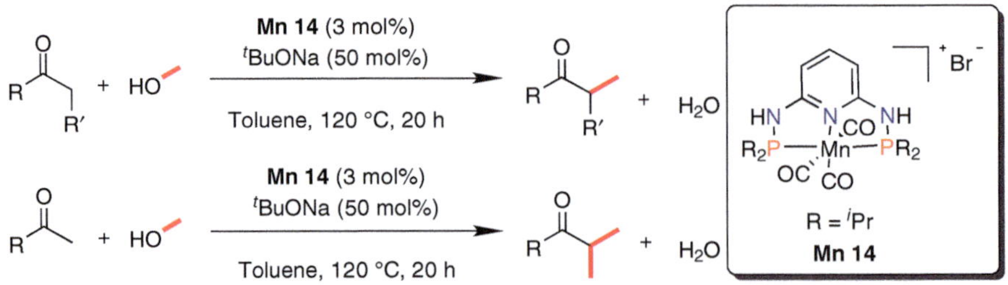

Scheme 3.25 α-Methylation of ketones catalyzed by **Mn 14**.

excellent yields. For these transformations, 4 equiv of esters and 2 equiv of amides were needed to achieve reasonable yields.

El-Sepelgy and Rueping used PNP- and PNN-based systems for the alkylation of aliphatic esters and amides (Scheme 3.26b) [45]. The role of an aliphatic PNP (**L3**), a pyridine-based PNP (**L1**), and an aliphatic PNN was investigated, whereas high reactivity of the PNN system (**Mn 17**) could be observed. Under the same conditions, the PNP systems performed poorly. A variety of amides could be selectively alkylated with benzylic alcohols containing halides, ethers, and heterocycles and aliphatic primary alcohols. In case of esters, the catalyst loading and base content had to be increased to 5 mol% **Mn 17** and 2 equiv of base.

Deuterium-labeling was used to elucidate the reaction mechanism. This proved that the alcohol proton was incorporated in the α-position of the amide group, whereas the benzylic deuterium atoms of the former alcohol substrate were not scrabbled over the product. These examinations indicate the presence of a HB reaction, involving a metal hydride species and participation of the non-innocent ligand.

Balaraman and coworkers used aliphatic PNP for the alkylation of amides and esters under HB conditions [46]. Different alkyl groups connected to the phosphine donor were investigated. Cyclohexyl groups (**Mn 9′**) gave slightly better yields than *iso*propyl groups, where phenyl groups performed only moderate under the given reaction conditions.

The alkylation reaction of amides proceeded at 110 °C in toluene with a remarkably low catalyst loading of only 0.5 mol% as seen in Scheme 3.26c. Only 2 equiv of amide had to be used to achieve moderate to excellent yield with benzylic and aliphatic alcohols.

In case of ester as substrate, the reaction temperature could be decreased to 80 °C. Furthermore, this protocol also operated with only 0.5 mol% of catalyst, which is only one-tenth in comparison to all other protocols.

Sortais and coworkers reported the selective methylation of benzylic ester using methanol as feedstock using **Mn 14** (Scheme 3.26d) [44].

3.3.4 Alkylation of Nitriles and Sulfonamide

Maji and coworkers were the first to report an NN(S) ligand-based catalyst for the α-alkylation of benzylic nitriles with alcohols [47]. As seen in Scheme 3.27a, 2 mol% of an *in situ* generated catalyst was sufficient for this transformation. Elaborate reaction temperature (140 °C) were needed to achieve the desired reactivity in

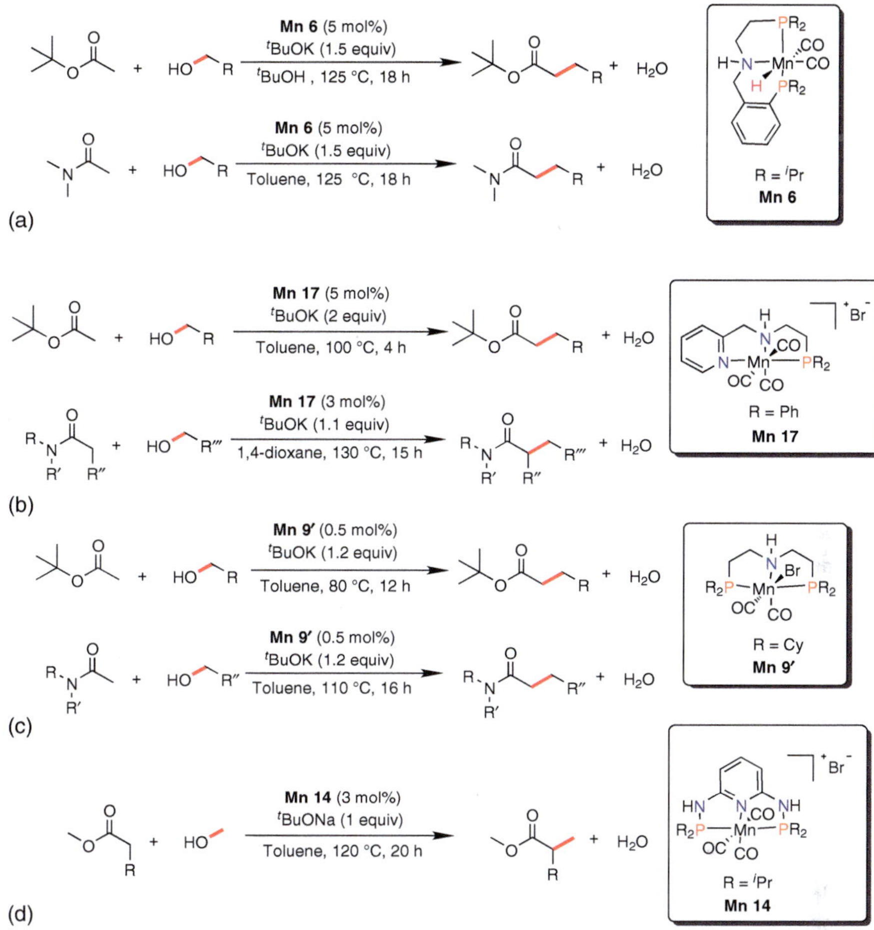

Scheme 3.26 Alkylation of esters and amides catalyzed by well-defined Mn(I) catalysts. (a) Alkylation of esters and amides via HB reaction (Milstein and coworkers [40]). (b) Alkylation of esters and amides via HB reaction (El-Sepelgy and coworkers [45]) (c) Alkylation of esters and amides via HB reaction (Balaraman and coworkers [46]). (d) Methylation of esters via HB reaction (Sortais and coworkers [44]).

^{t}amyl alcohol. Interestingly only a moderate amount of base (20 mol% ^{t}BuOK) were needed. A broad variety of benzylic alcohol including heteroaromatic, halide, and ether containing systems were used as alkyl surrogates, although 2 equiv were needed. In case of aliphatic alcohols, 5 equiv were used, whereas the attempt of methylation yielded only traces of product.

Mechanistic studies revealed a high content of incorporated deuterium atoms in the β-position. This indicated the presence of a manganese hydride species and consecutive reduction of the formed carbon–carbon double bound via metal–ligand bifunctionality mode.

El-Sepelgy and coworkers investigated the potential use of a neutral aliphatic PNP (**L3**)-, a cationic aromatic PNP (**L1**)-, and a cationic pyridine-based PNN ligand (**Mn**

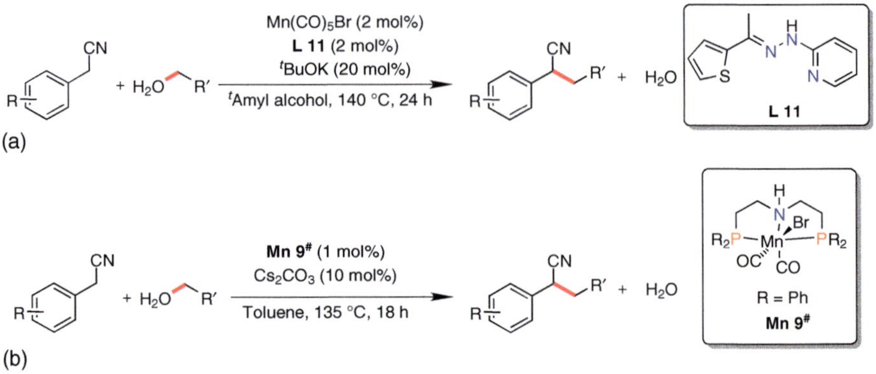

Scheme 3.27 Selective α-alkylation of nitriles catalyzed by Mn(I) complexes via HB. (a) Alkylation of nitriles via HB reaction (Maji and coworkers [47]). (b) Alkylation of nitriles via HB reaction.

17) with phenyl groups system for the alkylation of benzylic nitriles [48]. Under the given reaction conditions, the neutral PNP-based complex (Scheme 3.27b, **Mn 9#**) performed slightly better than the PNN-based system, whereas neglectable reactivity could be detected with the pyridine-based PNP system. With only 1 mol% catalyst at 135 °C in ᵗamyl alcohol and a low amount of Cs_2CO_3 (10 mol%) as base, 2 equiv of benzylic or aliphatic alcohols could be used as alkylation agent.

The functional group tolerance could be increased, tolerating heterocycles, ethers, halides, and acetals on the nitrile substrate. In the case of the alcohol as coupling partner, heterocycles, ethers, halides, and terminal alkenes were well tolerated. This protocol is also suitable for α-methylation of benzylic nitriles using methanol as alkylation agent.

In 2017, the group of Sortais reported the first methylation protocol for sulfonamides with a small substrate scope employing **Mn 14** (Scheme 3.28a) [31].

Morill and coworkers reported the use of **Mn 9** for the alpha alkylation of sulfonamides with primary alcohols in 2019 (Scheme 3.28b) [49]. The protocol operated with an equimolar ratio of alcohol and sulfonamide, resulting in a highly atom-efficient reaction. Unfortunately, 5 mol% catalyst and 150 °C in xylenes were required with a comparingly low amount of K_2CO_3 (10 mol%) as base.

The scope of alcohols resulted in a broad applicability of benzylic and aliphatic alcohols, including methanol as alkyl surrogate. Heterocycles, halides, ethers, terminal alkenes, and even the sensitive ester moiety were well tolerated. However, alkyne containing sterically hindered primary and generally secondary alcohols could not be used in this procedure.

Investigation of the amide substrates revealed tolerance of halides, ethers, terminal alkenes, and even sterically hindered aromatic and aliphatic systems. Substrates containing a pyridine-, nitro-, or nitrile-motive were unsuitable within this process.

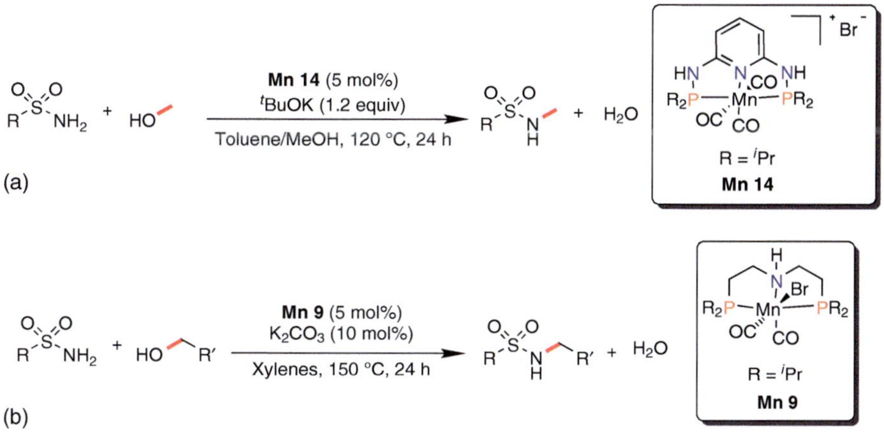

Scheme 3.28 Alkylation of sulfonamides catalyzed by **Mn 14** and **Mn 9**. (a) Methylation of sulfonamides via HB reaction (Sortais and coworkers [31]). (b) Alkylation of sulfonamides via HB reaction (Morill and coworkers [49]).

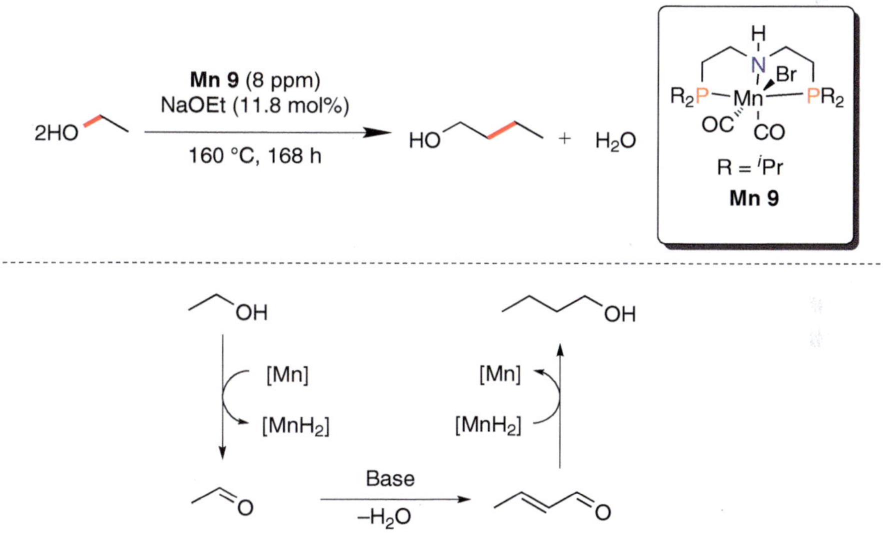

Scheme 3.29 Upgrading of ethanol into 1-butanol via HB catalyzed by **Mn 9**.

3.3.5 Upgrading of Ethanol into 1-Butanol

The group of Liu [50] and the group of Jones [51] independently studied the catalytic upgrading reaction of ethanol into 1-butanol using Guerbet reaction in 2018 using **Mn 9**. The general reaction pattern is displayed in Scheme 3.29. Liu and coworkers could report that 8 ppm of catalyst was sufficient for this reaction, whereas selectivity of 92% could be achieved.

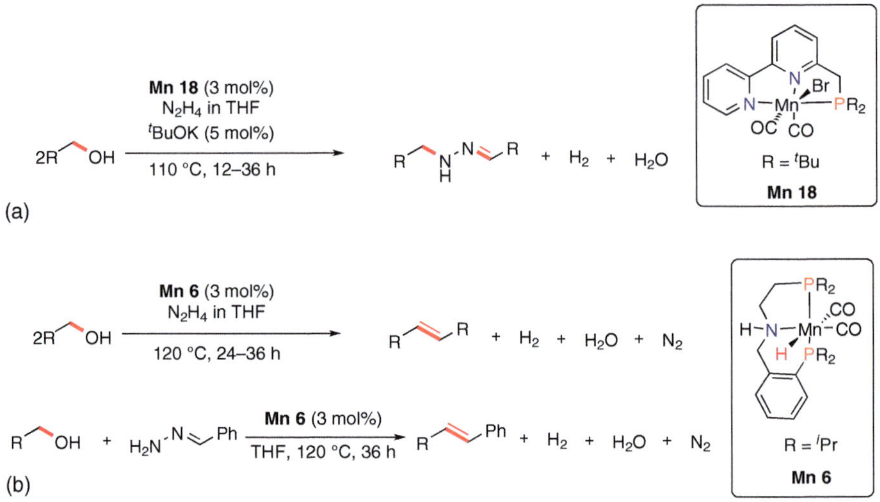

Scheme 3.30 Synthesis of N-substituted hydrazones and direct synthesis of alkene from alcohols catalyzed by **Mn 18** and **Mn 6**. (a) Synthesis of N-substituted hydrazones via ADC and HB reaction (Milstein and coworkers [52]). (b) Conversion of alcohols into alkenes with hydrazine/hydrazones via ADC (Milstein and coworkers [53]).

Detailed mechanistic studies underlined the essential role of the "*NH*" motive in this system, since very low reactivity could be observed when the "*NH*" moiety was substituted by "*NMe*."

3.3.6 Alkylation of Hydrazine

Milstein and coworker reported a partial HB reaction for the alkylation of hydrazine using a bipyridine-based PNN system (**Mn 18**) [52]. This unique reaction, operating via ADC and HB reactions, proceeded at 110 °C in THF with 3 mol% catalyst and 5 mol% ^{t}BuOK, as seen in Scheme 3.30a.

Investigation of the reaction mechanism indicated the formation of a coordinatively unsaturated 16 e^- complex. The fivefold coordinated species was proven to catalyze this reaction under base-free conditions.

Using a slightly different protocol Milstein and coworkers were able to convert alcohols directly in alkenes, employing **Mn 6** (Scheme 3.30b) [53]. Mechanistic investigations revealed the presence of a highly interesting deprotonated benzylidenehydrazine ligand, coordinated to the manganese center.

3.3.7 Combining Acceptorless Dehydrogenative Coupling with Hydrogen-Borrowing

Recently, Cavllo, El-Sepelgy, and Rueping reported a three-component reaction for the synthesis of pyrroles using **Mn 19** [54]. In this seminal work, a combination of ADC and HB was presented as seen in Scheme 3.31.

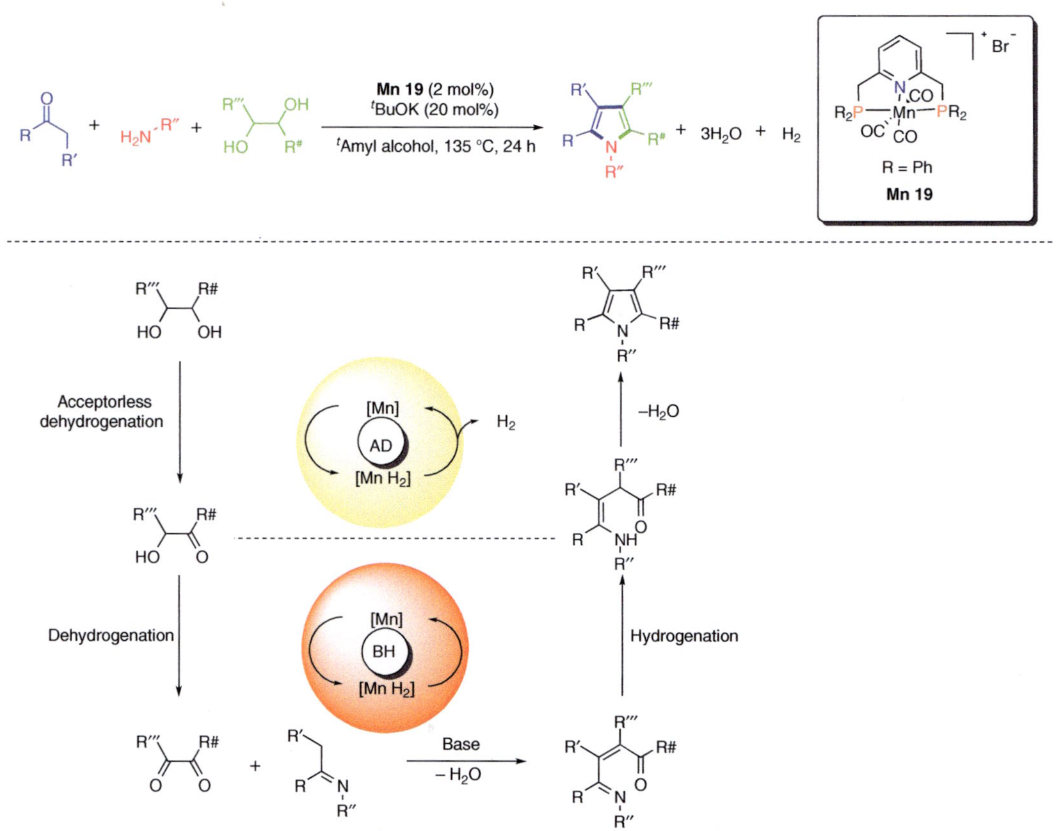

Scheme 3.31 Synthesis of pyrroles via ADC and HB catalyzed by **Mn 19**.

3.4 Conclusions and Perspectives

In conclusion, we describe the development of some distinct families of well-defined manganese-based catalysts that are active in a variety of dehydrogenation reactions involving also an HB process. Most of these successful examples originate from the discovery of somehow privileged metal–ligand PNP pincer platforms that can efficiently support and stabilize the manganese metal centers or even participate in elementary steps of the reaction. The key aspects in the development of these catalysts have been the recognition of decisive structural features as well as the determination and optimization of proper reaction conditions that permit catalytic turnover. Moreover, mechanistic insights including catalyst activation and the identification of the reactive species revealed specific similarities and differences between the described systems. These catalysts already cover a broad scope of catalytic applications in some cases, exhibiting activities and productivities that are competitive to well-established noble metal catalysts. However, the current knowledge about the nature and characteristics of active manganese-based systems paved the way for conceptually and mechanistically well-founded research, which might lead to further developments and the discovery of novel catalysts extending the current scope and limitations of reactivity. One aspect is the replacement of CO by other π-accepting ligands such as NO^+, CN^-, and isocyanides or the partial or complete removal of CO ligands that may provide new reactivities. Another challenge in Mn(I) chemistry will be the development of catalysts that originate from inexpensive Mn(II) materials, which is as yet restricted to the expensive precursor $[Mn(CO)_5X]$ (X = Cl, Br). In view of the many inspiring discoveries and achievements in homogeneous manganese catalysis in the last few years, it seems that research into this field is merely at the beginning of its development and many breakthroughs are still ahead, rewarding those who do not mind great challenges.

References

1 Bullock, R.M. (2013). Abundant metals give precious hydrogenation performance. *Science* 342: 1054–1055.

2 (a) Christianson, D.W. (1997). Structural chemistry and biology of manganese metalloenzymes. *Prog. Biophys. Mol. Biol.* 67: 217. (b) Wieghardt, K. (1989). The active sites in manganese-containing metalloproteins and inorganic model complexes. *Angew. Chem. Int. Ed.* 28: 1153.

3 (a) Valyaev, D.A., Lavigne, G., and Lugan, N. (2016). Manganese organometallic compounds in homogeneous catalysis: past, present, and prospects. *Coord. Chem. Rev.* 308: 191–235. (b) Carney, J.R., Dillon, B.R., and Thomas, S.P. (2016). Recent advances of manganese catalysis for organic synthesis. *Eur. J. Org. Chem.*: 3912–3929. (c) Garbe, M., Junge, K., and Beller, M. (2017). Homogeneous catalysis by manganese-based pincer complexes. *Eur. J. Org. Chem.*: 4344–4362. (d) Maji, B. and Barman, M. (2017). Recent developments of manganese complexes for catalytic hydrogenation and dehydrogenation reactions. *Synthesis*

49: 3377–3393. (e) Filonenko, G.A., van Putten, R., Hensen, E.J.M., and Pidko, E.A. (2018). Catalytic (De)hydrogenation promoted by non-precious metals - Co, Fe and Mn: recent advances in an emerging field. *Chem. Soc. Rev.* 47: 1484–1515. (f) Gorgas, N. and Kirchner, K. (2018). Isoelectronic manganese and iron hydrogenation/dehydrogenation catalysts - similarities and divergences. *Acc. Chem. Res.* 51: 1558–1569. (g) Zell, T. and Langer, R. (2018). From ruthenium to iron and manganese - a mechanistic view on challenges and design principles of base metal hydrogenation catalysts. *ChemCatChem* 10: 1930–1940. (h)Mukherjee, A. and Milstein, D. (2018). Homogeneous catalysis by cobalt and manganese pincer complexes. *ACS Catal.* 8: 11435–11469.

4 Gunanathan, C. and Milstein, D. (2013). Applications of acceptorless dehydrogenation and related transformations in chemical synthesis. *Science* 341: 1229712.

5 de Vries, J.G. and Elsevier, C.J. (eds.) (2006). *The Handbook of Homogeneous Hydrogenation*. Weinheim: Wiley-VCH Verlag GmbH.

6 Khusnutdinova, J.R. and Milstein, D. (2015). Metal-ligand cooperation. *Angew. Chem. Int. Ed.* 54: 12236–12273.

7 Zell, T. and Milstein, D. (2015). Hydrogenation and dehydrogenation iron pincer catalysts capable of metal-ligand cooperation by aromatization/dearomatization. *Acc. Chem. Res.* 48: 1979–1994.

8 Mukherjee, A., Nerush, A., Leitus, G. et al. (2016). Manganese-catalyzed environmentally benign dehydrogenative coupling of alcohols and amines to form aldimines and H_2: a catalytic and mechanistic study. *J. Am. Chem. Soc.* 138: 4298–4301.

9 Mastalir, M., Glatz, M., Gorgas, N. et al. (2016). Divergent coupling of alcohols and amines catalyzed by isoelectronic hydride MnI and FeII PNP pincer complexes. *Chem. Eur. J.* 22: 12316–12320.

10 Samuelsen, S.V., Santilli, C., Ahlquist, M.S.G., and Madsen, R. (2019). Development and mechanistic investigation of the manganese(III) salen-catalyzed dehydrogenation of alcohols. *Chem. Sci.* 10: 1150–1157.

11 Azizi, K., Akrami, S., and Madsen, R. (2019). Manganese(III) porphyrin-catalyzed dehydrogenation of alcohols to form imines, tertiary amines and quinolines. *Chem. Eur. J.* 25: 6439–6446.

12 Espinosa-Jalapa, N.A., Kumar, A., and Milstein, D. (2017). Synthesis of cyclic imides by acceptorless dehydrogenative coupling of diols and amines catalyzed by a manganese pincer complex. *J. Am. Chem. Soc.* 139: 11722–11725.

13 Chakraborty, S., Gellrich, U., Diskin-Posner, Y. et al. (2017). Manganese-catalyzed N-formylation of amines by methanol liberating H_2: a catalytic and mechanistic study. *Angew. Chem. Int. Ed.* 56: 4229–4233.

14 Daw, P., Kumar, A., Espinosa-Jalapa, N.A. et al. (2019). Direct synthesis of amides by acceptorless dehydrogenative coupling of benzyl alcohols and ammonia catalyzed by manganese pincer complex: unexpected crucial role of base. *J. Am. Chem. Soc.* 141: 12202–12206.

15 Nguyen, D.H., Trivelli, X., Capet, F. et al. (2017). Manganese pincer complexes for the base-free, acceptorless dehydrogenative coupling of alcohols to esters: development, scope, and understanding. *ACS Catal.* 7: 2022–2032.

16 Wang, Z., Shao, K., and Zhang, Q.L. (2018). Manganese-catalyzed dual-deoxygenative coupling of primary alcohols with 2-arylethanols. *Angew. Chem. Int. Ed.* 57: 15143–15147.

17 Chakraborty, S., Das, U.K., Ben-David, Y., and Milstein, D. (2017). Manganese catalyzed α-olefination of nitriles by primary alcohols. *J. Am. Chem. Soc.* 139: 1710–11713.

18 Luque-Urrutia, J.A., Sola, M., Milstein, D., and Poarter, A. (2019). Mechanism of the manganese-pincer-catalyzed acceptorless dehydrogenative coupling of nitriles and alcohols. *J. Am. Chem. Soc.* 141: 2398–2403.

19 Zhang, G., Irrgang, T., Dietel, T. et al. (2018). Manganese-catalyzed dehydrogenative alkylation or α-olefination of alkyl-N-heteroaromatics by alcohols. *Angew. Chem. Int. Ed.* 57: 9131–9135.

20 Barman, M.K., Waiba, S., and Maji, B. (2018). Manganese-catalyzed diect olefination of methyl-substituted heteroarenes with primary alcohols. *Angew. Chem. Int. Ed.* 57: 9126–9130.

21 Mastalir, M., Glatz, M., Pittenauer, E. et al. (2016). Sustainable synthesis of quinolines and pyrimidines catalyzed by manganese PNP pincer complexes. *J. Am. Chem. Soc.* 138: 15543–15546.

22 Deibl, N. and Kempe, R. (2017). Manganese-catalyzed multicomponent synthesis of pyrimidines from alcohols and amidines. *Angew. Chem. Int. Ed.* 56: 1663–1666.

23 Kallmeier, F., Dudziec, B., Irrgang, T., and Kempe, R. (2017). Manganese-catalyzed sustainable synthesis of pyrroles from alcohols and amino alcohols. *Angew. Chem. Int. Ed.* 56: 7261–7265.

24 Mastalir, M., Pittenauer, E., Allmaier, G., and Kirchner, K. (2017). Manganese-catalyzed aminomethylation of aromatic compounds with methanol as sustainable C1 building block. *J. Am. Chem. Soc.* 139: 8812–8815.

25 Daw, P., Kumar, A., Espinosa-Jalapa, N.A. et al. (2018). Synthesis of pyrazines and quinoxalines via acceptorless dehydrogenative coupling routes catalyzed by manganese pincer complexes. *ACS Catal.* 8: 7734–7741.

26 Das, K., Mondal, A., and Srimani, D. (2018). Selective synthesis of 2-substituted and 1,2-disubstituted benzimidazoles directly from aromatic diamines and alcohols catalyzed by molecular defined nonphosphine manganese(I) complex. *J. Org. Chem.* 83: 9553–9560.

27 Das, K., Mondal, A., and Srimani, D. (2018). Phosphine free Mn-complex catalysed dehydrogenative C—C and C-heteroatom bond formation: a sustainable approach to synthesize quinoline, pyrazine, benzothiozole and quinoline derivatives. *Chem. Commun.* 54: 10582–10585.

28 Elangovan, S., Neumann, J., Sortais, J.-B. et al. (2016). Efficient and selective N-alkylation of amines with alcohols catalysed by manganese pincer complexes. *Nat. Commun.* 7: 12641.

29 Huang, M., Li, Y., Liu, J. et al. (2019). Room temperature N-heterocyclic carbene manganese catalyzed selective N-alkylation of anilines with alcohols. *Chem. Commun.* 55: 6213–6216.

30 Neumann, J., Elangovan, S., Spannenberg, A. et al. (2017). Improved and general manganese-catalyzed N-methylation of aromatic amines using methanol. *Chem. Eur. J.* 23: 5410–5413.

31 Bruneau-Voisine, A., Wang, D., Dorcet, V. et al. (2017). Mono-N-methylation of anilines with methanol catalyzed by a manganese pincer-complex. *J. Catal.* 347: 57–62.

32 Landge, V.G., Mondal, A., Kumar, V. et al. (2018). Manganese catalyzed N-alkylation of anilines with alcohols: ligand enabled selectivity. *Org. Biomol. Chem.* 16: 8175–8180.

33 Homberg, L., Roller, A., and Hultzsch, K.C. (2019). A highly active PN^3 manganese pincer complex performing N-alkylation of amines under mild conditions. *Org. Lett.* 21: 3142–3147.

34 Fertig, R., Irrgang, T., Freitag, F. et al. (2018). Manganese-catalyzed and base-switchable synthesis of amines or imines via borrowing hydrogen or dehydrogenative condensation. *ACS Catal.* 8: 8525–8530.

35 Liu, T., Wang, L., Wu, K., and Yu, Z. (2018). Manganese-catalyzed β-alkylation of secondary alcohols with primary alcohols under phosphine-free conditions. *ACS Catal.* 8: 7201–7207.

36 El-Sepelgy, O., Matabor, E., Brzozowska, A., and Rueping, M. (2019). C-Alkylation of secondary alcohols by primary alcohols through manganese-catalyzed double hydrogen autotransfer. *ChemSusChem* 12: 3099–3102.

37 Kaithal, A., Gracia, L.-L., Camp, C. et al. (2019). Direct synthesis of cycloalkanes from diols and secondary alcohols or ketones using a homogeneous manganese catalyst. *J. Am. Chem. Soc.* 141: 17487–17492.

38 Kaithal, A., van Bonn, P., Hölscher, M., and Leitner, W. (2020). Manganese(I)-catalyzed β-methylation of alcohols using methanol as C_1 source. *Angew. Chem. Int. Ed.* 59: 215–220.

39 Peña-López, M., Piehl, P., Elangovan, S. et al. (2016). manganese-catalyzed hydrogen-autotransfer C—C bond formation: α-alkylation of ketones with primary alcohols. *Angew. Chem. Int. Ed.* 55: 14967–14971.

40 Chakraborty, S., Daw, P., Ben David, Y., and Milstein, D. (2018). Manganese-catalyzed α-alkylation of ketones, esters, and amides using alcohols. *ACS Catal.* 8: 10300–10305.

41 Lan, X.-B., Ye, Z., Huang, M. et al. (2019). Nonbifunctional outer-sphere strategy archived a highly active α-alkylation of ketones with alcohols by *N*-heterocyclic carbene manganese (NHC-Mn). *Org. Lett.* 21: 8065–8070.

42 Barman, M.K., Jana, A., and Maji, B. (2018). Phosphine-free NNN-manganese complex catalyzed α-alkylation of ketones with primary alcohols and Friedländer quinoline synthesis. *Adv. Synth. Catal.* 360: 3233–3238.

43 Gawil, S.S., Pandia, B.K., and Gunanathan, C. (2019). Manganese(I)-catalyzed α-alkenylation of ketones using primary alcohols. *Org. Lett.* 21: 3842–3847.

44 Bruneau-Voisine, A., Pallova, L., Bastin, S. et al. (2019). Manganese catalyzed α-methylation of ketones with methanol as a C1 source. *Chem. Commun.* 55: 314–317.

45 Jang, Y.K., Krückel, T., Rueping, M., and El-Sepelgy, O. (2018). Sustainable alkylation of unactivated esters and amides with alcohols enables by manganese catalysis. *Org. Lett.* 20: 7779–7783.

46 Rana, J., Gupta, V., and Balaraman, E. (2019). Manganese-catalyzed direct C—C coupling of α-C—H bonds of amides and esters with alcohols via hydrogen autotransfer. *Dalton Trans.* 48: 7094–7099.

47 Jana, A., Reddy, C.B., and Maji, B. (2018). Manganese catalyzed α-alkylation of nitriles with primary alcohols. *ACS Catal.* 8: 9229–9231.

48 Borghs, J.C., Tran, M.A., Sklyaruk, J. et al. (2019). Sustainable alkylation of nitriles with alcohols by manganese catalysis. *J. Org. Chem.* 84: 7927–7935.

49 Reed-Berendt, B.G. and Morill, L.C. (2019). Manganese-catalyzed N-alkylation of sulfonamides using alcohols. *J. Org. Chem.* 84: 3715–3724.

50 Fu, S., Shao, Z., Wang, Y., and Liu, Q. (2017). Manganese-catalyzed upgrading of ethanol into 1-butanol. *J. Am. Soc.* 139: 11941–11948.

51 Kulkarni, N.V., Brennessel, W.W., and Jones, W.D. (2017). Catalytic upgrading of ethanol to *n*-butanol via manganese-mediated Guerbet reaction. *ACS Catal.* 8: 997–1002.

52 Das, K., Ben-David, Y., Diskin-Posner, Y., and Milstein, D. (2018). N-Substituted hydrazones by manganese catalyzed coupling of alcohols with hydrazine; 'Borrowing Hydrogen' and acceptorless dehydrogenation in one system. *Angew. Chem. Int. Ed.* 57: 2179–2182.

53 Das, K.U., Chakraborty, S., Diskin-Posner, Y., and Milstein, D. (2018). Direct conversion of alcohols into alkenes by dehydrogenative coupling with hydrazine/hydrazone catalyzed by manganese. *Angew. Chem. Int. Ed.* 57: 13444–13448.

54 Borghs, J.C., Azofra, L.M., Biberger, T. et al. (2019). Maganese-catalyzed multicomponent synthesis of pyrroles through acceptorless dehydrogenation hydrogen autotransfer catalysis: experiment and computation. *ChemSusChem* 12: 3083–3088.

4

Manganese-Catalyzed Hydrosilylation and Hydroboration Reactions

Thao T. Nguyen and Ryan J. Trovitch

Arizona State University, School of Molecular Sciences, Physical Sciences Building D-102, Tempe, AZ 85287, USA

4.1 Introduction

Hydrofunctionalization, which involves H—X bond addition across an unsaturated carbon–carbon or carbon–heteroatom bond, unlocks possibilities for atom-economical synthesis. Metal catalysts play a crucial role in this transformation in terms of achieving regio-, stereo-, and enantioselectivity. This chapter covers two important manganese-catalyzed hydrofunctionalization reactions, hydrosilylation and hydroboration, parts of which have been covered in recent reviews [1–4].

Hydrosilylation, the addition of a Si—H bond across an unsaturated bond, has attracted widespread interest for its industrial and laboratory-scale applications. The hydrosilylation of olefins to prepare silicones still relies on catalysts derived from platinum. Speier first reported using $[H_2PtCl_6]\cdot H_2O$ to mediate alkene hydrosilylation with high selectivity in 1957 [5], and a significant advance was made by Karstedt in 1973, who developed a soluble platinum(0) catalyst that exhibits desirable activity and selectivity for olefin hydrosilylation (Figure 4.1a) [6]. The hydrosilylation of carbonyl-containing compounds has become a popular way to prepare alcohols on a small scale. In 1972, Ojima's group first reported that Wilkinson's rhodium(I) catalyst (Figure 4.1b) mediates carbonyl hydrosilylation under mild conditions to give silyl ethers in high yield [7].

Hydroboration, the process of adding a B—H bond across a double or triple bond, also provides useful reagents for organic synthesis. Herbert C. Brown was recognized with a Nobel Prize in Chemistry for the synthesis of organoboranes, and he also discovered a convenient route to prepare alcohols following direct alkene hydroboration [8]. This method involved the use of pyrophoric boranes such as BH_3 and B_2H_6; therefore, metal-catalyzed routes that use milder boranes have been targeted. In 1975, Kono et al. found that Wilkinson's catalyst oxidatively adds catecholborane [9], and, a decade later, Männig and Nöth reported the first metal-catalyzed alkene hydroboration using Wilkinson's catalyst under mild conditions (0.05 mol% cat., 20 °C, 25 minutes) [10]. Interestingly, the Rh-catalyzed

Manganese Catalysis in Organic Synthesis, First Edition. Edited by Jean-Baptiste Sortais.

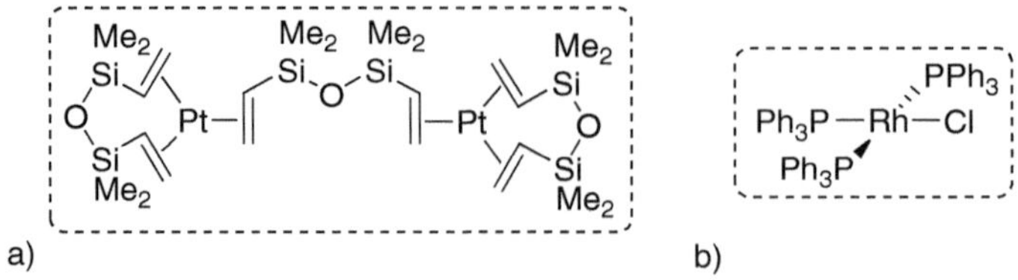

Figure 4.1 (a) Karstedt's catalyst. (b) Wilkinson's catalyst.

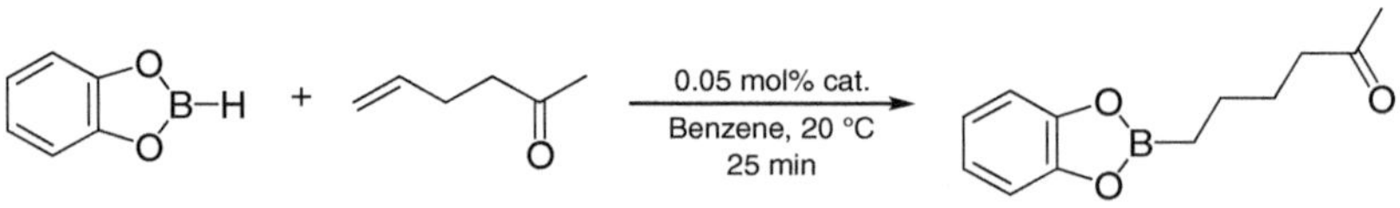

Scheme 4.1 Selective alkene hydroboration catalyzed by Wilkinson's catalyst.

hydroboration of 5-hexene-2-one using catecholborane resulted in selective C=C bond hydroboration (Scheme 4.1).

Although precious metal catalysts exhibit distinctive reactivity and stability, the evaluation of nontoxic, environmentally benign, and economical alternatives that feature base metals is critical for sustainable catalyst development. Manganese is a promising alternative to precious metals in hydrofunctionalization catalysis due to its low cost, Earth abundance, and lack of toxicity. Considering that its common oxidation states vary from 0 to +7, manganese has immense potential to generate useful catalysts. In particular, this chapter highlights the recent development of manganese-based hydrosilylation and hydroboration catalysts. In addition to catalytic activity, mechanistic insight is discussed.

4.2 Hydrosilylation

4.2.1 Hydrosilylation of Carbonyls and Carboxylates

The first manganese-catalyzed hydrosilylation of a ketone was reported in 1982 by Yates under UV irradiation (Scheme 4.2); however, $Mn_2(CO)_{10}$ turned out to be a poor photocatalyst for acetone and 2-heptanone hydrosilylation, with just 5% yield of silyl ether product obtained after acetone photolysis (Scheme 4.2) [11].

In the 1990s, the Cutler group made significant advances in the use of manganese acyl complexes for the hydrosilylation of C=O bonds, including the ketones and esters of organoiron carbonyl complexes. In 1991, they evaluated the hydrosilylation of FpC(O)R compounds [R = CH_3, Ph; Fp = $Fe(CO)_2(\eta^5\text{-}C_5H_5)$] with mono-, di-, and trihydrosilanes to yield the corresponding Fp(α-siloxyalkyl) complexes. Eighteen examples were demonstrated using a series of manganese complexes including $Mn_2(CO)_{10}$, $(CO)_5MnR$ [R = CH_3, $SiMe_3$, and $CHPh(OSiHR'_2)$] and $L(CO)_4MnC(O)R$ as the precatalysts [L = CO, R = CH_3, Ph; L = PPh_3, PEt_3, R = CH_3] (Scheme 4.3a) [12]. All of these catalysts were effective for the hydrosilylation of $FpC(O)CH_3$ with near-complete conversion observed, and the hydrosilylation of

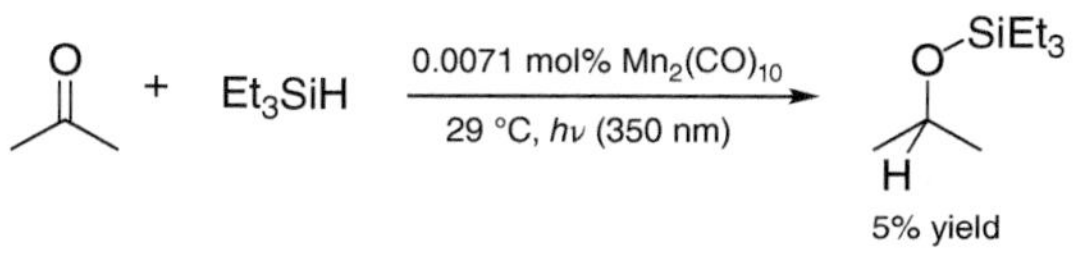

Scheme 4.2 Hydrosilylation of acetone catalyzed by $Mn_2(CO)_{10}$.

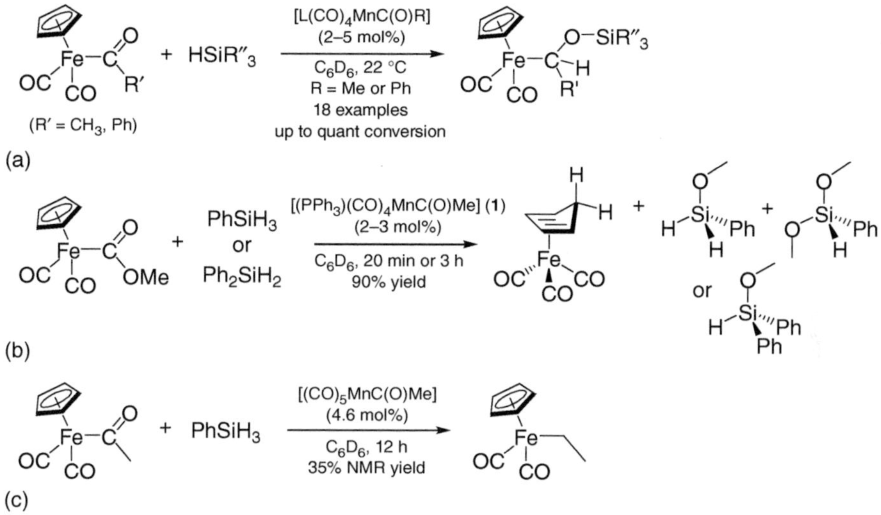

Scheme 4.3 Hydrosilylation of organoiron complexes mediated by $L(CO)_4MnC(O)R$. (a) Formation of $FpCH(R')OSiR''_3$. (b) Hydrosilylation of $FpC(O)OCH_3$. (c) Generation of iron alkyl product.

FpC(O)Ph with Ph_2SiH_2 was achieved using 2.4 mol% of $PPh_3(CO)_4MnC(O)CH_3$ to afford $FpCH(Ph)OSiHPh_2$ after 0.5 hours.

This method was extended to the hydrosilylation of $FpC(O)OCH_3$ with $PhSiH_3$ or Ph_2SiH_2 in the presence of 2–3 mol% $(Ph_3P)(CO)_4MnC(O)CH_3$ (**1**) to afford $(\eta^4\text{-}C_5H_6)Fe(CO)_3$, which features an η^4-cyclopentadiene ligand. The silane moiety was determined to be a mixture of unreacted $PhSiH_3$ and methoxysilanes (Scheme 4.3b). Note that a longer reaction time was required when using Ph_2SiH_2 (3 hours compared to 20 minutes when using $PhSiH_3$) [13]. Interestingly, the acyl manganese catalysts were able to mediate deoxygenative iron acyl hydrosilylation to afford the corresponding alkyl compounds (Scheme 4.3c) [14].

The manganese acetyl complexes were also utilized for the hydrosilylation of organic ketones and esters. The activity of **1** for the hydrosilylation of acetone with $PhMe_2SiH$ was much higher than what was achieved using $(PPh_3)_3RhCl$, with turnover frequencies (TOFs) of $27.2 \pm 5.0\,\text{min}^{-1}$ (Scheme 4.4a) [15]. The hydrosilylation of other ketones including acetophenone, cyclohexanone, and 2-cyclohexene-1-one was also evaluated with $PhMe_2SiH$ and Ph_2SiH_2 using 2.4 mol% of **1**, which resulted in up to >95% NMR yield after 4 minutes. The proposed mechanism for ketone hydrosilylation (Scheme 4.4b) begins with the generation of active catalyst **2** from manganese acyl complex **1** via σ-bond metathesis

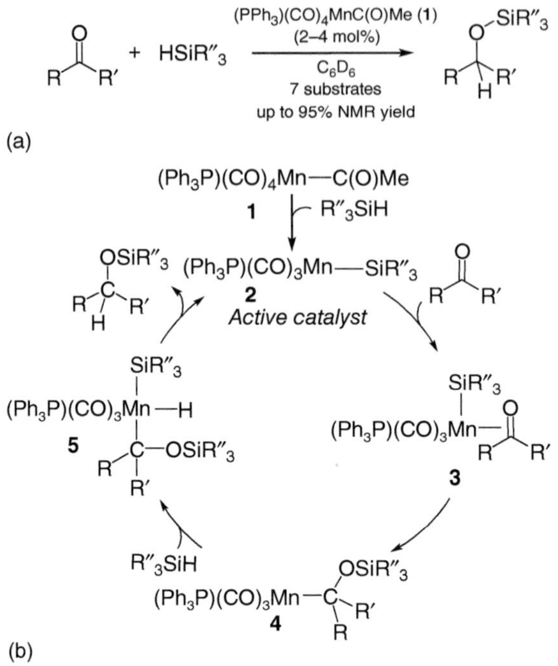

Scheme 4.4 (a) Hydrosilylation of ketones catalyzed by **1**. (b) Mechanism of hydrosilylation catalyzed by **1**.

with phenylsilane. Next, alkyl complex **4** forms following C=O bond coordination and insertion into the Mn—Si bond. Catalysis continues with Si—H oxidative addition and reductive elimination of the silyl ether product to regenerate **2**.

A large scope of 13 ester substrates was also evaluated for hydrosilylation to yield the corresponding ether products (Scheme 4.5a) [16]. Deoxygenation was proposed to proceed through two successive hydrosilylation events that are catalyzed by manganese (Scheme 4.5b). In particular, the hydrosilylation process was similar to that of ketone hydrosilylation, but the silyl ether product is reduced by intermediate **6** to afford the corresponding ether and siloxane.

In 1999, Chung's group found that the naphthalene-derived manganese complexes, (η^5-$C_{10}H_9$)$Mn(CO)_3$ (**7**) and [(η^6-$C_{10}H_8$)$Mn(CO)_3$][BF_4] (**8**), catalyze ketone hydrosilylation to give alcohols with high yields following hydrolysis [17]. In both cases, facile ring slippage was believed to play a vital role in the catalytic activity of the catalysts (Scheme 4.6). A scope of 10 substrates was evaluated for ketone hydrosilylation using 5 mol% of **7** under ambient temperature to give up to 99% NMR yield after 3 hours. Substitution patterns were found to significantly affect the conversion of aryl ketones. In fact, no transformation was observed for acetophenones containing electron-withdrawing cyano and nitro groups, while 28% NMR yield was obtained for 4-methoxyacetophenone after 7 hours due to competing methoxy group coordination. The η^6-coordinated cation was found to be more reactive for ketone hydrosilylation than the neutral η^5-coordinated complex (Scheme 4.6) [17, 18].

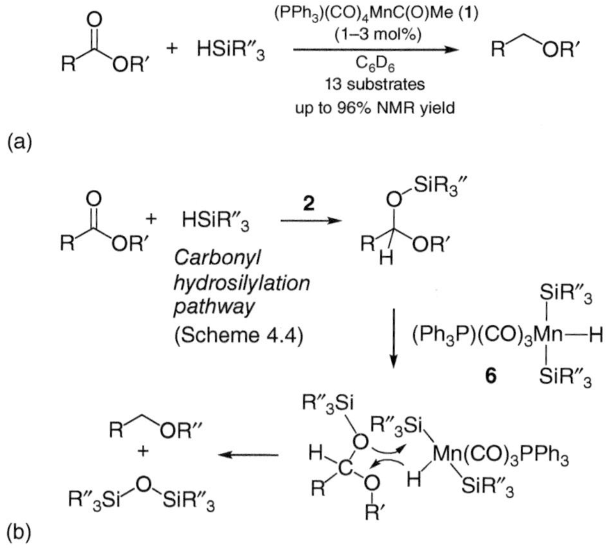

Scheme 4.5 (a) Ester hydrosilylation catalyzed by **1**. (b) Mechanism of ester hydrosilylation mediated by active catalyst **2**.

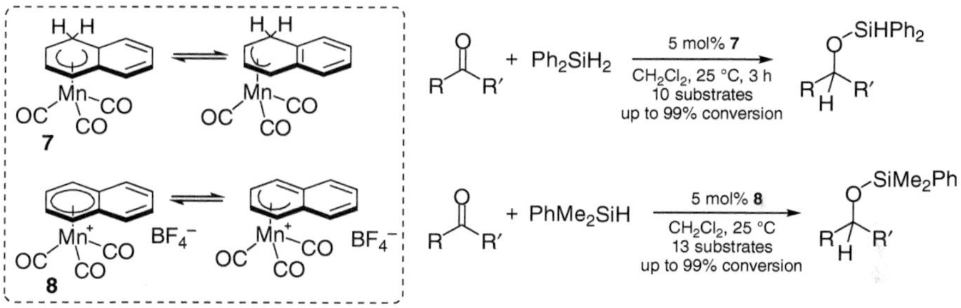

Scheme 4.6 Ring slippage and ketone hydrosilylation catalyzed by **7** and **8**.

In 2013, Du and coworker developed an air-stable manganese catalyst featuring a salen ligand framework [19]. Their catalyst, (salen-3,5-^{t}Bu$_2$)MnN (**9**, Figure 4.2), was able to mediate the hydrosilylation of aldehydes and ketones with impressive activity (maximum TOF of 11 760 h^{-1}) at 80 °C. The hydrosilylation of 14 aliphatic and aromatic carbonyl substrates with phenylsilane was evaluated using 0.5 mol% of **9** (Scheme 4.7a). This catalyst was found to exhibit good functional group tolerance, yielding a diverse set of primary and secondary alcohols following hydrolysis. Most aldehyde substrates were completely reduced after 20 minutes, while more time was required for ketones (up to 3 hours). However, substitution did have a significant impact on the hydrosilylation of aryl carbonyl compounds. In particular, the electron-withdrawing nitro group of *p*-nitrobenzaldehyde accelerated the hydrosilylation compared with benzaldehydes featuring electron-donating substituents such as *p*-OCH_3 or *p*-Cl. Nitro substitution was unfavorable for ketone hydrosilylation; complete reduction was not achieved for *p*-nitroacetophenone even after two days.

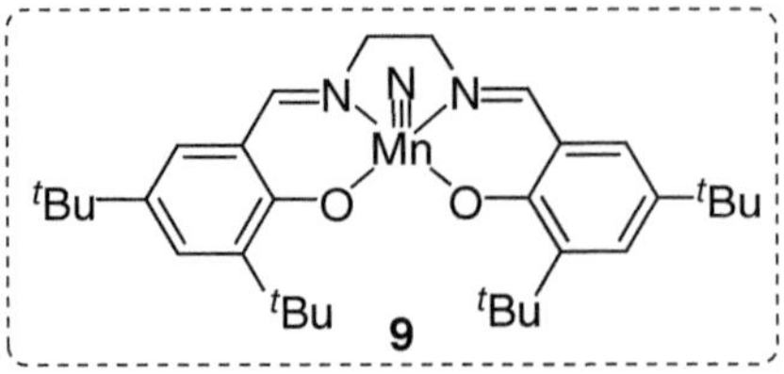

Figure 4.2 Structure of catalyst **9**.

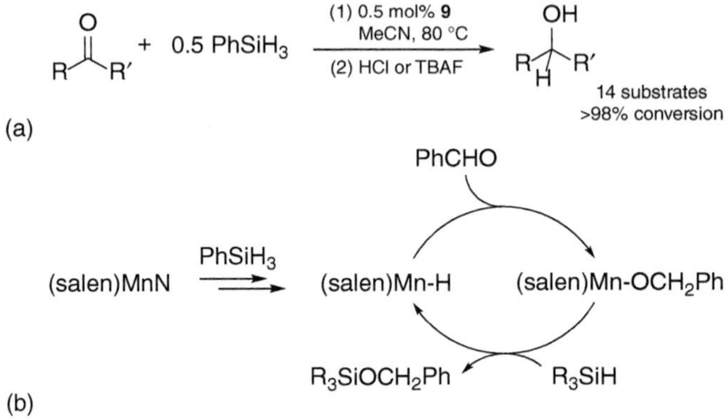

Scheme 4.7 (a) Carbonyl hydrosilylation catalyzed by **9**. (b) Proposed mechanism for **9**-catalyzed carbonyl hydrosilylation.

Notably, 1,3-diphenylpropan-1-one was obtained for the hydrosilylation of chalcone – an α,β-unsaturated ketone, which was believed to go through a silyl enol ether intermediate. The proposed mechanism involves the reduction of catalyst **9** with silane to form a manganese(III) hydride that participates in carbonyl insertion to generate the desired silyl ether product (Scheme 4.7b).

With an interest in preparing renewable materials to replace plastic, subsequent studies by Du and coworkers demonstrated that **9** can enable the polymerization of diols, dicarbonyls, and cellulosic furans to prepare poly(silylether)s (PSEs) following dehydrogenative coupling and carbonyl hydrosilylation [20, 21]. Note that PSEs are promising alternative materials since their Si—O—C linkages can be easily degraded through hydrolysis. Using hydrosilanes to produce PSEs can overcome the disadvantages of using chlorosilanes, including air and moisture sensitivity and the formation of by-products such as HCl. Several diol and dicarbonyl compounds including biorenewable furans such as 2,5-diformylfuran (DFF), 5,5′-[oxybis(methylene)]di(2-furaldehyde) (OBMF), and 5-(hydroxymethyl)furfural (HMF) (Scheme 4.8) were evaluated for manganese-catalyzed polymerization.

The reaction conditions were optimized to use toluene as the solvent, refluxing temperature (110 °C), and a catalyst loading of 1 mol% **9**. Polymerization of dialdehydes (terephthalaldehyde, adipaldehyde) with Ph_2SiH_2 took 24 hours to reach 90% conversion, while 1,4-cyclohexandione took 48 hours. However, the isolated yield and molecular weight of the corresponding polymers were lower than those prepared from the dehydrogenative coupling of diols and silanes (yield = 40–58%, M_n = 1800–2400, Table 4.1). This result was explained by the stronger nucleophilic

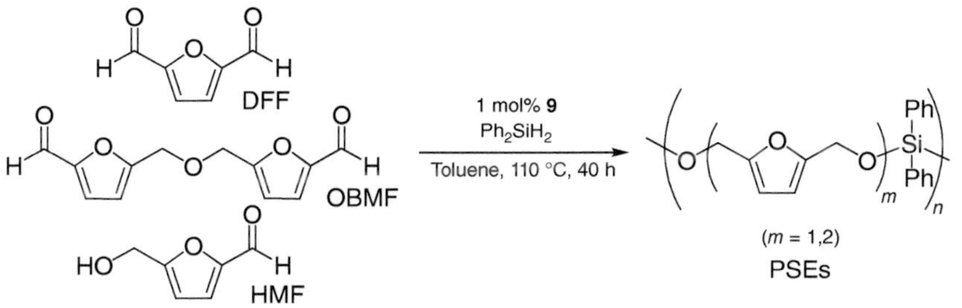

Scheme 4.8 PSE preparation by way of carbonyl hydrosilylation catalyzed by **9**.

Table 4.1 Polymerization of carbonyl compounds with Ph_2SiH_2.

Substrate	*t* (h)	M_n (g/mol)	Yield (%)
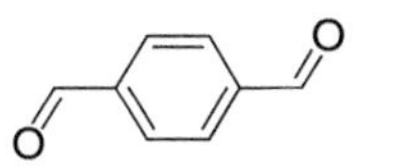	24	2400	58
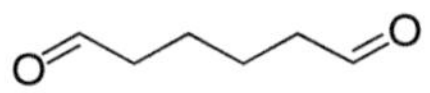	24	1800	40
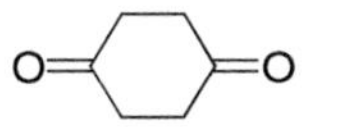	48	2100	49
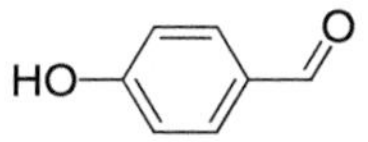	18	3800	69
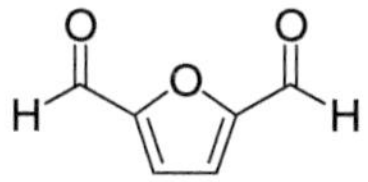	40	5500	76
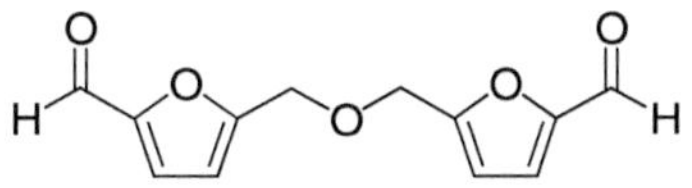	40	6400	78
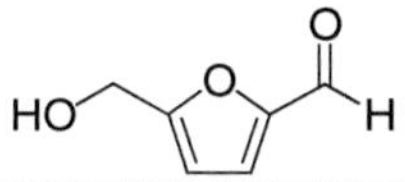	40	8000	74

properties of hydroxyls than carbonyls in the reaction with silanes. Silyl ether polymers with varying connectivity were observed when using *p*-hydroxybenzaldehyde, which possesses both alcohol and aldehyde functional groups. This protocol was extended to the hydrosilylation of DFF, OBMF, and HMF under the same catalytic conditions (Scheme 4.8), and the corresponding high molar-weight PSE polymers were collected after 40 hours in modest yield (Table 4.1). The mechanism of **9**-catalyzed polymer formation was proposed to start with the reduction of **9** by Ph_2SiH_2 to form a low-valent complex [either Mn(II) or Mn(III)]. Then Si—H coordination to the metal center in an η^1- or η^2-fashion affords a silane complex that undergoes nucleophilic attack by a carbonyl group to form an Si—O bond.

Scheme 4.9 Carboxylic acid hydrosilylation mediated by $Mn_2(CO)_{10}$.

The groups of Darcel and Sortais reported the selective reduction of carboxylic acids with Et_3SiH under UV irradiation followed by acidic hydrolysis to yield aldehydes using $Mn_2(CO)_{10}$ (Scheme 4.9) [22]. The catalytic conditions were optimized (5 mol% cat., 3.3 equiv of Et_3SiH, toluene or Et_2O, UV irradiation at 350 nm, 25 °C), and conversion was determined after 3 hours. In particular, the electronic properties and positions of aryl substituents had a considerable effect on the catalytic activity. Most *para*-substituted phenylacetic acids including *p*-alkyl, *p*-halide, and *p*-methoxy variants underwent complete reduction (>97% conversion observed). The amine group of *p*-aminophenylacetic acid did not inhibit hydrosilylation, affording the corresponding amino aldehyde in good isolated yield (85%). Extra considerations were required for some substrates. For example, 5 equiv of silane were required for 4-hydroxyphenylacetic acid due to silylation of the hydroxyl group, while halide-substituted substrates took 24 hours to reach up to 91% conversion. However, nitrobenzene was believed to form an adduct with $Mn_2(CO)_{10}$ under catalytic conditions; therefore, it did not undergo hydrosilylation. Both *meta*- and *ortho*-methyl substitution was found to hinder hydrosilylation (89% and 61% conversion, respectively). Notably, carboxylic acids featuring heteroaryl groups were also reduced successfully. In the case of saturated aliphatic substrates, extended carbon chains did not influence the reduction of palmitic and tetradecanoic acids, and the C=C bond in oleic acid was retained following reduction. The terminal double bond in hept-6-enoic acid was observed to participate in hydrosilylation, which resulted in a mixture of products. Finally, formic acid and acetic acid were hydrosilylated in good yield, while trifluoroacetic acid was not successfully reduced.

In subsequent papers, several half-sandwich Mn(I) complexes featuring an *N*-heterocyclic carbene (NHC) ligand were synthesized and evaluated for photocatalyzed carbonyl hydrosilylation (Scheme 4.10a) [23, 24]. Among six $Cp(CO)_2Mn(NHC)$ derivatives, $Cp(CO)_2Mn(I^{Mes})$ (**10**) was found to exhibit the highest photocatalytic activity for carbonyl hydrosilylation (maximum TON of 960) [24]. Aldehyde hydrosilylation was carried out under mild conditions (25 °C in toluene) for 13 substrates in the presence of 1 mol% **10** under 350 nm irradiation (Scheme 4.10b) [23]. Good to excellent conversion was observed for most substrates including aliphatic, aromatic, and heterocyclic aldehydes after 1 hour and the corresponding primary alcohols were collected following hydrolysis with NaOH (2 M) solution and extraction into MeOH. The electronic properties of the *para*-substituent did not affect the transformation of aromatic aldehydes. However, *m*-fluoro substitution was found to reduce the catalytic activity, requiring 8 hours to reach 92% conversion. Importantly, **10** selectively catalyzed the hydrosilylation of alkenyl and alkynyl substrates with >97% conversion, without reduction or

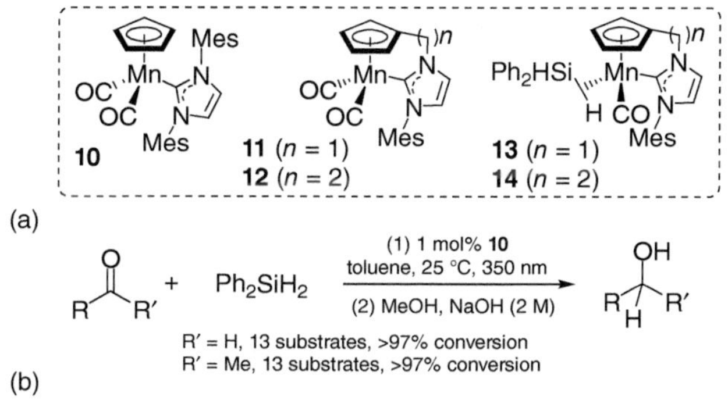

Scheme 4.10 (a) Structures of **10–14**. (b) Carbonyl hydrosilylation mediated by **10**.

polymerization of the unsaturated C—C bonds. In the case of ketone hydrosilylation, a scope of 13 substrates was effectively reduced under identical catalytic conditions. Many aliphatic and substituted aromatic ketones were hydrosilylated within 4 hours; however, less active ketones such as *o*-methylacetophenone, acetylferrocene, dialkyl, and cyclic substrates required up to 24 hours to be fully converted. Chloro- and bromo-substituents contributed to the decomposition of **10** under UV irradiation, hence no conversion was observed for these substrates.

The synthesis of piano-stool Mn(I) complexes, $(Cp(CH_2)_nNHC^{Mes})Mn(CO)_2$ (**11**, **12**; $n = 1$, 2; Scheme 4.10a), was then reported and their hydrosilylation activity was compared with **10** by hydrosilylating 2-acetonaphthone with Ph_2SiH_2 [24]. Under identical conditions, **11** and **12** exhibited less activity with just 56% and 66% NMR yield observed in the presence of 1 mol% cat., respectively. However, the synthesis and characterization of these complexes gave access to the monocarbonyl η^2-silane intermediates, $(Cp(CH_2)_nNHC^{Mes})Mn(CO)(\eta^2\text{-H–SiHPh}_2)$ (**13**, **14**; $n = 1$, 2; Scheme 4.10a), suggesting that the catalytic mechanism is consistent with traditional Ojima ketone hydrosilylation [25].

In 2018, Royo's group reported the development of a manganese complex featuring a bis-NHC ligand, $(\text{bis-NHC}^{Me})Mn(CO)_3Br$ (**15**), which is not only air stable but is also an effective catalyst for ketone hydrosilylation [26]. Compound **16**, which was prepared by treating **15** with $AgBF_4$, was also found to be stable in air (Scheme 4.11a). To compare the catalytic activity of these complexes, 14 ketones were separately evaluated for hydrosilylation using phenylsilane in the presence of either **15** or **16** at elevated temperature under air (Scheme 4.11b). In particular, aromatic substrates were effectively reduced except for *p*-iodobenzaldehyde, which afforded the dehalogenated product 1-phenylethanol. Aliphatic ketones, including alkyl, alkenyl, and cyclic substrates, also underwent hydrosilylation, and *trans*-4-phenyl-3-buten-2-one was selectively reduced without reduction or polymerization of the C=C bond. The silyl ether products were collected following hydrolysis with 2 M NaOH and extracted with dichloromethane to afford the corresponding secondary alcohols in good yield. It was determined that **16** is more active than **15** since the former hydrosilylated all substrates within 5 hours, whereas more time was required using **15** under identical conditions. In fact, the maximum

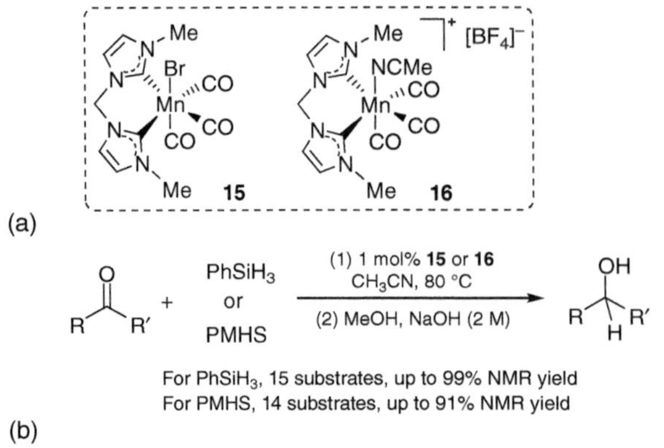

Scheme 4.11 (a) Structures of **15** and **16**. (b) Carbonyl hydrosilylation mediated by **15** or **16**.

TON observed using 0.1 mol% **16** for the hydrosilylation of acetophenone and phenylsilane was 1000 (TOFs of up to 320 h^{-1}). Surprisingly, when polymethylhydrosiloxane (PMHS) was employed as the silane source instead of $PhSiH_3$, **15** was found to be significantly more effective. Therefore, ketone hydrosilylation using PMHS was evaluated for a large scope of aliphatic and aromatic ketones under identical conditions. Except for substrates featuring iodo-substitution, hydrosilylation occurred in 56–91% NMR yield after 8 hours.

The same group recently employed **15** to achieve the hydrosilylation of esters to yield alcohols [27]. A scope of 14 esters featuring a variety of functional groups was screened for hydrosilylation in the presence of $PhSiH_3$. Excellent conversions and yields were observed; however, limitations were found in the hydrosilylation of benzoates featuring NO_2, NH_2, and CN groups. Notably, this study presented the first examples of Mn-catalyzed ester hydrosilylation that utilize cheap and readily available PMHS as the silane source. In the presence of this reagent, **15** was found to catalyze methyl and ethyl ester hydrosilylation to yield the corresponding high molecular weight alcohol in moderate to excellent yields, excluding NO_2- and NH_2-functionalized benzoates.

Very recently, Bagh and coworkers described a Mn(I) catalyst for ester hydrosilylation under neat conditions at 100 °C [28]. Loadings of *fac*-[Mn-(xantphos)$(CO)_3$Br] as low as 1 mol% were used for the hydrosilylation of 17 different esters to yield the corresponding alcohols in up to 97% isolated yield after basic workup. The substrate scope was expanded to include popular industrial alcohols; 11 fatty acid esters were hydrosilylated under neat conditions to generate the desired alcohols (up to 97% isolated yield).

Since 2014, our group has been working to develop manganese hydrosilylation catalysts featuring pyridine diimine (PDI) and pyridine di(eneamide) (PDEA) ligands (Scheme 4.12, **17**–**20**). Hydrosilylation has not only been evaluated for ketones [29–31] but also been extended to esters [29], aldehydes [30–32], and formates [31, 32] under mild conditions with low catalyst loadings (Scheme 4.12). Note that (^{Ph_2PPr}PDI)Mn (**17**) [29], (^{PyEt}PDEA)Mn (**19**) [30], and [(^{Ph_2PEt}PDI)Mn]$_2$

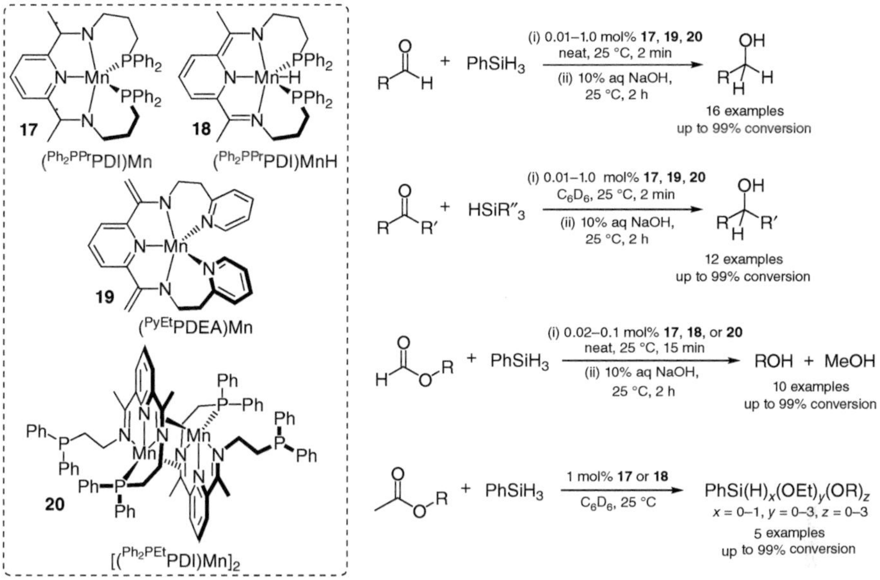

Scheme 4.12 Structures of **17–20**, carbonyl hydrosilylation, and carboxylate dihydrosilylation catalyzed by **17–20**.

(**20**) [31] show extraordinary reactivity for aldehyde hydrosilylation with respective TOFs of up to 4900, 2475, and 4950 min^{-1}. Notably, catalysts **17** and **20** were the leading transition metal catalysts for carbonyl hydrosilylation under ambient conditions when published. Although the hydride complex (^{Ph_2PPr}PDI)MnH (**18**) [32] is slightly less active for aldehyde and ketone hydrosilylation, it turned out to be highly effective for formate dihydrosilylation with a maximum observed TOF of 330 min^{-1}. This activity is orders of magnitude higher than what had been reported for carboxylate dihydrosilylation catalysts prior to 2017.

The mechanism of **17**-catalyzed carbonyl hydrosilylation was proposed to follow the modified Ojima mechanism shown in Scheme 4.13 [32]. The catalytic process begins with phosphine dissociation and the reversible oxidative addition of a Si—H bond to generate **21**. Next, carbonyl substrates reversibly coordinate to the metal center to afford intermediate **22** and then insert into the Mn—H bond to generate **23**. Insertion was believed to be the rate-determining step (RDS) due to kinetic studies. For carbonyl hydrosilylation, **23** undergoes fast reductive elimination to release the desired hydrosilylated product and regenerate catalyst **17**. In the case of di- and trihydrosilanes, the remaining Si—H bonds can participate in hydrosilylation to form doubly and triply condensed products. Regarding formate and ester dihydrosilylation, **23** undergoes fast β-alkoxide elimination to form **24**, followed by reductive elimination of silyl ether and loss of aldehyde to regenerate **17**. Interestingly, the redox non-innocent ^{Ph_2PPr}PDI ligand stabilizes the intermediates, allowing Mn(II) to remain throughout the catalytic process.

Although **18** showed lower activity than **17** for carbonyl hydrosilylation, its activity for carboxylate dihydrosilylation was higher, and the proposed mechanism for

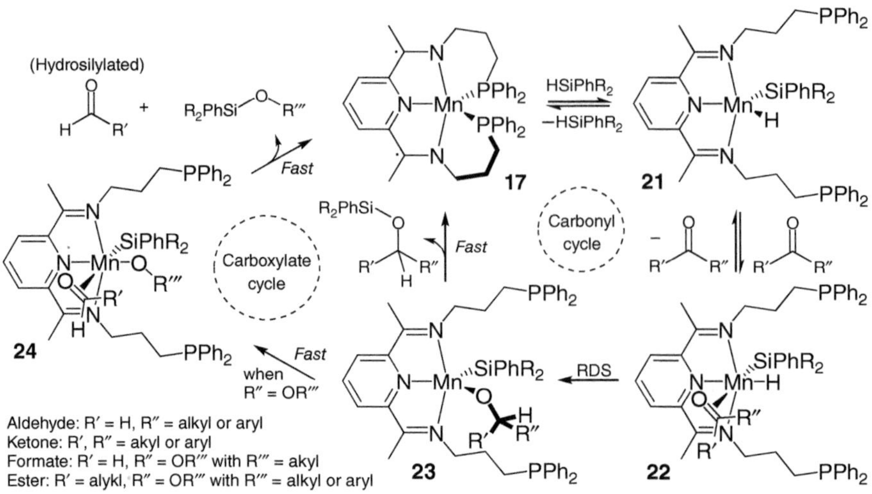

Scheme 4.13 Modified Ojima mechanism for aldehyde, ketone, formate, and ester hydrosilylation catalyzed by **17**.

18-catalyzed transformations is shown in Scheme 4.14 [32]. For carbonyl hydrosilylation, coordination of the substrate C=O bond to Mn is followed by insertion to form **26**. The RDS was proposed to be the σ-bond metathesis reaction between **26** and silane to release the hydrosilylated product and regenerate catalyst **18**. In the case of carboxylate dihydrosilylation, **26** undergoes fast β-alkoxide elimination to afford **28** and aldehyde. Then σ-bond metathesis between the Mn—O bond of **28** and Si—H yields the silyl ether product and **18**. Note that the aldehyde released in both mechanisms (Schemes 4.13 and 4.14) was believed to be hydrosilylated during the catalytic reaction to form the corresponding silyl ether.

In 2017, Huang's group synthesized a series of Mn(II) complexes featuring chiral iminopyridine oxazoline (IPO) ligands that were found to mediate the asymmetric hydrosilylation of aryl ketones (**29–32**, Scheme 4.15a, left) [33]. Interestingly, the catalyst possessing the bulkiest IPO ligand (**32**) was found to exhibit the highest activity for ketone hydrosilylation. Catalysis required the use of $NaEt_3BH$ as an activating reagent, and the reactions were conducted with 0.1 M ketone since concentration was found to significantly influence the rate of hydrosilylation (Scheme 4.15a). A broad scope of aromatic ketones were selectively hydrosilylated to obtain secondary alcohols in excellent yield and enantioselectivity following hydrolysis of the corresponding silyl ethers. Substrate substitution patterns had a considerable influence on the conversion and enantioselectivity of the transformation. For example, acetophenones featuring *o*-chloro or *o*-methoxy groups required heating to 60 °C for selective conversion. *m*-Methoxyacetophenone underwent hydrosilylation with higher enantioselectivity than *p*-methoxyacetophenone, while different positions of chloro-substitution did not affect the results.

In 2017, Gade and coworkers developed manganese catalysts bearing [2,5-bis(2-oxazolinyldimethylmethyl)pyrrole] pincer ligands (PdmBOX) [34]. In comparison to iron and cobalt sources, the addition of ligand **33** and **34** to $Mn(CH_2TMS)_2$ allowed for efficient acetophenone hydrosilylation using $(Et_2O)_2SiMeH$. Greater than 95% conversion was observed for both catalysts;

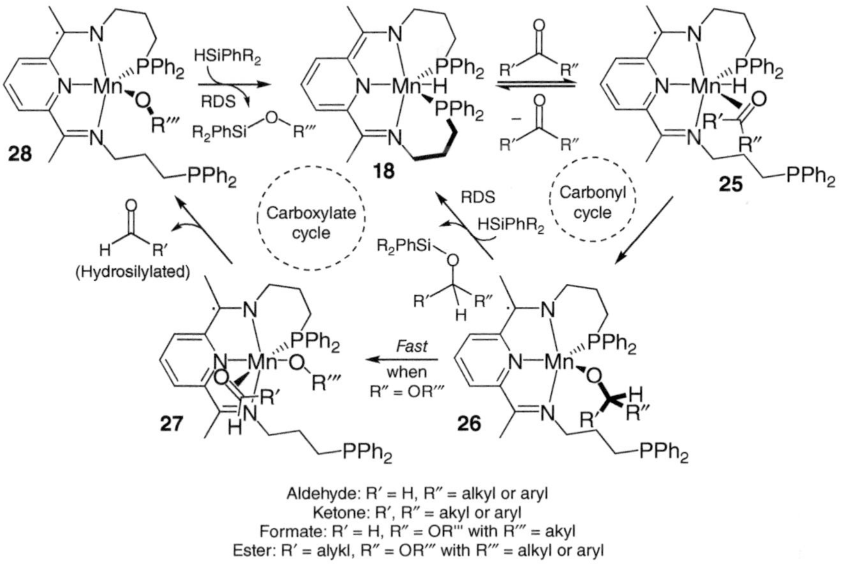

Scheme 4.14 Mechanism for aldehyde, ketone, formate, and ester hydrosilylation catalyzed by **18**.

however, the products were obtained with only modest enantioselectivity (ee = 9% (*R*) and 50% (*R*), respectively) (Scheme 4.15b).

The Werlé and Leitner groups recently found that Mn(I) complexes bearing triazole ligands [35] are active for the hydrosilylation of carbonyl and carboxyl groups. Of the evaluated catalysts, **35** and **36** exhibited the best performance (Scheme 4.16) [36]. In particular, 19 ketones were screened for hydrosilylation catalyzed by 0.25–1 mol% **35** at 80 °C (Scheme 4.16a). Moderate to high yield was observed for most substrates, with the notable exceptions of 2-acetylanisole and 2-acetylfuran. Interestingly, cationic complex **36** exhibited activity for carbonyl hydrosilylation and the hydrosilylative reduction of carboxyl groups. Using 2 mol% **36**, a variety of esters underwent hydrosilylation with phenyl silane, and hydrolysis yielded the corresponding alcohols and ethers (Scheme 4.16b). Notably, **36** represents the first manganese catalyst capable of hydrosilylating carboxylic acids to generate alcohols (Scheme 4.16c). The independent reaction of **36** with $PhSiH_3$ supported the formation of a catalytically relevant hydride species.

Solid-phase Mn catalysts have recently been shown to mediate carbonyl hydrosilylation. Copéret and coworkers described silica-supported Mn(II) sites for the reduction of carbonyl compounds and transesterification [37]. Their catalysts, $Mn_2[OSi(O^tBu)_3]_4$@$SiO_{2\text{-}400}$ and $Mn\{N(SiMe_3)_2\}_2$·THF@$SiO_{2\text{-}400}$, were well defined, and the latter was found to effectively mediate benzaldehyde hydrosilylation using either $PhSiH_3$ (>99% NMR yield after 10 minutes) or triethoxysilane (97% NMR yield after 1 hour) under mild conditions (0.5 mol% cat., 25 °C). Very recently, the groups of Zhang and Zheng described the synthesis of an air-stable 1D Mn(II) coordination polymer featuring a 4′-pyridyl-2,2′;6′,2″-terpyridine (pytpy) ligand,

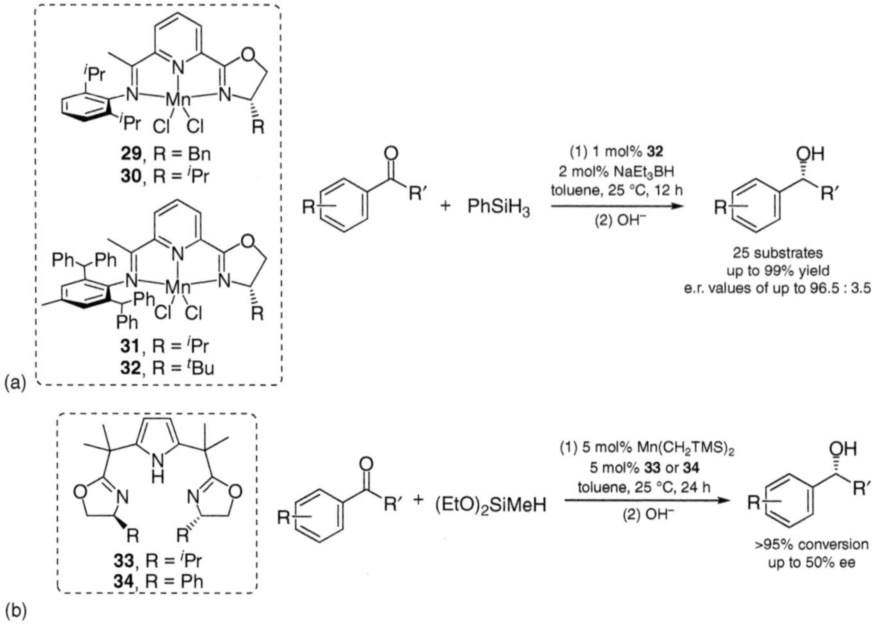

Scheme 4.15 Asymmetric ketone hydrosilylation mediated by manganese catalysts. (a) Hydrosilylation using IPO Mn precursors. (b) PdmBOX ligands for Mn-based hydrosilylation.

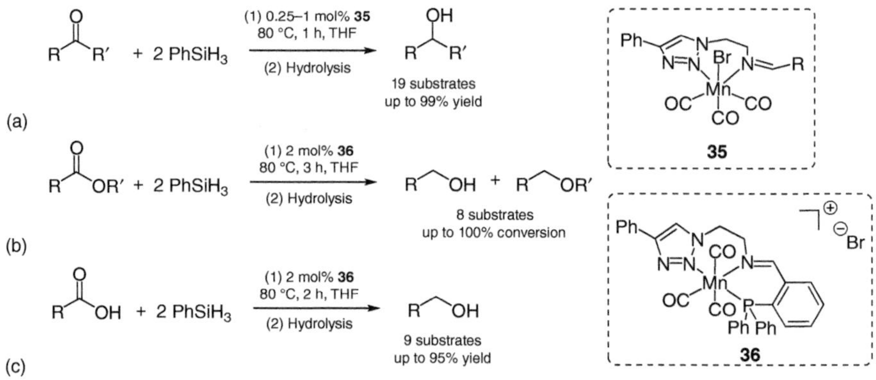

Scheme 4.16 (a) Ketone hydrosilylation mediated by **35**. (b) Ester hydrosilylation mediated by **36**. (c) Carboxylic acid hydrosilylation mediated by **36**.

which is an effective precatalyst for hydrofunctionalization reactions including hydroboration and hydrosilylation [38]. Using KO^{t}Bu as an activator and $PhSiH_3$, 10 aldehydes and ketones were successfully reduced and worked up to yield the corresponding alcohols (complete carbonyl reduction was noted and isolated yields of up to 95% were achieved).

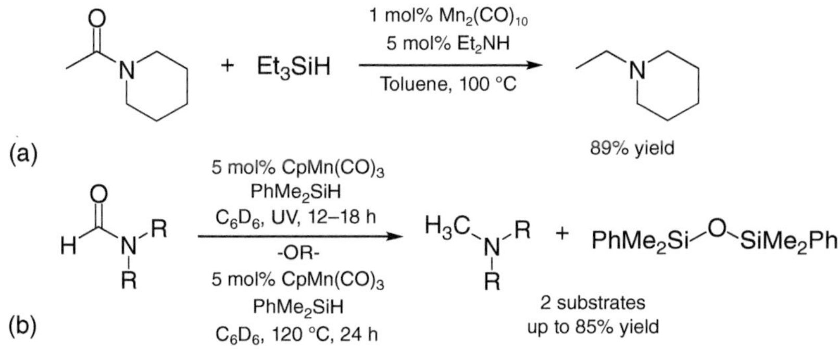

Scheme 4.17 Amide hydrosilylation catalyzed by manganese carbonyl compounds. (a) Hydrosilylation of *N*-acetylpiperidine. (b) Formamide hydrosilylation.

Figure 4.3 Structure of **37**.

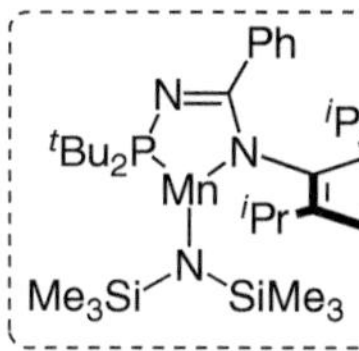

4.2.2 Hydrosilylation of Amides

So far, relatively few examples of manganese-catalyzed amide reduction have been reported. In 2001, the reduction of amides with hydrosilanes using a number of transition metal complexes was reported by the Fuchikami group [39]. In the presence of $Mn_2(CO)_{10}$ and Et_2NH, *N*-acetylpiperidine was reduced to *N*-ethylpiperidine in good yield (Scheme 4.17a). In 2012, Pannell and coworkers reported the photocatalytic activity of $CpMn(CO)_3$ for the hydrosilylation of dimethylformamide (DMF) and *N,N*-diethylformamide under UV irradiation (Scheme 4.17b). They also found this catalyst to be active for the same reactions at 120 °C [40].

In 2017, Sydora and coworkers developed a Mn precatalyst [(κ^2-P,N) $Mn(N(SiMe_3)_2)$] that deoxygenates tertiary amides to give the corresponding amines in the presence of silane reductant (Figure 4.3) [41]. A large scope of aliphatic and aromatic tertiary amides were reduced with phenylsilane under mild conditions (Scheme 4.18a). Using 5 mol% **37** and heating to 75 °C allowed amide hydrosilylation to reach completion after 1 hour. The reduction of substituted benzyl amides, as well as sterically bulky tertiary dibenzyl amides containing (hetero)benzyl and even *N*-benzylpiperidine functionalities, were effectively deoxygenated with >90% conversion observed. Aliphatic substrates were also reduced to tertiary alkyl amines; however, lower catalyst loadings and temperature (2 mol% **37**, 25 °C) reduced the rate of the reaction (18 hours needed for completion). In addition, this catalyst was found to mediate the reduction of aldehydes, ketones (Scheme 4.18b), and esters (Scheme 4.18c) using phenylsilane under mild conditions (1 mol% **37**, benzene, 25 °C).

Acetophenones and benzaldehydes featuring electron-donating and electron-withdrawing groups underwent hydrosilylation (>90% conversion observed after

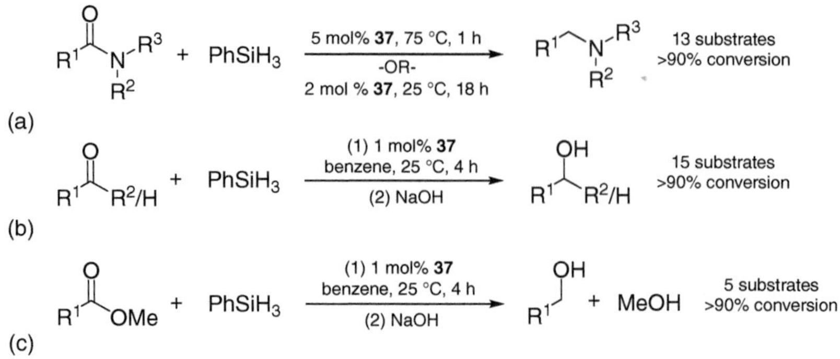

Scheme 4.18 Hydrosilylation of (a) amides, (b) ketones/aldehydes, and (c) esters using **37**.

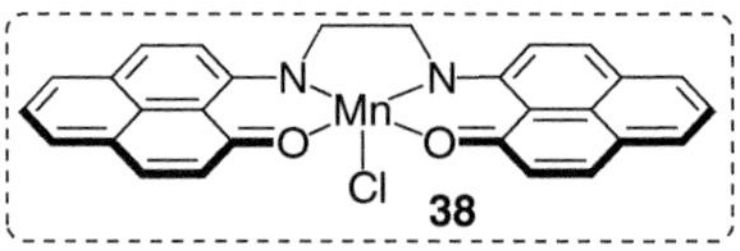

Figure 4.4 Structure of **38**.

4 hours, Scheme 4.18b) and afforded the corresponding alcohols. Saturated and unsaturated aliphatic ketones were also selectively hydrosilylated under the catalytic conditions. A maximum TON of 400 and TOFs of up to 200 min^{-1} were achieved upon evaluating the reduction of 4-heptanone at low catalyst loading (0.25 mol%). The hydrosilylation of esters using **37** was assessed under the same conditions as carbonyl hydrosilylation. Methyl benzoates featuring varied substitution and methyl heptanoate were successfully reduced to obtain the corresponding primary alcohols and MeOH upon hydrolysis.

Mandal's group recently reported a Mn(III) complex featuring a phenalenyl ligand (**38**, Figure 4.4) that converts primary amides to either amines or nitriles following hydrosilylation (Scheme 4.19a) [42]. In the presence of KO^tBu and $PhSiH_3$ at 50 °C, 5 mol% of **38** enabled the hydrosilylation of aryl, heteroaryl, and aliphatic amides to yield primary amines as their hydrochloride salts in moderate to excellent yield (61–95%). In particular, benzylamides with varied substitution were reduced and worked up to afford benzylamine salts in moderate to excellent yield (64–95%), while aromatic amides with hetero rings such as 3-pyridinecarboxamide and 2-thiophenecarboxamide underwent the transformation in moderate yield (61–63%). Aliphatic amides were successfully converted to the corresponding aliphatic amine salts. In addition, the protocol was extended to yield nitriles under identical conditions in the presence of 20 mol% of *N*-methylbenzamide as the inhibitor to prevent the nitrile from further reduction. A scope of 16 substrates with various aryl and aliphatic amides were assessed for deoxygenation under these conditions (Scheme 4.19b). Benzylamides were converted to the corresponding nitriles in good to excellent yield (74–96%), as were heteroaryl and aliphatic substrates. Interestingly, selective reduction of the primary amides in 4-cyanobenzamide and *N*-(4-carbamoylphenyl)benzamide afforded the corresponding nitrile products, with the existing nitrile or secondary amide groups retained.

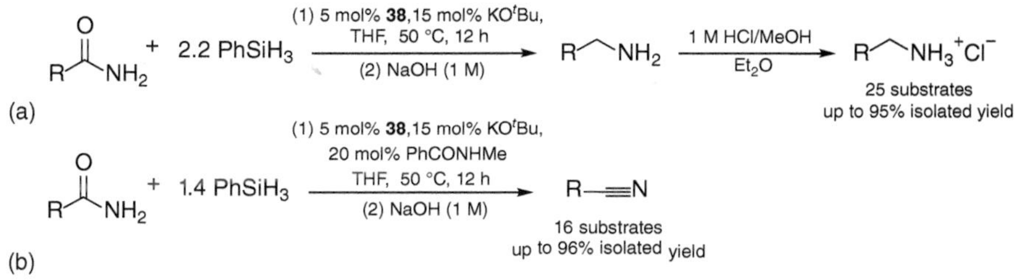

Scheme 4.19 (a) Amide hydrosilylation to amines catalyzed by **38**. (b) Amide hydrosilylation to nitriles catalyzed by **38**.

Figure 4.5 Structure of **39**, **40**.

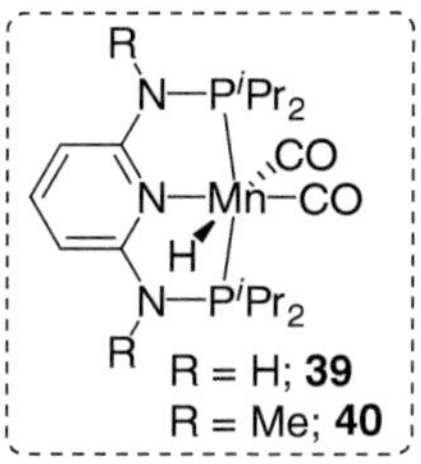

4.2.3 Hydrosilylation of Carbon Dioxide

Carbon dioxide, an abundant greenhouse gas, is also a convenient C_1 source for small-scale synthetic chemistry. Gonsalvi and coworkers reported that their Mn(I) PNP pincer complexes, $(PNP^{NR}{-}^iPr)Mn(CO)_2H$ (**39**: R = H, **40**: R = Me; Figure 4.5), effectively enabled CO_2 (1 bar) hydrosilylation under mild conditions to form methoxysilyl products and MeOH upon simple hydrolysis [43]. Screening experiments indicated that **39** exhibits higher CO_2 hydrosilylation activity than catalyst **40**. The activity of **39** for CO_2 reduction was evaluated using $PhSiH_3$ (5 equiv with respect to CO_2) and 2 mol% cat. (with respect to Si—H) at 25 and 80 °C in DMSO-d_6 (Scheme 4.20), and product formation was monitored by NMR spectroscopy. Reactions at room temperature took up to 24 hours to reach the methoxysilyl product, while CO_2 hydrosilylation at elevated temperature afforded reduced products in high yield (89%) after just 1 hour. Note that the formation of methane was not observed in this work. Hydrosilylation using an excess amount of CO_2 (1 : 1 M ratio vs. Si—H) was also conducted at 80 °C. Complete reduction of CO_2 to the methoxysilyl product was observed after 24 hours without the presence of any by-products. A maximum TON of 100 was achieved by lowering the catalyst loading to 1 mol% under identical conditions.

The mechanism for **39**-catalyzed CO_2 hydrosilylation was proposed based on NMR experiments, DFT calculations, and the isolation of key intermediates (Scheme 4.20b). The first step involves the insertion of CO_2 into the Mn—H bond to generate intermediate **41**. Next, σ-bond metathesis between **41** and an incoming Si—H bond yields **39** and silylformate, which undergoes insertion to afford **42**. Subsequent σ-bond metathesis between another Si—H equivalent and the pendent C—O bond yields intermediate **43** and liberates siloxane. The final step also

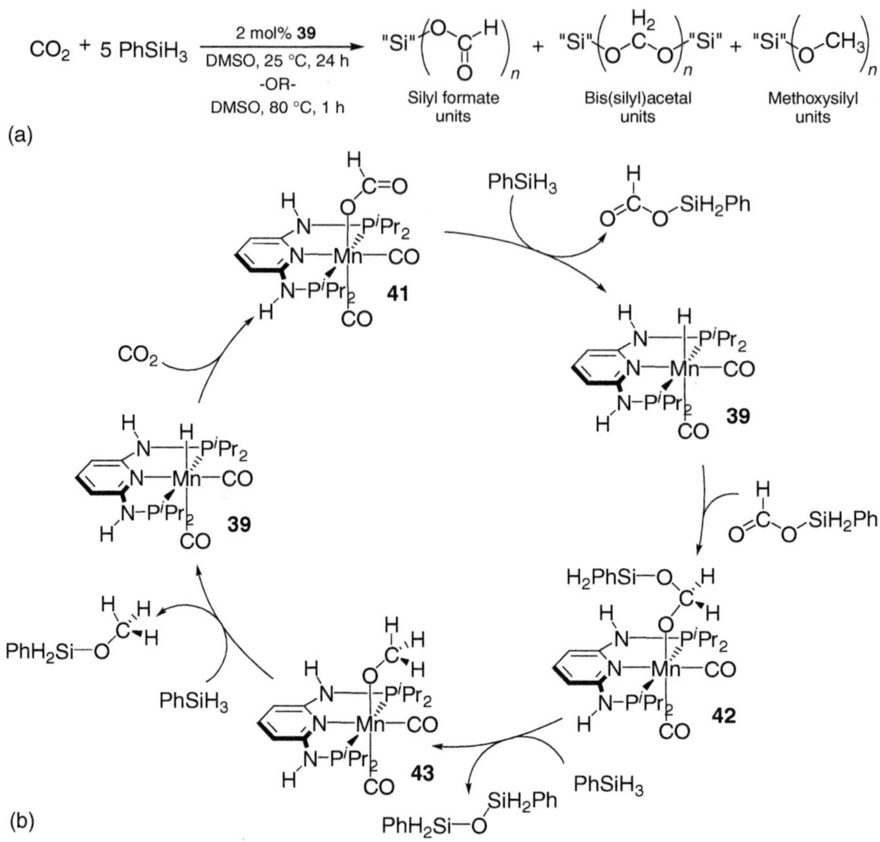

Scheme 4.20 (a) Hydrosilylation of CO_2 catalyzed by **39**. (b) Proposed mechanism for **39**-catalyzed CO_2 hydrosilylation.

involves σ-bond metathesis between Si—H and Mn—O to regenerate **39** and afford methoxyphenylsilane.

4.2.4 Hydrosilylation of C=C and C≡C Bonds

In comparison with carbonyl hydrosilylation, the development of manganese catalysts for alkene and alkyne hydrosilylation remains under- investigated. The chemoselectivity of carbon–carbon multiple bond hydrosilylation is readily complicated by side reactions such as isomerization, hydrogenation, dehydrogenative silylation, and dihydrosilylation. Given the lower polarity of C=C and C≡C bonds compared with C=O bonds, only a few examples of manganese-based alkene and alkyne hydrosilylation have been reported.

4.2.4.1 Olefin Hydrosilylation

In 1983, an early investigation by Faltynek utilized the triphenylsilyl complex, $(CO)_5MnSiPh_3$ (Figure 4.6, **44**), for the hydrosilylation of terminal alkenes following both thermal and photochemical activation [44]. The reactions were carried

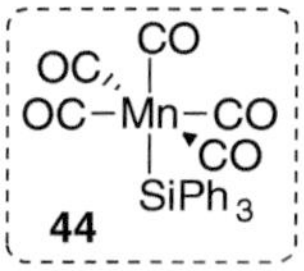

Figure 4.6 Structure of **44**.

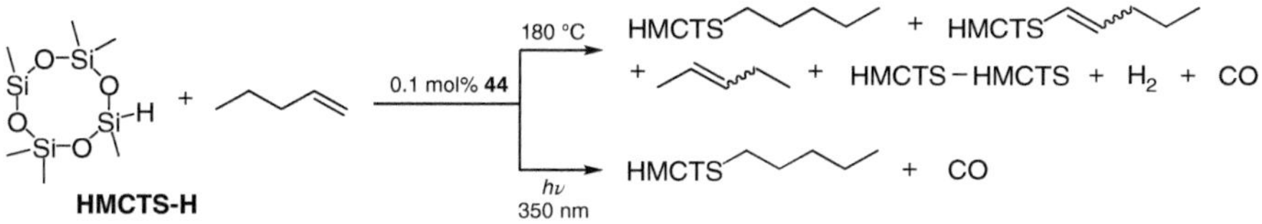

Scheme 4.21 Hydrosilylation of 1-pentene with HMCTS-H catalyzed by **44**.

out using heptamethylcyclotetrasiloxane (HMCTS-H) and 1-pentene. Interestingly, heating at 180 °C afforded a mixture of products including the desired alkylsilane and by-products arising from dehydrogenative silylation and alkene isomerization. In contrast, activation of **44** with UV light allowed for sole preparation of the hydrosilylation product in high yield (Scheme 4.21).

In 1987, Hilal and coworkers reported the use of $Mn_2(CO)_{10}$ to hydrosilylate 1-hexene with Et_3SiH or $(EtO)_3SiH$ [45]. The reaction catalyzed by $Mn_2(CO)_{10}$ was slightly slower than the one using $Co_2(CO)_8$; however, the manganese complex was found to exhibit better selectivity for alkene hydrosilylation despite the low yield observed (Scheme 4.22a). In 1999, they developed a protocol to mediate the hydrosilylation of 1-octene using triethoxysilane and $Mn_2(CO)_{10}$ supported by an aminated poly(siloxane) surface (Figure 4.7). The alkylsilane product was collected in high yield with no evidence of by-product formation due to alkene isomerization (Scheme 4.22b) [46]. Continuing work on the development of supported manganese catalysts, Hilal studied the activity of tetra(4-pyridyl)porphyrinato-manganese(III) cation $[Mn^{III}(TPyP)]^+$ for alkene hydrosilylation from three different sources: homogeneous $[Mn(TPyP)]^+$ ions; the microparticle-supported catalyst, MnTPyP@Micro-Clay; and the nanoparticle supported catalyst, MnTPyP@Nano-Clay [47]. Comparison was made by evaluating the hydrosilylation of 1-octene with triethoxysilane (Scheme 4.22c). Interestingly, intercalation was believed to enhance both the activity and the selectivity of the catalyst, especially in the case of the nanoparticle-supported system, for which conversion of 85% and TOFs of up to 1200 min^{-1} were observed. However, the activity decreased as leaching out of $MnTPyP^+$ occurred.

Figure 4.7 Structure of supported $Mn_2(CO)_{10}$ catalyst.

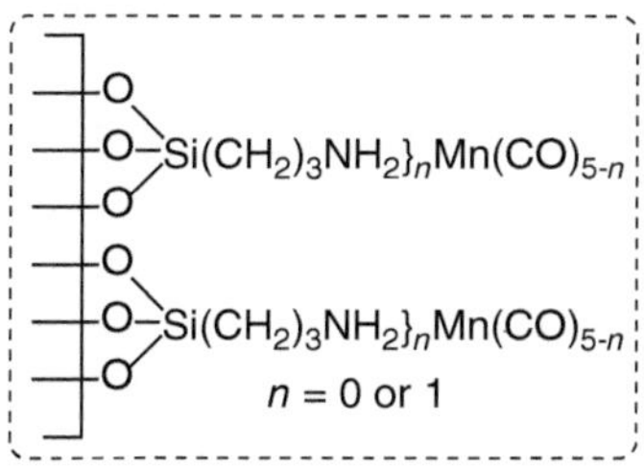

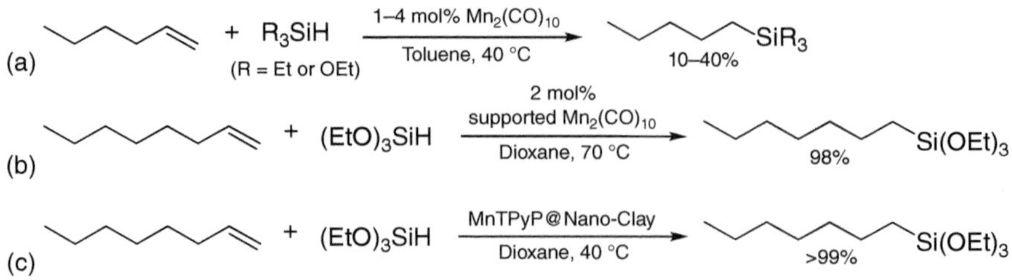

Scheme 4.22 (a) Hydrosilylation of 1-hexene catalyzed by $Mn_2(CO)_{10}$. (b) Hydrosilylation of 1-octene catalyzed by supported $Mn_2(CO)_{10}$. (c) Hydrosilylation of 1-octene catalyzed by MnTPyP@Nano-Clay.

In 2017, Thomas's group made a significant contribution by evaluating base metal catalyst activation in the presence of sodium *tert*-butoxide for a series of reactions including hydrosilylation, hydroboration, hydrovinylation, hydrogenation, and [2π + 2π] alkene cycloaddition [48]. This study included the related iron and cobalt complexes; however, the manganese precatalyst $(^{2,6\text{-}Et_2Ph}PDI)MnBr_2$ (**45**) was also evaluated for alkene hydrosilylation and found to exhibit moderate activity (Scheme 4.23a). After reporting this preliminary result, they sought to develop improved alkene hydrosilylation catalysts by employing different ligand frameworks, alternating the steric bulk of the ligand, and screening activators [49]. The combination of $(^{2,6\text{-}iPr_2Ph}PDI)MnBr_2$ (**46**) and NaO^tBu in a 1 : 3 M ratio was found to exhibit the best activity, and a wide scope of alkenes were hydrosilylated under mild conditions. Three different silane reductants were used ($HSi(OEt)_3$, $HSiMe(OEt)_2$, and $PhSiH_3$), and high yields of the linear silane products were obtained with excellent regioselectivity (Scheme 4.23b). Concurrently, Emslie and coworkers reported the synthesis of manganese hydride complexes derived from $(dmpe)_2MnH(C_2H_4)$ [50]. Preliminary results showed that 2 mol% of *cis*-$[(dmpe)_2MnH(Et_2Si{=}CHMe)]$ mediated ethylene hydrosilylation using diethylsilane at 60 °C to give triethylsilane with 60% conversion after 24 hours, suggesting that compounds of this type can be used as precatalysts for future alkene hydrosilylation studies.

In 2018, our group reported the synthesis of a dimeric manganese compound featuring β-diketiminate ligands, $[(^{2,6\text{-}iPr_2Ph}BDI)Mn(\mu\text{-}H)]_2$ (**47**), which was found to mediate olefin hydrosilylation at 130 °C [51]. In particular, aliphatic alkenes underwent hydrosilylation with phenylsilane to yield anti-Markovnikov products, while styrenes were found to participate in Markovnikov hydrosilylation (Scheme 4.24a). Despite the non-ideal conditions required for alkene hydrosilylation, the hydride complex was found to catalyze the cross-linking of PMHS with vinyl-terminated poly(dimethylsiloxane) to prepare silicones that are commonly used in household consumer products and as coatings. A silicone used for oxygen permeable contact lenses and transparent LED screen coatings was also prepared by hydrosilylating 1,2,4-trivinylcyclohexane with PMHS in the presence of **47** (Scheme 4.24b). Notably, these are the first silicones featuring non-hydrolysable Si—C bonds prepared by a Mn catalyst.

The mechanisms for aliphatic and aromatic alkene hydrosilylation were proposed and are shown in Scheme 4.25. Notice that these transformations do not follow the

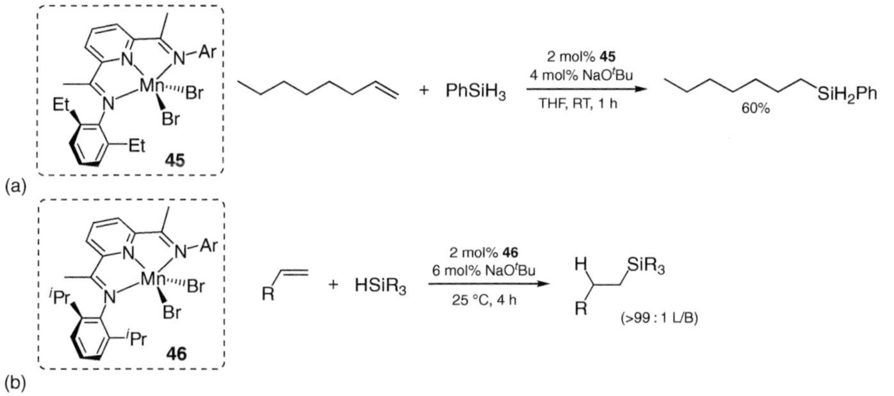

Scheme 4.23 (a) Hydrosilylation of 1-octene with phenylsilane catalyzed by **45** and NaO^tBu. (b) Regioselective hydrosilylation of alkenes catalyzed by **46** and NaO^tBu.

commonly involved Chalk–Harrod mechanism for alkene hydrosilylation, which involves Si—H oxidative addition [52]. Instead, the hydrosilylation of aliphatic alkenes begins with alkene insertion into the Mn—H bond and concurrent dimer dissociation to generate primary alkyl intermediate **48**. Then σ-bond metathesis between the Si—H and Mn—C bonds releases the linear silane product and regenerates **47** (Scheme 4.25a). In contrast, styrenes other than α-methylstyrene underwent insertion to form the secondary alkyl intermediate **49**, which results in preparation of the anti-Markovnikov product.

Soon after, Yang and coworkers reported chemo- and regioselective olefin hydrosilylation using $Mn(CO)_5Br$ [53]. Hydrosilylation was evaluated for a wide substrate scope including functionalized styrenes and aliphatic alkenes (Scheme 4.26a). Most substrates were effectively hydrosilylated using benzyldimethylsilane under the optimized conditions (5 mol% cat., hexane, 60 °C, 4 hours) to give the corresponding anti-Markovnikov products with good to excellent isolated yield. In particular, styrene substitution did not affect the transformation and aryl derivatives such as 2-naphthalenyl and 2-pyridyl underwent hydrosilylation in modest yield at 100 °C. Interestingly, $Mn_2(CO)_{10}$ was also found to be active for alkene hydrosilylation under UV irradiation. Mechanistic insight was gained by conducting deuterium labeling experiments and isolating key intermediates, which implicated a radical pathway for anti-Markovnikov hydrosilylation. The proposed mechanism for $Mn(CO)_5Br$-catalyzed alkene hydrosilylation is shown in Scheme 4.26b, which begins with the formation of $HMn(CO)_5$ via σ-bond metathesis between $Mn(CO)_5Br$ and silane, followed by the generation of $(CO)_5Mn\cdot$ (**50**). The reaction of **50** and silane affords $HMn(CO)_5$ and silyl radical **51**, which then reacts with alkene to yield the β-silyl alkyl radical **52**. Hydrogen atom transfer between $HMn(CO)_5$ and **52** affords the desired anti-Markovnikov product and regenerates **50**.

As part of Zhang and coworkers recent demonstration of carbonyl hydrosilylation using 1D Mn(II) coordination polymers, KO^tBu activation was also found to enable ambient temperature alkene hydrosilylation [38]. The preliminary screening

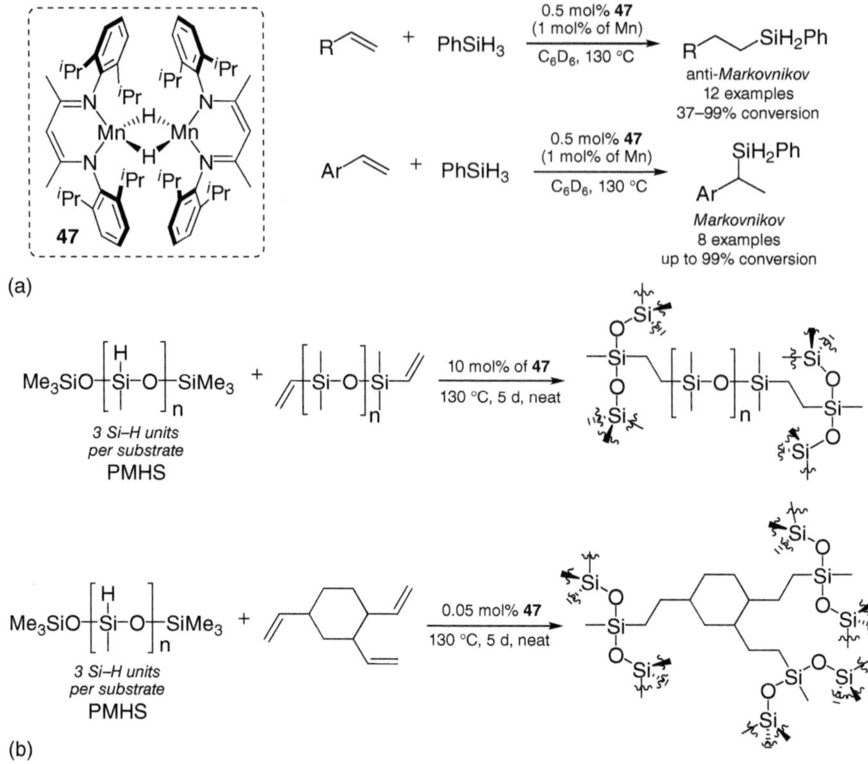

Scheme 4.24 (a) Hydrosilylation of alkenes and styrenes catalyzed by **47**. (b) Silicone preparation by **47**.

of five aromatic alkenes indicated that the catalyst system effectively and regioselectively hydrosilylates styrene to yield the branched product in 90% isolated yield. The hydrosilylation of functionalized styrenes was modestly successful (up to 58% isolated yield was found). Notably, these results were achieved using 0.1 mol% of catalyst under ambient conditions in the absence of solvent.

4.2.4.2 Alkyne Hydrosilylation

It was not until 2018 that Wang and coworkers reported the first examples of Mn-mediated alkyne hydrosilylation [54]. A variety of alkynes were evaluated for hydrosilylation with mono-, di-, and tri-substituted aliphatic and aromatic silanes. In particular, the use of mononuclear $MnBr(CO)_5$ in the presence of $AsPh_3$ afforded the *E*-products in moderate to good yields (Scheme 4.27a). In contrast, dinuclear $Mn_2(CO)_{10}$ in the presence of dilauroyl peroxide (LPO) allowed for the formation of *Z*-products with good regioselectivity (Scheme 4.27b).

A series of experiments to isolate the key catalytic components were conducted to study the mechanism of hydrosilylation catalyzed by $MnBr(CO)_5$. Substitution of CO with $AsPh_3$ is believed to occur, followed by silane activation and HBr loss to yield intermediate **54** (Scheme 4.28). Next, **54** undergoes alkyne substitution then

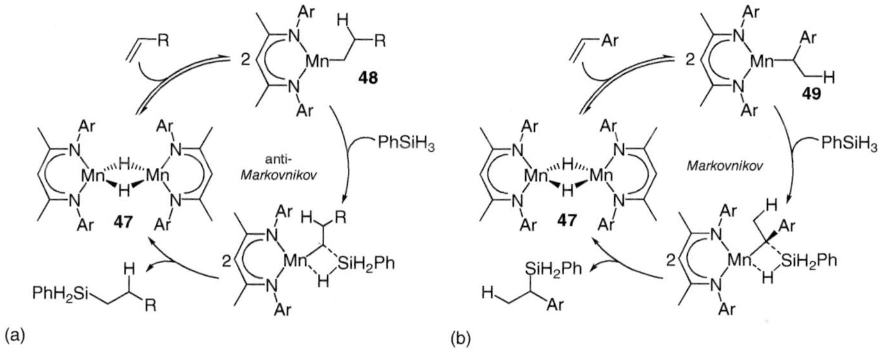

Scheme 4.25 Proposed mechanisms of alkene hydrosilylation catalyzed by **47**: (a) anti-Markovnikov pathway and (b) Markovnikov pathway.

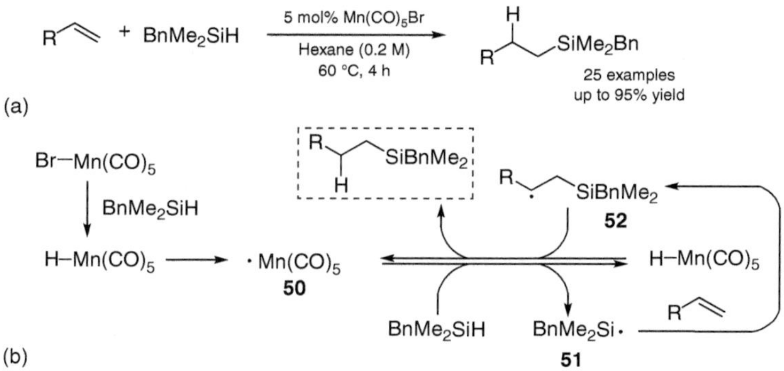

Scheme 4.26 (a) Alkene hydrosilylation catalyzed by $Mn(CO)_5Br$. (b) Proposed mechanism for anti-Markovnikov alkene hydrosilylation.

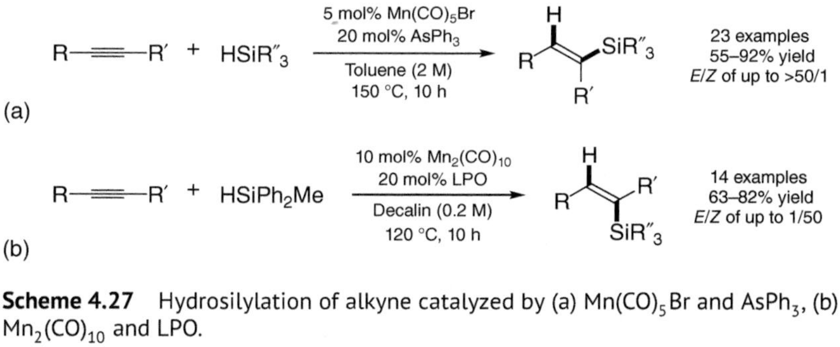

Scheme 4.27 Hydrosilylation of alkyne catalyzed by (a) $Mn(CO)_5Br$ and $AsPh_3$, (b) $Mn_2(CO)_{10}$ and LPO.

insertion into the Mn—Si bond to afford the (β-silyl)alkenyl complex **56**. In the presence of silane, Si—H coordination and σ-bond metathesis releases the *E*-configured product to regenerate **54** (Path A). Alternatively, **57** can undergo σ-bond metathesis to release the product and directly coordinate an alkyne to generate **55** (Path B).

Experimental work suggested that *Z*-selective alkyne hydrosilylation catalyzed by $Mn_2(CO)_{10}$ does not follow an organometallic mechanism, but rather a radical

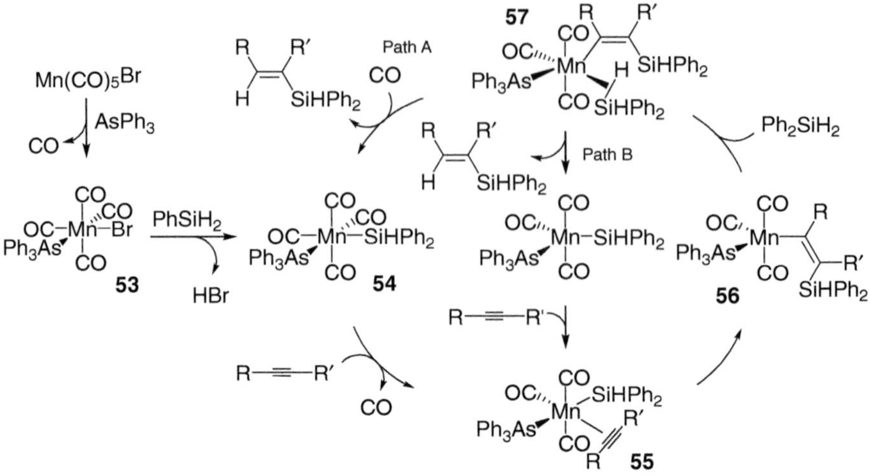

Scheme 4.28 Proposed mechanism for *E*-selective hydrosilylation catalyzed by $Mn(CO)_5Br$ and $AsPh_3$.

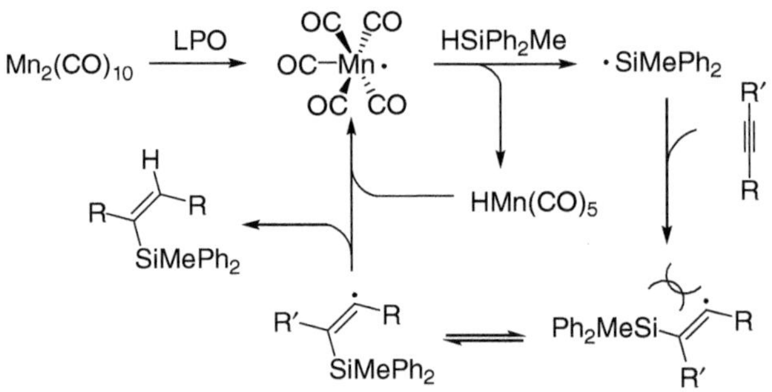

Scheme 4.29 Proposed mechanism for *Z*-selective hydrosilylation catalyzed by $Mn_2(CO)_{10}$ and LPO.

mechanism (Scheme 4.29). The catalytic process starts with the generation of $(CO)_5Mn\cdot$ from $Mn_2(CO)_{10}$ via homolysis under UV irradiation, which then reacts with silane to produce a silyl radical and $HMn(CO)_5$. The reaction of silyl radical and alkyne affords *E*- and *Z*-configured alkenyl radicals. Next, H-atom transfer to these alkenyl radicals from $HMn(CO)_5$ yields the vinyl silane product and regenerates $(CO)_5Mn\cdot$. Notice that *Z*-selectivity is preferable to *E*-selectivity due to the steric hindrance of the diphenylmethylsilyl group.

4.3 Hydroboration

Organoboronate compounds are important reagents in synthetic chemistry due to their stability, ease of handling, impressive functional group compatibility, and ability to be applied in a variety of cross-coupling transformations. However, the

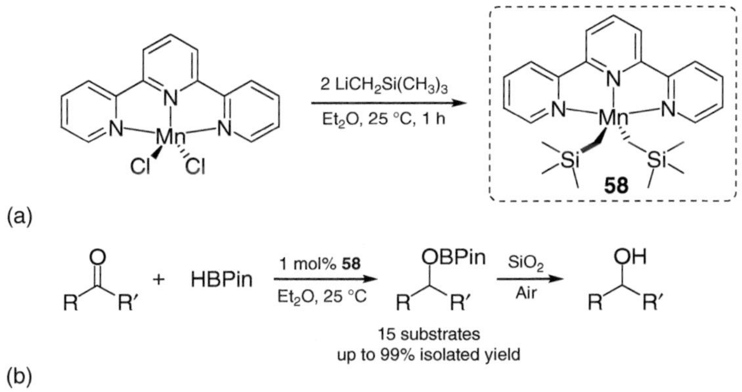

Scheme 4.30 (a) Synthesis of **58**. (b) Ketone hydrosilylation mediated by **58**.

development of manganese catalysts for hydroboration has lagged behind when compared with recent progress using other Earth-abundant transition metals.

4.3.1 Hydroboration of C=O Bonds

In 2016, Zhang and coworkers reported the synthesis of a well-defined Mn(II) complex bearing an NNN chelating ligand that effectively mediates the hydroboration of aldehydes and ketones [55]. The active catalyst, (tpy)Mn(CH_2SiMe_3)$_2$ (**58**), was synthesized by treating the dichloride precursor with 2.0 equiv of $LiCH_2SiMe_3$ in Et_2O at room temperature (Scheme 4.30a). Mn-catalyzed carbonyl hydroboration was evaluated under mild conditions using pinacolborane (HBPin) for a scope of 15 aldehyde and ketone substrates. Boryl products were hydrolyzed and purified by SiO_2 column chromatography to afford the corresponding primary and secondary alcohols in good to excellent yield (Scheme 4.30b). Interestingly, several competition reactions were conducted, and **58** was found to exhibit chemoselectivity for C=O over C=C bond hydroboration.

In 2017, Gade's group developed the first Mn catalyst for enantioselective ketone hydroboration, (^{Ph,H}boxmi)MnCH_2SiMe_3 (**59**, Scheme 4.31a), which features a chiral bis(oxazolinyl-methylidene)isoindoline pincer ligand (boxmi) [56]. This complex was prepared either by direct reaction of boxmi ligand with Mn(CH_2SiMe_3)$_2$ or by alkylation of the manganese chloride variant with $LiCH_2SiMe_3$. The ligand substitution pattern, the borane reductant, and solvent were evaluated, and 25 ketones were evaluated for hydroboration with HBPin under ambient conditions in the presence of 2.5 mol% **59** (Scheme 4.31b). Excellent conversion and enantioselectivity was observed for all substrates (up to >99% ee) and the chiral alcohol products were obtained in high yields via hydrolysis. In addition, C—C multiple bond hydroboration was not observed for vinyl and alkynyl ketones, which revealed that direct aryl group attachment to the carbonyl functionality was not essential for the selectivity. Importantly, TOFs of up to 80 min^{-1} were achieved by reducing the catalyst loading to 0.1 mol%.

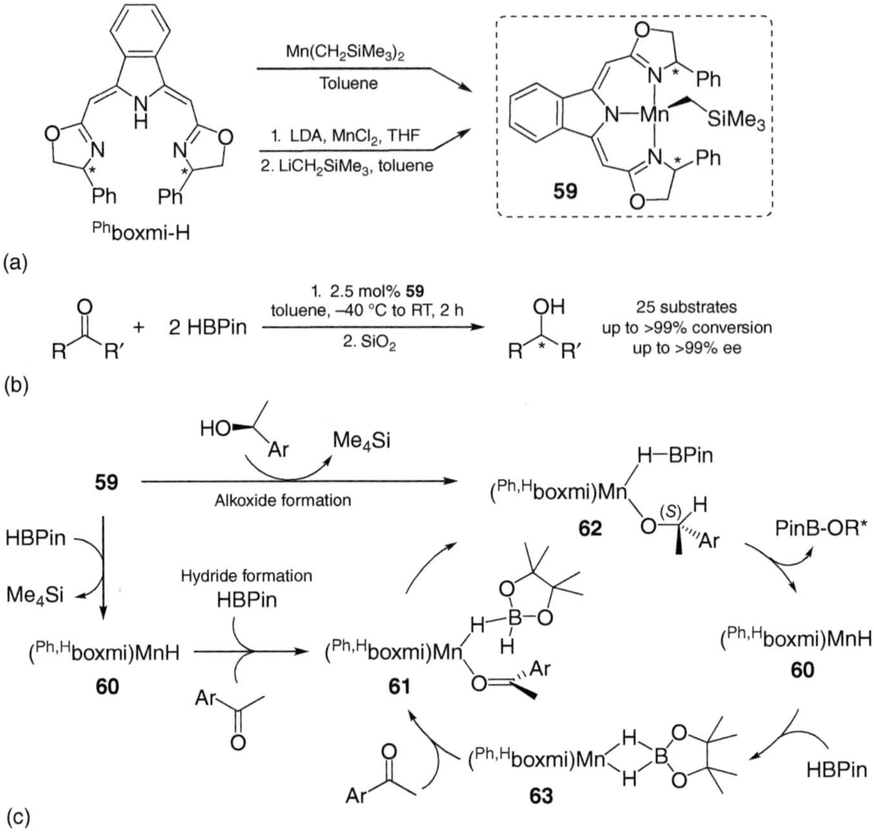

Scheme 4.31 (a) Synthesis of catalyst **59**. (b) Asymmetric ketone hydroboration mediated by **59**. (c) Proposed mechanism for **59**-catalyzed ketone hydroboration.

Experiments using a radical trap confirmed that this transformation does not occur via a radical process. The reaction of DBPin with 4-fluoroacetophenone yielded a deuterium-labeled alcohol after silica workup, which indicated borane as the hydrogen/deuterium source. Further experiments using pregenerated manganese hydride (^{Ph,H}boxmi)MnH (**60**) also suggested that hydrogen transfer takes place from the borane instead of the metal hydride. The proposed mechanism begins with the coordination of HBPin and ketone to **60** to form intermediate **61**, which is set up to stereoselectively transfer hydrogen from the borane to the carbonyl (Scheme 4.31c). Then, σ-bond metathesis releases the chiral boronic ester to regenerate **60**, which coordinates HBPin to form a borane–manganese adduct with bridging hydrides (**63**), restarting the catalytic cycle. A follow-up DFT study featuring stoichiometric and kinetic experiments suggested that **59** could enter the catalytic cycle as **60** following σ-bond metathesis with HBPin (Scheme 4.31c, left). In addition, **59** could undergo alcoholysis to form the catalytically active Mn(II) alkoxide **62** (Scheme 4.31c, top). The dominant pathway involves σ-bond metathesis between HBPin and the Mn alkoxide [57].

Figure 4.8 Structure of **64**.

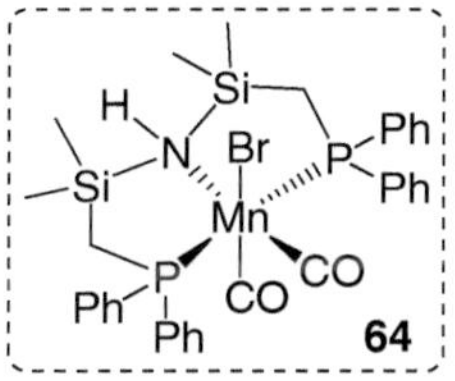

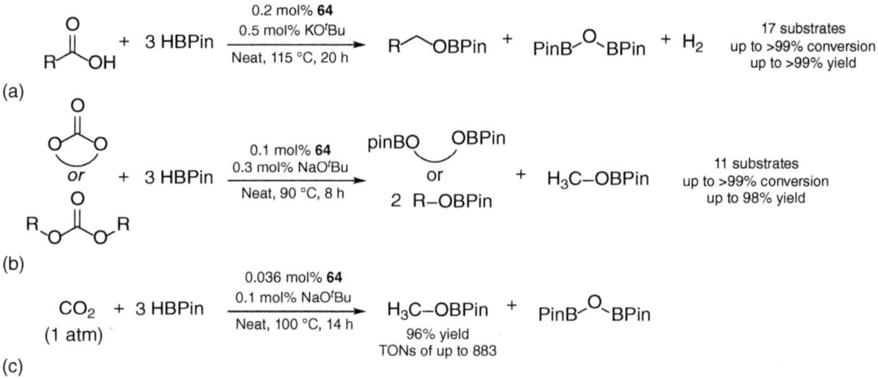

Scheme 4.32 Hydroboration of (a) carboxylic acids, (b) carbonates, and (c) CO_2 catalyzed by **64**.

In 2018, Leitner utilized the manganese pincer complex, $((Ph_2PCH_2SiMe_2)_2NH)Mn(CO)_2Br$ (**64**, Figure 4.8), for the hydroboration of carboxylates, carbonates, and carbon dioxide [58]. The reduction of these functionalities is important for sustainable chemical and pharmaceutical synthesis and the preparation of alternative non-fossil feedstocks derived from biomass or carbon dioxide. A wide range of 17 carboxylic acids was reduced with HBPin using 0.2 mol% of **64** in the presence of 0.5 mol% KO^{t}Bu (Scheme 4.32a). Good to excellent conversion was observed for most substrates including aliphatic acids, benzoic acids, the biomass-derived platform chemical succinic acid, and fatty acids. Notably, unsaturated substrates were selectively hydroborated in good yield without C=C bond reduction.

The selective reduction of organic carbonates has been a challenge since the compounds are kinetically stable toward hydrogenation and are often used as solvents for reductive reactions [59]. Additionally, the reduction of carbonates derived from CO and CO_2 can provide industrially important methanol derivatives. Therefore, cyclic and linear carbonates were assessed for hydroboration at elevated temperature with 3 equiv of HBPin using 0.1 mol% of **64** and 0.3 mol% of NaO^{t}Bu under neat conditions (Scheme 4.32b). This catalyst was found to be highly effective for the hydrosilylation of aromatic and cyclic carbonates featuring five-membered or six-membered rings as well as heterocyclic carbonates with up to 98% yield observed. Interestingly, the 1,2-carbonate of glycerol, which is a by-product of industrial vegetable oil production, was also hydroborated to the corresponding boryl ether in 88% yield. However, the hydroboration of linear carboxylates was slightly less effective.

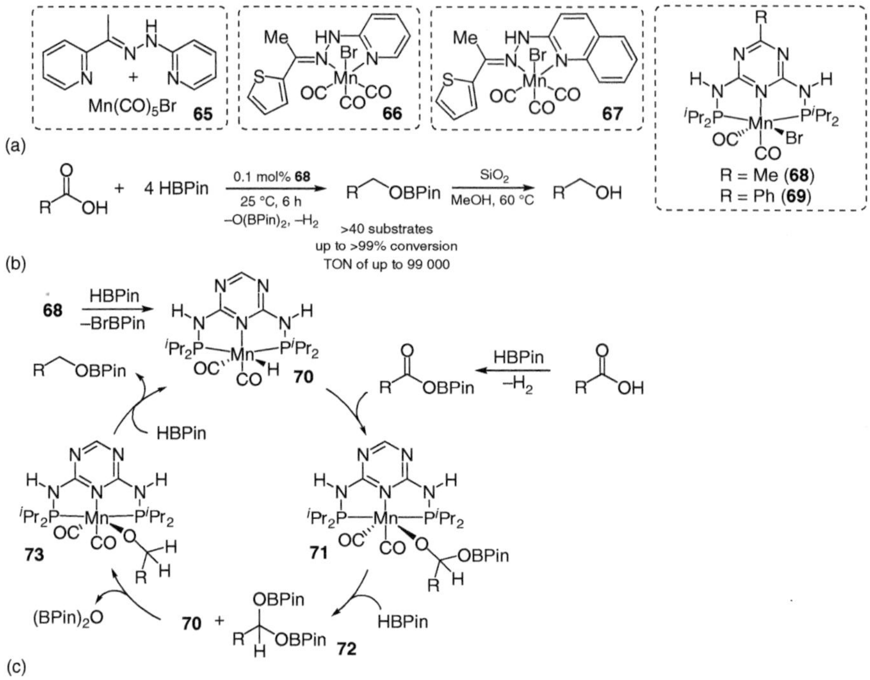

Scheme 4.33 (a) Structure of catalysts **65–69**. (b) Hydroboration of carboxylic acids catalyzed by **68**. (c) Proposed mechanism for carboxylic acid hydrosilylation catalyzed by **68**.

Catalyst **64** (0.036 mol%) was able to reduce CO_2 (1 atm) with 3 equiv of HBPin in the presence of NaO^{t}Bu at 100 °C to yield MeOBPin (96% by ^{1}H NMR). Turnover numbers of up to 883 were observed (Scheme 4.32c).

In 2019, Maji's group studied a series of hydrazine-based (**65–67**) and triazine-based catalysts (**68, 69**) for the deoxygenative hydroboration of carboxylic acids (Scheme 4.33a). Each catalyst was found to effectively catalyze carboxylate hydroboration under ambient conditions at low catalyst loadings (0.001–0.1 mol%) [60]. In particular, 0.1 mol% of **65–68** fully hydroborated benzoic acid to the corresponding boronate ester (TONs of up to 990). Catalyst **66** showed excellent reactivity for this transformation with a maximum TON of more than 99 000 at 25 °C (TOFs of up to 34 min^{-1}), which renders it the best homogeneous catalyst known for this process. The catalytic activity of **68** was evaluated using 0.1 mol% **68** (50 μl in a stock solution in toluene) with HBPin at 25 °C for a wide range of 42 substrates including aromatic and aliphatic acids, fatty acids, and even some bioactive carboxylic acids (Scheme 4.33b). Hydrolysis with SiO_2/methanol at 60 °C gave the corresponding alcohols in up to 96% yield.

The aromatic carboxylic acids screened were fully reduced (>99%), including derivatives that possess electron-donating or electron-withdrawing groups or heteroaromatic functionalities such as furan and thiophene. In the case of aliphatic substrates, conversions up to >99% were achieved, although by-products from C=C hydroboration were detected for cinnamic acid (22%). Cyclopropane rings were unmodified under such conditions, while hindered (pivalic acid), unhindered (acetic acid), and C_1 acids were reduced quantitatively. Long-chain fatty acids reacted smoothly with HBPin to afford the corresponding alcohols in excellent yield, while drug molecules and bile acid were also successfully converted under these conditions. Notably, **68** showed great chemoselectivity for the hydroboration of carboxylic acids in the presence of reducible functional groups. Halogenated aromatic and aliphatic substrates were tolerated and nitro, nitrile, and ester groups were retained.

The proposed mechanism is shown in Scheme 4.33c, which involves the addition of HBPin to precatalyst **68** to afford hydride complex **70**. Next, the insertion of RC(O)OBPin (derived from the noncatalytic reaction of carboxylic acid with HBPin) forms alkoxy intermediate **71**. This compound undergoes σ-bond metathesis with HBPin to regenerate **70** and intermediate **72**. Once formed, **72** could release $O(BPin)_2$ to liberate aldehyde, which can be inserted to yield **73**. Finally, σ-bond metathesis between **73** and HBPin affords the final product and **70**.

In Copéret's silica-supported Mn catalysis study, $Mn\{N(SiMe_3)_2\}_2 \cdot THF@SiO_{2-400}$ was also found to be an active catalyst for carbonyl hydroboration [37]. A scope of eight functionalized aldehydes and ketones underwent hydroboration under mild conditions (0.5 mol% cat., 25 °C, six hours) to achieve the desired products in up to 99% yield. Notably, this organic ligand-free system exhibited great recyclability with no loss in activity observed after three catalytic cycles. Most recently, the Mn coordination polymer developed by Zhang and Zheng was found to effectively mediate the hydroboration of carbonyl compounds with TONs of up to 990 following KO^tBu activation [38]. Using 0.1 mol% of Mn and 1 mol% of KO^tBu, a wide range of 16 functionalized ketones and 11 aldehydes were successfully transformed to the corresponding boronate esters under ambient conditions (up to 99% GC yields). The products were purified by column chromatography in up to 95% isolated yield.

4.3.2 Hydroboration of Alkenes

Zhang and Zheng's earlier study on $(tpy)Mn(CH_2SiMe_3)_2$ (**58**)-catalyzed carbonyl hydroboration also revealed that this compound mediates the ambient temperature hydroboration of alkenes using 1.1 equiv of HBPin (Scheme 4.34) [55]. Notably, styrenes underwent hydroboration to yield the Markovnikov products with up to 98% regioselectivity, while aliphatic alkenes tended to yield the anti-Markovnikov products. Electron-donating or electron-withdrawing substituents in either the *meta*- or *para*-positions of styrene did not affect the transformation, whereas *o*-chlorostyrene required more forcing conditions to reach 75% isolated yield (50 °C, 24 hours). Importantly, **58**-catalyzed alkene hydroboration was not selective for substrates containing nitro and cyano groups, which yielded a mixture of products.

In Thomas's 2018 investigation, it was found that the combination of $(^{2,6\text{-}iPr_2Ph}PDI)MnBr_2$ (**46**) and NaO^tBu in a 1 : 3 M ratio not only enabled alkene hydrosilylation but also effectively mediated alkene hydroboration under mild

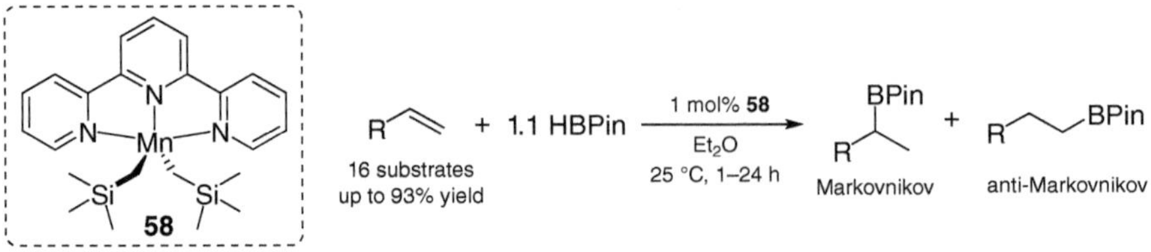

Scheme 4.34 Alkene hydroboration mediated by **58**.

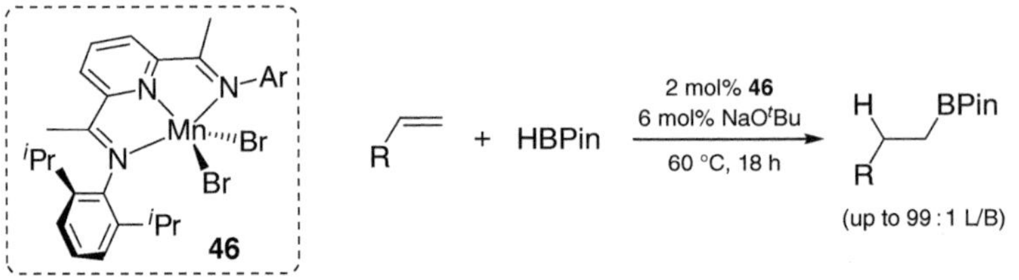

Scheme 4.35 Alkene hydroboration catalyzed by **46**.

conditions [49]. A scope of 10 alkenes and styrenes were evaluated in the presence of 3 equiv of HBPin (Scheme 4.35). Aliphatic alkenes such as 1-octene, 4-phenylbutene, and 3-triethoxypropene were hydroborated with good regioselectivity for the anti-Markovnikov product (>99 : 1 *L*/*B*). Styrenes containing different electronic and steric substitution patterns underwent hydroboration to provide alkylboranes in moderate to excellent yield with excellent regioselectivity (*L*/*B* of up to >99 : 1).

In 2018, Turculet and coworkers reported the synthesis and catalytic activity of structurally analogous first-row metal complexes featuring an *N*-phosphinoamidinate ligand [(PN)M; M = Mn, Fe, Co, and Ni] [61]. As a part of the investigation, DFT calculations were performed to evaluate the electronic structure of $[(\kappa^2\text{-P,N})Mn(N(SiMe_3)_2)]$ (**37**). The doublet, quartet, and sextet open-shell spin states were found to be accessible, which is consistent with its paramagnetic character. To compare the catalytic activity of **37** to other members of the series, the isomerization–hydroboration of 1-octene, *cis*-4-octene, and *trans*-4-octene was evaluated (Scheme 4.36). Negligible conversion was observed after one hour at 23 °C, while poor conversion (<50%) and selectivity were observed under more forcing conditions (65 °C, 18 hours). The product mixture determined by NMR spectroscopy is shown in Table 4.2 and DFT calculations revealed that the relative activity of the (PN)M catalysts is related to their spin-state energy gaps. Particularly, for the reaction of **37** with *trans*-4-octene, there is no viable spin crossover mechanism, and the migratory insertion was found to be slow, resulting in a poor catalytic outcome.

4.3.3 Hydroboration of Alkynes

In 2020, Rueping and coworkers reported the hydroboration of propargylic alcohols and amines, as well as symmetric internal alkynes, to yield the corresponding

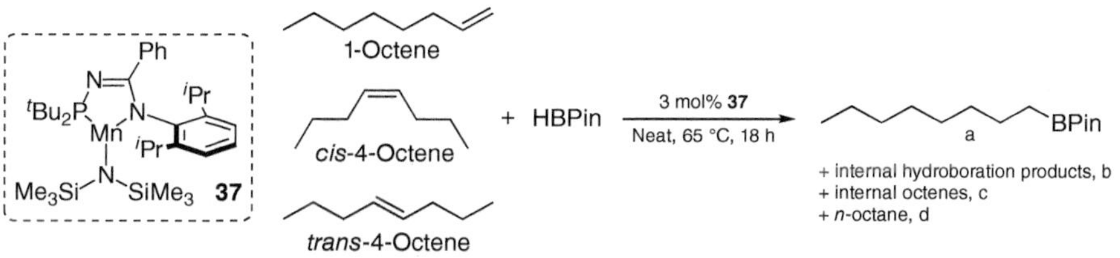

Scheme 4.36 Alkene hydroboration catalyzed by **37**.

Table 4.2 Partial isomerization–hydroboration of alkenes catalyzed by **37**.

Substrate	Residual substrate	a	b	c	d
1-Octene	55%	25%	<5%	15%	5%
cis-4-Octene	>95%	<5%	<5%	<5%	<5%
trans-4-Octene	>95%	<5%	<5%	<5%	<5%

alkenes using $NaEt_3BH$-activated Mn(II) precursors ($^{2,6\text{-}iPr_2Ph}$PDI)$MnCl_2$ and ($^{2,6\text{-}Et_2Ph}$PDI)$MnCl_2$ [62]. In particular, 14 TBDMS-protected hydroxyalkynes were screened for hydroboration using the former precatalyst under ambient conditions, which provided the corresponding vinyl borane in excellent yield (up to 99% NMR yield and 95% isolated yield after column chromatography) (Scheme 4.37a). Activated ($^{2,6\text{-}iPr_2Ph}$PDI)$MnCl_2$ also mediated the hydroboration of varied Bn-protected propargylic amines to yield the corresponding alkenes (up to 95% NMR yield and 85% isolated yield after column chromatography) (Scheme 4.37b). Interestingly, ($^{2,6\text{-}Et_2Ph}$PDI)$MnCl_2$ was found to show higher activity for the reduction of symmetric alkynes with HBPin in the presence of $NaEt_3BH$. A scope of nine alkynes were screened for hydroboration to yield the *syn*-addition products (Scheme 4.37c).

4.3.4 Hydroboration of Nitriles

Our group recently described the first examples of Mn-catalyzed nitrile dihydroboration using the hydride dimer, [($^{2,6\text{-}iPr_2Ph}$BDI)Mn(μ-H)]$_2$ (**47**) [63]. The reduction of 14 nitriles in the presence of HBPin was achieved using 0.5 mol% **47** at 80 °C (Scheme 4.38). In most cases, aromatic and aliphatic nitriles were transformed into the desired *N,N*-diborylamine products with excellent conversion (up to 99%) and modest isolated yield (up to 92%). Benzonitrile substitution did not adversely affect nitrile reduction, while the C=O functionality of *p*-acetylbenzonitrile was reduced in the presence of 3.3 equiv of HBPin (indicating that this catalyst is also active for carbonyl hydroboration). The mechanism of **47**-catalyzed nitrile hydroboration was explored using experimental and computational methods. Adding 2 equiv of benzonitrile to **47** allowed for the isolation of [($^{2,6\text{-}iPr_2Ph}$BDI)Mn(μ-NCHPh)]$_2$, and this intermediate was found to mediate benzonitrile dihydroboration under the same catalytic conditions. DFT calculations revealed that nitrile insertion into the Mn–H

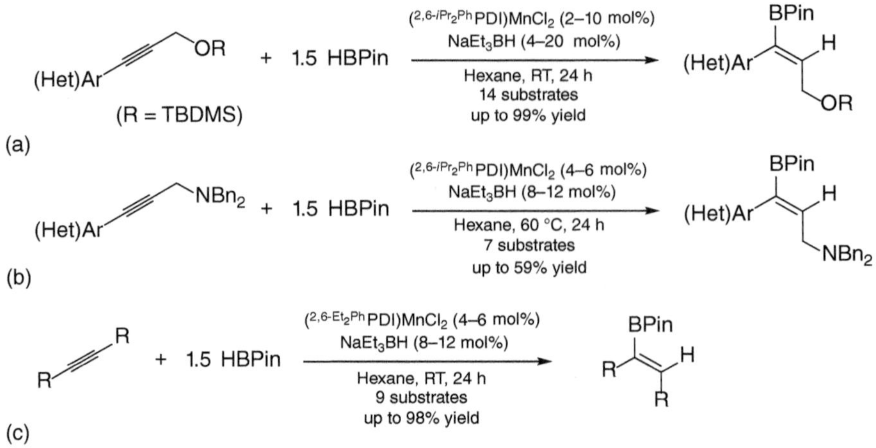

Scheme 4.37 (a) Propargylic alcohol hydroboration catalyzed by ($^{2,6\text{-}iPr_2Ph}$PDI)$MnCl_2$/$NaEt_3BH$. (b) Propargylic amine hydroboration catalyzed by ($^{2,6\text{-}iPr_2Ph}$PDI)$MnCl_2$/$NaEt_3BH$. (c) Symmetric alkyne hydroboration catalyzed by ($^{2,6\text{-}Et_2Ph}$PDI)$MnCl_2$/$NaEt_3BH$.

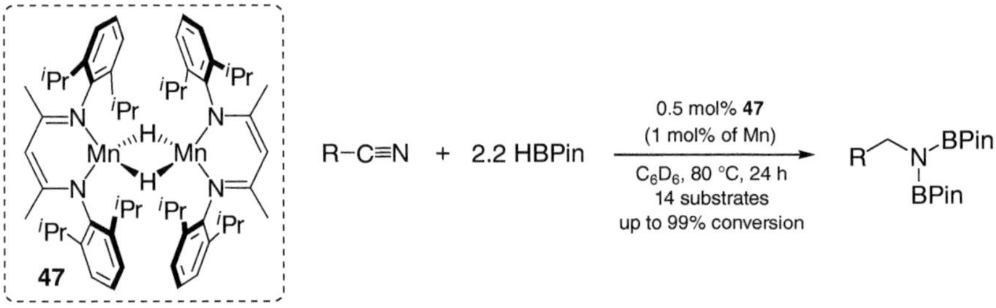

Scheme 4.38 Nitrile dihydroboration catalyzed by **47**.

moiety takes place, followed by σ-bond metathesis between Mn amide monomers and HBPin to regenerate the active catalyst and release the desired products.

4.4 Summary and Conclusions

This chapter highlighted major advances in manganese-based hydrofunctionalization catalysis. The pace of this development has accelerated dramatically in recent years. Prior to 2010, only a handful of studies had utilized manganese catalysts for hydrosilylation, and remarkably, the first report of manganese-catalyzed hydroboration was published in 2016. Over the last five years, many highly active manganese catalysts for these transformations have been described, and detailed mechanistic insight has been gathered. These hydrofunctionalization efforts have greatly contributed to the recent expansion of reductive manganese catalysis.

References

1 Trovitch, R. (2014). *Synlett* 25 (12): 1638–1642.
2 Garbe, M., Junge, K., and Beller, M. (2017). *Eur. J. Org. Chem.* 2017 (30): 4344–4362.
3 Trovitch, R.J. (2017). *Acc. Chem. Res.* 50 (11): 2842–2852.
4 Yang, X. and Wang, C. (2018). *Chem. Asian J.* 13 (17): 2307–2315.
5 Speier, J.L., Webster, J.A., and Barnes, G.H. (1957). *J. Am. Chem. Soc.* 79 (4): 974–979.
6 Karstedt, B.D. (1973). Platinum Complexes of Unsaturated Siloxanes and Platinum Containing Organopolysiloxanes. US Patent 3775452A, 27 November 1973.
7 Ojima, I., Nihonyanagi, M., and Nagai, Y. (1972). *J. Chem. Soc., Chem. Commun.* 16: 938a.
8 Brown, H.C. (1961). *Tetrahedron* 12 (3): 117–138.
9 Kono, H., Ito, K., and Nagai, Y. (1975). *Chem. Lett.* 4 (10): 1095–1096.
10 Männig, D. and Nöth, H. (1985). *Angew. Chem. Int. Ed.* 24 (10): 878–879.
11 Yates, R.L. (1982). *J. Catal.* 78 (1): 111–115.
12 Hanna, P.K., Gregg, B.T., and Cutler, A.R. (1991). *Organometallics* 10 (1): 31–33.
13 DiBiase Cavanaugh, M., Gregg, B.T., Chiulli, R.J., and Cutler, A.R. (1997). *J. Organomet. Chem.* 547 (1): 173–182.
14 Mao, Z., Gregg, B.T., and Cutler, A.R. (1998). *Organometallics* 17 (10): 1993–2002.
15 DiBiase Cavanaugh, M., Gregg, B.T., and Cutler, A.R. (1996). *Organometallics* 15 (12): 2764–2769.
16 Mao, Z., Gregg, B.T., and Cutler, A.R. (1995). *J. Am. Chem. Soc.* 117 (40): 10139–10140.
17 Son, S.U., Paik, S.-J., Lee, I.S. et al. (1999). *Organometallics* 18 (20): 4114–4118.
18 Son, S.U., Paik, S.-J., and Chung, Y.K. (2000). *J. Mol. Catal. A: Chem.* 151 (1): 87–90.
19 Chidara, V.K. and Du, G. (2013). *Organometallics* 32 (18): 5034–5037.
20 Vijjamarri, S., Chidara, V.K., and Du, G. (2017). *ACS Omega* 2 (2): 582–591.
21 Vijjamarri, S., Streed, S., Serum, E.M. et al. (2018). *ACS Sustainable Chem. Eng.* 6 (2): 2491–2497.
22 Zheng, J., Chevance, S., Darcel, C., and Sortais, J.-B. (2013). *Chem. Commun.* 49 (85): 10010–10012.
23 Zheng, J., Elangovan, S., Valyaev, D.A. et al. (2014). *Adv. Synth. Catal.* 356 (5): 1093–1097.
24 Valyaev, D.A., Wei, D., Elangovan, S. et al. (2016). *Organometallics* 35 (24): 4090–4098.
25 Ojima, I., Nihonyanagi, M., Kogure, T. et al. (1975). *J. Organomet. Chem.* 94 (3): 449–461.
26 Pinto, M., Friães, S., Franco, F. et al. (2018). *ChemCatChem* 10 (13): 2734–2740.
27 Sousa, S.C.A., Realista, S., and Royo, B. (2020). *Adv. Synth. Catal.* 362 (12): 2437–2443.

28 Behera, R.R., Ghosh, R., Panda, S. et al. (2020). *Org. Lett.* 22 (9): 3642–3648.
29 Mukhopadhyay, T.K., Flores, M., Groy, T.L., and Trovitch, R.J. (2014). *J. Am. Chem. Soc.* 136 (3): 882–885.
30 Ghosh, C., Mukhopadhyay, T.K., Flores, M. et al. (2015). *Inorg. Chem.* 54 (21): 10398–10406.
31 Mukhopadhyay, T.K., Ghosh, C., Flores, M. et al. (2017). *Organometallics* 36 (18): 3477–3483.
32 Mukhopadhyay, T.K., Rock, C.L., Hong, M. et al. (2017). *J. Am. Chem. Soc.* 139 (13): 4901–4915.
33 Ma, X., Zuo, Z., Liu, G., and Huang, Z. (2017). *ACS Omega* 2 (8): 4688–4692.
34 Wenz, J., Vasilenko, V., Kochan, A. et al. (2017). *Eur. J. Inorg. Chem.* 2017 (47): 5545–5556.
35 Martínez-Ferraté, O., Werlé, C., Franciò, G., and Leitner, W. (2018). *ChemCatChem* 10 (20): 4514–4518.
36 Martínez-Ferraté, O., Chatterjee, B., Werlé, C., and Leitner, W. (2019). *Catal. Sci. Technol.* 9 (22): 6370–6378.
37 Ghaffari, B., Mendes-Burak, J., Chan, K.W., and Copéret, C. (2019). *Chem. Eur. J.* 25 (61): 13869–13873.
38 Zhang, G., Zeng, H., Li, S. et al. (2020). *Dalton Trans.* 49 (8): 2610–2615.
39 Igarashi, M. and Fuchikami, T. (2001). *Tetrahedron Lett.* 42 (10): 1945–1947.
40 Arias-Ugarte, R., Sharma, H.K., Morris, A.L.C., and Pannell, K.H. (2012). *J. Am. Chem. Soc.* 134 (2): 848–851.
41 Kelly, C.M., McDonald, R., Sydora, O.L. et al. (2017). *Angew. Chem. Int. Ed.* 56 (50): 15901–15904.
42 Das, H.S., Das, S., Dey, K. et al. (2019). *Chem. Commun.* 55 (79): 11868–11871.
43 Bertini, F., Glatz, M., Stöger, B. et al. (2019). *ACS Catal.* 9 (1): 632–639.
44 Pratt, S.L. and Faltynek, R.A. (1983). *J. Organomet. Chem.* 258 (1): C5–C8.
45 Hilal, H.S., Abu-Eid, M., Al-Subu, M., and Khalaf, S. (1987). *J. Mol. Catal.* 39 (1): 1–11.
46 Hilal, H.S., Suleiman, M.A., Jondi, W.J. et al. (1999). *J. Mol. Catal. A: Chem.* 144 (1): 47–59.
47 Jondi, W., Zyoud, A., Mansour, W. et al. (2016). *React. Chem. Eng.* 1 (2): 194–203.
48 Docherty, J.H., Peng, J., Dominey, A.P., and Thomas, S.P. (2017). *Nat. Chem.* 9 (6): 595–600.
49 Carney, J.R., Dillon, B.R., Campbell, L., and Thomas, S.P. (2018). *Angew. Chem. Int. Ed.* 57 (33): 10620–10624.
50 Price, J.S., Emslie, D.J.H., and Britten, J.F. (2017). *Angew. Chem. Int. Ed.* 56 (22): 6223–6227.
51 Mukhopadhyay, T.K., Flores, M., Groy, T.L., and Trovitch, R.J. (2018). *Chem. Sci.* 9 (39): 7673–7680.
52 Chalk, A.J. and Harrod, J.F. (1965). *J. Am. Chem. Soc.* 87 (1): 16–21.
53 Yang, X. and Wang, C. (2018). *Chin. J. Chem.* 36 (11): 1047–1051.
54 Yang, X. and Wang, C. (2018). *Angew. Chem. Int. Ed.* 57 (4): 923–928.
55 Zhang, G., Zeng, H., Wu, J. et al. (2016). *Angew. Chem. Int. Ed.* 55 (46): 14369–14372.

56 Vasilenko, V., Blasius, C.K., Wadepohl, H., and Gade, L.H. (2017). *Angew. Chem. Int. Ed.* 56 (29): 8393–8397.

57 Vasilenko, V., Blasius, C.K., and Gade, L.H. (2018). *J. Am. Chem. Soc.* 140 (29): 9244–9254.

58 Erken, C., Kaithal, A., Sen, S. et al. (2018). *Nat. Commun.* 9 (1): 4521–4529.

59 Schäffner, B., Schäffner, F., Verevkin, S.P., and Börner, A. (2010). *Chem. Rev.* 110 (8): 4554–4581.

60 Barman, M.K., Das, K., and Maji, B. (2019). *J. Org. Chem.* 84 (3): 1570–1579.

61 Macaulay, C.M., Gustafson, S.J., Fuller, J.T. III, et al. (2018). *ACS Catal.* 8 (11): 9907–9925.

62 Brzozowska, A., Zubar, V., Ganardi, R.C., and Rueping, M. (2020). *Org. Lett.* 22 (10): 3765–3769.

63 Nguyen, T.T., Kim, J.H., Kim, S. et al. (2020). *Chem. Commun.* 56 (28): 3959–3962.

5

Manganese Complexes for Electro- and Photocatalytic Transformations

Sergio Fernandez[1], *Geyla C. Dubed Bandomo*[1], *and Julio Lloret-Fillol*[1,2]

[1] *Institute of Chemical Research of Catalonia (ICIQ), The Barcelona Institute of Science and Technology, Avinguda Països Catalans 16, 43007 Tarragona, Spain*
[2] *Catalan Institution for Research and Advanced Studies (ICREA), Passeig Lluís Companys 23, 08010 Barcelona, Spain*

5.1 Introduction

Manganese is one of the most abundant first-row transition metals in the crust of the Earth (0.1%). As it is the case for other 3d metals, the redox chemistry of manganese is very rich and takes a vital role as an enzyme cofactor in biochemical processes. Indeed, photosystem II, which is one of the essential enzyme machineries for life in Earth, depends on a Mn_4CaO_5 cluster to catalyze the splitting of the water molecule into dioxygen, protons, and electrons during the process of natural photosynthesis [1–4]. Besides the most common oxidation state of manganese in Nature is Mn^{2+}, manganese compounds have a rich redox chemistry in both, high and low oxidation states. Despite their available redox properties and the essential role in the natural biochemistry, Mn complexes did not call great attention of the organic synthetic community until recently [5]. Now, Mn chemistry is emerging as a powerful tool to solve challenging synthetic problems using photo- and electrochemistry [6–9]. Electron transfer (ET) processes are at the core of both photo- and electrochemical transformations relevant for energy conversion and synthetic methods to shorting synthetic routes [10–18].

Toward this direction, understanding the fundamental questions about ET is crucial. Photo- and electrocatalytic processes covered in this review are initiated by ET processes, which in turn triggers the desired chemical reactivity. Although the same initial ET processes can be achieved using chemical, photochemical, or electrochemical methods, the reaction conditions are of paramount importance to control the reactivity and selectivity [19–25]. And they are not necessarily easily interchangeable. For instance, while photoinduced ET processes are produced in the homogeneous phase, used electrochemical methods imply the utilization of electrodes; therefore, the diffusion effect of reduced, oxidized, and radical species is intrinsically different. Nevertheless, new concepts start to be developed to overcome these differences and limitations [26].

Manganese Catalysis in Organic Synthesis, First Edition. Edited by Jean-Baptiste Sortais.

To facilitate the comprehension of this chapter, we will focus first on the discussion of the main applications of Mn complexes as catalysts in photochemical and electrochemical organic transformations. Then, we will also highlight the most relevant examples of molecular catalysts based on Mn for the reduction of CO_2 due to its importance in the synthesis of renewable fuels such as CO and HCOOH [27].

5.2 Mn-Catalyzed Organic Oxidations

In this section, we will focus on the reactivity of Mn complexes in catalytic organic transformations in which high oxidation states of the metal center are involved. To stabilize these high oxidation states, most of the Mn coordination compounds reported as catalysts for organic oxidations bear strong donating ligands such as porphyrin and corrole derivatives or strong-chelating non-heme pyridylamino ligands (Figure 5.1). In particular, Mn complexes tend to form metal–oxo species in high oxidation states (IV, V) upon reaction with oxidants. These new Mn–oxo species are highly reactive and can undergo a wide variety of organic transformations such as C–H hydroxylation (P450-type activity) or alkene epoxidation among other biomimetic oxygenation reactions [28–32].

As an alternative to the classical chemical oxidation reactions, photocatalytic methods have recently appeared as interesting strategies in which H_2O is used as the oxygen source to form high valent metal–oxo intermediates (Eq. (5.1)) without the need of strong one-electron oxidants (e.g. cerium(IV) ammonium nitrate) [9].

$$M^{n+} + H_2O \rightarrow M^{(n+2)+}(O) + 2H^+ + 2e^- \quad (5.1)$$

Under photocatalytic conditions, strong sacrificial oxidants can be replaced by a milder oxidant together with the proper photosensitizer (PS). In this regard, Fukuzumi et al. reported for the first time a highly efficient photocatalytic system based on Mn porphyrin catalysts for oxygenation reactions in water [33]. In this catalytic system, $[Ru^{II}(bpy)_3]^{2+}$ (bpy = 2,2′-bipyridine) is the photosensitizer and $[Co^{III}(NH_3)_5Cl]^{2+}$ is the weak one-electron oxidant. Upon visible light irradiation ($\lambda > 430$ nm), the excited state $[Ru^{II}(bpy)_3]^{2+*}$ is formed and is then oxidatively quenched, by the sacrificial Co^{III} complex, giving the corresponding Co^{II} complex and $[Ru^{III}(bpy)_3]^{3+}$. Mechanistic studies suggest that $(Por)Mn^{IV}(O)$ is formed by one-electron oxidation of $(Por)Mn^{III}(OH)$ with the oxidized photosensitizer $[Ru(bpy)_3]^{3+}$. Then, the $Mn^{IV}{=}O$ species undergoes disproportionation in the presence of an acid, regenerating the starting $(Por)Mn^{III}(OH)$ and the catalytically active species $[(Por)Mn^{V}(O)]^+$ that can oxidize the corresponding organic substrate (Figure 5.2). The scope of these reactions is limited to the use of water-soluble substrates such as sodium *p*-styrenesulfonate (epoxidation), sodium 4-ethylbenzene sulfonate (hydroxylation), and 2-(methylthio)ethanol (sulfoxidation).

After this pioneer work, similar reactivity was observed with other non-heme Mn and Fe complexes [34, 35]. For instance, Sun and coworkers reported an aminoquinoline Mn complex $[Mn^{II}(BQCN)(OTf)_2]$ (BQCN = *N,N′*-dimethyl-*N,N′*-bis(8-quinolyl)cyclohexanediamine; OTf = $CF_3SO_3^-$) that can oxidize alcohols to

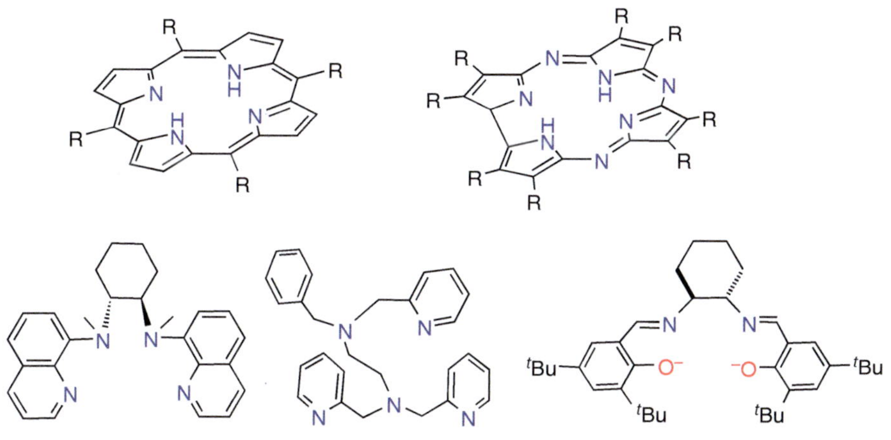

Figure 5.1 Typical ligands used in oxidation catalysis with Mn coordination compounds.

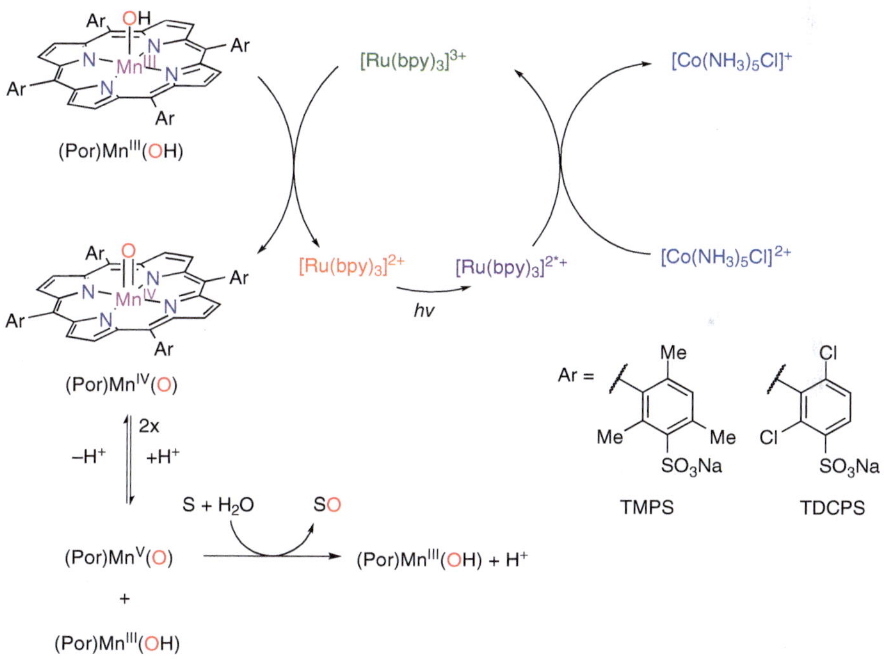

Figure 5.2 Proposed reaction mechanism of photocatalytic oxidation of organic substrates with a Mn porphyrin catalyst.

aldehydes and organic sulfides to sulfoxides under similar photocatalytic conditions [35]. In this case, the catalytically active intermediate is a Mn^{IV}=O species that is formed *in situ* after the sequential oxidation of Mn^{II} to Mn^{III}–OH and then to Mn^{IV}=O by two cycles of the $[Ru(bpy)_3]^{2+/3+}$ photosensitizer. The oxidation of sulfides to sulfoxides occurs by oxygen atom transfer (OAT) from Mn^{IV}=O, recovering the starting Mn^{II} complex. In contrast, the alcohol oxidation occurs by

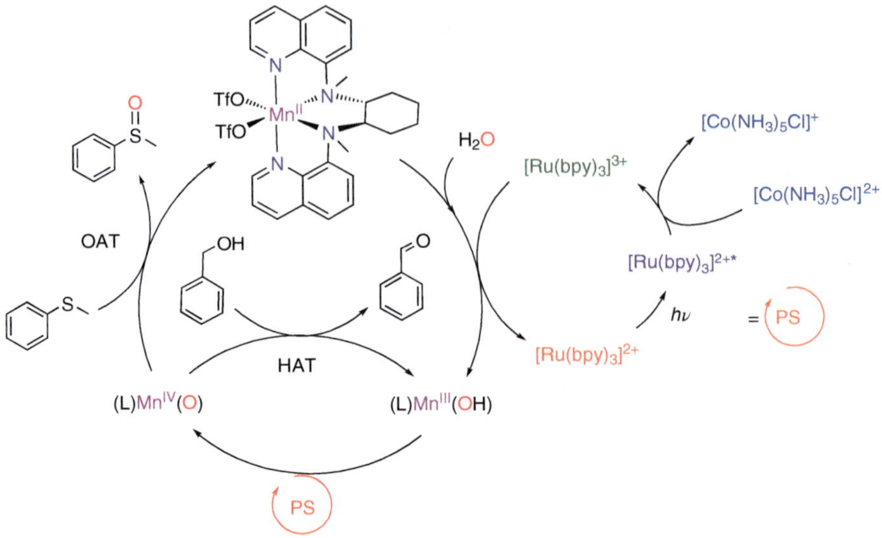

Figure 5.3 Proposed reaction mechanism for the light-driven oxidation of organic substrates by $[Mn^{II}(BQCN)(OTf)_2]^{2+}$ in aqueous solution. $[Ru^{II}(bpy)_3]^{2+}$ and $[Co^{III}(NH_3)_5Cl]^{2+}$ were used as a photosensitizer and sacrificial electron acceptor, respectively.

hydrogen atom transfer (HAT) to give the corresponding aldehyde and a Mn^{III}–OH intermediate (Figure 5.3).

Later, Fukuzumi and coworkers reported the asymmetric epoxidation of terminal olefins employing the chiral version of the previous methodology. In this case, the authors used the enantiomerically pure $[Mn^{II}(R,R\text{-}BQCN)(OTf)_2]$ complex as a catalyst in the presence of acetic acid, $[Ru(bpy)_3]^{2+}$ as a photosensitizer, and $[Co^{III}(NH_3)_5Cl]^{2+}$ as a sacrificial electron acceptor [36]. This methodology allowed for the asymmetric epoxidation of different terminal olefins (up to 60% ee) under visible light irradiation using water as the main oxygen source (Figure 5.4). Mechanistic studies showed for the first time that the chemically generated $Mn^{IV}{=}O$ intermediate is responsible for the asymmetric epoxidation of olefins.

The same year, Lau and coworkers reported a similar photocatalytic approach for the oxidation of olefins and alcohols to the corresponding epoxides and carbonyl compounds [37]. The authors used a Mn(V)–nitrido complex $(PPh_4)_2[Mn(N)(CN)_4]$ that was previously known to catalyze the oxidation of alkenes, alcohols, and alkanes using H_2O_2 as chemical oxidant (Figure 5.5) [38, 39]. Under visible light irradiation in the presence of $[Ru(bpy)_3]^{2+}$ and $[Co^{III}(NH_3)_5Cl]^{2+}$, complex $(PPh_4)_2[Mn(N)(CN)_4]$ can oxidize styrene, *cis*- and *trans*-β-methylstyrene, and cyclohexene to the corresponding epoxides with high TON (>90). A drop in the catalytic activity was observed with more difficult linear and sterically hindered alkenes. In the case of alcohol oxidation, this methodology presents high efficiency toward electron-rich primary benzyl alcohols to form the corresponding aldehydes. Although with lower TON, the catalytic system also oxidizes electron-poor benzyl

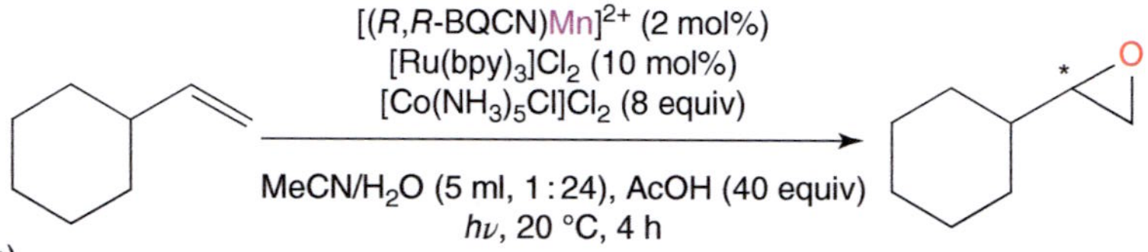

(a)

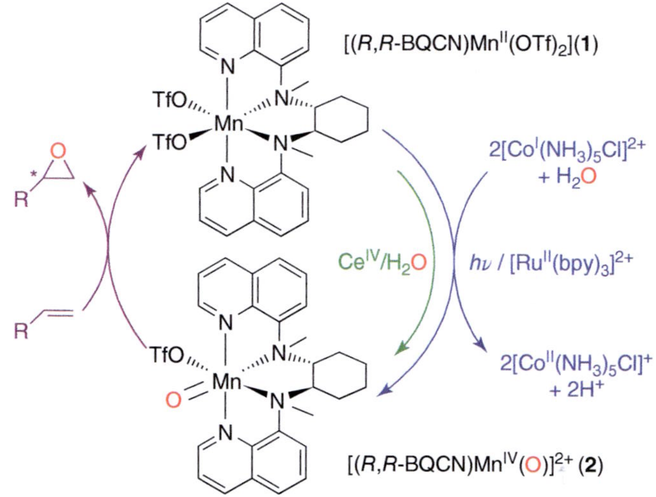

(b)

Entry	Substrate	Product	Yield (%)	*ee* (%)
1	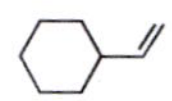	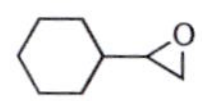	59 ± 4	53 ± 2
2	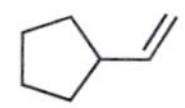	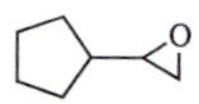	68 ± 5	50 ± 2
3	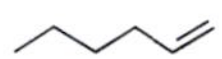	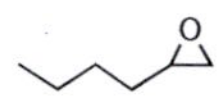	24 ± 3	57 ± 1
4		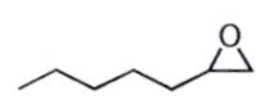	26 ± 3	60 ± 2
5		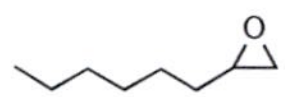	35 ± 4	56 ± 2
6	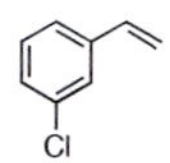	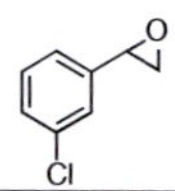	65 ± 5	43 ± 2

(c)

Figure 5.4 Photochemical asymmetric olefin epoxidation catalyzed by [Mn^{II}(*R,R*-BQCN)(OTf)$_2$]. General reaction conditions (a), simplified catalytic cycle (b), and scope table (c). Source: Shen et al. [36].

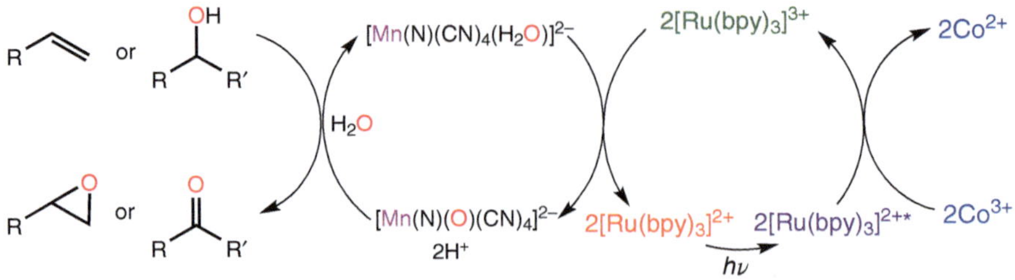

Figure 5.5 Simplified reaction scheme for the oxidation of olefins and alcohols (S) to the corresponding epoxides and carbonyl compounds using an Mn^{V}-nitrido complex $(PPh_4)_2[Mn(N)(CN)_4]$ as catalyst under visible light irradiation (458 nm LED).

alcohols and secondary benzyl alcohols such as α-methylbenzyl alcohol and cyclohexanol to the corresponding ketones. These results suggest that the active oxidant is an electrophilic oxo species. Indeed, mechanistic studies suggest that the active species is a $Mn^{IV}{=}O$ intermediate since a $[(PPh_4)[Mn^{VII}(N)(O)(CN)4]{\cdot}CF_3CH_2OH]^-$ adduct could be detected by mass spectrometry. Labeling studies with $H_2{}^{18}O$ water revealed that the oxygen in the epoxide comes from water, and the $Mn^{IV}{=}O$ species is responsible for the OAT. In the case of the alcohol oxidation, studies based in the kinetic isotopic effect in the competitive oxidation of non-deuterated and deuterated cyclohexanol ($KIE_{H/D} = 3.5$), together with the clean oxidation of cyclobutanol to cyclobutanone, indicate that the two-electron hydride abstraction is the rate-determining step in the reaction mechanism.

More recently, the same groups reported the hydroxylation of benzene to phenol with water using a non-heme $Mn^{IV}{=}O$ complex bound to Sc^{3+} ions. Although the $[Mn^{IV}(Bn\text{-}TPEN)(O)]^{2+}$ (Bn-TPEN = *N*-benzyl-*N*,*N*′,*N*′-tris(2-pyridylmethyl)-1,2-diamino-ethane) is not photoactive, in the presence of $Sc(OTf)_3$, it forms an adduct $[Mn^{IV}(Bn\text{-}TPEN)(O)]^{2+}\cdots(Sc(OTf)_3)_2$ that results in a long-lived state (6.4 μs) upon irradiation at 355 nm [40]. The formation of the long-lived excited state of a high valent metal–oxo has a great potential to produce challenging oxidations. In this case, the $[Mn^{IV}(O)]^{2+}\cdots(Sc^{3+})_2$ complex is a powerful photosensitizer that can even oxidize benzene in one electron. The benzene radical cation formed after reductive quenching undergoes hydroxylation in the presence of water to produce a neutral radical intermediate that restores the aromaticity after a second one-electron reduction and protonation (Figure 5.6). The (sub)stoichiometric nature of the reaction is the main limitation of the methodology since a maximum 50% yield concerning the starting $[Mn^{IV}(O)]^{2+}\cdots(Sc^{3+})_2$ adduct can be achieved. Moreover, an inactive $[Mn^{III}OMn^{III}]^{4+}$ decay product is accumulated after the reaction.

Another approach has been the use of molecular O_2 as an oxygen donor for the oxidation of organic substrates under photochemical or electrochemical conditions. In the late 1990s, Wong and Guo reported the first electrochemical enantioselective epoxidation of olefins using immobilized Mn–Schiff base complexes, also known as Mn–salen complexes, over glassy carbon electrodes [41]. In this example, a negative potential is applied to a solution of the complex in an oxygen-saturated acetonitrile electrolyte solution in order to reduce the starting Mn^{III} complex to Mn^{II}, which

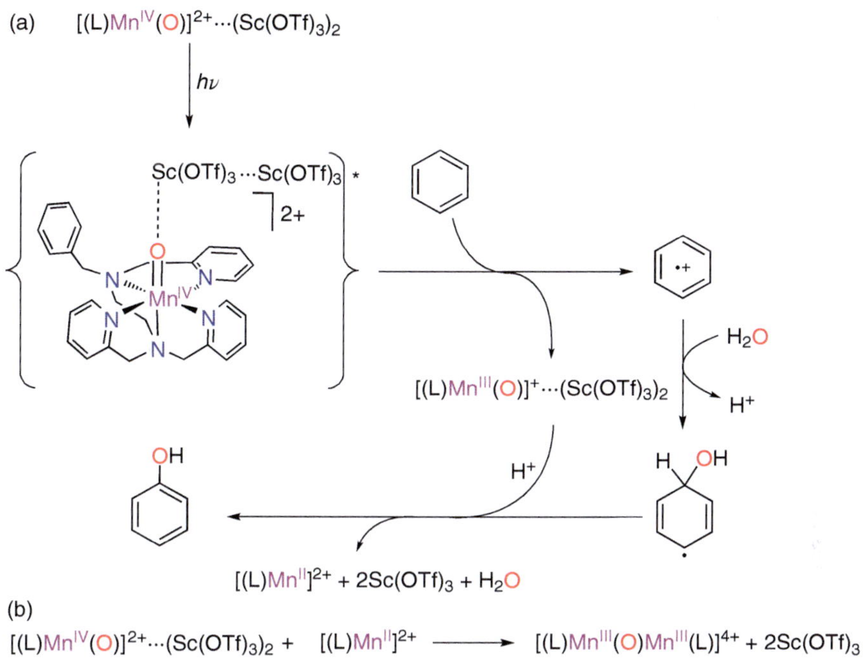

Figure 5.6 (a) Structure of the $[Mn^{IV}(O)]^{2+}\cdots(Sc^{3+})_2$ photoactive complex and reaction mechanism of the benzene photohydroxylation promoted by $[Mn^{IV}(O)]^{2+}\cdots(Sc^{3+})_2$. (b) Reaction of $[Mn^{IV}(O)]^{2+}\cdots(Sc^{3+})_2$ and $[Mn^{II}]^{2+}$ to form $[Mn^{III}OMn^{III}]^{4+}$.

is able to bind O_2, forming a Mn^{III}–superoxo species that can be further reduced to the corresponding Mn–oxo active species. Under controlled potential electrolysis (CPE) conditions, the formation of an inactive Mn^{IV}, Mn^{IV} bis(μ-oxo) dimer from Mn^{II} and O_2 is described as the main deactivation pathway. Later on, Torii and coworkers reported the asymmetric epoxidation of styrenes using the same type of Mn–salen complexes as catalysts in a CH_2Cl_2 aqueous NaCl two-phase solution at 0 °C within an undivided cell setup [42]. However, the authors propose that the formation of the Mn–oxo intermediate is mediated by the anodic oxidation of Cl^- to ClO^- (Figure 5.7). Instead of water or molecular O_2, ClO^- is the oxygen donor in the formation of the catalytically active Mn–oxo species that will be responsible for the olefin epoxidation.

Nonetheless, most of the work done in the field of Mn-catalyzed oxidation reactions using O_2 as oxidant take advantage of the photophysical and photochemical properties of Mn heme compounds (porphyrins and corrole derivatives) to form the high valent metal–oxo intermediate (Eq. (5.2)) responsible for the oxidation of organic substrates.

$$M^{n+} + O_2 + 2H^+ + 2e^- \rightarrow M^{(n+2)} + (O) + H_2O \qquad (5.2)$$

Goldberg and Prokop reported the photochemical activation of O_2 with Mn^{III} corrolazine complex $Mn^{III}(TBP_8Cz)$ [TBP_8Cz = (octakis(*p*-*tert*-butylphenyl)-corrolazinato)$^{3-}$] to give a Mn^V–oxo complex $Mn^V(TBP_8Cz)(O)$ through a free

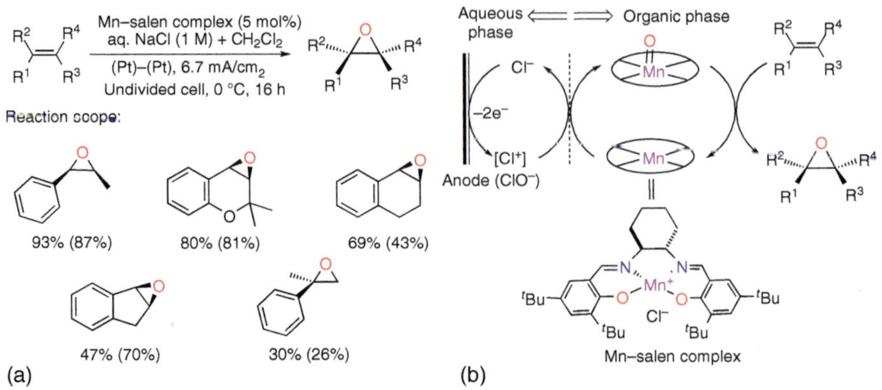

Figure 5.7 (a) General conditions for the Mn-catalyzed electrochemical epoxidation and reaction scope with yields (ee in parenthesis). (b) Electro-epoxidation mechanism using the Cl^-/Mn double mediatory system.

radical chain process [43]. This complex is able to catalyze the oxygenation of PPh_3 to $OPPh_3$ under visible light irradiation. The same system can oxidize the benzylic C—H bond of toluene derivatives selectively to the corresponding benzyl alcohols [44]. Indeed, the first evidence for the photocatalytic oxidation of C—H bond using O_2 as oxidant was obtained by using 10-methyl-9,10-dihydroacridine ($AcrH_2$) as substrate (TON = 11 in five hours), which is a stronger hydrogen donor than toluene or hexamethylbenzene (HMB) [45]. Mechanistic studies based on flash photolysis and KIEs show that the excited state of $Mn^{III}(TBP_8Cz)$ reacts with O_2 to form a superoxo species $Mn^{IV}(TBP_8Cz)(O_2^{\cdot-})$. The superoxo species abstracts a hydrogen atom from the substrate to form $Mn^{IV}(TBP_8Cz)(OOH)$ together with the corresponding benzyl radical. Finally, the benzyl radical undergoes oxygen rebounding with $Mn^{IV}(TBP_8Cz)(OOH)$ to form the benzyl alcohol derivative and $Mn^{V}(TBP_8Cz)(O)$. In the case of $AcrH_2$, the starting Mn^{III} species is recovered by the further $2e^-$ oxidation of the dihydroanthracene derivative (BDFE = 76 kcal/mol) by the Mn^V=O intermediate to form the corresponding ketone (Figure 5.8). Instead, the Mn^V=O species is not able to oxidize stronger benzylic C—H bonds like the ones of toluene (BDFE = 87 kcal/mol), which prevents the catalytic oxidation of these family of substrates.

Later in 2015, Goldberg and coworkers reported the photocatalytic aerobic C–H oxidation of HMB to the corresponding benzyl alcohol (18 TON) and aldehyde (9 TON) using $Mn^{III}(TBP_8Cz)$ as catalyst under visible light irradiation (Figure 5.9) [46]. In this work, the authors show that the strict control of the reaction medium acidity is the key to achieve catalytic turnovers. In the presence of $H^+[B(C_6F_5)_4]^-$, the starting $Mn^{III}(TBP_8Cz)$ gets protonated in one or two of the *meso*-N atoms as shown by single-crystal X-ray diffraction, UV–Vis spectrometry, and ^{1}H-NMR. The protonation of the ligand in a remote position allows the formation of a more reactive Mn^V–oxo intermediate that can now oxidize benzylic C—H bonds. The substitution of $H^+[B(C_6F_5)_4]^-$ by a coordinating acid such as HOTf boosted the catalytic activity and selectivity of the system toward the oxidation of HMB to the corresponding

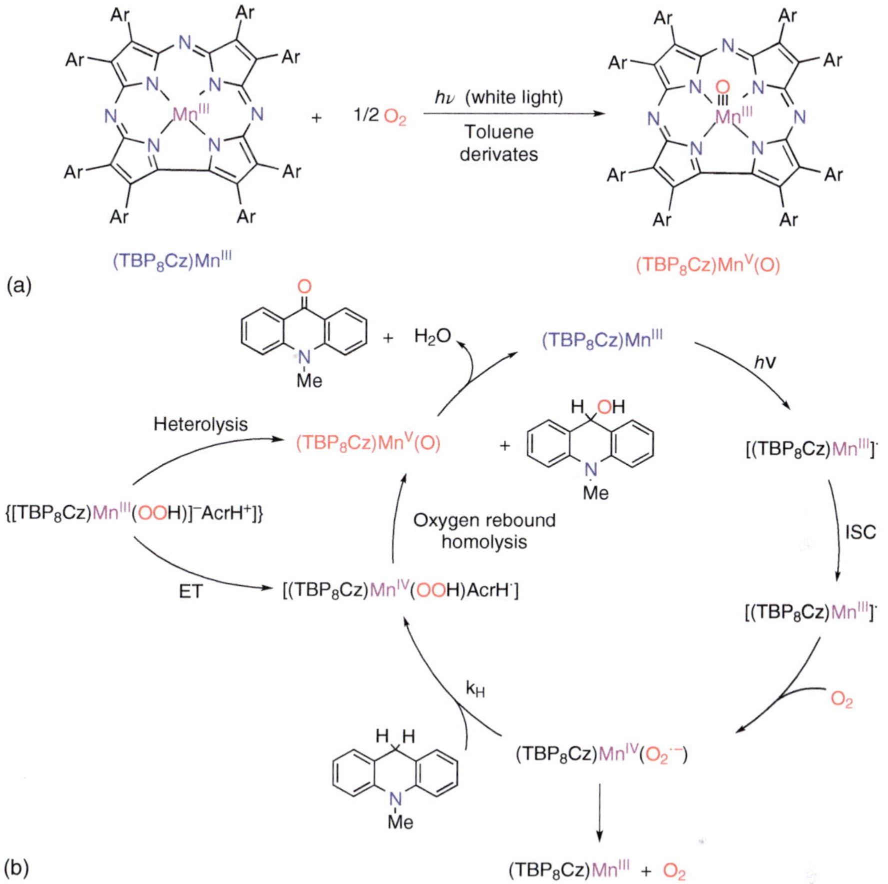

Figure 5.8 (a) Photochemical activation of O_2 by a Mn^{III} corrolazine complex to form the Mn^{V}–oxo species. (b) Proposed mechanism for the photocatalytic oxidation of $AcrH_2$ with $Mn^{III}(TBP_8Cz)$ and O_2.

alcohol (563 TON vs. 9 TON of aldehyde) and the sulfoxidation of thioanisole (902 TON) under visible light irradiation [47]. The coordination of the conjugate base to the monoprotonated Mn^{III} complex gives Mn^{III}(OTf)(TBP8Cz(H)), which is responsible for the improved catalytic properties. Instead, in the presence of a large excess of acid, the reaction stops due to the formation of the catalytically inactive double-protonated Mn^{III} corrolazine complex.

5.3 Mn-Catalyzed Reductions and Other Organic Transformations

In contrast to the high valent heme and non-heme Mn–oxo complexes used as oxidation catalysts, most of the catalytic reduction reactions involve Mn carbonyl complexes. Under reducing conditions, e.g. organic reductive transformations and CO_2

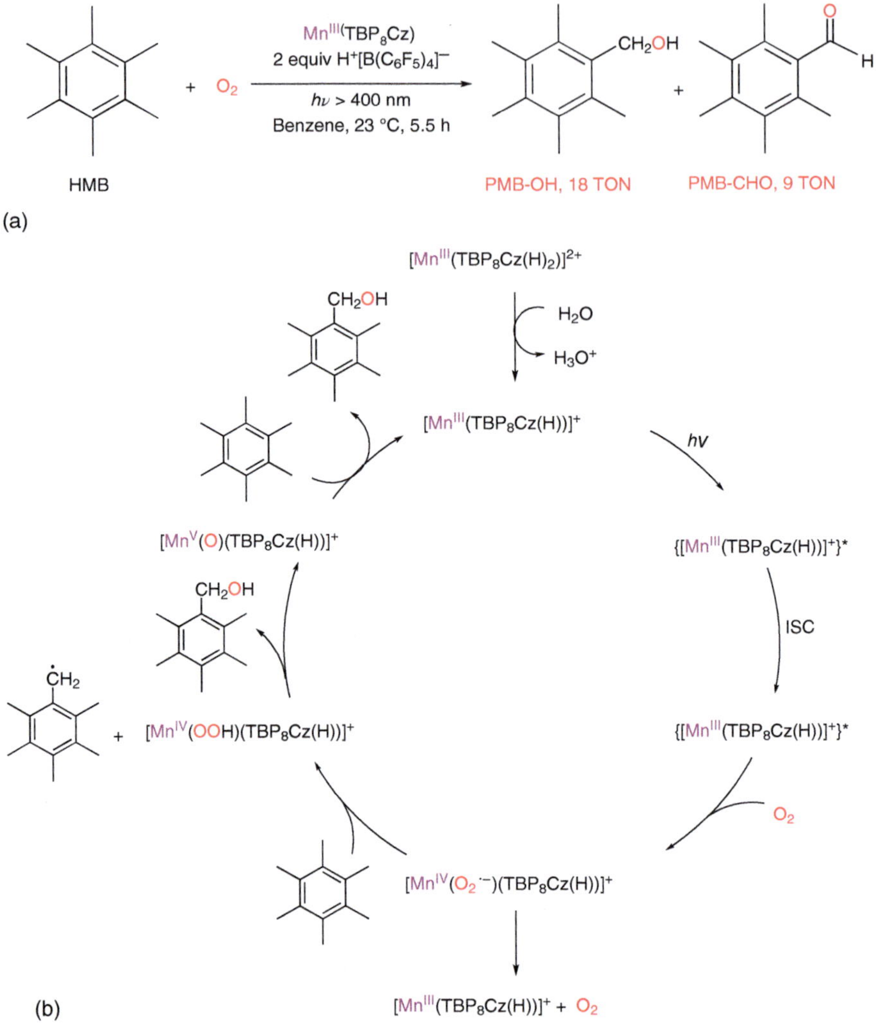

Figure 5.9 (a) Reaction conditions for the photocatalytic aerobic C–H oxidation of HMB by a Mn^{III} corrolazine complex. (b) Proposed reaction mechanism.

reduction (*vide infra*), low valent Mn intermediates that are strongly stabilized by π-acceptor ligands such as CO are involved. It is well known that upon light irradiation, the metal–carbonyl bond can be activated through an electronic transition in which an electron from the metal is promoted to a π* orbital of the ligand (MLCT). This photoexcitation drives the formation of the catalytically active species, through an isomerization reaction, ligand substitution, or the generation of a coordination vacancy [48].

In 1983 Pratt and Faltynek reported the hydrosilylation of pent-1-ene using an equimolar amount of heptamethylcyclotetrasiloxane in the presence of $Ph_3SiMn(CO)_5$ (0.1 mol%) [49]. Under light irradiation conditions, the main

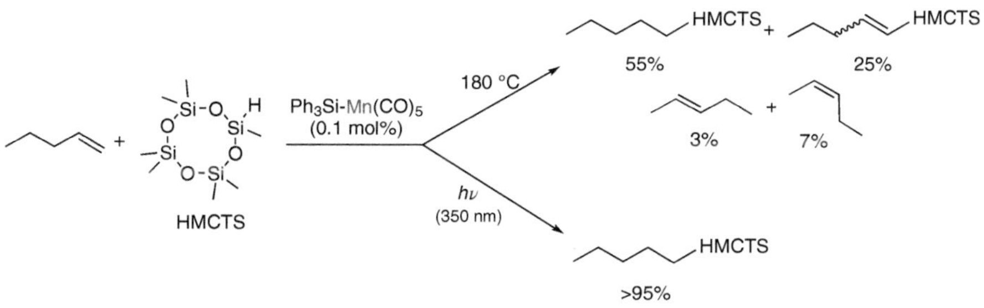

Figure 5.10 Different results of the thermal and photochemical hydrosilylation of alkenes catalyzed by $Ph_3SiMn(CO)_5$.

product of the reaction corresponds to the terminal addition at the double bond, whereas the thermal hydrosilylation at 180 °C in the dark produces a mixture of the terminal addition product, isomers from dehydrogenative silylation, isomers of pent-2-ene, and bis(heptamethylcyclotetrasiloxane) (Figure 5.10). The authors hypothesize that this difference in selectivity has a mechanistic origin. While the thermal reaction proceeds through a radical process, under light irradiation the substrate coordinates to the metal center following a Chalk–Harrod mechanism [5].

More efforts have been recently focused on the development of catalytic systems based on Mn carbonyl compounds for the reduction of organic carbonyl compounds using silanes under photoirradiation. For example, under UV irradiation, $CpMn(CO)_3$ is able to catalyze the reduction of dimethylformamide (DMF) and diethylformamide (DEF) to the corresponding tertiary amines and disiloxanes (Figure 5.11a) [50]. Darcel and coworkers reported the reduction of carboxylic acids (phenylacetic acids, alkenyl, and alkyl carboxylic acids) to the corresponding aldehydes in good yields by using $Mn_2(CO)_{10}$ as catalyst under similar conditions. Although this methodology tolerates non-conjugated internal olefins, conjugated and terminal non-conjugated olefins are fully and partially hydrosilylated, respectively. In comparison, benzylic acids were reduced in moderate yields. In this case, the reaction takes place in two steps: (i) hydrosilylation of the acid to the disilylacetal followed by (ii) its hydrolysis in acidic media to give the aldehyde (Figure 5.11b) [51]. Lugan and coworkers expanded the scope of this photocatalytic reaction to the hydrosilylation of aldehydes and ketones by using a family of half-sandwich Mn^I *N*-heterocyclic carbene (NHC) complexes as catalysts. The best results were obtained using $CpMn(CO)_2(^{Mes}NHC)$ as catalyst and Ph_2SiH_2 as reducing agent. With this methodology, a wide range of aliphatic and aromatic aldehydes and ketones were reduced in good to excellent yields. As it was observed previously with carboxylic acids, the reduction of α,β-unsaturated ketones gave a mixture of the desired allyl alcohol and the corresponding aliphatic alcohol. Moreover, a hydrolysis step under basic conditions was required after the light-driven hydrosilylation to obtain the alcohol as the final product (Figure 5.11c) [52].

More recently, several research groups have focused their efforts on the development of new organic methodology to address more challenging organic

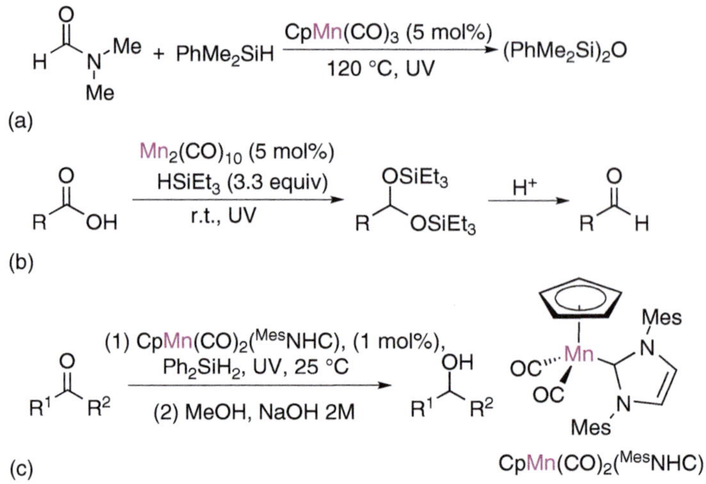

Figure 5.11 Selected Mn-catalyzed hydrosilylation of carbonyl compounds under photoirradiation. (a) Reduction of DMF-catalyzed bt $CpMn(CO)_3$. (b) Hydrosilylation of carboxylic acids catalyzed by $Mn_2(CO)_{10}$. (c) Reduction of ketones catalyzed by $CpMn(CO)_2(^{Mes}NHC)$.

transformations such as C—C bond formation, C–H activation, and cyclization reactions using either photocatalysis or electrocatalysis.

In 2018, Nagib and coworkers reported a new visible light-driven strategy for the synthesis of vinyl iodides via ketyl radical reactivity using the readily available $Mn_2(CO)_{10}$ catalyst [53]. This methodology allows obtaining a broad scope of synthetically useful α-acetoxy iodides with high Z-selectivity in one step by the coupling of an aldehyde and alkyne in the presence of an acyl iodide under mild conditions. Contrary to the traditional ketyl radical reactivity, this new strategy is based on a redox-neutral atom transfer catalysis, in which visible light plays a role in the homolysis of $Mn_2(CO)_{10}$ to give the radical intermediate $Mn(CO)_5$. This $17e^-$ species can abstract an iodine atom ($I^{\cdot}$) through an atom transfer radical addition (ATRA) mechanism from the *in situ* generated α-acetoxy iodide ($BDE_{C\text{-}I}$ = 58 kcal/mol), giving $Mn(CO)_5I$ (Figure 5.12). The resulting protected ketyl radical intermediate couples with the alkyne reagent affording a vinyl radical. The final product ($BDE_{C\text{-}I}$ = 61–68 kcal/mol) is formed by one-electron oxidation of the vinyl radical with $Mn(CO)_5I$ regenerating the active $Mn(CO)_5$ radical intermediate. If we consider the Bond Dissociation Energy (BDEs) of bonds cleaved and formed, the reaction is exothermic. The scope of the reaction includes the coupling of aliphatic aldehydes with substituted alkynes with broad electronic nature including silanes, boronates, aryl alkynes, 1,3-enynes, and propargyl esters. This methodology also tolerates alcohols, ketones, and unsubstituted triple bonds tethered to the alkyne reagent. Substituted aldehydes containing diverse functionalities such as esters, amides, and alkyl halides also gave good results. Finally, the reactivity was expanded to ketones in place of the aldehyde coupling partner and alkenes instead of alkynes, and even more complex molecules such as a vitamin E derivative could be successfully functionalized. The photochemical reactivity of $Mn_2(CO)_{10}$ was used later by Wang and coworkers in a photo-initiated Mn-catalyzed Giese addition of alkyl iodides to alkenes (*vide infra*).

Lynam and coworkers demonstrated that light irradiation is a key factor in the Mn-catalyzed C–H functionalization of 2-phenylpyridine [54]. To this end, they performed a mechanistic study of the C—C bond formation step based on time-resolved IR spectroscopy starting with $Mn(ppy)(CO)_4$ as a model catalyst. Photolysis of the metallated tetracarbonyl complex under UV light irradiation (355 nm) promotes the ultrafast (<0.5 ps) liberation of one of the CO ligands generating a coordination vacancy in cis-position to the aryl ligand. In the presence of a coordinating substrate (S), such as phenylacetylene, this vacancy is occupied, forming a *fac*-$[Mn(ppy)(CO)_3(S)]$. Coordination of the substrate to the Mn by the triple bond leads to the alkyne insertion to the $Mn—C_{ppy}$ bond resulting in the formation of a new C—C bond. A similar reaction mechanism is also operative when using acrylates, isocyanates, and benzaldehyde as substrates.

The same year, Ackerman and coworkers reported a flow-chemistry approach for the Mn-catalyzed C–H arylation of benzene and other (hetero)arene substrates at room temperature under visible light irradiation [55]. The inexpensive $CpMn(CO)_3$ complex displayed good catalytic performance in the C–H arylation of benzenes, furanes, thiophenes, and pyrroles using different substituted aryldiazonium salts as coupling partners. Mechanistic studies suggest that blue light is needed for the generation of an excited-state adduct between an Mn-arene intermediate and the aryl diazonium salt. This excited-state adduct undergoes intramolecular ET delivering aryl radicals that can attack the (hetero)arene to yield the final product after one-electron oxidation and deprotonation. The methodology proved useful in the gram-scale synthesis of a biaryl product (64% yield). It was also applied for the preparation of the commercial drug dantrolene.

Later, Wang and coworkers reported the Mn-catalyzed addition of alkyl iodides to Michael acceptors initiated by visible light [56]. As in the previous example by Nagib and coworkers, the authors take advantage of the photochemical reactivity of $Mn_2(CO)_{10}$, which dissociates upon visible light irradiation forming the $Mn(CO)_5$ radical intermediate. In this reaction, the authors propose the use of a Hantzsch ester (HE) as a sacrificial reducing agent. The reaction methodology performs efficiently (up to 92% isolated yield) with primary, secondary, and tertiary alkyl iodides. It tolerates different functional groups such as alkyl chloride, alkyl and aryl iodide, CF_3, CO_2Et, and TMS. The yield decreased with benzyl iodides due to the lower reactivity of the resulting benzyl radical. Other electrophilic olefins apart from acrylates (e.g. α,β-unsaturated malonate, maleimide, acrylamide, and phenyl vinyl sulfone) delivered the desired coupling product in synthetically useful yields. Simple but very informative reactivity experiments allowed for the indirect detection of C-centered radicals. Moreover, the incorporation of deuterium in the final product was observed only when a deuterated Hantzsch ester was used, suggesting that this is the main hydrogen atom source in the reaction mechanism (Figure 5.13).

Nonetheless, the electrochemical organic reactivity has emerged as a promising strategy for the functionalization of organic molecules under mild conditions [7, 25].

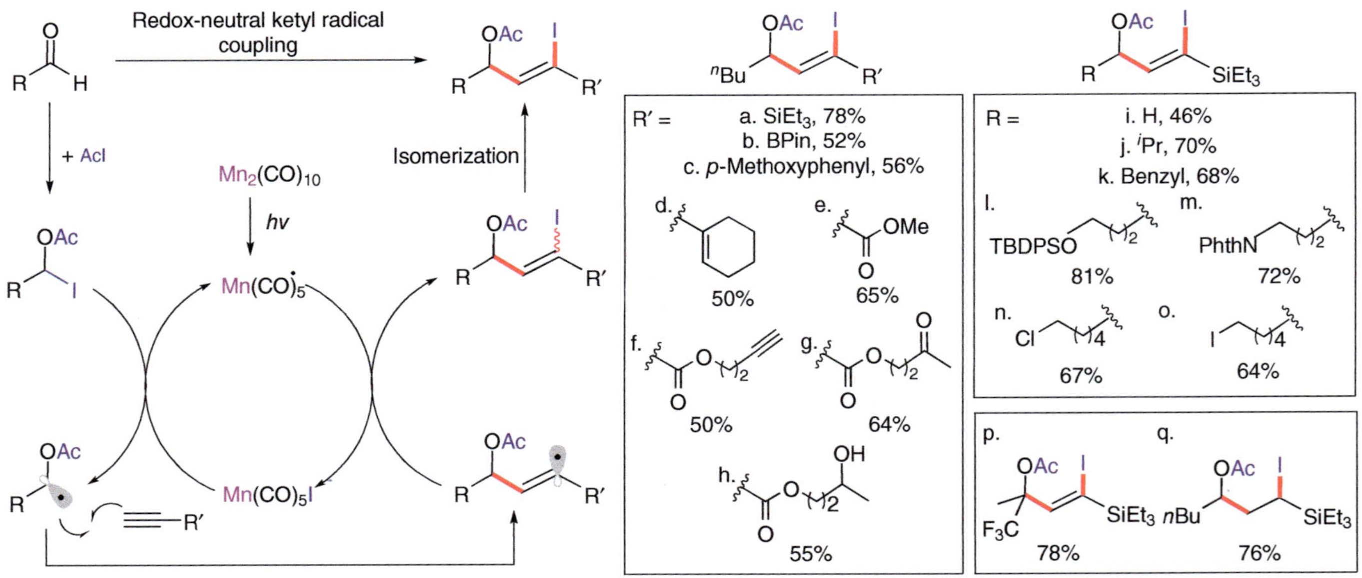

Figure 5.12 (right) Photochemical ketyl radical reactivity via atom transfer catalysis with $Mn_2(CO)_{10}$. (top) Reaction mechanism. (left) Selected examples and yields.

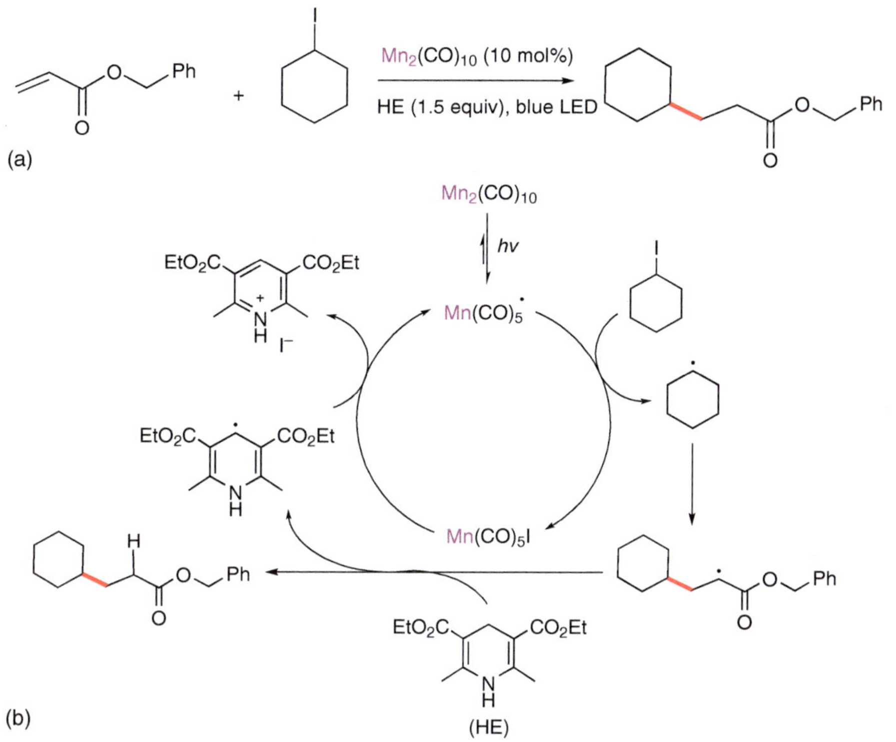

Figure 5.13 Light-initiated Giese addition of alkyl iodides to electron-poor olefins catalyzed by $Mn_2(CO)_{10}$. (a) Reaction conditions. (b) Mechanistic proposal.

In this regard, the group of Lin and coworkers have reported a large variety of useful electrocatalytic organic transformations. Their approach is relevant in terms of synthetic complexity, but, at the same time, they use inexpensive and straightforward catalysts, primarily based on Mn [57]. In 2017, they reported a new electrochemical methodology for the diazidation of alkenes using the inexpensive $MnBr_2{\cdot}4H_2O$ and NaN_3 as catalyst and azide source, respectively [58]. The diazides obtained can give access to vicinal diamines that are useful precursors for the synthesis of bioactive products. This methodology has an excellent tolerance (up to 98% yield) to different functional groups such as alcohols, aldehydes, tertiary amines, sulfides, epoxides, esters, and carboxylic acids, among others. Moreover, it demonstrated the utility in the gram-scale syntheses of 2,3-dimethylbutane-2,3-diazide with high yields and with no need from a chromatographic purification. The reaction takes place in the anode at $E_{anode} \approx 0.5$ V vs. $Fc^{+/0}$ (U_{cell} = 2.3 V). First, the Mn center catalyzed the oxidation of N_3^- to $N_3^{\cdot}$ (in the form of $[Mn^{III}]$-N_3), following by the consecutive addition of $N_3^{\cdot}$ to the olefin to give a C-centered radical. The subsequent addition of $N_3^{\cdot}$ to generated C-centered radical gives the final product (Figure 5.14).

Taking the diazidation mechanistic proposal as a starting point, the Lin group developed other difunctionalization reactions. For instance, the same year, they reported a new chemoselective and stereoselective electrochemical method for

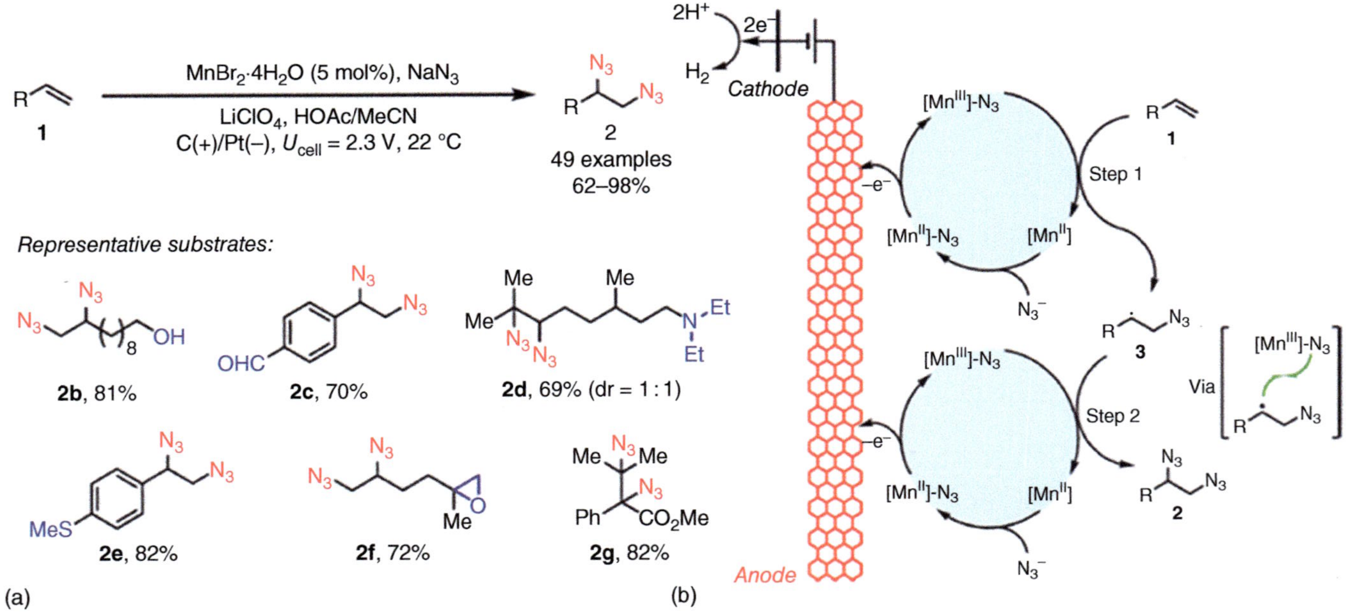

Figure 5.14 Mn-catalyzed electrochemical diazidation of alkenes. (a) Reaction conditions and selected examples of the reaction scope. (b) Mechanistic proposal. Source: Siu et al. [57].

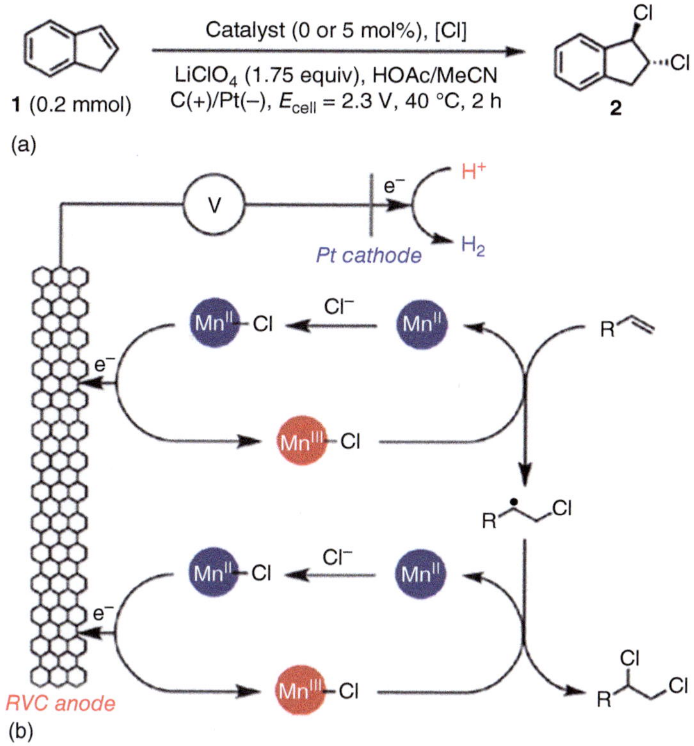

Figure 5.15 Mn-catalyzed electrochemical dichlorination of alkenes. (a) Reaction conditions. (b) Mechanistic proposal. Source: Fu et al. [59].

the dichlorination of alkenes, using $Mn(OTf)_2$ as catalyst (Figure 5.15) [59]. The chloride source in this transformation is $MgCl_2$ and, therefore, toxic electrophilic chloride sources such as Cl_2 are avoided. Moreover, alternative selectivity patterns are obtained to the ones obtained by traditional electrophilic chlorination reactions. As the previously reported diazidation reaction, the electrochemical dichlorination method is based on the anodic generation of radicals at a cell potential of 2.3 V vs. $Fc^{+/0}$ (E_{anode} = 0.94 V). This methodology gave excellent results (up to 97% yield) for the trans-dichlorination of styrene derivatives with different substitution patterns as well as for the difunctionalization of aliphatic olefins. A wide variety of functional groups were tolerated such as alkyl bromides, esters, and oxidatively sensitive groups, e.g. alcohols, aldehydes, carboxylic acids, sulfides, pyridines, and tertiary amines.

To increase the structural and functional complexity of the reaction products, the same group developed a strategy called anodically coupled electrolysis (ACE). This strategy allows to perform the heterodifunctionalization of alkenes under electrochemical conditions. This process is based on the generation of two different types of radical intermediates in the anode surface: a transient radical and a persistent radical intermediate in combination with the appropriate metal catalyst. The first transformation developed using the ACE strategy was the chlorotrifluoromethylation of alkenes [60]. The introduction of the CF_3 functionality has a broad interest in both material science and medicinal chemistry.

Moreover, the incorporation of a vicinal C_{sp3}—Cl bond in the final molecule allows the concomitant functionalization of the final product. In this particular reaction, $CF_3^{\cdot}$ is the transient radical, whereas the persistent one is [Mn^{III}]-Cl. According to the mechanistic proposal (Figure 5.16), the transient $CF_3^{\cdot}$ species is formed in the anodic surface. Then, $CF_3^{\cdot}$ attacks to the olefin generating a C-centered radical intermediate that undergoes chlorination after a bimolecular reaction with [Mn^{III}-Cl]. This methodology provided good to excellent yields for the heterodifunctionalization of styrenes, linear, and cyclic alkenes (in this case favoring the trans-isomer). Substrates containing nitrogen heterocycles were compatible with the reaction conditions. As a proof of concept, the ACE methodology was expanded to the alkynes chlorotrifluoromethylation and styrenes bromotrifluoromethylation.

Further efforts were focused on the development of a tandem cyclization–difunctionalization of ene–ynes to synthesize chlorotrifluoromethylated pyrrolidines using very similar reaction conditions (Figure 5.17) [61]. However, the addition of 2,2′-bipyridine was crucial to obtain good stereoselectivity. The chlorination at the less hindered face of the C_{sp2}-centered radical is favored.

Later on, the same group reported the use of the ACE strategy to forge a C–C vicinal to a C–halogen bond in a single step via a Mn-catalyzed chloroalkylation of alkenes [62]. Under very similar conditions to the previous ACE reactivity, $Mn(OTf)_2$ was employed as precatalyst, 2,2′-bpy as a ligand, malononitrile as alkylating agents, and NaCl as chlorine source. In this reaction scheme, the persistent radical is the carbon-centered obtained from the anodic oxidation of the activated carbonyl compound, and the persistent radical is [Mn^{III}]-Cl (Figure 5.18), as in the previous chlorotrifluoromethylation reaction. The use of NaCl as a chlorine source instead of $MgCl_2$ was the key to obtain good selectivity of the desired product vs. the competing 1,2-dichlorination of the alkene. This electrochemical methodology allowed for the effective chloroalkylation of styrene derivatives functionalized with different groups such as formyl or benzyl chloride. Aliphatic alkenes were suitable substrates for this heterodifunctionalization reaction. In the case of cyclic internal alkenes, the trans-chloroalkylated product was preferentially obtained. Replacing NaCl by NaBr, the corresponding bromoalkylation product was obtained in moderate yields. Different substituted malononitriles, and also methyl 2-benzylcyanoacetate, proved to be useful alkylating agents in the chloroalkylation of styrenes. Acetylacetone and acetylacetates also engage reactivity as alkylating agents with indene delivering 1,2-dihydrofuran from the oxyalkylation reaction even in the absence of NaCl.

Finally, the same type of reactivity was observed in the presence of a diphenylphosphine oxide to obtain the chlorophosphinoylation product [63]. In this process, the same conditions were used as in the previous heterodifunctionalization reactions, i.e. $Mn(OTf)_2$ as a catalyst and chloride salt as chlorine source, that is, LiCl. In this case, the transient radical is the one generated after the anodic oxidation of diphenylphosphine oxide and [Mn^{III}]-Cl is the persistent radical. The method tolerates the same kind of alkene substrates as in the other heterodifunctionalization reactions going from moderate to excellent yields.

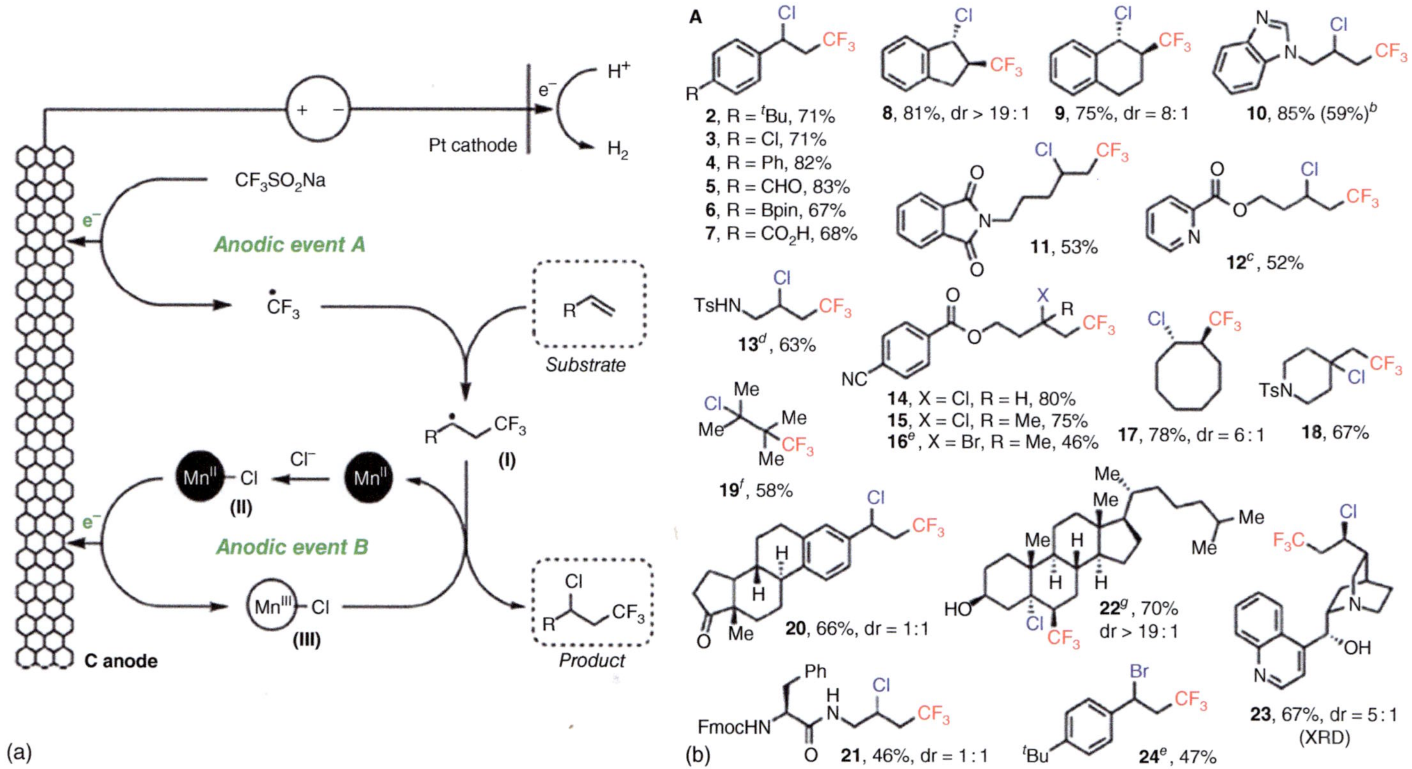

Figure 5.16 Mn-catalyzed chlorotrifluoromethylation of alkenes. (a) Mechanistic proposal. (b) Reaction scope. Source: Ye et al. [60].

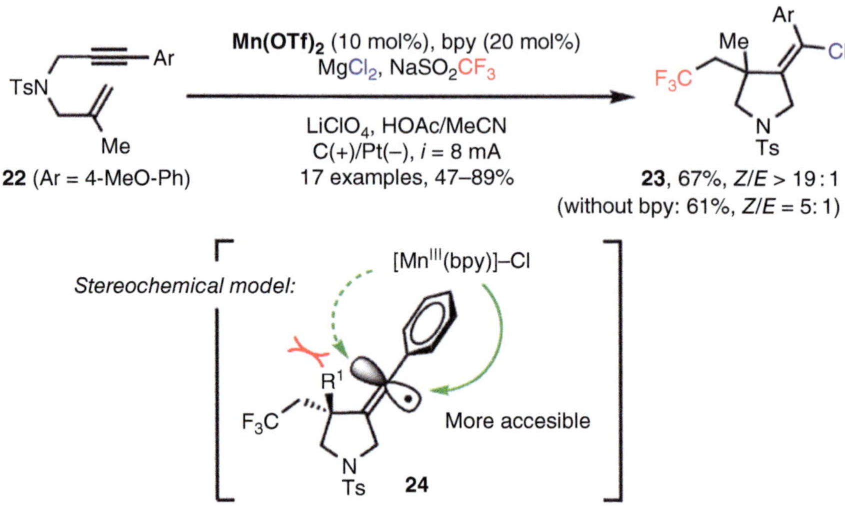

Figure 5.17 Mn-catalyzed tandem cyclization–chlorotrifluoromethylation of ene–ynes. Source: Siu et al. [57].

Other groups have also contributed with valuable electrochemical organic transformations catalyzed by Mn compounds [64]. For instance, similar conditions to the ones already mentioned (hetero)difunctionalization reactions were also suitable for the deconstructive chlorination of cycloalkanols [65]. In this work, Morrill and coworkers proposed the transformation of cyclopropanols and cyclobutanols (40 examples) to synthetically useful β- and γ-chlorinated ketones [65]. This reaction is based in the anodic generation of transient alkoxy radicals using $Mn(OTf)_2$ and $MgCl_2$ as catalyst and chlorine source, respectively. The strained alkoxy radical undergoes ring opening to form the corresponding carbon-centered radical, which reacts with $[Mn^{III}]$-Cl (persistent radical), giving the desired product (Figure 5.19). Electrochemistry in flow coupled to a continuous inline purification allowed to perform a gram-scale reaction in almost 80% yield.

An alternative strategy to the ACE is the paired electrolysis in which the products from the two half-reactions (anodic oxidation and cathodic reduction) collapse to form the target molecule. In this regard, Chen and coworkers reported a practical synthesis of chloroacetophenones by the oxychlorination of styrenes using $MnCl_2$ and $MgCl_2$ as electrocatalyst and chlorine source, respectively [66]. This methodology is based in the coupling of the Mn-catalyzed anodic chlorination of olefins and the O_2 reduction over a carbon cathode. Following a similar scheme as in the ACE strategy, the $[Mn^{III}]$-Cl persistent radical formed in the anode reacts with styrene to yield a transient chlorinated benzyl radical. At the same time, superoxo radicals (persistent) are formed in the cathode surface. A bimolecular reaction between the transient benzyl radical and the superoxo species gives a benzyl alcohol intermediate that evolves to the desired chloroacetophenone after a second oxidation event (Figure 5.20).

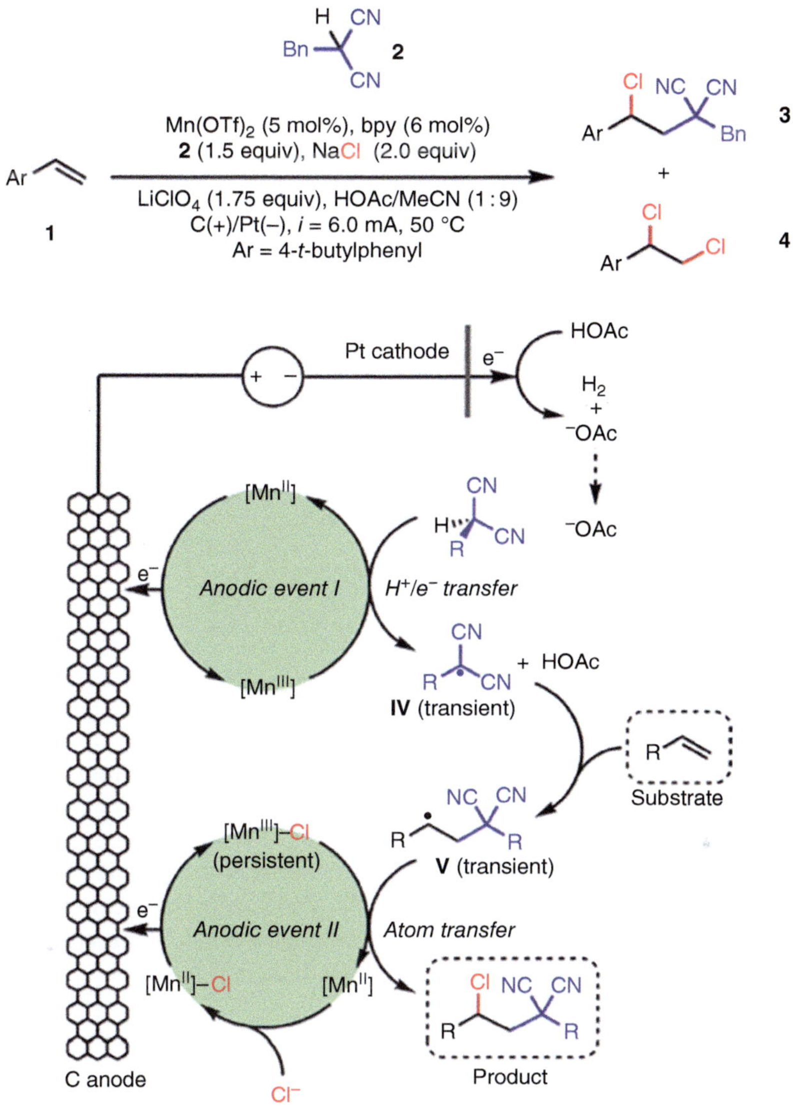

Figure 5.18 Mn-catalyzed chloroalkylation of alkenes under ACE conditions. Source: Fu et al. [62].

5.4 Mn-Catalyzed CO_2 Reduction

While CO_2 is utilized industrially for some transformations, it is generally viewed as a waste product and greenhouse gas [67, 68]. The catalytic conversion of CO_2 into valuable products such as liquid fuels not only will decrease the CO_2 release into the atmosphere and reduce its environmental impacts but will also lead to renewable chemicals and fuels from CO_2, if the energy used for CO_2 transformation is coming from renewable sources, such as solar power [69]. Electrochemical CO_2 reduction

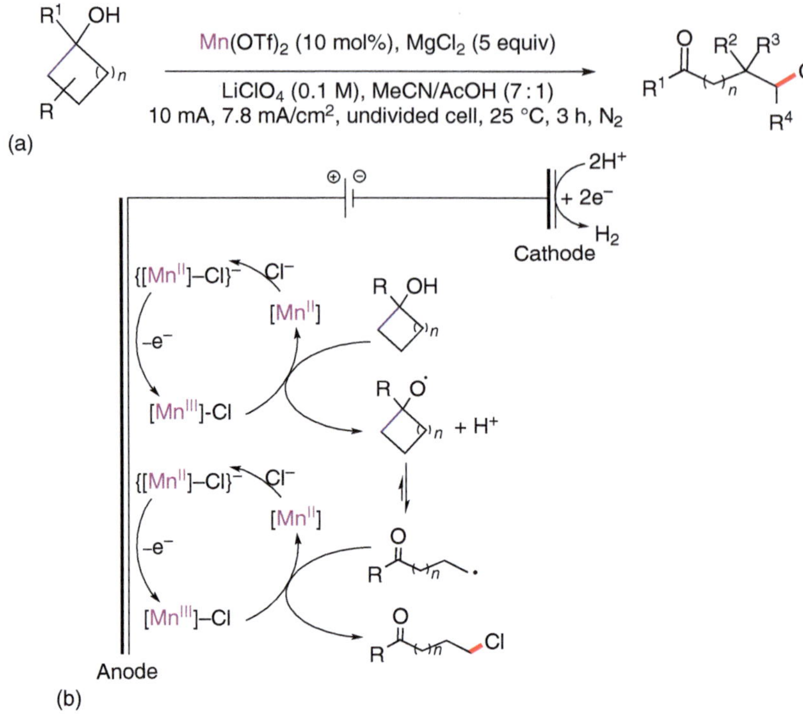

Figure 5.19 Manganese-catalyzed electrochemical deconstructive chlorination of cycloalkanols via alkoxy radicals. (a) General batch reaction conditions. (b) Mechanistic proposal.

is particularly attractive because CO_2 could be transformed into valuable products using electricity as an input. At the same time, CO_2 photocatalysis reduction is an appealing approach to accomplish its sunlight conversion. Thus, simultaneously, abundant, clean, and renewable solar energy is stored in chemicals, helping the transition toward more sustainable energy sources [70].

5.4.1 Electrocatalytic CO_2 Reduction with Manganese-Based Catalysts

The electrocatalytic reduction of CO_2 has less thermodynamic and kinetic constraints when it is coupled to proton transfer reactions (Table 5.1) [71]. However, competitive reduction of protons to H_2 can occur at similar potentials to the CO_2 reduction pathways. For instance, at pH 7, the thermodynamic potential for reducing protons is only 100 mV less negative than the reduction of CO_2 to CO. Consequently, a suitable catalyst must be highly selective.

In general, metal complexes are good catalyst candidates for the CO_2 reduction because of their redox and coordination flexibility. The reduction of a metal complex, at the right low oxidation state, facilitates the appearance of coordination vacancies, leading access to binding positions for oxidizing substances such as CO_2. In general, the CO_2 binding to the reduced metal center triggers its activation [72].

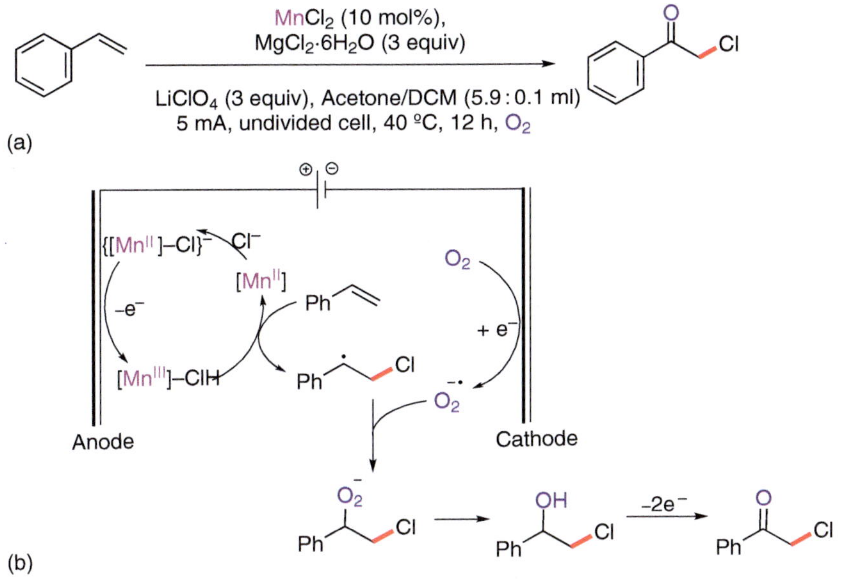

Figure 5.20 Manganese-catalyzed electrochemical oxychlorination of styrenes. (a) General batch reaction conditions. (b) Mechanistic proposal.

Table 5.1 Reduction potentials for selective CO_2 reduction products.

	$E°$ (*V* vs. NHE) at pH 7
$CO_2 + 2H^+ + 2e^- \rightarrow CO + H_2O$	−0.53
$CO_2 + 2H^+ + 2e^- \rightarrow HCO_2H$	−0.61
$CO_2 + 4H^+ + 4e^- \rightarrow H_2CO + H_2O$	−0.48
$CO_2 + 6H^+ + 6e^- \rightarrow CH_3OH + H_2O$	−0.38
$CO_2 + 8H^+ + 8e^- \rightarrow CH_4 + H_2O$	−0.24
$CO_2 + e^- \rightarrow CO_2^{\bullet -} + H_2O$	−1.90
$2H^+ + 2e^- \rightarrow H_2$	−0.42

Subsequent ET event to the metal complex and proton transfer to the CO_2 moiety bonded to the metal drives the C–O cleavage. There are several reviews on electro- and photo-catalytic CO_2 reduction with coordination complexes [73–77].

Among the reported metallic complexes, Mn carbonyl ones are excellent catalysts for CO_2 reduction. Moreover, Mn is attractive for large-scale applications, like CO_2 reduction to fuels, since it is one of the most abundant metals in Earth's crust and is biocompatible [78]. There are already some reviews that cover part of the Mn-based electro- [79, 80] or photocatalytic activity [6, 81]. However, herein, we will focus on a selection of Mn-based catalysts for CO_2 reduction to illustrate their versatility in terms of ligand framework and access to mechanistic information and to develop heterogenized systems.

5.4.1.1 Pyridine-Based Complexes

Chardon-Noblat and coworkers firstly reported that highly active carbonyl polypyridyl manganese complexes act as electrocatalysts for the reduction of CO_2 in the presence of a proton source [82, 83]. They demonstrated that *fac*-$[Mn(L)(CO)_3X]^n$ complexes (L = 2,2′-bipyridine (bpy) or 4,4′-dimethyl-2,2′-bipyridine (dmbpy) for X = Br^-, $n = 0$) (**1a** and **1b**, Figure 5.21A) exhibit an excellent efficiency and selectivity of CO_2-to-CO electroreduction in the presence of protons [82]. Since the first report, a large collection of Mn systems has been described. In the *fac*-$[Mn(L)(CO)_3X]^n$ family of complexes, the Mn center is typically in the Mn^I oxidation state, L is a bidentate polypyridyl or related ligand, and X is a neutral or anionic monodentate ligand, e.g. CH_3CN ($n = 1$) or Br^- ($n = 0$).

The cyclic voltammograms (CVs) for the *fac*-$[Mn(bpy)(CO)_3Br]$ showed two single irreversible electron reduction waves that were assigned to the $Mn^{I/0}$ and $Mn^{0/-I}$ redox systems (Figure 5.21B). In the presence of CO_2, the CV slightly changes, while the subsequent addition of water, as a proton source, produced an intense current enhancement, corresponding to catalytic CO_2 reduction. CPE (four hours) in CH_3CN and 5% H_2O at −1.40 vs. SCE showed quantitative faradaic efficiency (FE) for production of CO. However, the FE dropped slightly at extended CPE (FE_{CO} = 85% in 22 hours), due to the formation of H_2 (FE_{H2} 15%). UV–Vis recorded during the bulk reductive electrolysis showed the formation of the $[Mn^0(bpy)(CO)_3]_2$ dimer and the unstable $[Mn^{-I}(bpy)(CO)_3]^-$, in agreement with spectroscopic results previously reported [86–88]. The proposed mechanism involves the formation of the $[Mn^{-I}(bpy)(CO)_3]^-$ that reacts with CO_2. Later on, based on pulsed-EPR complemented with DFT studies, the authors also proposed that the Mn^0–Mn^0 dimer could also react with CO_2, although at slower reaction rates [83]. The oxidative addition of CO_2 and H^+ over the Mn^0–Mn^0 dimer can lead to a Mn^{II}-carboxylic acid intermediate, which finally gets reduced with a second electron and protonated to evolve a CO and an H_2O molecule (Figure 5.22) [89, 90].

Kubiak's group studied the influence of weak acids on the catalytic activity of $[Mn(bpy\textit{t}Bu)(CO)_3X]^n$ (where X: Br^- [n = 0], CH_3CN [n = 1]) (**2a** and **2b**, Figure 5.21A) [84]. They found a direct correlation between acid concentration and its strength (H_2O, MeOH, and TFE) with the increase of the CV catalytic current. Bulk electrolysis of **2a** in CH_3CN with TFE (1.4 M at −2.2 V vs. SCE), gave a turnover frequency (TOF) value for CO of 340 s^{-1} with quantitative FE. To gain further insights on the mechanism, they performed IR-SEC of $[Mn(bpy\textit{t}Bu)(CO)_3Br]$, suggesting that upon the first reduction, the manganese complex rapidly dimerizes to form a $[Mn^0(bpy\text{-}\textit{t}Bu)(CO)_3]_2$, which breaks upon further reduction to yield the catalytically active anionic species (Figure 5.21C) [84, 91].

To prevent dimerization, Sampson et al. introduced bulky substituents at the ortho-positions of the bpy ligand [85]. In particular, they reported the $[Mn(mesbpy)(CO)_3Br]$ (mesbpy = 6,6′-dimesityl-2,2′-bipyridine) complex (**3**, Figure 5.23). Interestingly, the Mn complex exhibited a single two-electron

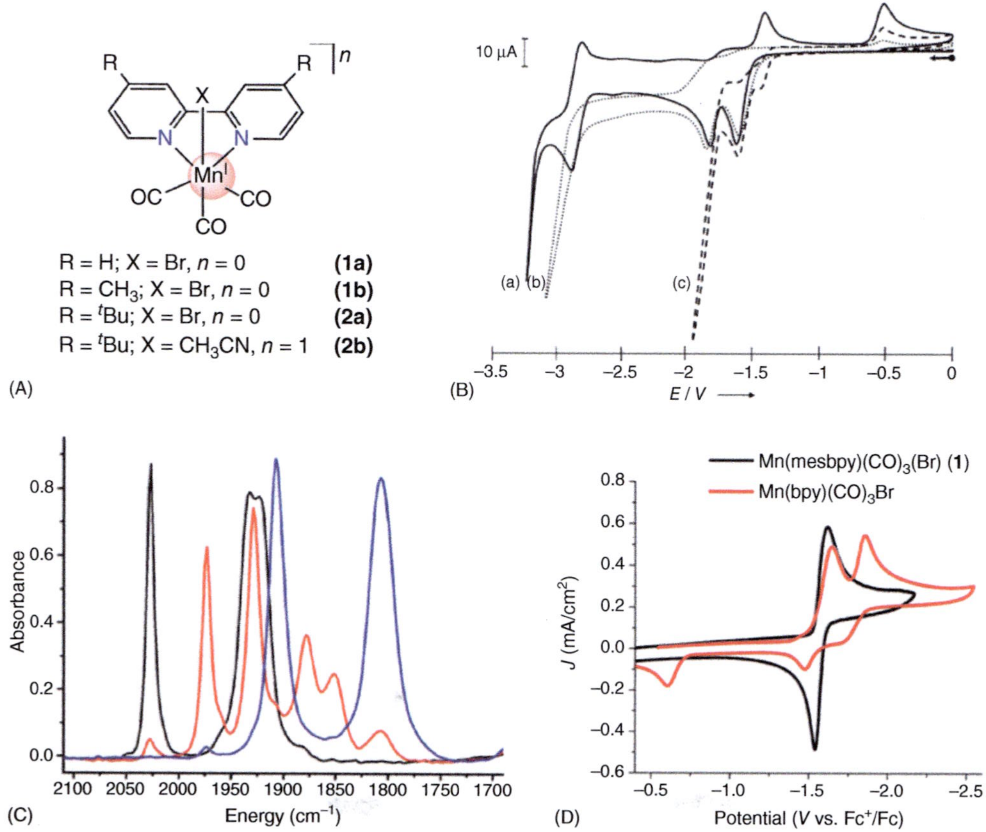

Figure 5.21 (A) Selected carbonyl bipyridine manganese complexes. (B) Cyclic voltammetry of **1a** (1 mM) in CH_3CN/TBAP (0.1 M) at $v = 100$ mV/s vs. Ag/$AgNO_3$ (a) under Ar and under CO_2 (b) without and (c) with 5% of H_2O. Source: (a, b) Based on Bourrez et al. [82]. (C) Overlap of the species observed in IR-SEC of complex [Mn(bpy*t*Bu)$(CO)_3$Br] (black) in CH_3CN under Ar, the monoreduced [Mn(bpy*t*Bu)$(CO)_3]_2$ (red), and double reduced [Mn(bpy*t*Bu)$(CO)_3]^-$ (blue) species. Source: Smieja et al. [84]. (D) Comparison of the CV of [Mn(bpy)$(CO)_3$Br] (red) and [Mn(mesbpy)$(CO)_3$Br] (black) (1 mM) in CH_3CN/TBAP (0.1 M) at $v = 100$ mV/s under N_2. Source: Sampson et al. [85].

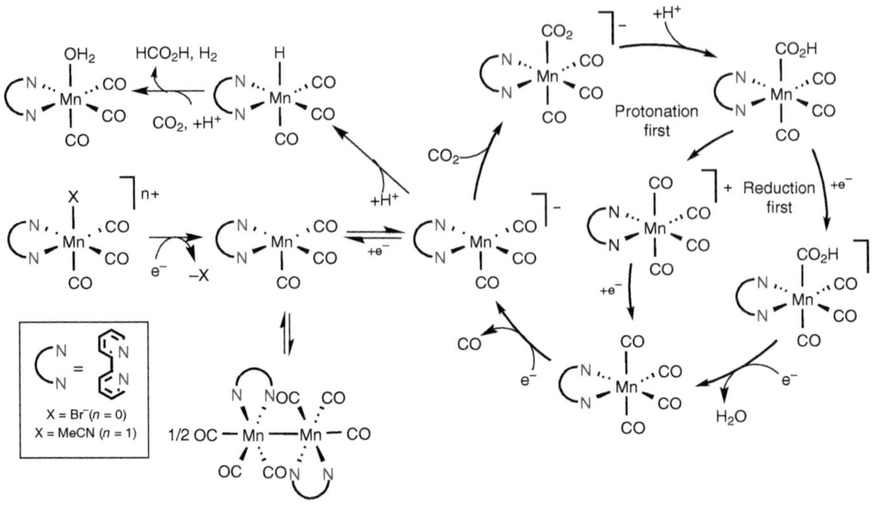

Figure 5.22 Proposed mechanisms for the electrocatalytic CO_2 reduction to CO and HCOOH.

reversible reduction wave under N_2 (Figure 5.21D). The anionic complex, $[Mn(mesbpy)(CO)_3]^-$, was formed at 300 mV less negative potential in comparison with analogous $[Mn(bpy)(CO)_3Br]$. However, the reaction with CO_2 does not occur until 400 mV regarding the redox feature. The catalytic activity was remarkable with a maximum TOF of $5 \times 10^3\ s^{-1}$ (CH_3CN with 1.4 M TFE), and high FE for CO_2-to-CO reduction. The IR-SEC and chemical prepared reaction intermediates and corroborated the formation of $[Mn(mesbpy)(CO)_3]^0$, $[Mn(mesbpy)(CO)_3]^-$, with no evidence of dimerization. Besides, $[Mn(mesbpy)(CO)_3][K(18\text{-crown-}6)(THF)]$ was characterized by single-crystal X-ray crystallography (Figure 5.24). The corresponding cationic $[Mn(mesbpy)(CO)_3(MeCN)]^+$ was afterwards successfully used in H_2 evolution reactions [85].

Replacing the Brønsted acid by a Lewis acid [84, 85, 92] induces a modification of the catalytic mechanism for the CO_2 reduction. Therefore, when using Mg cations, $[Mn(mesbpy)(CO)_3Br]$ binds CO_2 at significantly reduced overpotentials (η = 0.3–0.45 V) and increases the rate of catalysis by 10-fold. The catalytic mechanism based on variable concentration, cyclic voltammetry, infrared spectroelectrochemistry, and bulk electrolysis studies indicated that, in this case, the catalysis proceeds via a reductive disproportionation reaction of $2CO_2 + 2e^-$ to give CO and CO_3^{2-} (Figure 5.25). Besides the first CO_2 molecule that binds the Mn, a second CO_2 molecule plays the role of acid as oxygen abstractor and helps to cleave the C—O bond in the postulated Mn^I–CO_2–Mg intermediate adduct. A TOF of $630\ s^{-1}$ for CO could be obtained at an overpotential as small as c. 350 mV, showing the boosting effect of the Lewis acid on the catalytic process.

In addition to the steric effect of the mesbpy ligand that prevents the dimerization [85], Grills and coworkers tested {*fac*-$Mn^I([(MeO)_2Ph]_2bpy)(CO)_3(CH_3CN)$}(OTf) (where $[(MeO)_2Ph]_2bpy$ = 6,6′-bis(2,6-dimethoxyphenyl)-2,2′-bipyridine) complex

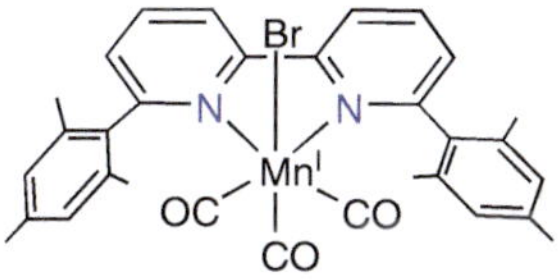

(3)

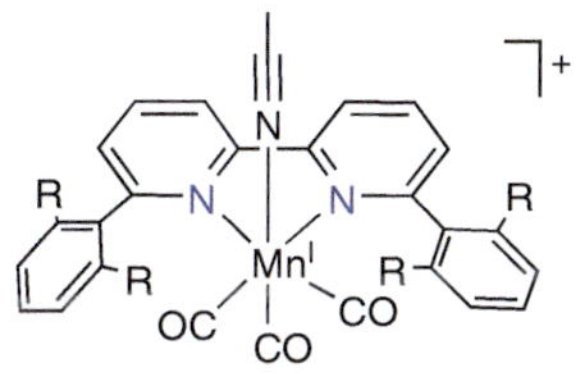

R = OCH_3

(4)

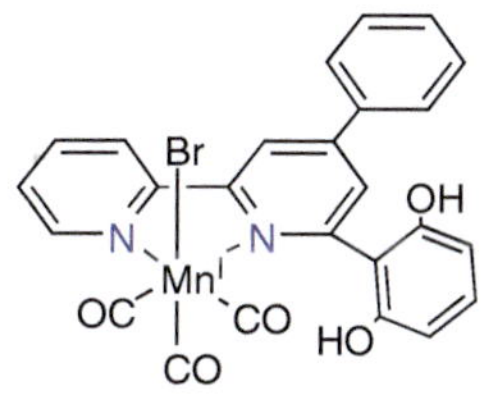

(5)

(6)

R = OH **(7a)**
R = Me **(7b)**
R = NEt_2 **(7c)**

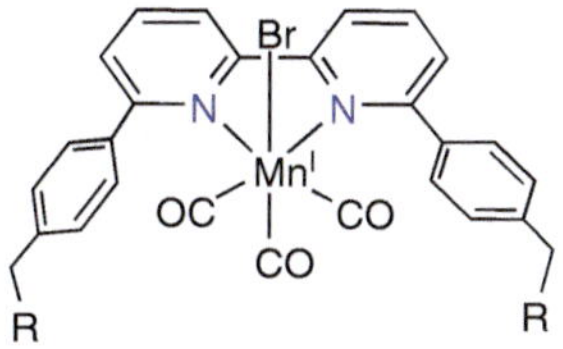

R = NEt_2 **(7d)**

Figure 5.23 Selected carbonyl bipyridine manganese complexes with ortho substituents introducing a steric influence (**3, 7b**), weak hydrogen bonding interactions (**4**), and pendant base (**5, 6, 7a, 7c, 7d**).

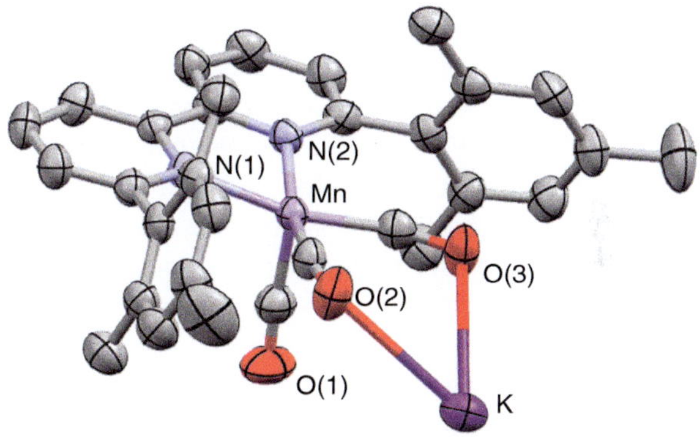

Figure 5.24 Molecular structure of [Mn(mesbpy)$(CO)_3$][K(18-crown-6)(THF)], hydrogens, 18-crown-6, and THF have been omitted for clarity. Ellipsoids are set at the 50% probability level. Source: Sampson et al. [85].

with an extra functionality – *ortho*-MeO groups at the phenyl rings (**4**, Figure 5.23) [93]. The four pendant methoxy groups of $[(MeO)_2Ph]_2$bpy ligand introduce an electronic influence combined with a weak allosteric hydrogen-bonding interaction that significantly lowers the activation barrier for C—OH bond cleavage from the metallocarboxylic acid intermediate. This system presents the first example of a protonation-first pathway and comprises the protonation of the $\{Mn^I\text{-}CO_2\}$ to give $\{Mn^I\text{-}COOH\}$, followed by its reduction, thus minimizing the overpotential requirement. PhOH and TFE as proton sources exhibited the largest protonation-first catalytic currents saving up to 0.55 V in overpotential with respect to the thermodynamically demanding reduction.

Another example of the effect of a local proton source on the activity of bipyridine Mn complexes was investigated by Gobetto and coworkers [94]. The electrochemical

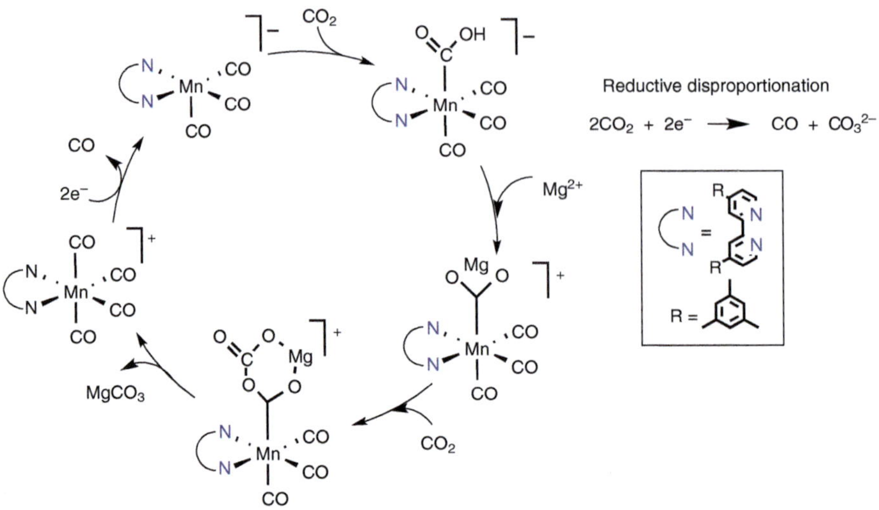

Figure 5.25 Proposed mechanism for the electrocatalytic reductive disproportionation reaction of CO_2.

behavior of *fac*-[Mn(dhbpy)(CO)$_3$Br] (dhbpy = 4-phenyl-6-(1,3-dihydroxybenzen-2-yl) 2,2′-bipyridine) complex (**5**, Figure 5.23) contains two acidic OH groups in the proximity of the metal center that act as an effective local proton source. This study was the first experimental evidence of an intramolecular proton-assisted catalytic process for a Mn^{I}-carbonyl catalyst. Complex **5** showed substantial catalytic activity, and CPE at −1.8 V vs. SCE under catalytic conditions showed an unusual change in selectivity for CO_2 reduction, giving a mixture of CO (FE = 70%) and HCOOH (FE = 22%), suggesting an alternative reaction pathway in catalysis. The mechanism involves the formation of a hydride intermediate (Figure 5.22), issued from intramolecular protonation in the reduced catalyst anion that would further insert CO_2 and evolve HCOOH.

Bocarsly and coworkers [95] reported a ligand framework with a phenolic proton near the CO_2 binding site, Mn(HOPh-bpy)(CO)$_3$Br (**6**, Figure 5.23), which allows for facile proton-assisted C—O bond cleavage. The phenolic pendant group, in the second coordination sphere, stabilizes the O atom of CO_2 through hydrogen bonding. In this system, the substitution reduced the overpotential and increased sevenfold the electrocatalytic current compared with the classic [Mn(bpy)(CO)$_3$Br], even with the presence of an added Brønsted acid. CPE produced CO with a FE of 76% after four hours of electrolysis. Experimental and computational studies suggested a key mechanistic role for the phenolic proton from the entropic contribution to the activation of free energy for the increased catalytic activity. In the proposed mechanism, the Mn–CO_2 adduct is stabilized via hydrogen bonding between the phenolic proton and an O atom of the CO_2.

Recently, Daasbjerg and coworkers reported a series of manganese carbonyl complexes with bipyridine and phenanthroline ligands with various pendant groups around the coordination sphere of the Mn (**7**, Figure 5.23) [96].

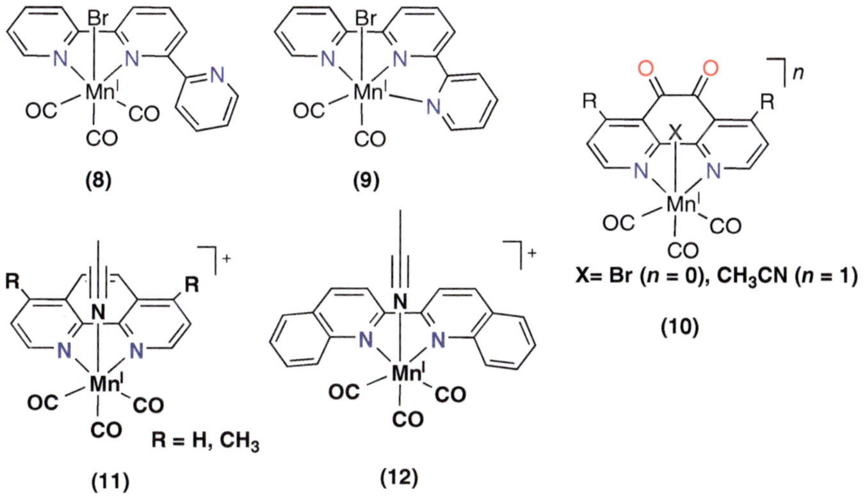

Figure 5.26 Carbonyl pyridine-based manganese complexes as electrocatalyst for the CO_2 reduction.

Among others, they positioned an alcohol (**7a**), a methyl (**7b**), and a dimethylamine group (**7c**) at an equivalent localization at the methoxy groups in the {*fac*-Mn^I([$(MeO)_2Ph]_2$bpy)$(CO)_3(CH_3CN)$}(OTf) complex (**4**). Interestingly, the substituent had a significant effect on the products formed, yielding CO for **7a** and **7b** and HCOOH for **7c**. The strategically positioned tertiary amines close to the metal center were essential for obtaining HCOOH; CO was the main product if there was no amine group (**7a**, **7b**) or amines placed far from the metal center (**7d**). The amine-modified complexes are among the most active catalysts for reducing CO_2 to formic acid, with a TOF of up to $5500\,s^{-1}$ at 630 mV of overpotential. Experimental and theoretical studies revealed the role of the amines as proton shuttles, leading to the formation of Mn–hydride species.

Based on some initial reports on photo- and electrochemical behavior of Mn with a 2,2′ : 6,2″-terpyridine (tpy) ligand [97], Machan and Kubiak reported the electrocatalytic characterization of Mn(tpy) carbonyl complexes with κ^2-*N,N′* (κ^2-tpy) and κ^3-*N,N′,N″*(κ^3-tpy) coordination modes, Mn(κ^2-tpy)$(CO)_3$Br and Mn(κ^3-tpy)$(CO)_2$Br (**8**, **9**, Figure 5.26) [98]. Although the κ^3 form is active for CO_2 reduction, it is susceptible to fast degradation, achieving a maximum TON of 4 before a loss of selectivity. Results suggested that CO ligand replacement requires the use of more strongly binding ligands to stabilize the Mn center with bulky functional groups to inhibit dimerization.

New manganese tricarbonyl complex bearing the redox-active ligand based on the *N,N′*-coordinated 1,10-phenanthroline-5,6-dione (phen-dione), and *fac*-[Mn(phen-dione)$(CO)_3X]^n$ (where X: Br^- [$n = 0$] or CH_3CN [$n = 1$]) (**10**, Figure 5.26) was reported by Chardon-Noblat and coworkers [99]. After a first two-electron ligand-based reduction, the complex can trap two molecules of CO_2, by ECEC reactions. In CH_3CN, the catalyst performed excellent selectivity for catalytic CO_2 reduction to CO or HCOOH through electrochemical or photochemical

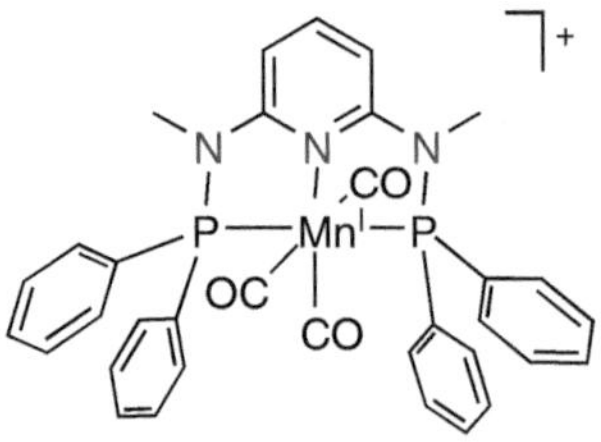

(13)

Figure 5.27 Mn{κ^3-[2,6-{Ph_2PNMe}v_2(NC_5H_3)]}$(CO)_3$Br complex.

activation, respectively. In their study, switching the solvent from DMF to CH_3CN enhanced the selectivity of the reduction reaction toward HCOOH. The presence of the π-accepting redox-active phen-dione ligand led to different reduced intermediates compared with standard 2,2′-bipyridine (bpy, 1a) and 1,1′-phenanthroline (phen, **11**) complexes, with the first two redox processes remaining localized on the ligand.

Systematic variation of the polypyridyl ligand from bpy to phen, to dmphen, and finally to bqn was reported by Rochford and coworkers [100]. The authors concluded that both steric and electronic-based ligand influence the activity of *fac*-[Mn(N^N)$(CO)_3(CH_3CN)]^+$ catalysts for CO_2 reduction. In addition, they highlighted the *fac*-[Mn(bqn)$(CO)_3(CH_3CN)]^+$ (bqn = 2,2′-biquinoline) complexes (**12**, Figure 5.26). The bqn ligand exhibits both steric and electronic influence over the fundamental redox properties of the complex, and, consequently, it is related to the catalytic properties with respect to the activation of CO_2.

In search for new type of Mn catalysts, Richeson and coworkers studied the planar coordinating tridentate PNP ligand κ^3-[2,6-{Ph_2PNMe}v_2(NC_5H_3)] (**13**, Figure 5.27) [101]. The complex [Mn{κ^3-[2,6-{Ph_2PNMe}v_2(NC_5H_3)]}$(CO)_3$Br] presents a unique metal geometry for CO production showing a FE of 96% in dry CH_3CN during more than five hours electrolysis at 960 mV of overpotential. The selectivity for CO dropped during CPE in the presence of 5% of water with a FE of 41% for H_2 evolution, suggesting a competitive pathway in the catalytic mechanism. In contrast to reported Mn electrocatalysts, **13** does not possess a coordinated halide; it has a *mer*-tricarbonyl geometry and a PNP neutral pincer ligand. The presence of carbonate points to a reductive disproportionation of the CO_2.

On the other hand, taking inspiration by the supramolecular assembly of bimetallic species, Kubiak and coworkers studied the co-catalytic effects of a Re–Mn heterobimetallic supramolecular system (**14**, Figure 5.28) [98]. They combined Re- and Mnbpy-based complexes functionalized with methyl acetamidomethyl groups [(M(dacbpy)$(CO)_3$Br)] (M = Re, Mn). The formation of hydrogen bonds between the catalysts generated active bimetallic species that produced ~10% more current under catalytic conditions than the individual Re and Mn complexes. The FE for CO production was 86% through 4 TONs, based on the total moles of added catalyst. FT-IR spectroelectrochemical results supported a reaction pathway where a hydrogen-bonded heterobimetallic dimer is initially formed between [Re^0(dacbpy)$(CO)_3$] and [Mn^0(dacbpy)$(CO)_3$] (Figure 5.28). Under CO_2 and a proton source, Mn binds and reduces CO_2 to yield a hydroxycarbonyl species that can undergo either a protonation or reduction to initiate C—O bond cleavage. The reduced Re species can reduce the Mn hydroxycarbonyl through an ET and then the spontaneous CO release allows the dimer to re-enter the catalytic cycle.

5.4.1.2 Diimine-Based Complexes

Hartl and coworkers evaluated a Mn-diimine complexes containing non-aromatic ligand [Mn(^{i}Pr-DAB)$(CO)_3$X] (DAB = *N*,*N*′-di-R-1,4-diazabuta-1,3-diene, X = halide) (**15**, Figure 5.29) [102]. [Mn_3(^{i}Pr-DAB)(CO)Br] catalytically reduced CO_2 to CO in CH_3CN with 5% of water. In the light of spectroelectrochemical results, they suggested that the electrochemical reduction of the complex produces the 2e$^-$ reduced anion [Mn(^{i}Pr-DAB)$(CO)_3$]$^-$ instead of the dimer [Mn(^{i}Pr-DAB)$(CO)_3$]$_2$. The lower overpotential needed to reduce CO_2 was attributed to the difference in the reduction paths in the catalytic mechanism with respect to the analogous [Mn(R-bpy)$(CO)_3$Br] (R = H, *t*Bu, Me).

Kubiak and coworkers investigated additional *fac*-[(α-diimine)Mn(DAB)$(CO)_3$Br] (DAB = 1,4-diazabutadiene), (**16**–**19**, Figure 5.29) together with their Re analogues [103]. The complexes described poor electrochemical CO_2 conversion to CO but provide insights into the role of redox-active ligands in catalysis. Experimental data suggested that Mn–DAB complexes undergo two sequential one-electron reductions to form the active radical, anionic catalyst. Following the reaction, 1 equiv of CO_2 generates a Mn^I-COOH species, where a second CO_2 equivalent resulted in a disproportionation reaction to form CO and CO_3^{2-}. In contrast, for the Zeng's DAB complexes [102], the formation of [Mn$(CO)_5$]$^-$ was not observed [103].

5.4.1.3 *N*-Heterocyclic Carbene-Based Complexes

NHCs are attractive for catalysis [104] owing to their strong σ-donor and weak π-acceptor properties, which lead to exceptional stability. Modification of the coordinating atom would significantly affect the electronic properties of the whole complex [105]. However, ligands with alternative coordinating moieties have received comparatively little attention in designing new manganese-based catalysts. In this regard, imine–NHC ligands have been successfully integrated to afford *fac*-[Mn(N^C)$(CO)_3$Br] catalysts.

The first examples of manganese catalysts bearing NHC ligands were reported by Agarwal et al. [106]. They replaced one pyridine ring in bpy with an NHC moiety to afford a *fac*-[MnBr(NC)$(CO)_3$]-type catalysts. [Mn(*N*-methyl-*N*′-2-pyridylbenzimidazol-2-ylidine)$(CO)_3$Br] (**20**, Figure 5.30) and [Mn(*N*-methyl-*N*′-2-pyridylimidazol-2-ylidine)$(CO)_3$Br] (**21**, Figure 5.30) reduce CO_2 to CO following two-electron reduction of the Mn^I center, with no H_2 production observed. The 1e$^-$ and 2e$^-$ reduction processes for both catalysts were reported at approximately the same potential (about −1.4 and −1.5 V vs. SCE, respectively). However, the catalytic current was observed at potentials, similar to the classic [Mn(bpy)$(CO)_3$Br].

Complex **20** was used by Agarwal and coworkers [107] as a benchmark for understanding the effect of increased π-acidity of pyrimidine compared with pyridine (**22**,

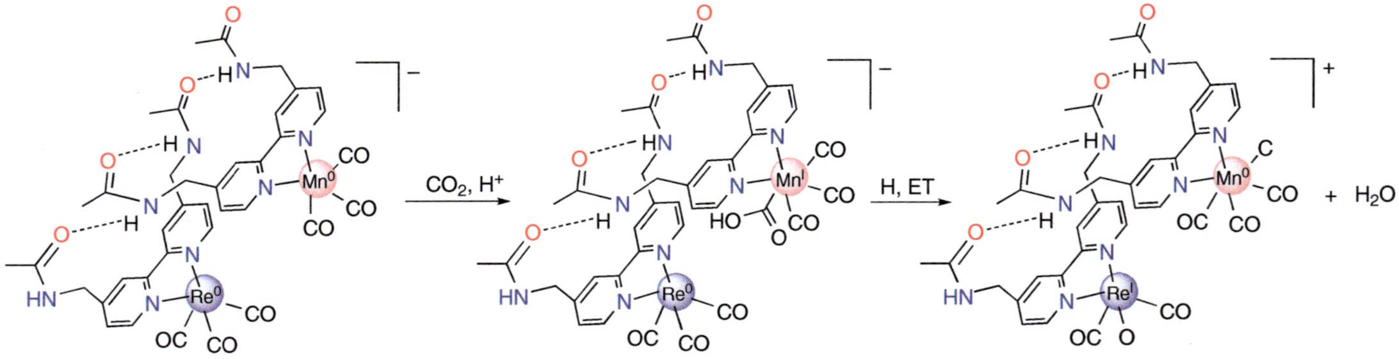

(14)

Figure 5.28 Hydrogen-bonded heterobimetallic Mn–Re active species for the electrocatalytic CO_2 reduction.

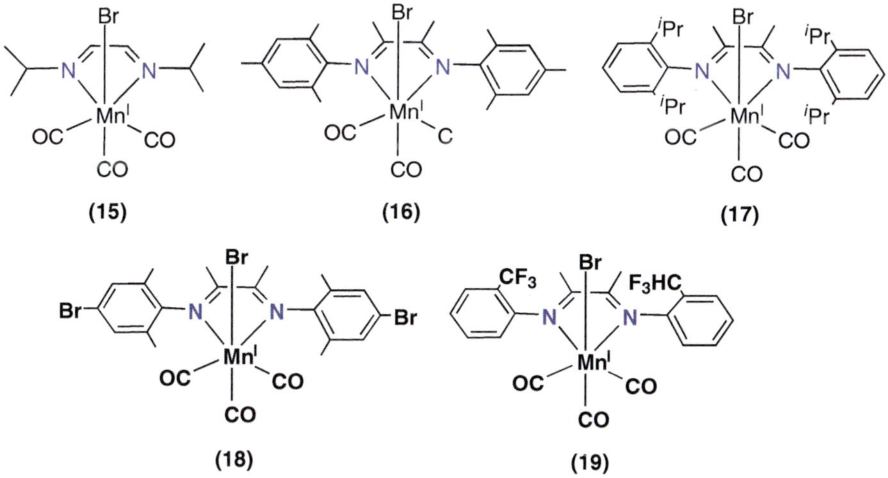

Figure 5.29 Diimine-based Mn complexes for electrocatalytic CO_2 reduction.

Figure 5.30). They further investigated the effect of the axial ligand on the electrocatalytic activity of [Mn(N^C)$(CO)_3$X] (X = Br^-, NCS^-, CN^-) complexes [108]. Bromide was replaced by cyanide and isothiocyanate, respectively (**23** and **24**, Figure 5.30). Even though all compounds exhibited a catalytic current enhancement under CO_2, those with a pseudohalogen (NCS^- and CN^-) displayed poor FE for the conversion of CO_2 to CO, favoring proton reduction to form H_2.

In Mn-NHC catalysts, the first reduction is more preferred over the halide dissociation. The Schafer group also investigated the mechanisms of CO_2 reduction on complexes **23**, **24**, and **25** (Figure 5.30) [109]. They demonstrated that CO_2 binding is strongly preferred over H^+ binding. Therefore, strong acids could be used safely without significantly increasing H_2 production. The H_2 evolution was not competing with the CO_2 reduction even when the catalyst included a pendant phenol in the second coordination sphere (**25**). The effect of the inclusion of a pendant phenol was positive, showing a decrease of the C—O bond cleavage barrier. Interestingly, they reported a dissociation of the pyridyl fragment upon reduction of the [Mn(N^C)$(CO)_4$]$^+$ species, potentially leading to a catalyst degradation pathway.

The first purely organometallic *fac*-[Mn(bis-MeNHC)$(CO)_3$Br] complex [110] (**26**, Figure 5.30) presented unprecedented activity for the selective electrocatalytic reduction of CO_2 to CO, exceeding 100 TONs with a FE of 95% in anhydrous CH_3CN. The TOF in CH_3CN was of 2100 s^{-1}, and in the presence of protons, the catalyst leads to 320 000 s^{-1}. The detection of Mn^I tetracarbonyl intermediate was also reported under catalytic conditions for the first time. FTIR-SEC measurements and computational data supported that the strongly nucleophilic character of the Mn atom is likely responsible for the positive ligand effect on catalysis.

Recently, two new Mn complexes based on a 2,6-bis(*N*-heterocyclic carbene)pyridine ligand framework, [MnCNC^R$(CO)_3$]Br (R = Me or benzyl) (**27a**, **27b**, Figure 5.30), were evaluated for the electrocatalytic CO_2 reduction [111]. CV

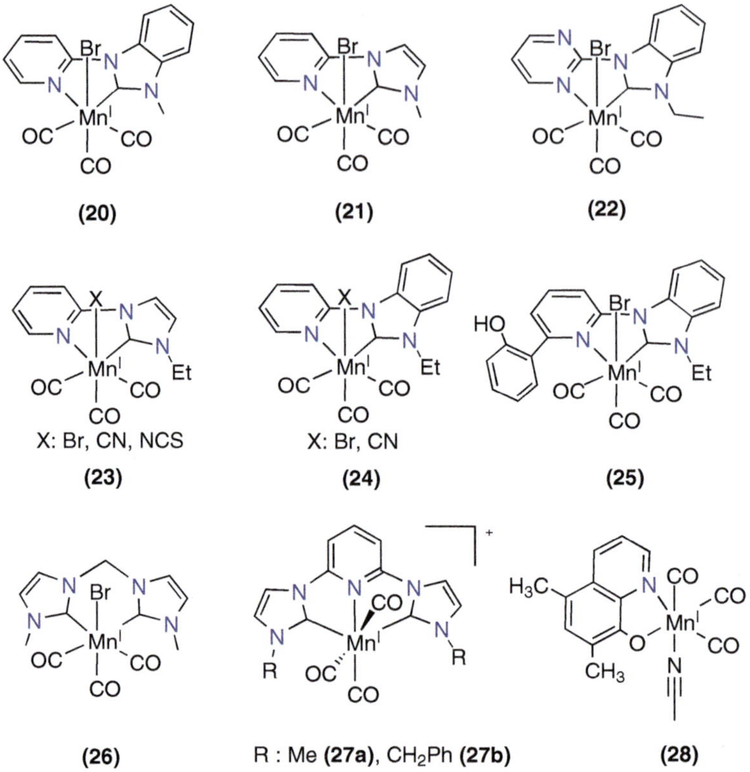

Figure 5.30 Selected Mn catalysts for electrocatalytic CO_2 reduction with NHC ligands.

analysis, under CO_2 in the absence of protons, produced a substantial increase in the current density at a 1e$^-$ reduction, indicating that catalysis may occur in the absence of protons. The FE for CO production in 0.229 M TFE in CH_3CN were 84% and 50% for **27a** and **27b** with TOF values of 237 and 97 s^{-1}, respectively. DOSY experiments coupled with normal pulse voltammetry studies suggested that the reduction of complexes occurs irreversibly as a 1e$^-$ step, highlighting the unique electronic properties of these metal complexes.

The Mn tricarbonyl complexes, [Mn(Me_2OQN)$(CO)_3$(CH_3CN)] containing the 5,7-dimethyl-8-oxyquinolate (Me_2OQN$^-$) ligand (**28**, Figure 5.30), were reported by Rochford and coworkers [112]. The non-innocent character of the Me_2OQN ligand precludes the need for an additional overpotential, typically associated with the reduction-first pathway in the catalytic cycle of the similar [Mn(bpy)(CO_3)Br], facilitating a protonation-first CO_2 reduction pathway for the precatalyst upon their two-electron activation. The addition of a Brønsted acid yielded a CO_2 reduction TOF of 14 s^{-1} and FEs of 88% and 5% for CO and H_2, respectively. The metal–ligand synergy was determined by combining studies on theoretical modeling, electrochemistry, spectroelectrochemistry (FTIR and EPR), and pulse radiolysis coupled with time-resolved infrared spectroscopy (PR-TRIR), providing structural insight into the electronic properties of the reduced species.

5.4.2 Photochemical CO_2 Reduction

In contrast to electrocatalysis, the photocatalytic CO_2 reduction mediated by Mn complexes requires a photosensitizer (PS) along with a sacrificial electron donor to produce the catalysis. In the first step, the photosensitizer (usually used $[Ru(dmbpy)_3]^{2+}$; dmbpy = 4,4′-dimethyl-2,2′-bipyridine) absorbs the light, populating the ^{3}MLCT excited state. Then, the excited photosensitizer undergoes a reductive quenching with the sacrificial electron donor. The reduced photosensitizer, in turn, reduces the Mn^I precatalyst to the Mn^0 species, which dimerizes. As a difference regarding electrochemical conditions, the photoexcitation of the Mn–Mn dimer efficiently induced homolysis of the metal–metal bond, yielding the corresponding Mn radical species [113]. Importantly, photocatalytic CO_2 reduction with Mn-based complexes results mainly in HCOOH, whereas CO is produced in minor quantities.

Takeda et al. reported the first of these studies in 2014 [114], where *fac*-[Mn(bpy)$(CO)_3$Br] (Figure 5.31) was employed in combination with $[Ru(dmbpy)_3]^{2+}$ and led to the formation of a mixture of CO (TON = 12) and formate (TON = 149) in DMF/TEOA solution with 1-benzyl-1,4-dihydronicotinamide (BNAH) as a sacrificial reagent. Spectroscopic studies suggested that the photocatalytic reduction follows the reduction of the PS ^{3}MLCT excited-state by BNAH to generate the $[Mn(bpy)(CO)_3]_2$ dimer. In concordance, the CO evolved decreased concomitant with the disappearance of the $[Mn(bpy)(CO)_3]_2$ dimer, whereas a steady-state formation of HCOOH was observed.

Kubiak and coworkers reported the photocatalytic CO_2 reduction with $[Mn(bpy)(CO)_3(CN)]$ (**29**, Figure 5.32) employing $[Ru(dmbpy)_3]^{2+}$ as a photosensitizer (PS2) in mixtures of dry *N*,*N*-DMF-TEOA or MeCN-TEOA [113, 115]. After 15 hours of irradiation, the TON for CO and HCOOH was 21 and 127, respectively, with the product preference dependent on the solvent, shifting the selectivity from CO to HCOOH when changing the solvent from MeCN to *N*,*N*-DMF. The electrochemical and spectroelectrochemical results supported that the catalyst, similar to the previously described electrochemical mechanism [116], undergoes a disproportionation reaction involving two singly reduced species, $[Mn(bpy)(CO)_3(CN)]^{\cdot-}$, to produce 1 equiv of the catalytically active species $[Mn(bpy)(CO)_3]^-$ and 1 equiv of $[Mn(bpy)(CO)_3(CN)]$.

The catalyst *fac*-$[Mn(phen)(CO)_3Br]$ (**30**, Figure 5.32) was studied in combination with the photosensitizer ZnTPP (PS3) [117]. Again, the photocatalytic system efficiently produced CO and HCOOH in an aqueous CH_3CN solution. They concluded that the Mn/Zn ratio affects the efficiency of CO_2 reduction. After 180 minutes irradiation with a Mn/Zn ratio of 2 : 1, the TONs reached 64 and 16 for the formation of CO and HCOOH, respectively. When PS3 was used, no significant formation of H_2 was observed.

Photocatalysts using $[Cu_2(P_2bph)_2]^{2+}$ (P_2bph = 4,7-diphenyl-2,9-di(diphenyl phosphinotetramethylene)-1,10-phenanthroline) as a redox photosensitizer (PS4) and *fac*-$[Mn(R_2bpy)(CO)_3Br]$ (R_2bpy = 4,4′-X2-2,2′-bipyridine [R = H, OMe (**31**)] or 6mes = 6,6′-(mesityl)2-2,2′-bipyridine (**32**)) as catalyst (Figure 5.32) were evaluated

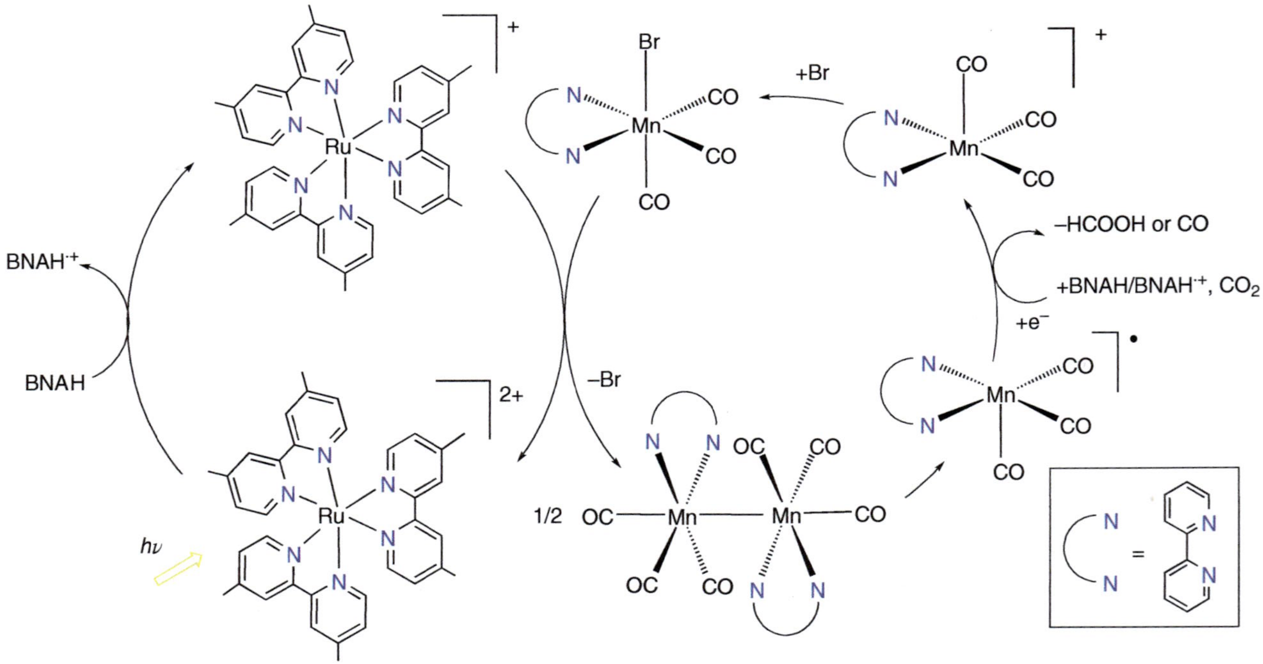

Figure 5.31 General photoinduced CO_2 reduction mechanism by Mn complexes.

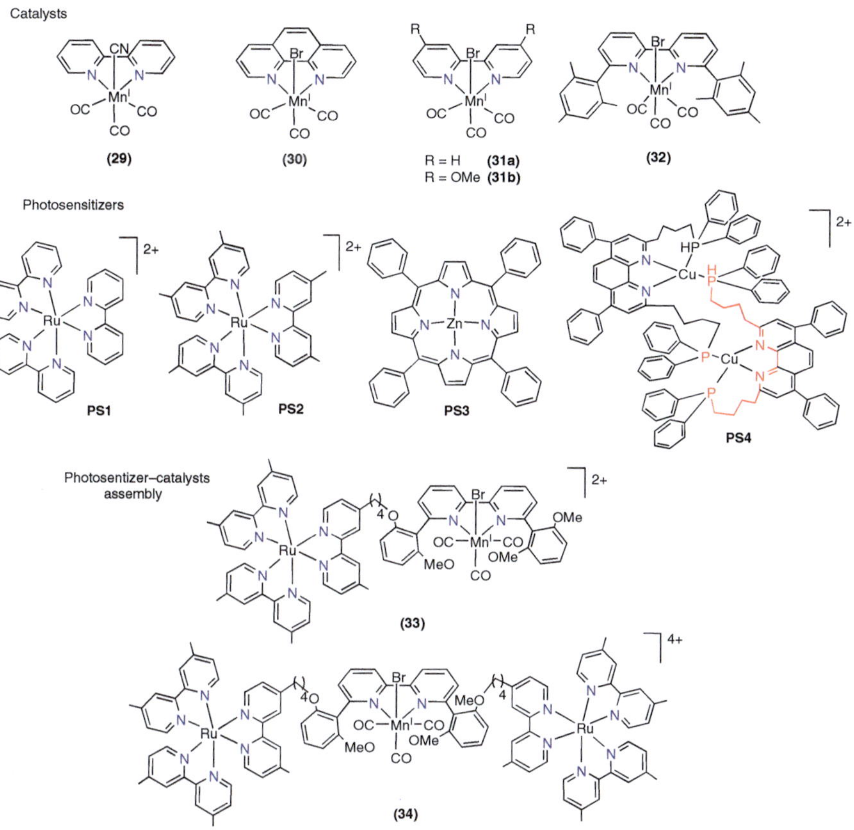

Figure 5.32 Catalysts and photosensitizers applied for the photochemical conversion of CO_2.

by Ishitani and coworkers [118]. The highest selectivity of CO formation was achieved by using **32** with a selectivity of 96% and TON based on the Mn catalyst over 1300. In contrast, **31a** yielded HCOOH as the main product (74%) with CO and H_2 as minor products (CO = 23%, H_2 = 2%). Electronic and steric effects of the substituents in the Mn complexes largely affected the photocatalytic efficiency and the product selectivity. The mechanistic studies suggested that the intermediates for CO_2 reduction are the 5-coordinate 18-electron species for **31** and **32**, whereas in the case of **31** the Mn–Mn dimer is also proposed.

Recently, new supramolecular photocatalysts with bulky groups featuring Mn catalyst unit for dimer suppression were reported by Ishitani and coworkers [119]. The photocatalysts Ru_x–Mn (x = 1, 2) were based on a Mn^I unit {$Mn(CO)_3Br$} coordinated to the bridging ligand (BL) present in the photosensitizer unit(s) $[Ru(dmb)_2(BL)]^{2+}$ (dmb = 4,4′-dimethyl-2,2′-bipyridine). As result, two supramolecular photocatalysts were evaluated, Ru_1–Mn (with BL_1 = 4-(4-(2-(6′-(2,6-dimethoxyphenyl)[2,2′-bipyridin]-6-yl)-3-methoxyphenoxy) butyl)-4′-methyl-2,2′-bipyridine) (**33**, Figure 5.32) and Ru_2–Mn (with BL_2 = 6,6′-

Bis(2-methoxy-6-(4-(4′-methyl[2,2′-bipyridin]-4-yl)butoxy)- phenyl)-2,2′-bipyridine) (**34**, Figure 5.32). Photocatalytic reactions were conducted by using Ru_1–Mn, Ru_2–Mn, and a 1 : 1 mixture of Ru and Mn molecular complexes in the presence of a mixture of DMA/TEA and BIH as an electron donor. The photocatalysts **33** reduced CO_2 selectively to HCOOH and CO as a minor product, with higher HCOOH formation than the 1 : 1 mixed system of the mononuclear complexes. In contrast, photocatalysis of the trinuclear complex **34** did not work as an efficient photocatalyst for CO_2 reduction due to slow intramolecular ET.

5.4.3 Heterogenization of Molecular Mn Catalysts for the CO_2 Reduction Reaction

Catalysts heterogenization has been a long-standing strategy to improve the stability of molecular catalysts that operate in homogeneous phase [120]. Heterogenization of well-defined catalysts could be prepared by immobilizing them in reticular materials, such as metal–organic frameworks, covalent organic frameworks, or polyoxometalates, on metal-oxide nanoparticles, and carbon-based conducting substrates, among others [121]. Understanding the relation between structure–activity correlations is the key to improve robustness, performance, and selectivity, especially in CO_2 reduction reactions [122, 123]. Active manganese complexes for CO_2 reduction have been immobilized following different strategies (Figure 5.33) including the simple immobilization of unfunctionalized [Mn(bpy)$(CO)_3$Br] or [Mn(bpy$(t\text{Bu})_2$)$(CO)_3$Br] complexes, mixed with Nafion® or Nafion/multi-walled carbon nanotubes (MWCNTs), by drop-cast on top glassy carbon (**35**) [125, 126]. They can also be immobilized by more elaborated strategies such as using π-stacking interactions (**36**) [127], covalently attached to glassy carbon (**37**) [128], metal-organic conjugated microporous polymers (**38**, **39**) [124, 129], metal-organic frameworks (**40**) [115], functionalized periodic mesoporous organosilicas (**41**) [130], or on top of silicon nanowires (SiNWs) photocathodes (**41**) [131], allowing to work with supported catalysts in water [132].

Molecular manganese catalysts immobilized on mesoporous TiO_2 (**42**, **43**) have been a successful strategy to enhance both the electrocatalytic and solar-driven CO_2 reduction, and to remark their use to understand the CO_2 reduction mechanism further [133], the formation of key intermediate catalytic species at the electrode surface such as the Mn^0–Mn^0 dimer, and its relation with the CO_2 reduction selectivity, has been identified. From these studies, the product selectivity ($HCOO^-$/CO) was controlled by changing the loading amount of the supported Mn catalyst on the TiO_2 surface. At low Mn concentrations, the supported catalyst has low mobility preventing the dimerization and the Mn–H hydride species, which are associated with the formation of the HCOOH that are dominant. While at higher concentrations, the Mn^0–Mn^0 prevents the formation of the Mn–H, leading to the CO production [133–136]. These conclusions are similar to the previously obtained by the group of Reisner when [Mn(bpy)$(CO)_3$Br] complex was anchored to a carbon nanotube electrode via a pyrene unit (**37**) [127].

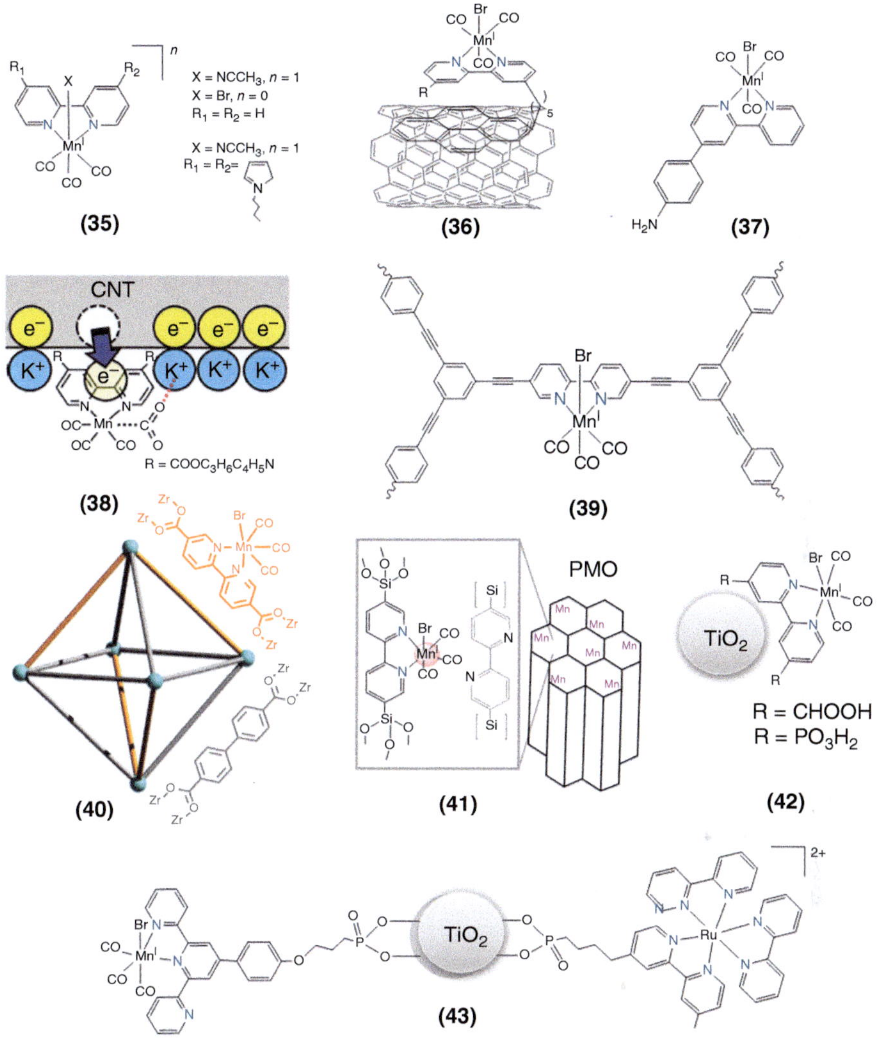

Figure 5.33 Strategies to improve the stability of molecular catalysts that operate in homogeneous phase. Source: Sato et al. [124]; Fei et al. [115].

5.5 Conclusions

The photo- and electro-induced catalytic processes mediated by manganese complexes are wide in reactivity, from catalyzing challenging oxidation reaction to CO_2 reduction. This broad window of reactivity can be ascribed to their redox flexibility and coordination chemistry. Both photo- and electrocatalysis have allowed the access of new reactivity, otherwise difficult to achieve under thermal conditions. In this regard, photo- and electro-assisted catalysis can serve as alternatives to classical synthetic methods, not only to increase the toolbox of chemical transformations but also to exploit renewable energy as energy input to trigger reactivity. From reported examples, it is expected to have an impact in shorting some synthetic routes.

On the other hand, there is also increasing attention to the catalytic reduction of CO_2. From the previous studies, mechanistic understanding and new ligand designs have brought progress on the control of the reaction selectivity. Secondary sphere interactions have emerged as a powerful tool, shifting the product selectivity from CO to HCOOH. Ligand design can tune the TOF and overpotential. In this regard, moving from organic solvents to water reduced the overpotential significantly but compromised the selectivity. Still, a full exploration of the capacity of manganese complexes in photo- and electrocatalytic transformations is at its infancy, and there is still a long way to go toward sustainability.

References

1 Cox, N., Pantazis, D.A., Neese, F., and Lubitz, W. (2015). *Interface Focus* 5 (3): 20150009.

2 Najafpour, M.M., Zaharieva, I., Zand, Z. et al. (2020). *Coord. Chem. Rev.* 409: 213183.

3 Yano, J. and Yachandra, V. (2014). *Chem. Rev.* 114 (8): 4175–4205.

4 Cox, N., Pantazis, D.A., and Lubitz, W. (2020). *Annu. Rev. Biochem.* 89 (1): 795–820.

5 Carney, J.R., Dillon, B.R., and Thomas, S.P. (2016). *Eur. J. Org. Chem.* 2016 (23): 3912–3929.

6 Hilt, G. (2020). *ChemElectroChem* 7 (2): 395–405.

7 Wang, P., Gao, X., Huang, P., and Lei, A. (2020). *ChemCatChem* 12 (1): 27–40.

8 Sauer, G.S. and Lin, S. (2018). *ACS Catal.* 8 (6): 5175–5187.

9 Fukuzumi, S., Lee, Y.-M., and Nam, W. (2019). *ChemSusChem* 12 (17): 3931–3940.

10 Ciamician, G. (1912). *Science* 36 (926): 385–394.

11 Vlček, A., Kvapilová, H., Towrie, M., and Záliš, S. (2015). *Acc. Chem. Res.* 48 (3): 868–876.

12 Escudero, D. (2016). *Acc. Chem. Res.* 49 (9): 1816–1824.

13 Zhang, W. and Gaffney, K.J. (2015). *Acc. Chem. Res.* 48 (4): 1140–1148.

14 Prier, C.K., Rankic, D.A., and MacMillan, D.W.C. (2013). *Chem. Rev.* 113 (7): 5322–5363.

15 Reisner, E. (2019). *Angew. Chem. Int. Ed.* 58 (12): 3656–3657.

16 Dalle, K.E., Warnan, J., Leung, J.J. et al. (2019). *Chem. Rev.* 119 (4): 2752–2875.

17 Armaroli, N. and Balzani, V. (2016). *Chem. Eur. J.* 22 (1): 32–57.

18 Stephenson, C.R.J., Yoon, T.P., and MacMillan, D.W.C. (eds.) (2018). *Visible Light Photocatalysis in Organic Chemistry*. Weinheim, Germany: Wiley-VCH Verlag GmbH & Co.KGaA.

19 Schultz, D.M. and Yoon, T.P. (2014). *Science* 343: 6174, 1239176.

20 Narayanam, J.M.R. and Stephenson, C.R.J. (2011). *Chem. Soc. Rev.* 40 (1): 102–113.

21 Corrigan, N., Shanmugam, S., Xu, J., and Boyer, C. (2016). *Chem. Soc. Rev.* 45 (22): 6165–6212.

22 Shatskiy, A., Lundberg, H., and Kärkäs, M.D. (2019). *ChemElectroChem* 6 (16): 4067–4092.

23 Atobe, M., Tateno, H., and Matsumura, Y. (2018). *Chem. Rev.* 118 (9): 4541–4572.

24 Kawamata, Y. and Baran, P.S. (2020). *Joule* 4 (4): 701–704.

25 Yan, M., Kawamata, Y., and Baran, P.S. (2017). *Chem. Rev.* 117 (21): 13230–13319.

26 Mo, Y., Lu, Z., Rughoobur, G. et al. (2020). *Science* 368 (6497): 1352.

27 Franco, F., Fernández, S., and Lloret-Fillol, J. (2019). *Curr. Opin. Electrochem.* 15: 109–117.

28 Meunier, B. (1992). *Chem. Rev.* 92 (6): 1411–1456.

29 Groves, J.T., Lee, J., and Marla, S.S. (1997). *J. Am. Chem. Soc.* 119 (27): 6269–6273.

30 Nam, W., Kim, I., Lim, M.H. et al. (2002). *Chem. Eur. J.* 8 (9): 2067–2071.

31 Lane, B.S. and Burgess, K. (2003). *Chem. Rev.* 103 (7): 2457–2474.

32 Vicens, L., Olivo, G., and Costas, M. (2020). *ACS Catal.* 10 (15): 8611–8631.

33 Fukuzumi, S., Kishi, T., Kotani, H. et al. (2011). *Nat. Chem.* 3 (1): 38–41.

34 Company, A., Sabenya, G., González-Béjar, M. et al. (2014). *J. Am. Chem. Soc.* 136 (12): 4624–4633.

35 Wu, X., Yang, X., Lee, Y.-M. et al. (2015). *Chem. Commun.* 51 (19): 4013–4016.

36 Shen, D., Saracini, C., Lee, Y.-M. et al. (2016). *J. Am. Chem. Soc.* 138 (49): 15857–15860.

37 Chen, G., Chen, L., Ma, L. et al. (2016). *Chem. Commun.* 52 (59): 9271–9274.

38 Kwong, H.-K., Lo, P.-K., Lau, K.-C., and Lau, T.-C. (2011). *Chem. Commun.* 47 (14): 4273–4275.

39 Ma, L., Pan, Y., Man, W.-L. et al. (2014). *J. Am. Chem. Soc.* 136 (21): 7680–7687.

40 Sharma, N., Jung, J., Ohkubo, K. et al. (2018). *J. Am. Chem. Soc.* 140 (27): 8405–8409.

41 Guo, P. and Wong, K.-Y. (1999). *Electrochem. Commun.* 1 (11): 559–563.

42 Tanaka, H., Kuroboshi, M., Takeda, H. et al. (2001). *J. Electroanal. Chem.* 507 (1): 75–81.

43 Prokop, K.A. and Goldberg, D.P. (2012). *J. Am. Chem. Soc.* 134 (19): 8014–8017.

44 Jung, J., Ohkubo, K., Prokop-Prigge, K.A. et al. (2013). *Inorg. Chem.* 52 (23): 13594–13604.

45 Jung, J., Ohkubo, K., Goldberg, D.P., and Fukuzumi, S. (2014). *J. Phys. Chem. A* 118 (32): 6223–6229.

46 Neu, H.M., Jung, J., Baglia, R.A. et al. (2015). *J. Am. Chem. Soc.* 137 (14): 4614–4617.

47 Jung, J., Neu, H.M., Leeladee, P. et al. (2016). *Inorg. Chem.* 55 (7): 3218–3228.

48 Kunz, K., Vitze, H., Bolte, M. et al. (2007). *Organometallics* 26 (18): 4663–4672.

49 Pratt, S.L. and Faltynek, R.A. (1983). *J. Organomet. Chem.* 258 (1): C5–C8.

50 Arias-Ugarte, R., Sharma, H.K., Morris, A.L.C., and Pannell, K.H. (2012). *J. Am. Chem. Soc.* 134 (2): 848–851.

51 Zheng, J., Chevance, S., Darcel, C., and Sortais, J.-B. (2013). *Chem. Commun.* 49 (85): 10010–10012.

52 Zheng, J., Elangovan, S., Valyaev, D.A. et al. (2014). *Adv. Synth. Catal.* 356 (5): 1093–1097.

53 Wang, L., Lear, J.M., Rafferty, S.M. et al. (2018). *Science* 362 (6411): 225–229.

54 Hammarback, L.A., Clark, I.P., Sazanovich, I.V. et al. (2018). *Nat. Catal.* 1 (11): 830–840.

55 Liang, Y.-F., Steinbock, R., Yang, L., and Ackermann, L. (2018). *Angew. Chem. Int. Ed.* 57 (33): 10625–10629.

56 Dong, J., Wang, X., Wang, Z. et al. (2019). *Chem. Commun.* 55 (78): 11707–11710.

57 Siu, J.C., Fu, N., and Lin, S. (2020). *Acc. Chem. Res.* 53 (3): 547–560.

58 Fu, N., Sauer, G.S., Saha, A. et al. (2017). *Science* 357 (6351): 575–579.

59 Fu, N., Sauer, G.S., and Lin, S. (2017). *J. Am. Chem. Soc.* 139 (43): 15548–15553.

60 Ye, K.-Y., Pombar, G., Fu, N. et al. (2018). *J. Am. Chem. Soc.* 140 (7): 2438–2441.

61 Ye, K.-Y., Song, Z., Sauer, G.S. et al. (2018). *Chem. Eur. J.* 24 (47): 12274–12279.

62 Fu, N., Shen, Y., Allen, A.R. et al. (2019). *ACS Catal.* 9 (1): 746–754.

63 Lu, L., Fu, N., and Lin, S. (2019). *Synlett* 30 (10): 1199–1203.

64 Strehl, J. and Hilt, G. (2019). *Org. Lett.* 21 (13): 5259–5263.

65 Allen, B.D.W., Hareram, M.D., Seastram, A.C. et al. (2019). *Org. Lett.* 21 (22): 9241–9246.

66 Tian, S., Jia, X., Wang, L. et al. (2019). *Chem. Commun.* 55 (80): 12104–12107.

67 Beer, C., Reichstein, M., Tomelleri, E. et al. (2010). *Science* 329 (5993): 834.

68 Lu, Q. and Jiao, F. (2016). *Nano Energy* 29: 439–456.

69 Centi, G. and Perathoner, S. (2009). *Catal. Today* 148 (3): 191–205.

70 Mikkelsen, M., Jørgensen, M., and Krebs, F.C. (2010). *Energy Environ. Sci.* 3 (1): 43–81.

71 Bard, A.J., Parsons, R., and Jordan, J. (eds.) (1985). *Standard Potentials in Aqueous Solution*. New York: CRC Press.

72 Collin, J.P. and Sauvage, J.P. (1989). *Coord. Chem. Rev.* 93 (2): 245–268.

73 Elgrishi, N., Chambers, M.B., Wang, X., and Fontecave, M. (2017). *Chem. Soc. Rev.* 46 (3): 761–796.

74 Takeda, H., Cometto, C., Ishitani, O., and Robert, M. (2017). *ACS Catal.* 7 (1): 70–88.

75 Bonin, J., Maurin, A., and Robert, M. (2017). *Coord. Chem. Rev.* 334: 184–198.

76 Chen, L., Guo, Z., Wei, X.-G. et al. (2015). *J. Am. Chem. Soc.* 137 (34): 10918–10921.

77 Liu, D.-C., Zhong, D.-C., and Lu, T.-B. (2020). *EnergyChem* 2 (3): 100034.

78 Yaroshevsky, A.A. (2006). *Geochem. Int.* 44 (1): 48–55.

79 Grills, D.C., Ertem, M.Z., McKinnon, M. et al. (2018). *Coord. Chem. Rev.* 374: 173–217.

80 Stanbury, M., Compain, J.-D., and Chardon-Noblat, S. (2018). *Coord. Chem. Rev.* 361: 120–137.

81 Sinopoli, A., La Porte, N.T., Martinez, J.F. et al. (2018). *Coord. Chem. Rev.* 365: 60–74.

82 Bourrez, M., Molton, F., Chardon-Noblat, S., and Deronzier, A. (2011). *Angew. Chem. Int. Ed.* 50 (42): 9903–9906.
83 Bourrez, M., Orio, M., Molton, F. et al. (2014). *Angew. Chem. Int. Ed.* 53 (1): 240–243.
84 Smieja, J.M., Sampson, M.D., Grice, K.A. et al. (2013). *Inorg. Chem.* 52 (5): 2484–2491.
85 Sampson, M.D., Nguyen, A.D., Grice, K.A. et al. (2014). *J. Am. Chem. Soc.* 136 (14): 5460–5471.
86 Hartl, F., Rossenaar, B.D., Stor, G.J., and Stufkens, D.J. (1995). *Recl. Trav. Chim. Pays-Bas* 114 (11–12): 565–570.
87 Stor, G.J., Morrison, S.L., Stufkens, D.J., and Oskam, A. (1994). *Organometallics* 13 (7): 2641–2650.
88 Kokkes, M.W., De Lange, W.G.J., Stufkens, D.J., and Oskam, A. (1985). *J. Organomet. Chem.* 294 (1): 59–73.
89 Riplinger, C., Sampson, M.D., Ritzmann, A.M. et al. (2014). *J. Am. Chem. Soc.* 136 (46): 16285–16298.
90 Neri, G., Walsh, J.J., Teobaldi, G. et al. (2018). *Nat. Catal.* 1 (12): 952–959.
91 Grills, D.C., Farrington, J.A., Layne, B.H. et al. (2014). *J. Am. Chem. Soc.* 136 (15): 5563–5566.
92 Sampson, M.D. and Kubiak, C.P. (2016). *J. Am. Chem. Soc.* 138 (4): 1386–1393.
93 Ngo, K.T., McKinnon, M., Mahanti, B. et al. (2017). *J. Am. Chem. Soc.* 139 (7): 2604–2618.
94 Franco, F., Cometto, C., Ferrero Vallana, F. et al. (2014). *Chem. Commun.* 50 (93): 14670–14673.
95 Agarwal, J., Shaw, T.W., Schaefer, H.F., and Bocarsly, A.B. (2015). *Inorg. Chem.* 54 (11): 5285–5294.
96 Rønne, M.H., Cho, D., Madsen, M.R. et al. (2020). *J. Am. Chem. Soc.* 142 (9): 4265–4275.
97 Compain, J.-D., Bourrez, M., Haukka, M. et al. (2014). *Chem. Commun.* 50 (19): 2539–2542.
98 Machan, C.W. and Kubiak, C.P. (2016). *Dalton Trans.* 45 (43): 17179–17186.
99 Stanbury, M., Compain, J.-D., Trejo, M. et al. (2017). *Electrochim. Acta* 240: 288–299.
100 McKinnon, M., Belkina, V., Ngo, K.T. et al. (2019). *Front. Chem.* 7: 628.
101 Rao, G.K., Pell, W., Korobkov, I., and Richeson, D. (2016). *Chem. Commun.* 52 (51): 8010–8013.
102 Zeng, Q., Tory, J., and Hartl, F. (2014). *Organometallics* 33 (18): 5002–5008.
103 Vollmer, M.V., Machan, C.W., Clark, M.L. et al. (2015). *Organometallics* 34 (1): 3–12.
104 Nelson, D.J. and Nolan, S.P. (2013). *Chem. Soc. Rev.* 42 (16): 6723–6753.
105 Asay, M., Jones, C., and Driess, M. (2011). *Chem. Rev.* 111 (2): 354–396.
106 Agarwal, J., Shaw, T.W., Stanton, C.J. III, et al. (2014). *Angew. Chem. Int. Ed.* 53 (20): 5152–5155.
107 Stanton, C.J., Vandezande, J.E., Majetich, G.F. et al. (2016). *Inorg. Chem.* 55 (19): 9509–9512.

108 Agarwal, J., Stanton Iii, C.J., Shaw, T.W. et al. (2015). *Dalton Trans.* 44 (5): 2122–2131.

109 Vandezande, J.E. and Schaefer, H.F. (2018). *Organometallics* 37 (3): 337–342.

110 Franco, F., Pinto, M.F., Royo, B., and Lloret-Fillol, J. (2018). *Angew. Chem. Int. Ed.* 57 (17): 4603–4606.

111 Myren, T.H.T., Lilio, A.M., Huntzinger, C.G. et al. (2019). *Organometallics* 38 (6): 1248–1253.

112 McKinnon, M., Ngo, K.T., Sobottka, S. et al. (2019). *Organometallics* 38 (6): 1317–1329.

113 Cheung, P.L., Machan, C.W., Malkhasian, A.Y.S. et al. (2016). *Inorg. Chem.* 55 (6): 3192–3198.

114 Takeda, H., Koizumi, H., Okamoto, K., and Ishitani, O. (2014). *Chem. Commun.* 50 (12): 1491–1493.

115 Fei, H., Sampson, M.D., Lee, Y. et al. (2015). *Inorg. Chem.* 54 (14): 6821–6828.

116 Machan, C.W., Stanton, C.J., Vandezande, J.E. et al. (2015). *Inorg. Chem.* 54 (17): 8849–8856.

117 Zhang, J.-X., Hu, C.-Y., Wang, W. et al. (2016). *Appl. Catal., A* 522: 145–151.

118 Takeda, H., Kamiyama, H., Okamoto, K. et al. (2018). *J. Am. Chem. Soc.* 140 (49): 17241–17254.

119 Fabry, D.C., Koizumi, H., Ghosh, D. et al. (2020). *Organometallics* 39 (9): 1511–1518.

120 Pittman, C.U. and Smith, L.R. (1975). *J. Am. Chem. Soc.* 97 (2): 341–344.

121 Piermatti, O., Abu-Reziq, R., and Vaccaro, L. (2020). *Catalyst Immobilization* (eds. M. Benaglia and A. Puglisi), 1–22. Germany: Wiley-VCH Verlag GmbH & Co. KGaA.

122 Yang, D., Ni, B., and Wang, X. (2020). *Adv. Energy Mater.* 10 (25): 2001142.

123 Wagner, A., Sahm, C.D., and Reisner, E. (2020). *Nat. Catal.* https://doi.org/10.1038/s41929-020-00512-x.

124 Sato, S., Saita, K., Sekizawa, K. et al. (2018). *ACS Catal.* 8 (5): 4452–4458.

125 Walsh, J.J., Smith, C.L., Neri, G. et al. (2015). *Faraday Discuss.* 183: 147–160.

126 Neri, G., Donaldson, P.M., and Cowan, A.J. (2019). *Phys. Chem. Chem. Phys.* 21: 7389–7397.

127 Reuillard, B., Ly, K.H., Rosser, T.E. et al. (2017). *J. Am. Chem. Soc.* 139 (41): 14425–14435.

128 Sun, C., Rotundo, L., Garino, C. et al. (2017). *ChemPhysChem* 18 (22): 3219–3229.

129 Smith, C.L., Clowes, R., Sprick, R.S. et al. (2019). *Sustainable Energy Fuels* 3 (11): 2990–2994.

130 Wang, X., Thiel, I., Fedorov, A. et al. (2017). *Chem. Sci.* 8 (12): 8204–8213.

131 Torralba-Peñalver, E., Luo, Y., Compain, J.-D. et al. (2015). *ACS Catal.* 5 (10): 6138–6147.

132 Walsh, J.J., Neri, G., Smith, C.L., and Cowan, A.J. (2019). *Organometallics* 38 (6): 1224–1229.

133 Walsh, J.J., Forster, M., Smith, C.L. et al. (2018). *Phys. Chem. Chem. Phys.* 20 (10): 6811–6816.

134 Rosser, T.E., Windle, C.D., and Reisner, E. (2016). *Angew. Chem. Int. Ed.* 55 (26): 7388–7392.

135 Le-Quang, L., Stanbury, M., Chardon-Noblat, S. et al. (2019). *Chem. Commun.* 55 (90): 13598–13601.

136 Woo, S.-J., Choi, S., Kim, S.-Y. et al. (2019). *ACS Catal.* 9 (3): 2580–2593.

6

Manganese-Catalyzed C—H Oxygenation Reactions

Konstantin P. Bryliakov[1,2]

[1] *Novosibirsk State University, Pirogova 2, Novosibirsk, 630090, Russian Federation*
[2] *Boreskov Institute of Catalysis, Pr. Lavrentieva 5, Novosibirsk, 630090, Russian Federation*

6.1 Introduction

Developing powerful and efficient general synthetic methods for the chemo- and stereoselective oxidative functionalization of C—H bonds of complex organic molecules has been a challenging goal for decades [1]. The difficulties faced by chemists stem from the inherent chemical inertness of organic C–H functionalities that are neither basic nor acidic, and poorly reactive toward electrophiles and nucleophiles. In turn, the high C—H bond dissociation energies of organic substances, mostly falling within the range 80–110 kcal/mol, require high energy reactants, capable of C–H activation via homolytic pathway, which is often disadvantageous from the selectivity perspective. The importance and relevance of the area was acknowledged by nominating J. Bercaw, R. Bergman, and G. Shul'pin 2017 Clarivate Analytics Citation Laureates in chemistry "for critical contributions to C–H functionalization" [2].

Existing approaches to selective C–H oxidative functionalization, using transition metal (mostly Fe and Mn)-based catalysts [3–6], in most cases rely on the Groves "rebound mechanism" [7, 8], proceeding via hydrogen atom abstraction, followed by oxygen rebound step, resulting in C—O bond formation (Scheme 6.1). This sequence sets severe restrictions on the catalytic pattern of the transition metal complex, as well as on the overall process organization. First of all, such catalytic processes require two-electron stoichiometric oxidants, such as iodosylarenes, peroxycarboxylic acids, sodium hypochlorite, or, preferably, "green" hydrogen peroxide; the use of dioxygen, the most attractive oxidant, has not been achieved so far in transition metal-catalyzed selective C–H oxygenation processes.

The net H abstraction is typically the rate-limiting step; the resulting intermediate solvent-caged radical should be sufficiently short-lived in order to ensure faster oxygen rebound as compared with its out-diffusion into the solution, which would trigger non-rebound-type reactions [9] and lead to undesired chemical transformations of the carbon-centered radical. Moreover, the C-radical may be prone

Manganese Catalysis in Organic Synthesis, First Edition. Edited by Jean-Baptiste Sortais.

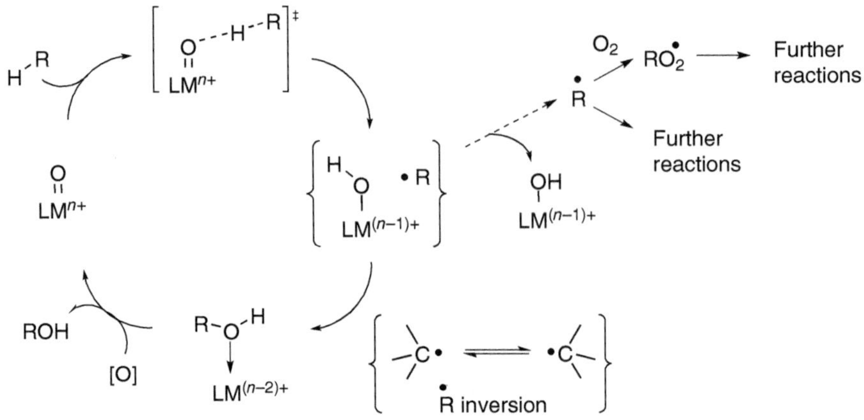

Scheme 6.1 Schematic representation of transition metal-mediated C–H functionalization.

to stereoisomerization (inversion), competitive to oxygen rebound, which would deteriorate the C–H hydroxylation stereospecificity or stereoselectivity. In principle, for the approaches focused on catalytic ketonization of methylene groups, the radical stereoisomerization is not critical since the initially formed alcohol contains weaker C—H bond than the substrate R–H itself and is therefore further rapidly oxidized to the ketone. On the other hand, these considerations highlight another serious problem, related to the catalytic hydroxylations of CH_2 groups: the latter reactions require special approaches in order to reduce the reactivity of the alcohol and thus suppress its undesired ketonization.

Nevertheless, the state of the art of transition metal-mediated C–H oxidations has demonstrated evident success in overcoming technical difficulties. Indeed, rational catalyst design in some cases paves the way to overriding the substrate bias and directing the oxidation to certain C–H fragments of a complex organic molecule. For instance, preferential oxidation of methylene groups in the presence of more reactive tertiary C–H groups can be achieved by manipulating the steric and electronic properties of the catalytic sites. To date, manganese-based catalysts have been the major protagonists in this area [5, 10–13], allowing effective tuning of the catalytic properties via ligand design; Fe-based catalysts, yet more extensively studied, seem to be less promising from the synthetic perspective, demonstrating generally lower catalytic efficiency and selectivity. In this chapter, we survey Mn-based systems for the selective C–H oxygenation processes, mostly focusing on the synthetic aspect of their catalytic properties.

6.2 Selective Manganese-Catalyzed C–H Oxidation Reactions: Early Studies

In 1990–1991, Grieco and coworkers reported the remote oxofunctionalization of inactivated C–H groups in steroids in the presence of a manganese porphyrin

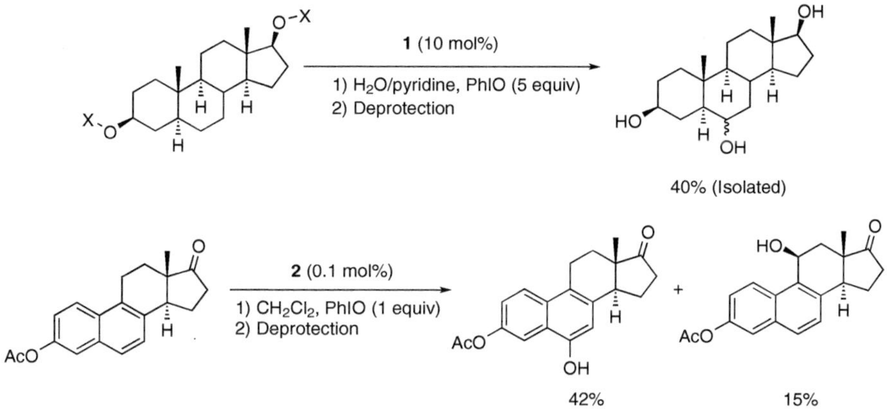

Scheme 6.2 Selective hydroxylation of steroid substrates in the presence of Mn porphyrins. X is protecting group.

complex, using iodosylbenzene as the terminal oxidant [14, 15]. Although the system clearly showed high site selectivity, its synthetic use would be problematic since the oxidation occurred in intramolecular fashion, the steroid substrate being covalently linked to the Mn porphyrin. The oxidation of C–H groups by various single oxygen donor oxidants (NaOCl, PhIO, $KHSO_5$, etc.) in the presence of Mn porphyrins was a focus of studies, too, mostly from the kinetic and mechanistic perspective [16, 17].

Later, Breslow et al. developed Mn porphyrin complex **1**, containing four β-cyclodextrin rings (Figure 6.1), which catalyzed the hydroxylation of protected androstane-3,17-diol into, after deprotection, the corresponding 3,6,17-triol, with iodosylbenzene (5 equiv) (Scheme 6.2) [18]. The system required 10 mol% catalyst loadings, and, to ensure sufficient substrate solubility in water, a hydrophilic SO_3H-containing protecting group was used. In subsequent publication, a broader set of substrates and porphyrin catalysts was systematically examined [19].

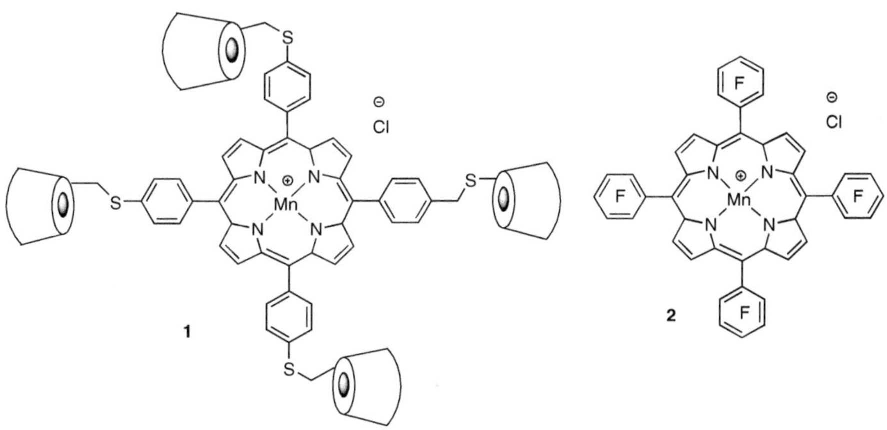

Figure 6.1 Examples of manganese porphyrin-based selective oxidation catalysts.

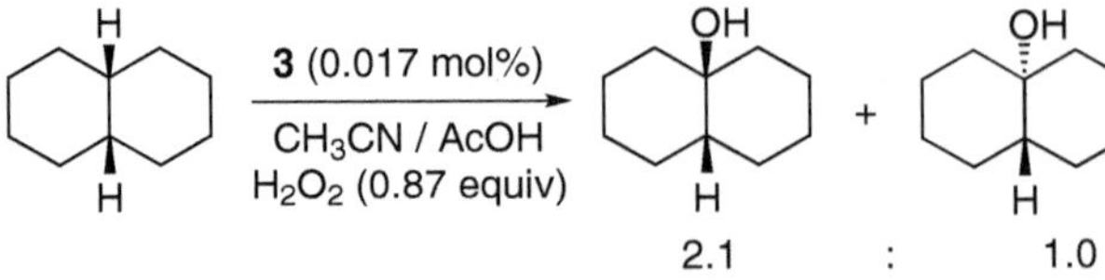

Scheme 6.3 Partial loss of stereochemistry in *cis*-decalin oxidation.

Alternatively, it was shown that even in the absence of substrate recognition motifs (e.g. cyclodextrins), manganese porphyrin catalyst can catalyze the oxidation of complex substrates with high selectivity. For example, fluorinated porphyrin **2** (0.1 mol%) mediated the oxidation of equilenin acetate to a mixture of C6-phenol (major) and C11-alcohol (minor) (Scheme 6.2) [20]. Surprisingly, in the presence of **2**, the C11-alcohol did not undergo further oxidation to ketone, which unusual reactivity of **2** was in sharp contrast to the catalytic pattern of the analogous ruthenium porphyrin [20].

Mohajer and Bagherzadeh reported the efficient oxidation of several cycloalkanes and ethylbenzene with tetrabutylammonium periodate in CH_2Cl_2 in the presence of various manganese(III) porphyrin catalysts (1.2 mol%) and imidazole additives (12 mol%). Ketones were the major reaction products in the benzylic C–H oxidations (of tetralin, indane, and ethylbenzene). On the other hand, the oxidation of cyclooctane and cyclohexane afforded nearly 1 : 1 mixtures of alcohol:ketone products, thus pointing to a free radical-driven oxidation mechanism [21]. Other manganese porphyrin-mediated C–H oxidations were reported [22–24].

A notable breakthrough in the area was the discovery of the efficient catalyst system for alkane oxygenation with H_2O_2 in the presence of binuclear manganese(IV) complex **3**, $[LMn^{IV}(O)_3Mn^{IV}L](PF_6)_2$, where L = 1,4,7-trimethyl-1,4,7-triazacyclononane, Me_3tacn (Figure 6.2) [25, 26]. The catalyst could use peroxycarboxylic acids (peracetic, *m*-chloroperbenzoic) as oxidants, as well as PhIO, *tert*-butyl hydroperoxide, and, crucially, hydrogen peroxide: in the latter case, carboxylic acid additives, preferentially acetic acid, were necessary. The catalysts exhibited very high efficiency, up to 1350 turnovers (TN). The alcohol/ketone ratios were close to 1 : 1 (typically 1.3–1.4), and partial loss of stereochemistry in the course of oxidation of *cis*- and *trans*-1,2-dimethylcyclohexanes and *cis*- and *trans*-decalins (Scheme 6.3) was reported, which facts point to substantial contribution of free radical-driven oxidation process. This catalyst system was subsequently reported to efficiently oxidize light alkanes, including methane, with hydrogen peroxide [27, 28]. Recent mechanistic study confirmed the involvement of atmospheric dioxygen (which is captured by the carbon-centered radical) into the oxidative transformations [29].

A number of variations of the Mn-triazacyclononane-based catalyst system were studied in the last 20 years [4], mostly from the mechanistic point of view, while its synthetic potential in the selective oxidation of complex organic molecules was not systematically examined. Nevertheless, the Mn-Me_3tacn catalyst system has been the inspiration for a number of next-generation bioinspired Mn-based catalyst systems for the chemo- and stereoselective oxidation of C–H groups with “green”

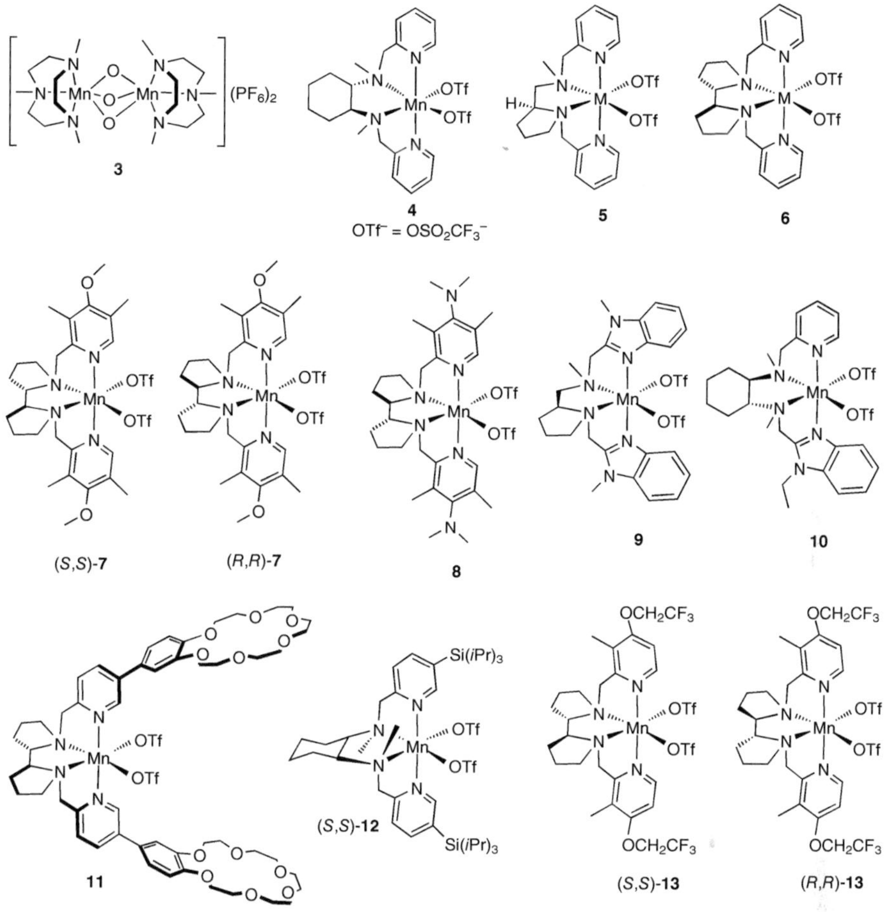

Figure 6.2 Structures of manganese complexes discussed here.

oxidant hydrogen peroxide. In Section 6.3, the catalytic properties of such systems are surveyed.

6.3 Manganese-Catalyzed Chemo- and Regioselective C–H Oxidations with Hydrogen Peroxide

In 2012, it was reported that manganese complexes (**4–6**, Figure 6.2), bearing bis-amine-bis-pyridine ligands, are able to efficiently conduct the oxidation of inactivated aliphatic C–H groups with H_2O_2, performing up to 1000 TN [30]. Using these catalysts (0.1 mol%) and 2.5 equiv of H_2O_2, cyclohexane was oxidized with 78–87% conversion and up to 97% selectivity for cyclohexanone. The oxidation of tertiary C–H groups of (−)-acetoxy-*p*-menthane, *trans*-1,2-dimethylcyclohexane, and *cis*-4-methylcyclohexyl-1-pivalate (Scheme 6.4) occurred stereospecifically, thus indicating fast oxygen rebound without meaningful stereoinversion of the

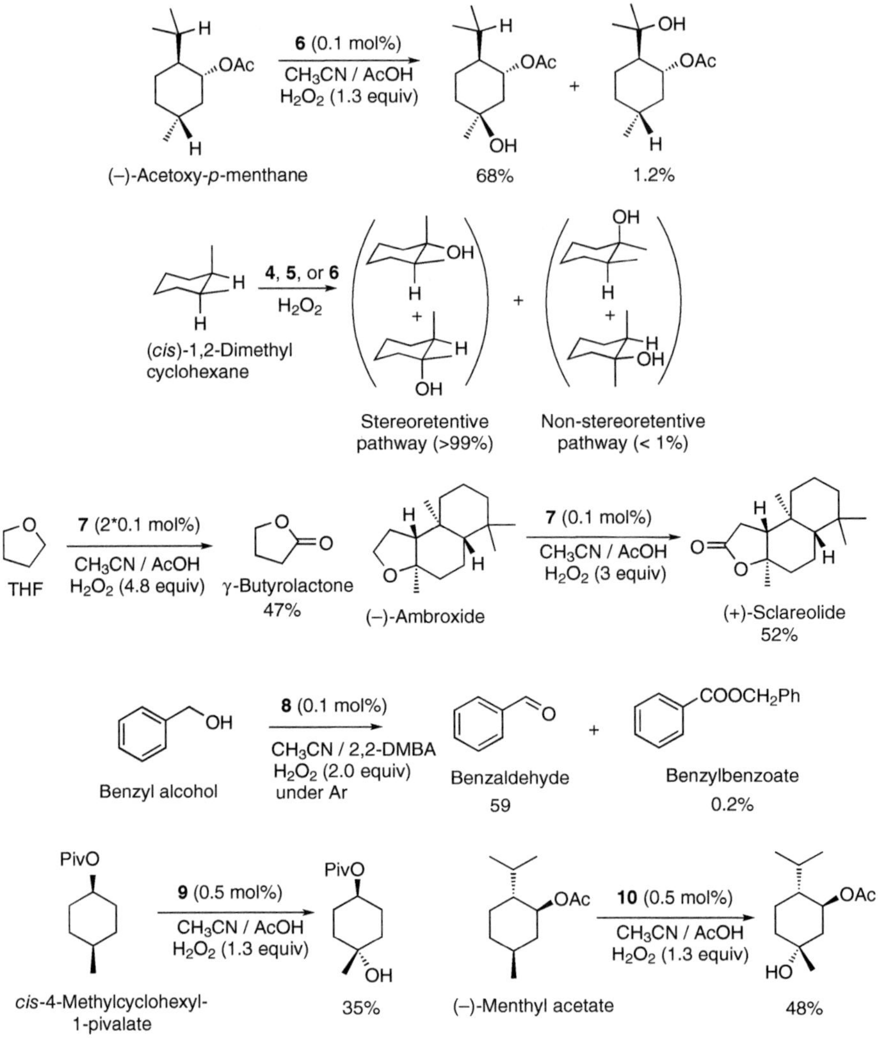

Scheme 6.4 Selective oxidation of complex substrates in the presence of Mn complexes **4–10**. 2,2-DMBA, 2,2-dimethylbutyric acid.

intermediate radical. For the reaction to occur, the use of catalytic additive, acetic acid, was necessary.

Subsequently, the reactivity pattern of analogous manganese complexes was examined. Using complex **7**, tetrahydrofuran and (–)-ambroxide were oxidized to the corresponding lactones, γ-butyrolactone [5] and (+)-sclareolide [8], respectively (Scheme 6.4). The presence of electron-donating substituents, attenuating the catalyst electrophilicity, is crucial for achieving high oxidation chemoselectivity. In a similar manner, in the presence of complex **8**, benzyl alcohol could be oxidized to benzaldehyde with very high selectivity (Scheme 6.4) [31].

Sun and coworkers developed C_1-symmetric complex **9**, which efficiently catalyzed the oxidation of benzylic and aliphatic C–H groups with H_2O_2, with preferential formation of ketones [32]. Complex **9** was also shown to catalyze the oxidation of a series of benzylic alcohols at only 0.05 mol% loadings, demonstrating up to 99% conversions [33]. Subsequently, the same group developed complex **10**, which ensured moderate to good yields of oxidized products in the course of oxidation of benzylic and aliphatic C–H groups (Scheme 6.4) [34].

Using complex **7** (Figure 6.2), Costas and coworkers examined the substrate electronic properties on the oxidation regioselectivity of *N*-1-(3-methylbutyl) derivatives (Scheme 6.5) [35]. The authors confirmed high sensitivity of catalysts of the type **4–10** to electronic effects, while in the presence of alkyl substituent R, the "activated" methylene position, adjacent to the nitrogen atom, oxidized preferentially; the oxidation of trifluoromethyl-substituted derivative afforded predominantly the tertiary alcohol, the remote oxidation product.

Manganese complexes with tetradentate ligands preferentially catalyze the oxidation of more electron-rich methylene (2°) and tertiary (3°) C–H groups [36]. Costas and coworkers suggested an interesting bioinspired approach to overriding the substrate bias, predetermined by its electronic properties, and directing the oxidation toward particular methylene groups [37, 38]. It was shown that, using crown ether appended catalyst **11**, the innate reactivity of undecylammonium tetrafluoroborate can be diverted via reversible "substrate recognition" by the catalyst. In effect, while reference catalyst **6** preferentially catalyzed the oxidation of substrates (in a mixture) with more oxidation-prone 3 ° C–H groups, supramolecular catalyst **11** exhibited high substrate selectivity toward undecyl ammonium tetrafluoroborate, together with high site selectivity, up to 92% C8 + C9 (Scheme 6.5). Linear free-energy relationships witnessed the same transition state for **6** and **11**, with rate-limiting hydrogen atom transfer, thus leaving supramolecular substrate–catalyst interactions, controlling the substrate orientation, as the only apparent mechanism of diverting the reactivity of the active (presumably manganyl) species toward the oxidation of methylene sites in the presence of far weaker 3° sites. The efficiency of this mechanism was demonstrated in a number of competitive oxidation experiments [38].

A serious limitation of the manganese-catalyzed hydroxylation protocols has been low alcohol yields, since the C–H group of the initially formed alcohol is weaker than in the starting substrate and is therefore more prone to further oxidation, affording the corresponding ketones. Costas with coworkers suggested a new synthetic approach, based on the use of hydrogen bond donor solvents (β-polyfluorosubstituted alcohols), capable of forming hydrogen bonds with the alcohol. This reduces its nucleophilicity, thus suppressing its oxidation to ketone [39]. In such a way, the alcohol can be obtained as the major reaction product, see e.g. cyclohexane oxidation in the presence of catalyst **12** (Scheme 6.5).

In a similar manner, by changing the reaction solvent (CH_3CN or trifluoroethanol), it has been possible to improve the selectivity of oxidation of estrone

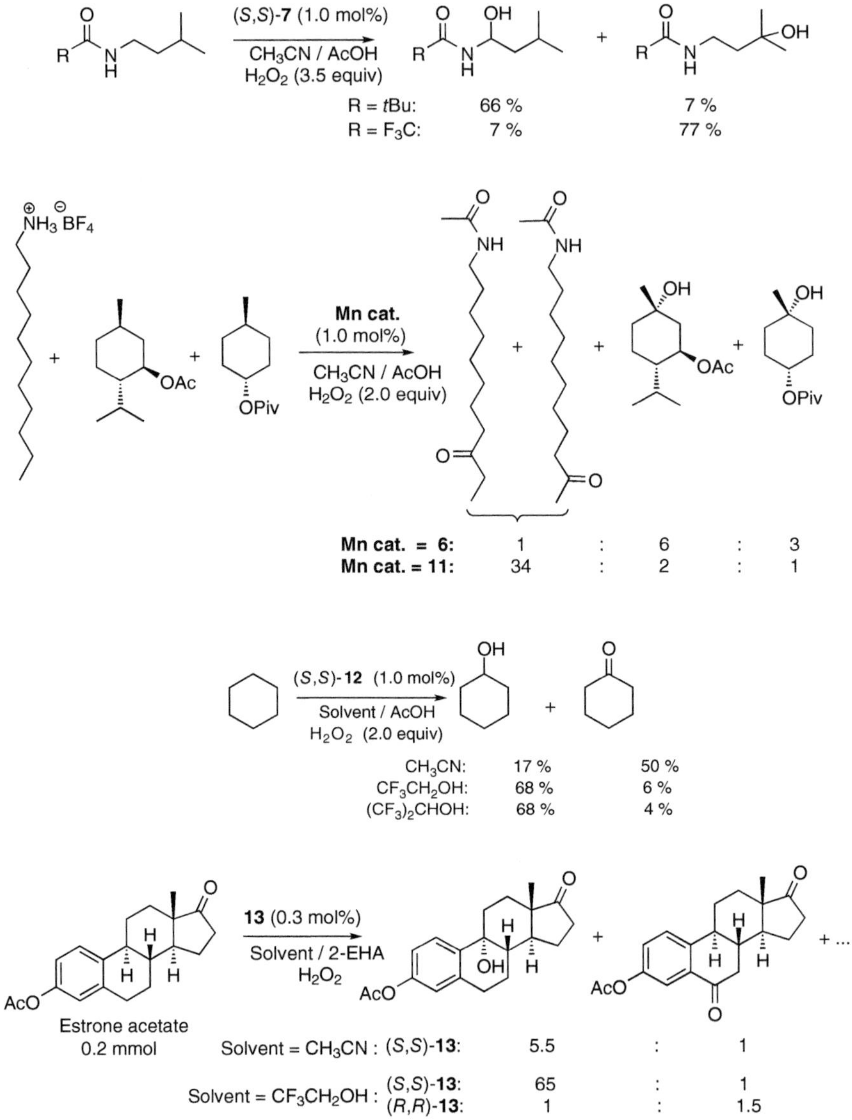

Scheme 6.5 Probing and exploiting steric and electronic effects in the oxidation of *N*-1-(3-methylbutyl) derivatives. 2-EHA, 2-ethylhexanoic acid.

acetate to the corresponding C9 alcohol (Scheme 6.5) [40]. The reason of this selectivity improvement is suppression of the overoxidation of the intermediate C6-alcohol to the C6-ketone. On the other hand, using catalyst with opposite absolute chirality ((*R*,*R*)-**13** instead of (*S*,*S*)-**13**) results in the alteration of the oxidation regioselectivity, in favor of the C6-ketone product.

6.4 Manganese-Catalyzed Enantioselective C–H Oxidations

In 1996, Katsuki and coworkers discovered that chiral (salen)manganese complexes of the type **14** (Figure 6.3), bearing elements of both central and axial chirality, can mediate the asymmetric oxidation of prochiral CH_2 groups with iodosylbenzene to the corresponding chiral secondary alcohols in moderate to good enantioselectivities (Scheme 6.6) [41]. The alcohol yields were low to moderate, which may indicate the presence of oxidative kinetic resolution, reducing the alcohol yield over the course of the catalytic reaction. On the other hand, according to the results of Murahashi et al., (salen)manganese complex **15**, bearing a chiral binaphthyl strapping unit, led to only moderate enantioselectivity of 34% ee (Scheme 6.6) in indane oxidation [42]. Under the conditions of high substrate excess (10 equiv vs. PhIO), reasonably high alcohol/ketone ratio of 4.1 apparently ruled out any significant contribution of oxidative kinetic resolution to the observed ee; the substrate conversions were typically rather low, 18–41%. The synthetic protocol required the use of 1,5-dicyclohexylimidazole as catalytic additive.

In 2012, Simonneaux and coworkers contributed the first catalytic asymmetric hydroxylation of prochiral CH_2 groups with "green" hydrogen peroxide as the terminal oxidant. Several chiral manganese porphyrins [43, 44] were prepared and tested as catalysts in the oxidation of several arylalkanes (ethylbenzene homologs and derivatives) in water–methanol mixtures, using 5 equiv of H_2O_2. The best catalytic performance was demonstrated by the sulfonated water-soluble manganese porphyrin complex **16** (Figure 6.3). High substrate conversions (88–100%) and moderate enantioselectivity was documented (32–57% ee) (Scheme 6.6). Noteworthy, surprisingly good to high alcohol selectivity was reported (the alcohol/ketone ratios were between 1.3 for ethylbenzene and 13 for 4-ethyltoluene). Substituting the SO_3^- groups with H, NO_2, or NMe_2 deteriorated the catalysts enantioselectivity, as well as selectivity for alcohol formation [43, 44].

We found that the asymmetric oxidation of arylalkanes with H_2O_2 can be efficiently performed in the presence of chiral manganese aminopyridine complexes **8** and **13** (Figure 6.2) [45, 46]. Typically, catalyst loadings of only 0.05–0.2 mol% ensured completion of the oxidation in CH_3CN at 0 °C within three hours in the presence of additives (carboxylic acids). Moreover, the use of *chiral* catalytic additives (such as *N*-Boc protected chiral prolines) substantially enhanced the oxidation enantioselectivity, up to 86% ee (Scheme 6.6) [46].

The resulting alcohols were prone to overoxidation to ketones under the reaction conditions, which required using a large excess of substrate and stopping the reaction at low conversion levels; otherwise, the ketone became the major reaction product. Over the reaction course, an increase of the optical purity of 1-phenylethanol was documented, thus indicating the contribution of oxidative kinetic resolution to the observed enantioselectivity.

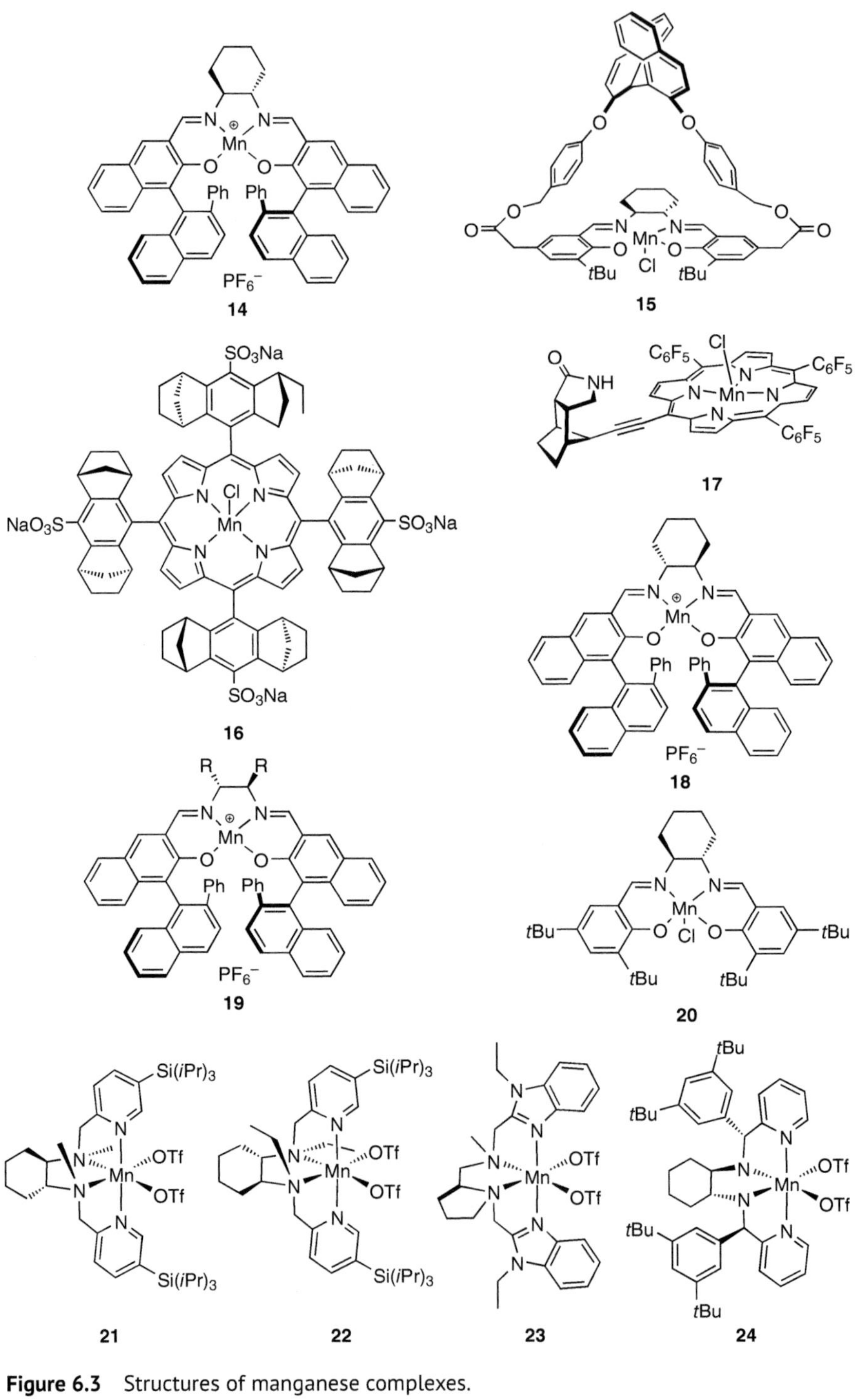

Figure 6.3 Structures of manganese complexes.

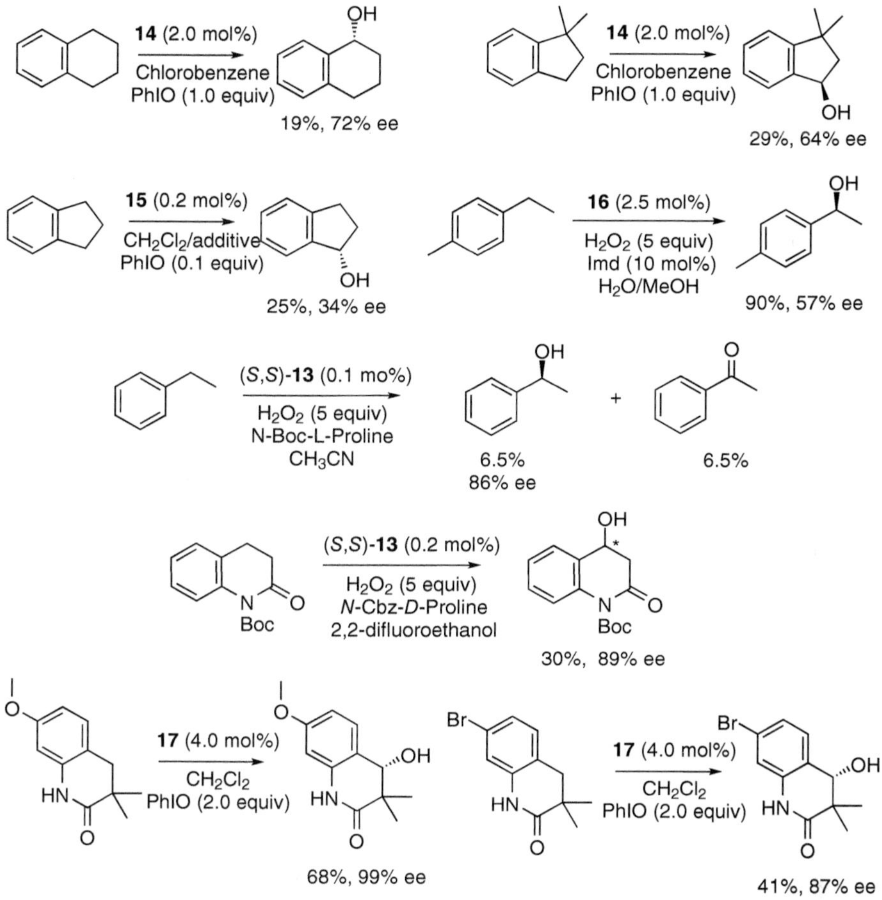

Scheme 6.6 Enantioselective C–H oxidations, mediated by manganese complexes. *N*-Cbz-*D*-proline, *N*-carbobenzyloxy-*D*-proline.

Replacing CH_3CN with β-fluorinated alcohols allowed substantial improvement of the alcohol formation selectivity (Scheme 6.6), 2,2-difluoroethanol ensuring the best combination of alcohol yield and enantioselectivity [40]. In such a way, several chiral benzylic alcohols having 56–89% ee were obtained in 20–30% chemical yield. In some cases, minor effect of kinetic resolution on the alcohol enantiomeric purity was documented.

Bach and coworkers developed the chiral porphyrin **17** (Figure 6.3), which was shown to catalyze the enantioselective oxidation of 3,3-disubstituted 3,4-dihydroquinolones with iodosylbenzene to the corresponding alcohols (Scheme 6.6) [47]. The authors ascribed the high regio- and enantioselectivity to hydrogen bonding between the substrates and the catalytically active porphyrin species, which apparently sets limitations on the substrate scope of this catalyst system.

6.5 Manganese-Catalyzed Oxidative Desymmetrizations

Another approach to chiral molecules through manganese-mediated hydrocarbon oxidation was pioneered by Katsuki and coworkers who in 1997 discovered that chiral (salen)manganese complexes (preferentially **18**, Figure 6.3) can mediate the oxidation of 4-(*tert*-butyl)tetrahydropyran to a mixture of (chiral) 1-alcohol and 1-ketone (Scheme 6.7), thus providing an example of oxidative desymmetrization of cyclic ethers [48]. In the cases of hexahydro-1*H*-cyclopenta[c]furan, 3,4-dimethyltetrahydrofuran, and 8-oxabicyclo[4.3.0]nonane, the reduction step was not necessary, chiral lactol being the major oxidation product (Scheme 6.7) [49, 50]. Subsequently, oxidative desymmetrization of several *meso*-pyrrolidine derivatives (with formation of the corresponding ketones) in the presence of several chiral (salen)manganese complexes, including **18** and **19**, was examined, and moderate to good yields (up to 70%) and enantioselectivities (up to 88% ee) were achieved; in some cases, conducting the reaction in acetonitrile gave better results [50, 51]. The developed oxidative desymmetrization synthetic protocol was utilized for the synthesis of (−)-alloyohimbane, one of the indole alkaloids of the yohimbane family [52]. Noticeably, the enantioselective benzylic hydroxylations with PhIO, mediated by catalysts of the type **14**, **18**, and **19**, exhibiting low yield and moderate to good enantioselectivity (up to 72% ee) were reported, too [53].

Murahashi and coworkers studied Jacobsen's catalyst **20** (both enantiomers) in the oxidative desymmetrization of 2-*tert*-butyldimethylsiloxyindane with iodosylbenzene as oxidant, in low yields (Scheme 6.7) [54]. The authors documented unusual nonmonotonic temperature dependence of enantioselectivity, with a maximum at 40 °C [54, 55].

Milan, Bietti, and Costas studied chiral manganese aminopyridine complexes **21** and **22** (Figure 6.3) as catalysts for the oxidation of mono-substituted cyclohexanes with H_2O_2 in the presence of carboxylic acid additives (Scheme 6.7) [56]. With cyclopropanecarboxylic acid additive, along with generally high C3-selectivity of oxidation (up to >99%), the resulting ketones demonstrated high enantioselectivities up to 96% ee. The highest yields of C3-ketone, as well as the highest enantioselectivities, were achieved for the oxidation of amide substituted cyclohexanes (Scheme 6.7). The crucial role of the amine moiety in achieving high regio- and enantioselectivity has remained unclear [56].

A related approach was proposed by Nam and Sun and coworkers who developed a synthetic protocol for the oxidative desymmetrization of spirocyclic β-ketones, mediated by chiral benzimidazole-derived manganese complexes [57]. With the best catalyst **23** (Figure 6.3), spirocyclic β,β′-diketones were obtained in moderate to good yields and good to excellent enantioselectivities (up to 98% ee, Scheme 6.7). The authors demonstrated scalability of the oxidation protocol to gram quantities and proposed further ketone reduction protocol (to alcohols), retaining the enantiomeric purity [58]. Subsequently, an analogous protocol for the oxidation of spirocyclic oxindoles to the corresponding ketones (with up to 91% ee) and alcohols (with up to 99% ee, Scheme 6.7, was developed) [58].

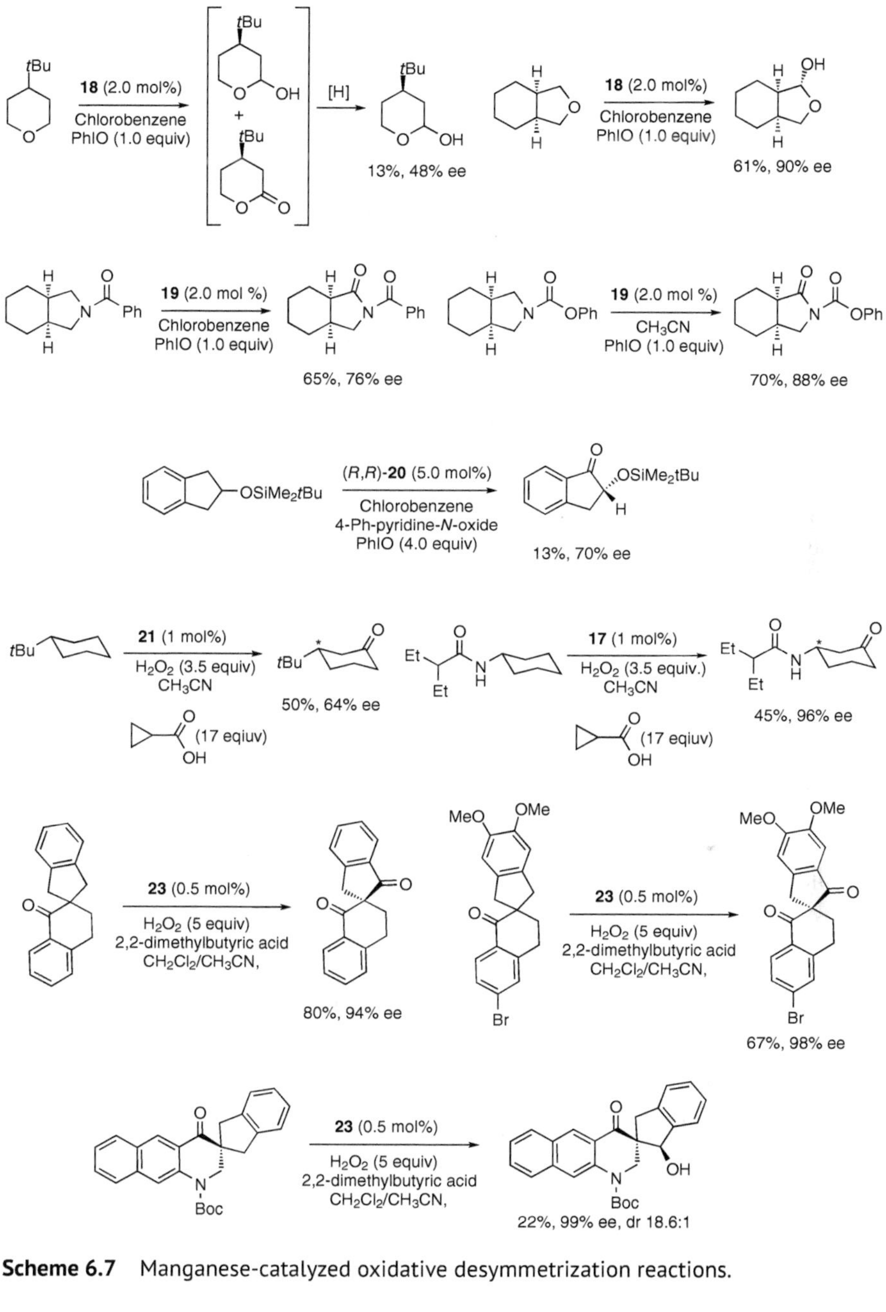

Scheme 6.7 Manganese-catalyzed oxidative desymmetrization reactions.

6.6 Mn-Catalyzed Oxidative Kinetic Resolution of *sec*-Alcohols

Oxidative kinetic resolution of racemic secondary alcohols provides an alternative synthetic approach to enantiomerically enriched alcohols. Until now, the majority of catalytic applications have relied on chiral (salen)manganese-based catalysts, either homogeneous or immobilized. In most cases, hazardous or high molecular weight oxidants were used, such as sodium hypochlorite, iodobenzene diacetate, or *N*-bromosuccinimide (with apparently the only exception when H_2O_2 was used [59]), which imposes constraints on their potential synthetic applications; an overview of this area can be found in Ref. [60]. In this section, we briefly survey the novel manganese-based catalysts (reported since 2017), capable of conducting the oxidative kinetic resolution of secondary alcohols, affording the unreacted desired enantiomerically enriched enantiomer and the C–H oxidation product, ketone.

In 2017 we found that chiral bis-amine-bis-pyridine manganese complexes of the types **7**, **8**, and **13** demonstrate different reactivities toward different enantiomers of 1-arylalkanols (to form ketones), thus allowing their kinetic resolution with H_2O_2 as the terminal oxidant [45]. The resolution selectivity factors (k_{rel}, the ratio of rate constants for the oxidation of the more reactive and the less reactive substrate enantiomers) were moderate, not exceeding 8.8 (in the case of *p*-tolylethanol). Remarkably, the use of carboxylic acid additives had detrimental effect on the selectivity factor, which eventually prompted us to get rid of additives (Scheme 6.8). Surprisingly, in the absence of additives, the observed k_{rel} values were not constant over the reaction course and increased as the conversions increased (Figure 6.4) [45]. This effect, explained by the coordination of the chiral substrate molecules to the catalytically active sites, which altered the instant catalyst stereoselectivity, was termed "asymmetric autoamplification" [61]. The mechanistic study provided evidence for the rate-limiting C–H atom abstraction, followed by oxygen rebound and *gem*-diol dehydration mechanism [46].

Alternatively, Sun and coworkers proposed an approach based on using chiral manganese complexes of the type **24** (Figure 6.3), which efficiently performed in the oxidative kinetic resolution of several secondary alcohols with H_2O_2 in the presence of small amounts of sulfuric acid additive (Scheme 6.8) [62]. Similarly, moderate selectivity factors (up to 9.3) were achieved in most cases, with the exception of *p*-NO_2-phenylethanol-1, in which case the reported conversion and ee values corresponded to k_{rel} of 17.

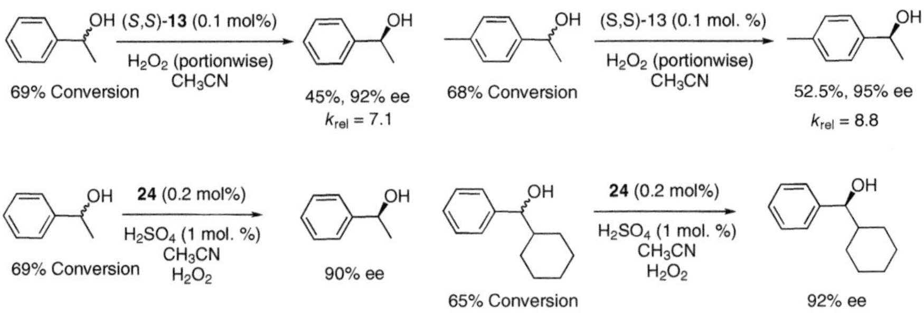

Scheme 6.8 Mn-catalyzed oxidative kinetic resolutions of *sec*-alcoholic substrates.

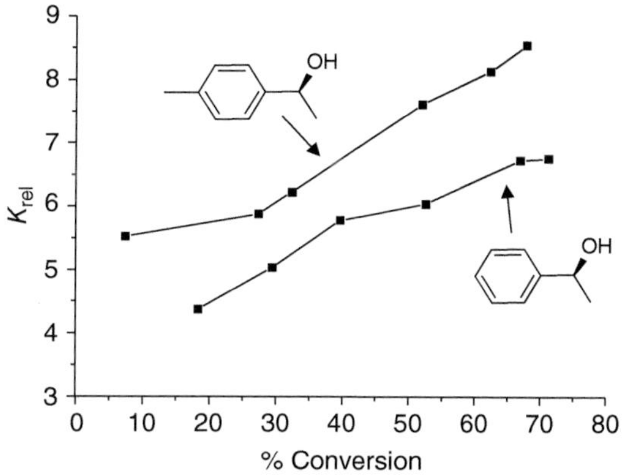

Figure 6.4 k_{rel} vs. conversion plot for the oxidative kinetic resolution of 1-arylethanols in the presence of (*S*,*S*)-**13**.

6.7 Mn-Catalyzed Oxidation of Aromatic Compounds

So far, examples of manganese-catalyzed aromatic C–H oxidations have been scarce, mainly focusing on the hydroxylation of naphthalenes in the presence of manganese porphyrins. Particularly, in 2000, Mansuy and coworkers examined the catalytic activities of β-polynitroporphyrins of the type **25** (Figure 6.5) in the oxidation of naphthalene with hydrogen peroxide in the presence of ammonium mandelate as cocatalyst [63]. Under high substrate excess conditions ([substrate]:[oxidant]:[Mn] = 500 : 50 : 1), 97% conversion (based on oxidant) was documented, with formation of α-naphthol (72%) and β-naphthol (7%) products (Scheme 6.9). A very similar products ratio (9,1) was documented in the kinetic study of naphthalene hydroxylation in the presence of manganese porphyrins of the type **26** (Figure 6.5), using peracetic acid as the terminal oxidant (Scheme 6.9) [64]. Guo and coworkers examined the oxidation of naphthalene and β-substituted naphthalenes with *m*-chloroperbenzoic acid in the presence of 0.1 mol% of manganese porphyrin complex **2** (Figure 6.1) and reported good selectivity for α-monohydroxylation of naphthalenes with electron-donating β-substituents

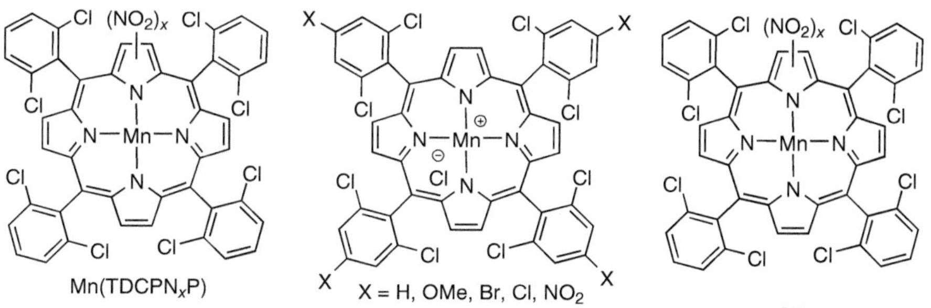

25 **26** **27**

Figure 6.5 Examples of manganese porphyrin-based selective aromatic C–H oxidation catalysts.

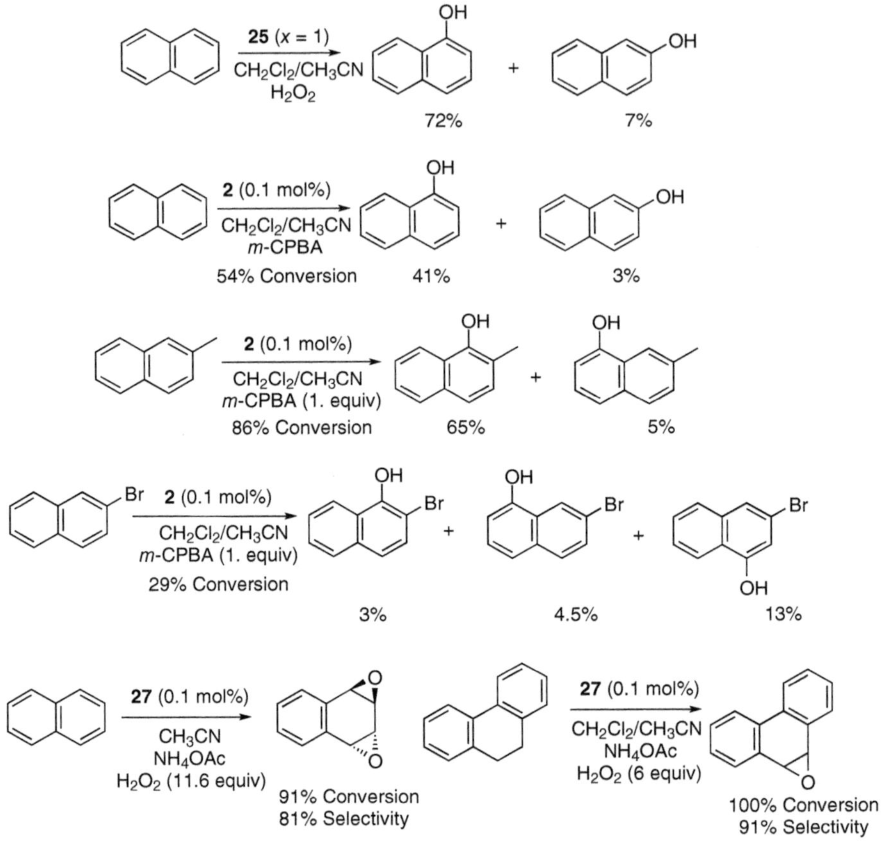

Scheme 6.9 Selective oxidation of aromatic compounds in the presence of manganese porphyrin complexes.

(Scheme 6.9) [65]. At the same time, the presence of electron-withdrawing β-substituents reduced the conversion and diverted the selectivity toward the oxidation at the δ-C–H site (Scheme 6.9). In general, it appears that the benzylic CH_2 groups in aromatic compounds are oxidized preferentially [63].

As distinct from the hydroxylation reactivity reported earlier, the oxidation of naphthalene and anthracene with excess of H_2O_2 in the presence of manganese porphyrin **27** (Figure 6.5) afforded predominantly the corresponding *anti*-1,2:3,4-arene diepoxides with 81% and 74% selectivity), respectively, while phenanthrene was converted to phenanthrene epoxide with 91% selectivity at 100% substrate conversion (Scheme 6.9) [66]. Due to the high catalytic efficiencies and selectivity, this system holds practical potential and could pave a new way to the synthesis of polycyclic aromatic hydrocarbon metabolites or corresponding precursors and to the transformation of polycyclic aromatic hydrocarbons into other functionalized compounds [66].

6.8 Summary and Conclusions

Since its emergence to the present, the area of chemo-, regio-, and stereoselective oxidations of aliphatic C–H groups in the presence of manganese catalysts has made spectacular progress. The current flourishing is mostly associated with the success of bioinspired, bis-amine-bis-pyridine manganese complexes that have dominated the field during the last decade. With the repertoire of so far reported manganese-based catalysts in hand, one can (i) in some cases conduct the oxidations with good to very high catalytic efficiencies (1000 TN or higher); (ii) use environmentally benign oxidant H_2O_2; (iii) selectively activate particular C–H groups in complex organic molecules, overcoming their innate reactivities; and (iv) conduct the oxidations in stereospecific and, most significantly, stereoselective fashion, in some cases with up to 99% enantioselectivity [67]. In all those respects, manganese-based catalysts have demonstrated superior characteristics as compared with their iron counterparts [5, 68].

While many manganese-based catalysts bear significant synthetic promise, practical aspects of their applications have remained significantly underdeveloped, although attempts to evaluate their synthetic potential are being undertaken [69]. To make the step from the laboratory to practice, the majority of existing catalyst systems need to broaden the substrate scopes, enhance the oxidation productivity (TN), and improve the selectivity of main product formation (especially alcohols), as well as oxidant consumption efficiency [70]. The subarea of manganese-catalyzed oxidative kinetic resolutions with H_2O_2, surprisingly prolific for the theory of asymmetric catalytic oxidations [71, 72], has so far been particularly underdeveloped from the practical point of view. To date, the oxidations of aromatic C–H groups in the presence of manganese complexes have been exemplified by only a handful of examples, in most cases focused on fundamental selectivity studies. Overall, there is much room for further developments that will likely follow in the near future.

More challenging could be the design of manganese-based catalyst systems, combining the capabilities of heterolytically activating atmospheric dioxygen in a cyclic process and ensuring the oxygen atom transfer to substrate with high level of regio- and stereocontrol. Solving this problem could make revolutionary impact on the approaches to the selective functionalization of complex organic molecules and is therefore highly tempting, yet this driving force so far has not produced synthetically feasible crops. Overall, since the milestone discoveries of Katsuki [73] and Jacobsen [74], manganese has been at the forefront of the modern oxidation catalysis, and the author would expect significant new advancements in near and medium-term future.

Acknowledgment

The author acknowledges support from the Russian Science Foundation (#20-13-00032).

References

1 Gunnoe, T.B. (2012). Introduction: Alkane C—H Activation by Single-Site Metal Catalysis. (ed. P.J. Pérez), 1–15. Dordrecht, Germany: Springer.
2 Krämer, K. (2017) *Chemistry World* https://www.scientificamerican.com/article/predictions-for-the-2017-chemistry-nobel-prize (accessed 15 January 2020).
3 Costas, M., Chen, K., and Que, L. Jr. (2000). *Chem. Rev.* 200: 517–544.
4 Talsi, E.P. and Bryliakov, K.P. (2012). *Coord. Chem. Rev.* 256: 1418–1434.
5 Ottenbacher, R.V., Talsi, E.P., and Bryliakov, K.P. (2016). *Molecules* 21: 1454.
6 Lubov, D.P., Talsi, E.P., and Bryliakov, L.P. (2020). *Russ. Chem. Rev.* 89: 587–628.
7 Groves, J.T. (2006). *J. Inorg. Biochem.* 100: 434–447.
8 Ottenbacher, R.V., Talsi, E.P., and Bryliakov, K.P. (2015). *ACS Catal.* 5: 39–44.
9 Cho, K.B., Hirao, H., Shaik, S., and Nam, W. (2016). *Chem. Soc. Rev.* 45: 1197–1210.
10 Talsi, E.P., Ottenbacher, R.V., and Bryliakov, K.P. (2015). *J. Organomet. Chem.* 793: 102–107.
11 Milan, M., Bietti, M., and Costas, M. (2018). *Chem. Commun.* 54: 9559–9570.
12 Milan, M., Salamone, M., Costas, M., and Bietti, M. (2018). *Acc. Chem. Res.* 51: 1984–1995.
13 Bryliakov, K.P. (2017). *Chem. Rev.* 117: 11406–11459.
14 Grieco, P.A. and Stuck, T.L. (1990). *J. Am. Chem. Soc.* 112: 7799–7801.
15 Stuck, T.L., Grieco, P.A., and Marsh, M.M. (1991). *J. Org. Chem.* 56: 2957–2959.
16 Sorokin, A.B., Robert, A., and Meunier, B.J. (1993). *Am. Chem. Soc.* 115: 7293–7299.
17 Meunier, B. (1992). *Chem. Rev.* 92: 1411–1456.
18 Breslow, R., Zhang, X., and Huang, Y. (1997). *J. Am. Chem. Soc.* 119: 4535–4536.
19 Yang, J., Gabriele, B., Belvedere, S. et al. (2002). *J. Org. Chem.* 67: 5057–5067.
20 Yang, J. and Breslow, R. (2000). *Tetrahedron Lett.* 41: 8063–8067.
21 Mohajer, D. and Bagherzadeh, M. (1998). *J. Chem. Res.*: 556–557.
22 Costas, M. (2011). *Coord. Chem. Rev.* 255: 2912–2932.
23 Da Silva, V.S., Nakagaki, S., Ucoski, G.M. et al. (2015). *RSC Adv.* 5: 106589–106598.
24 Bagherzadeh, M. and Mortazavi-Manesh, A. (2016). *RSC Adv.* 6: 41551–41560.
25 Lindsay Smith, J.R. and Shul'pin, G.B. (1998). *Tetrahedron Lett.* 39: 4909–4912.
26 Shul'pin, G.B. and Lindsay Smith, J.R. (1998). *Russ. Chem. Bull.* 47: 2379–2386.
27 Shul'pin, G.B., Süss-Fink, G., and Lindsay Smith, J.R. (1999). *Tetrahedron* 55: 5345–5358.
28 Shul'pin, G.B., Süss-Fink, G., and Shul'pina, L.S. (2001). *J. Mol Catal. A: Chem.* 170: 17–34.
29 Shul'pin, G.B., Nesterov, D.S., Shul'pina, L.S., and Pombeiro, A.J.L. (2017). *Inorg. Chim. Acta* 455: 666–676.
30 Ottenbacher, R.V., Samsonenko, D.G., Talis, E.P., and Bryliakov, K.P. (2012). *Org. Lett.* 14: 4310–4313.
31 Ottenbacher, R.V., Talsi, E.P., and Bryliakov, K.P. (2018). *Chem. Rec.* 18: 78–90.
32 Shen, D., Miao, C., Wang, S. et al. (2014). *Org. Lett.* 16: 1108–1111.

7

Manganese-Catalyzed Organometallic C—H Activation

Yunhui Yang and Congyang Wang

Institute of Chemistry, Chinese Academy of Sciences, Beijing National Laboratory for Molecular Sciences, CAS Key Laboratory of Molecular Recognition and Function, CAS Research/Education Center for Excellence in Molecular Sciences, Zhongguancun North First Street 2, Beijing 100190, China

7.1 Introduction

Transition metal-catalyzed activation of ubiquitous yet inert C—H bonds, which is thought to be the "holy grail" of organic chemistry, opens a new door to develop synthetic tools for biochemistry, material sciences, fine chemicals, and pharmaceutical industries. The inherent merits in atom- and step-economy of direct C—H bond functionalization make it more and more attractive in the area of chemical bond construction, especially in that of omnipresence C—C bond formation. The site-selective and catalytic transformations of C—H bonds to C—C bonds offer superior strategies to build up functional molecules compared to classical cross-coupling reactions, which involve pre-functionalized organohalides and organometallic reagents R_nM (M = Mg, Zn, B, Sn, Si, etc.). So far, precious and often toxic transition metals such as palladium, ruthenium, rhodium, and iridium have played a predominant role in the field of metal-catalyzed C—H activation. Nevertheless, the drawbacks of these metals limit their wide applications in industries and pharmaceuticals. Thus, it is highly desirable to develop more sustainable and economical alternative catalysts based on Earth-abundant metals for inert C—H transformations.

Manganese is the twelfth most abundant element and the third transition metal in the Earth crust (after iron and titanium). It is also an important trace nutrient in all forms of life. Without a doubt, manganese represents a highly attractive candidate for new catalyst development in accounts of its natural abundance, low cost, and good biocompatibility. Actually, various manganese complexes, in particular high-valance manganese species, have been extensively used in the catalysis of C—H nitrogenations [1], oxidation [2], and halogenations [3] reactions via an outer-sphere homolytic C—H cleavage regime over the last few decades, which have been well documented [4] and will not be considered herein. This chapter is limited to the catalytic processes promoted by manganese intermediates bearing Mn—C bonds, aiming to present a comprehensive overview on the latest and most representative

Manganese Catalysis in Organic Synthesis, First Edition. Edited by Jean-Baptiste Sortais.

contributions in the field of organomanganese-catalyzed C—H activation with focus on the construction of C—C bonds until August 2019.

7.2 Stoichiometric Cyclometalation with Manganese Complexes

As early as 1970, Stone, Bruce, and coworkers described the first stoichiometric manganese-mediated C—H activation of azobenzene with $MnMe(CO)_5$ leading to the formation of five-membered manganacycle, which could be isolated by column chromatography as a red solid in 93% yield (Scheme 7.1) [5]. After this pioneering work, a plethora of manganacycles have been successively synthesized from $MnR(CO)_5$ (R = Me, Bn, Ph, etc.) and the corresponding (hetero)arenes through similar protocols (Scheme 7.2). Generally, such inner-sphere manganese-mediated C—H activation for the formation of organomanganese complexes possessing well-defined Mn—C bonds usually required the assistance of coordinating functional groups.

Mechanistically, the cyclomanganation process may involve the first generation of a vacant site at the Mn center and then the coordination of arenes and cleavage of the C—H bond to yield a Mn—C bond. Specifically, as shown in

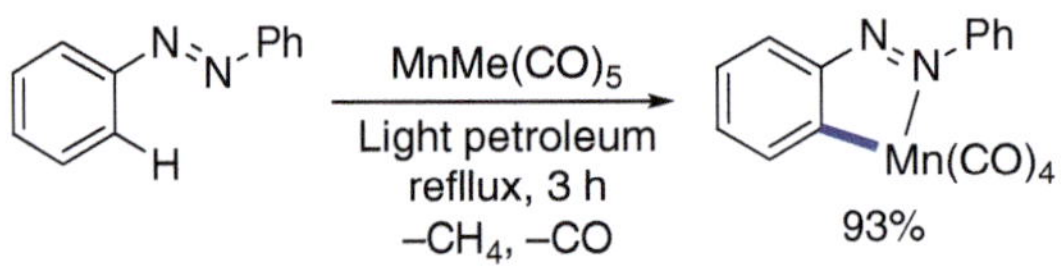

Scheme 7.1 Stoichiometric cyclometalation of azobenzene with $MnMe(CO)_5$. Source: Redrawn from Bruce et al. [5].

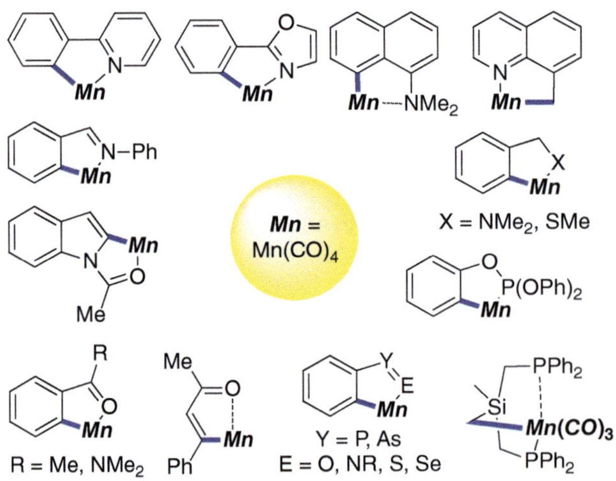

Scheme 7.2 Other representative manganacycles via cyclomanganation.

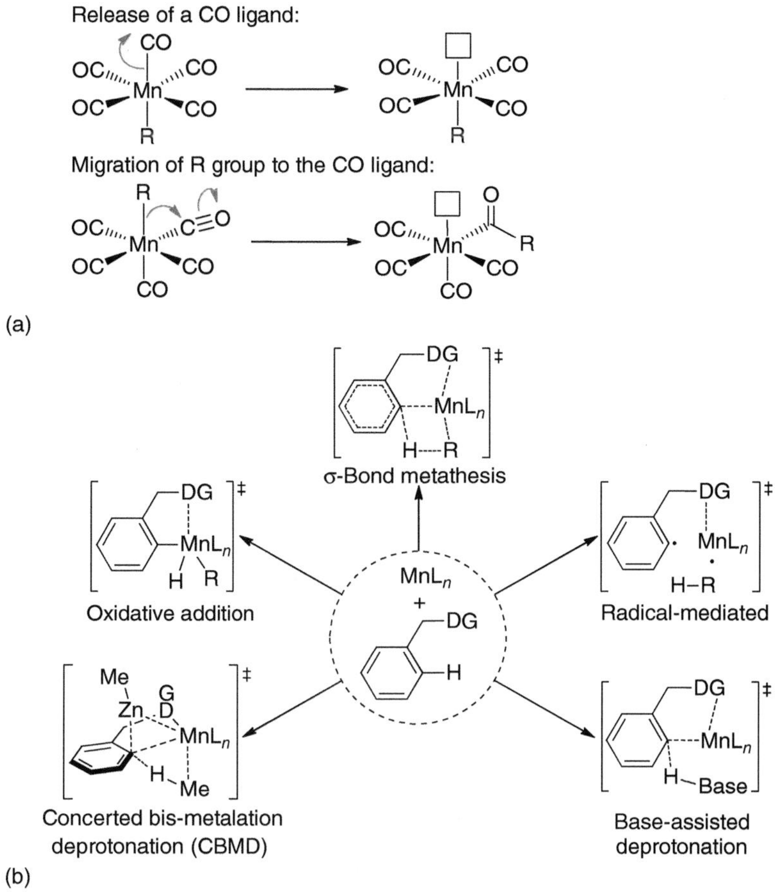

Scheme 7.3 Mechanistic pathways for manganese-promoted C—H bond cleavage. (a) Generation of the vacant site and (b) coordination and organometallic C—H activation.

Scheme 7.3a, release of a CO ligand from $MnR(CO)_5$ might generate the vacant site; the vacant site might be alternatively formed by migration of R group to the CO ligand [6]. With respect to the elementary step of organometallic C—H activation [7] (Scheme 7.3b), though controversial, it is proposed to proceed via an oxidative addition, but no solid evidences have yet been reported [8], a four-center transition state (σ-bond metathesis) with no change in the metal oxidation state [9], or a metallo-radical-mediated pathway [10]. In sharp contrast, more recent computational and experimental mechanistic studies revealed novel modes of C—H manganation through base-assisted deprotonation [11] and Mn–Zn-enabled concerted bis-metalation deprotonation (CBMD) mechanism [12].

As a follow-up on the successful preparation of various manganacycles, the reactivity of the Mn—C bonds in these complexes was further investigated by the group of Nicholson and Main [13], Liebeskind [14], Woodgate [15], etc. With the aid of a stoichiometric amount of $PdCl_2$ or trimethylamine *N*-oxide ($Me_3N^+O^-$), unsaturated molecules such as acrylates, alkenes, and alkynes inserted into the

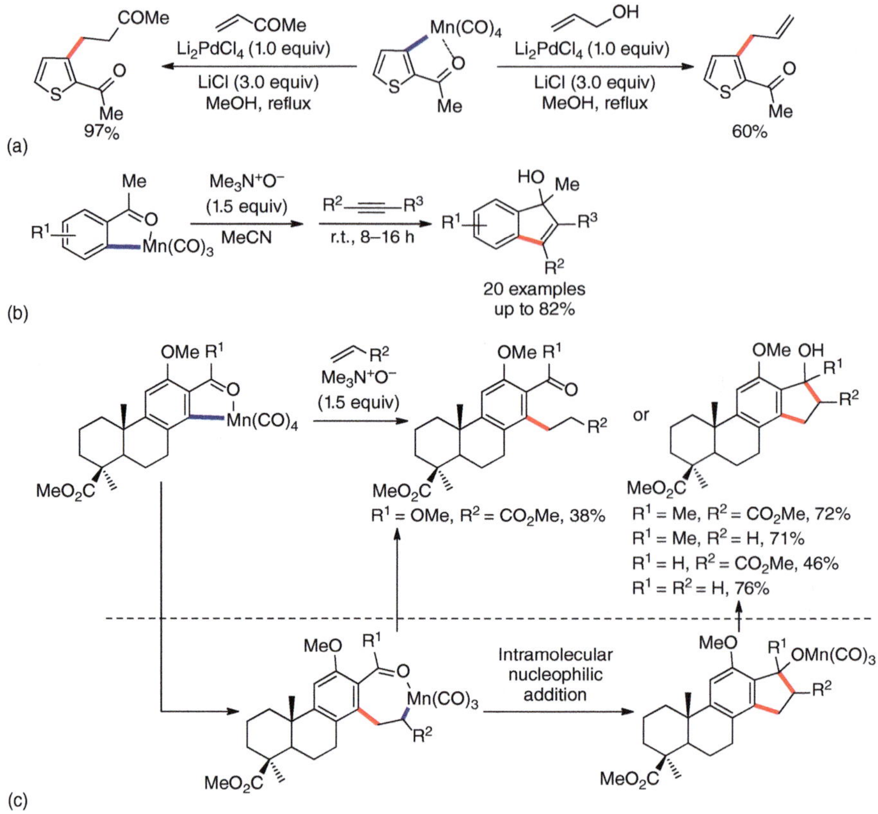

Scheme 7.4 C–C bond formation through insertion of unsaturated molecules into Mn–C bonds. (a) Nicholson and coworkers [13], (b) Liebeskind et al. [14], and (c) Woodgate and coworkers [15]. Sources: (a) Redrawn from Gommans et al. [13d]; (b) Redrawn from Liebeskind et al. [14]; and (c) Redrawn from Cambie et al. [15].

Mn—C bonds of manganacycles smoothly, resulting in the formation of new C—C bonds (Scheme 7.4).

7.3 Manganese-Catalyzed C–C Bond Formation via Directed C–H Activation

Insertion of unsaturated nonpolar or polar molecules such as alkynes, alkenes, allenes, aldehydes, imines, etc. into the Mn—C bonds of manganacyclic species are, by far, the most documented key steps among the Mn-catalyzed C—C bond formation reactions via organometallic C—H activation pathways.

7.3.1 Insertion of C≡C Bonds into Manganacycles

Inspiration for developing manganese-catalyzed C—H functionalization is largely based on the transformations of inert C_{sp^2}—H bonds into C—C bonds utilizing stoichiometric amounts of alkyl pentacarbonyl manganese reagent $MnR(CO)_5$. Though the groups of Liebeskind [14], Nicholson and Main [13b, 16], and Woodgate [15a], among others, have achieved manganese-mediated stoichiometric reactions between arenes and alkynes, analogous manganese-catalyzed aromatic C—H activation with alkynes remained elusive for more than 20 years, until the seminal work reported by Chen and Wang in 2013 [11a]. Therein, the authors described the first manganese-catalyzed C_{sp^2}–H alkenylation with terminal alkynes affording *E*-alkenylated products via an anti-Markovnikov addition pathway (Scheme 7.5). It is of note that the undesired di- or trimerization of terminal alkynes was inhibited in the presence of catalytic amounts of $MnBr(CO)_5$ and dicyclohexylamine (Cy_2NH). Remarkably, this protocol featured with a high regio-, stereo-, and monoselectivity, as well as good tolerance of various functional groups.

Experimental investigations on a string of bases demonstrated that Cy_2NH gave the best performance due to its moderate steric hindrance and basicity. The stoichiometric reaction of 2-phenylpyridine with $MnBr(CO)_5$ showed that Cy_2NH was essential to the C—H bond cleavage affording manganacycle in 31% isolated yield. DFT (Density Functional Theory) calculations further borne out the base-assisted deprotonation mechanism for the C—H activation step with an energy barrier of 12.5 kcal/mol (Scheme 7.6). Though the similar C—H activation mode was often observed in noble metal-catalyzed C—H activation reactions [17], it was the first time it was proven to be applicable in manganese catalytic systems.

On the basis of experimental and computational mechanistic studies, the authors put forward a plausible mechanism as shown in Scheme 7.7. The catalytic

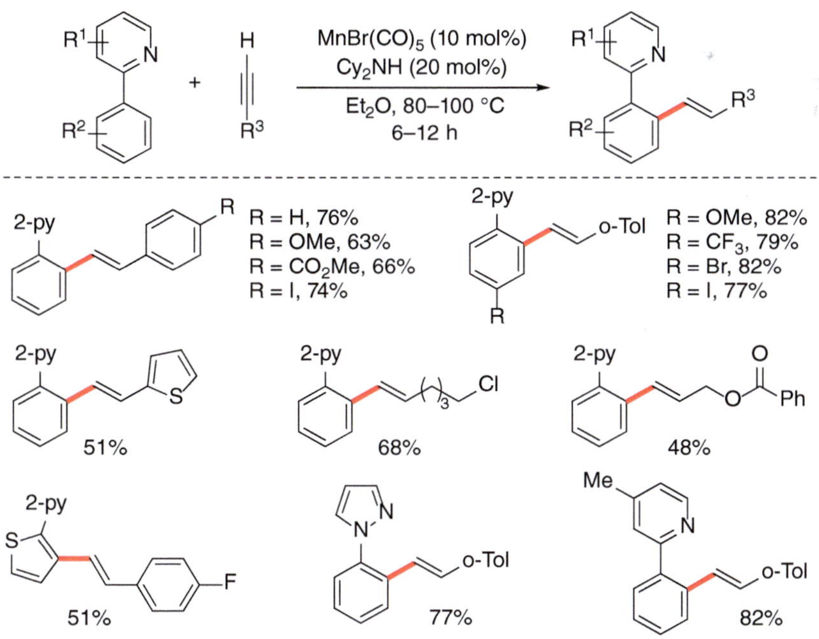

Scheme 7.5 The first manganese-catalyzed C_{sp^2}–H alkenylation with terminal alkynes.

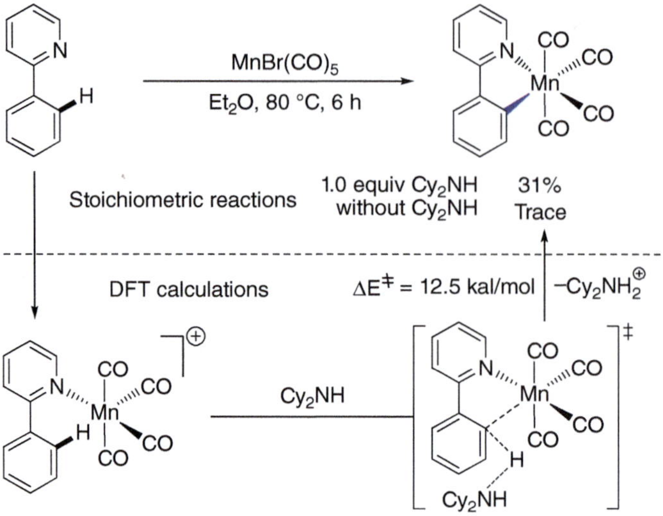

Scheme 7.6 Experimental and DFT studies for the synergetic effect of $MnBr(CO)_5$ and Cy_2NH in C_{sp^2}–H activation.

cycle is initiated by a base-assisted deprotonative C—H activation process of 2-phenylpyridine **1** to deliver manganacycle **Mn-1**. A coordination and migratory insertion of terminal alkyne **2** via intermediate **Mn-2** into the Mn—C bond of **Mn-1** gave the seven-membered manganacycle **Mn-3**. It then coordinated with a second alkyne and underwent a proto-demetalation step via ligand-to-ligand H-transfer (LLHT) to furnish species **Mn-5**. Ligand exchange occurred between **Mn-5** and **1** to afford the desired alkenylated product **3** and alkynyl manganese species **Mn-6**, followed by an alkyne-assisted C—H activation step via σ-bond metathesis regenerating **Mn-2** to complete the catalytic cycle of manganese.

In the previously mentioned catalytic cycle, the five-membered manganacycle **Mn-1** has been characterized unambiguously by Wang's group, and analogous seven-membered manganacycles have been synthesized independently by the groups of Nicholson and Main [16b] and Suárez [18]. While Wang's group failed to give more details about the seven-membered manganacycle **Mn-3** owing to its fast protonation with the terminal alkyne in catalysis, interestingly, Fairlamb, Lynam, and coworkers revealed a related transformation from five-membered manganacycles to seven-membered manganacycles [19], which provided strong endorsement for the mechanism as shown in Scheme 7.7.

Imine, one of simple yet synthetically useful functional groups, was used as a directing group (DG) for C—H activation in the stoichiometric cyclomanganation of benzylideneaniline described by Bruce and coworkers in 1971 [20]. However, the imine-directed manganese-catalyzed C—H activation reactions remain elusive for a long time. In 2014, Wang's group disclosed the first manganese-catalyzed dehydrogenative [4+2] annulation of *N*–H imines and alkynes via C—H/N–H activation, thereby providing expedient access to biologically meaningful isoquinolines in a step- and atom-economical manner (Scheme 7.8) [21]. Notably, this practically

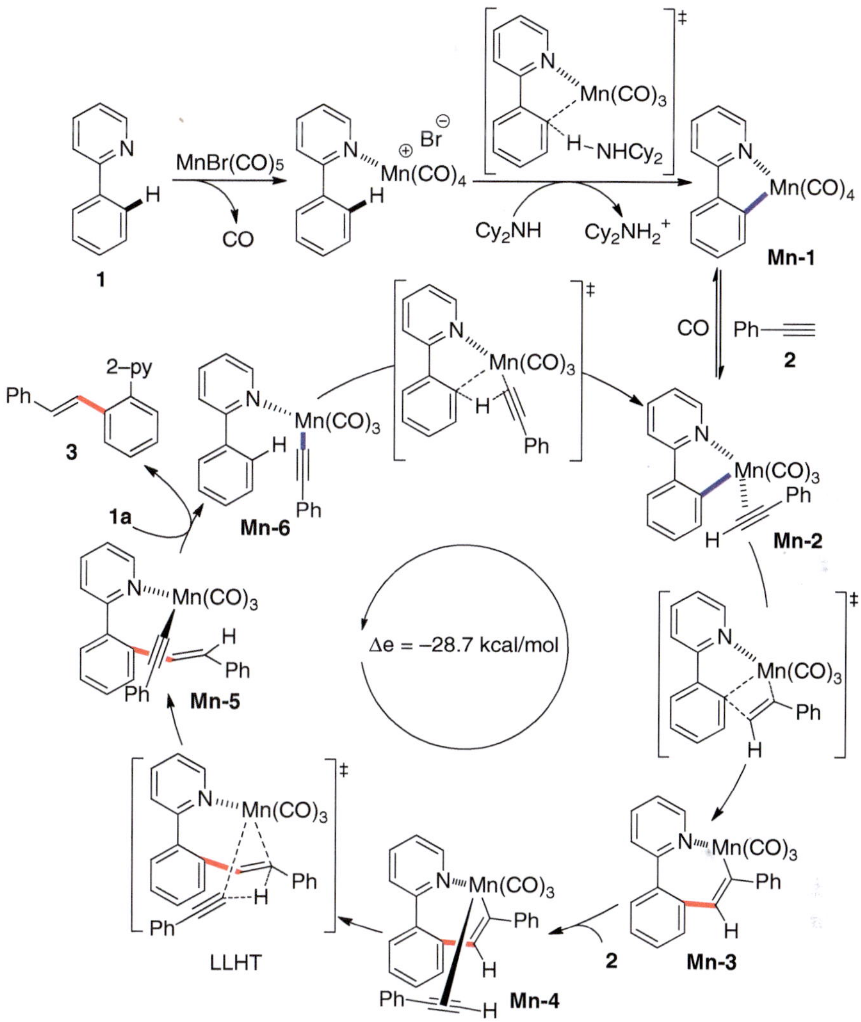

Scheme 7.7 Proposed mechanism for manganese-catalyzed C_{sp^2}–H alkenylation with terminal alkynes.

simple reaction eliminated the need of any oxidants, external ligands, or additives and extrudes H_2 as the major by-product, thus outcompeting known isoquinoline synthesis via other transition metal-catalyzed C—H annulations [22] in terms of atom-economy, sustainability, and tolerance of varied functional groups.

Based on the mechanistic studies of Wang's group [21] and the DFT calculations of Fu and Shi's group [23], a plausible catalytic cycle is shown in Scheme 7.9. The cyclomanganation of imine with $MnBr(CO)_5$ first generates five-membered manganacycle **Mn-7** with the assistance of bromine anion. Regioselective migratory insertion of alkyne into the Mn—C_{aryl} bond of **Mn-7** via a four-center transition state **TS-2** forms the seven-membered manganacycle **Mn-9**, which undergoes intramolecular

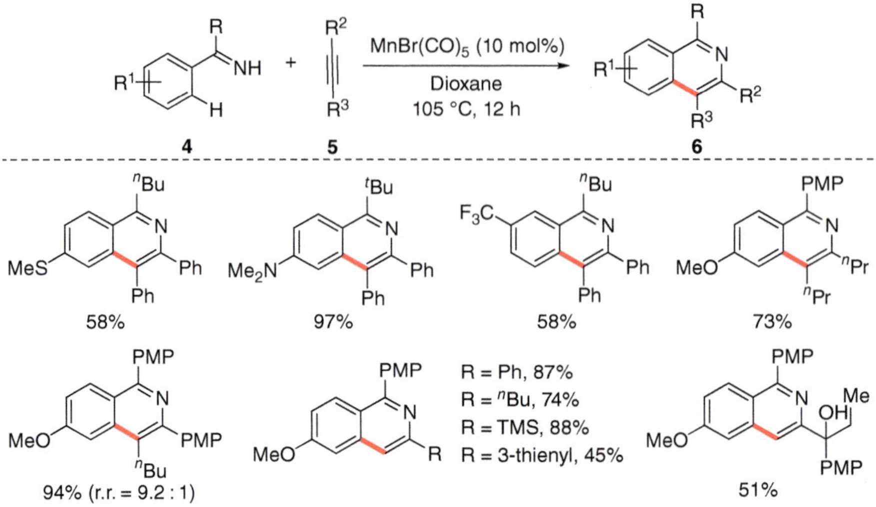

Scheme 7.8 Manganese-catalyzed dehydrogenative [4+2] annulation of *N*–H imines and alkynes. Source: Redrawn from He et al. [21].

cyclization, leading to a fused manganacycle **Mn-10** through transition state **TS-3**. The subsequent β–H elimination of **Mn-10** gives rise to product isoquinoline **6** and manganese hydride **Mn-11**. The coordination of unsaturated Mn–H species combines with imine **4** and subsequent C—H activation regenerates **Mn-7** along with evolution of H_2 through a concerted dehydrogenation process (**TS-4**).

In 2015, Li, Lei, and coworkers described a manganese/base co-catalyzed C—H alkenylation of indoles with terminal and internal alkynes (Scheme 7.10) [24]. It was found that a catalytic amount of benzoic acid, working as an H-transmitter in the C—H alkenylation process, was essential to control the reaction chemoselectivity. Otherwise, in the absence of benzoic acid, indoles were prone to undergo [2+2+2] cyclization with alkynes leading to carbazoles, involving H_2 evolution from Mn–H species and indoles, which echoed the work from Wang's group [21].

The direct *ortho*-C—H alkenylation of aromatic nitriles has largely lagged behind due to the linear configuration of the cyano group. To address this issue, Wang and coworkers devised a mask strategy to access *mono*-C—H alkenylated aromatic nitriles (Scheme 7.11) [25]. Specifically, alcohol was introduced as the mask for aromatic nitriles affording the corresponding imidates, which underwent C—H activation/insertion with alkynes through the co-catalysis of $Mn_2(CO)_{10}$ and NaOPiv·H_2O. The *in situ* elimination of ethanol unmasked the C—H alkenylated imidates to the expected nitriles.

A simple protocol of manganese-catalyzed C_{sp^2}–H alkenylation of aromatic amidines with alkynes was also achieved by Wang and coworkers (Scheme 7.12) [26]. Therefore, a catalytic amount of PhMgBr was essential to promoting the reaction efficiency. Of note, double bond migration was observed in the amidine moiety of the final C—H alkenylation product. This reaction proceeded with good tolerance of various functional groups as well as excellent *E*-stereoselectivity. Experimental

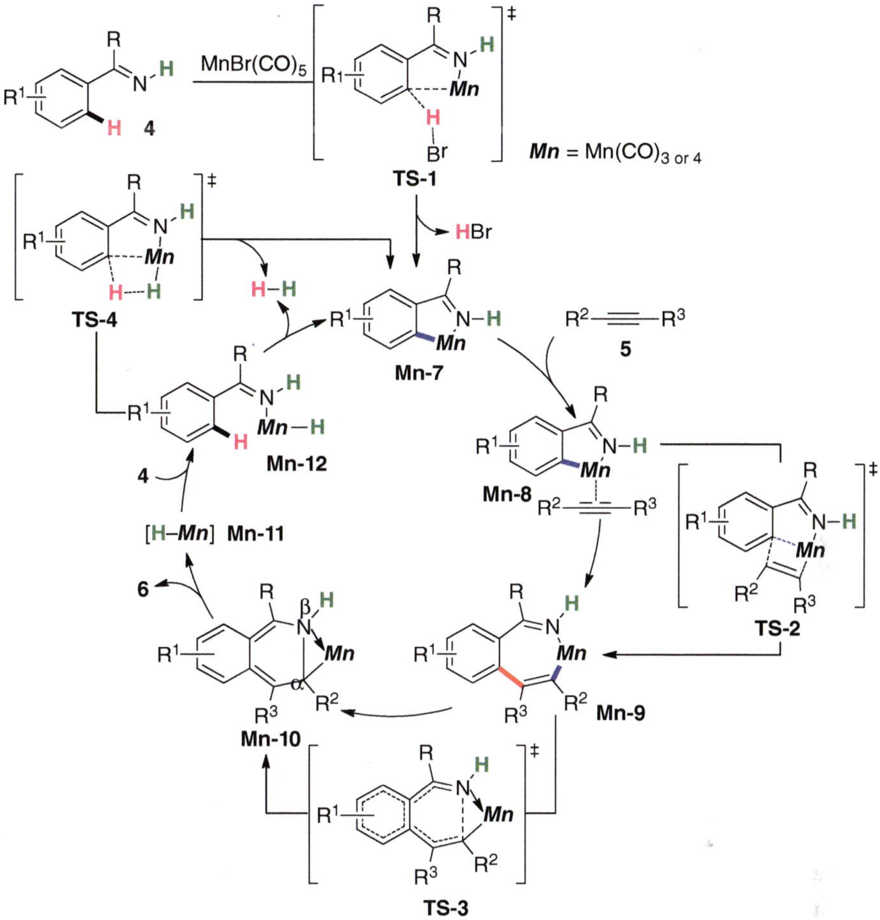

Scheme 7.9 A plausible catalytic cycle of manganese-catalyzed [4+2] annulations.

mechanistic studies revealed a similar base-assisted C—H activation/alkenylation sequence, analogous to previous manganese-catalyzed C—H alkenylation reactions with alkynes [11a, 25].

Ackermann and coworkers revealed a manganese-catalyzed C—H alkynylation of indoles and pyrroles with bromoalkynes (Scheme 7.13) [27]. The reaction was applicable to a wide range of haloalkynes bearing silyl, alkyl, alkenyl, and aryl moieties. Additionally, this synergistic catalysis regime set the late-stage diversification of cyclic and acyclic peptides under racemization-free conditions. Based on their mechanistic studies, the authors proposed a plausible reaction mechanism as depicted in Scheme 7.14. After a fast and reversible C—H bond cyclomanganation, a regioselective migratory insertion of alkynes into Mn—C bond delivered the seven-membered manganacycle **Mn-14**. Thereafter, the crucial β–Br elimination step, which was accelerated by BPh_3 additive, took place to deliver the desired product via the manganese species **Mn-15**.

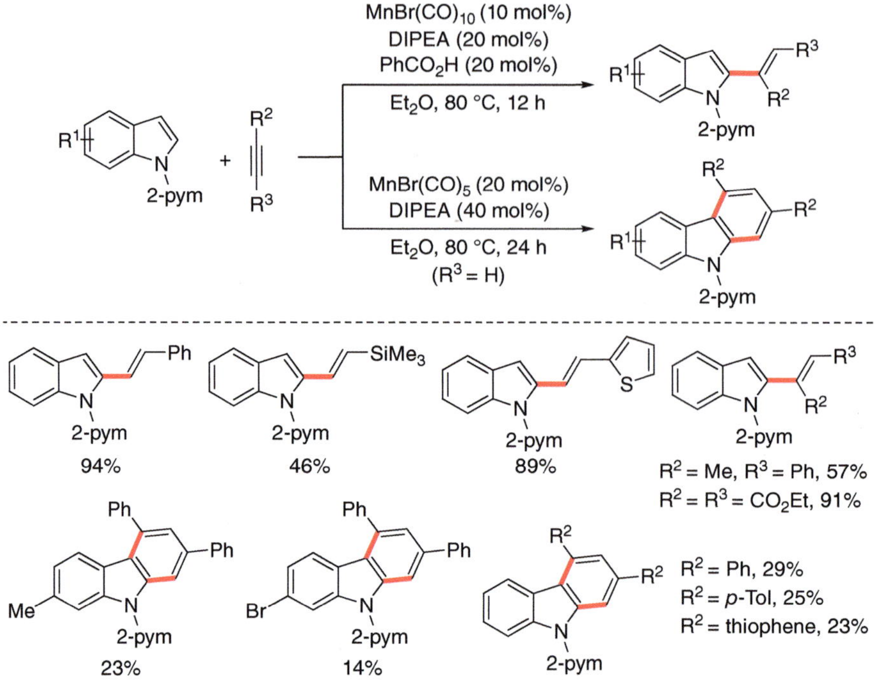

Scheme 7.10 Manganese/base co-catalyzed C–H alkenylation of indoles with alkynes. Source: Redrawn from Shi et al. [24].

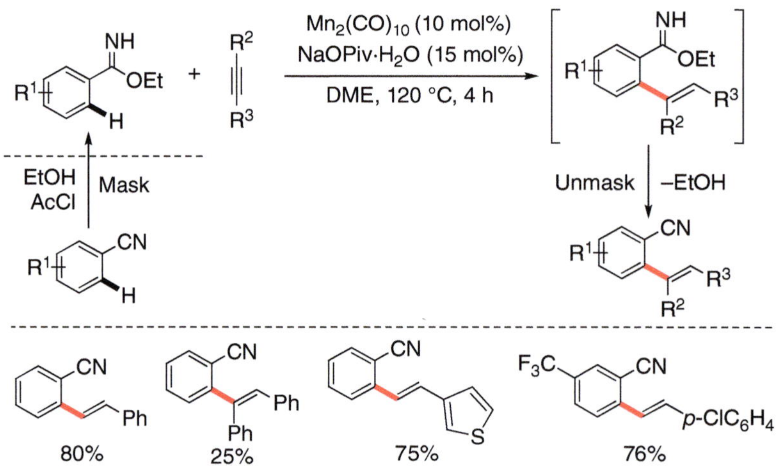

Scheme 7.11 Manganese-catalyzed *ortho*-C–H alkenylation of *N*–H imidates. Source: Redrawn from Yang et al. [25].

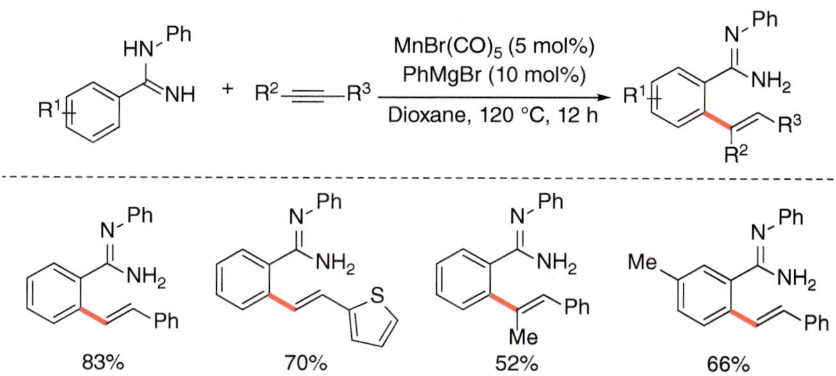

Scheme 7.12 Mn-catalyzed C–H alkenylation of aromatic amidines with alkynes. Source: Redrawn from Jia et al. [26].

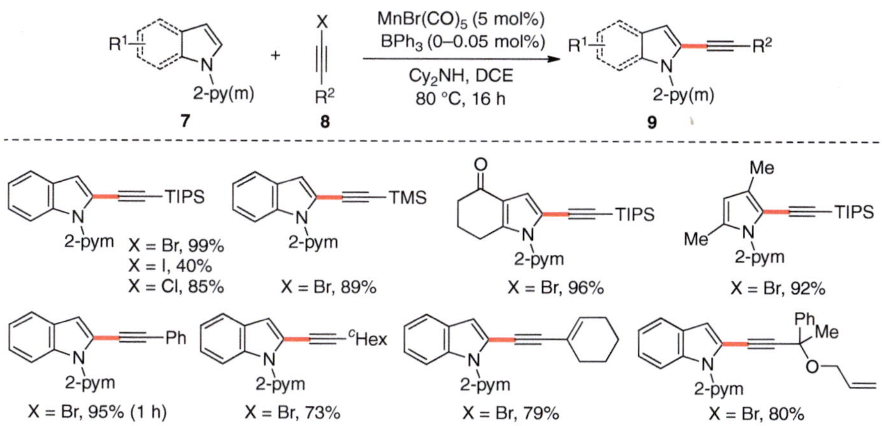

Scheme 7.13 Manganese-catalyzed C–H alkynylation with haloalkynes. Source: Redrawn from Ruan et al. [27].

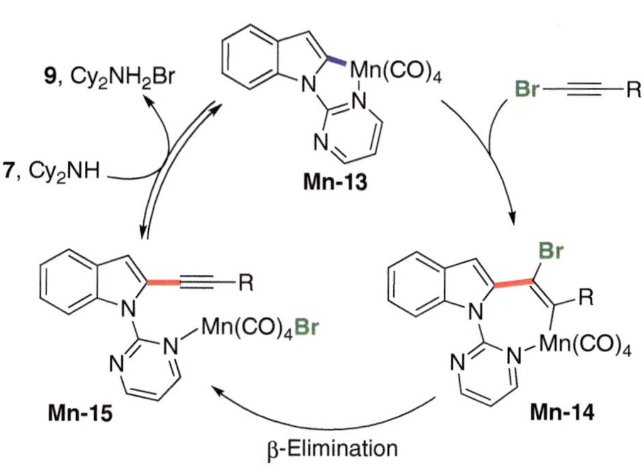

Scheme 7.14 A plausible reaction mechanism of manganese-catalyzed C–H alkynylation.

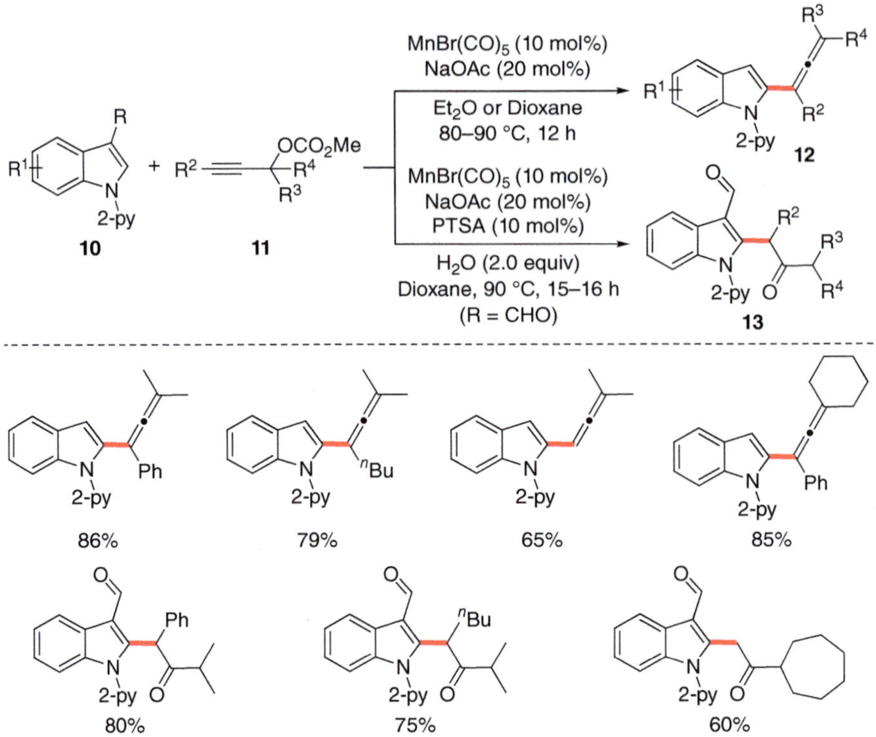

Scheme 7.15 Manganese-catalyzed C–H functionalization of indoles with propargylic carbonates.

Glorius and coworkers developed a manganese-catalyzed regioselective C—H allenylation of indoles with propargylic carbonates [28]. This reaction proceeded with good functional group tolerance to provide multisubstituted allenes bearing different aliphatic as well as aromatic substituents (Scheme 7.15). Moreover, aromatic ketones were prepared in some cases through the cooperation of manganese and *p*-toluenesulfonic acid.

The authors proposed that the catalytic cycle was initialed by the base-assisted cyclomanganation to furnish the five-membered manganacycle **Mn-17** (Scheme 7.16). Thereafter, the coordination of the carbonyl oxygen atom of the carbonate group to the manganese center through a ligand exchange delivered intermediate **Mn-18**. Subsequently, a regioselective migratory insertion of the alkyne moiety into Mn—C bond formed a bicyclic manganacycle **Mn-19**. Subsequently, a β-oxygen elimination, which was facilitated by the proximal tertiary center, occurred to release the desired product and regenerate the active species **Mn-16**. However, an alternative S_N2' attack of the Mn—C bond to the carbonate could not be ruled out.

As an extension of this approach, Glorius and coworkers subsequently reported a manganese-catalyzed C—H annulation of propargyl carbonates with *N*–H ketimines to produce isoquinolines (Scheme 7.17) [29]. Mechanistically, the use

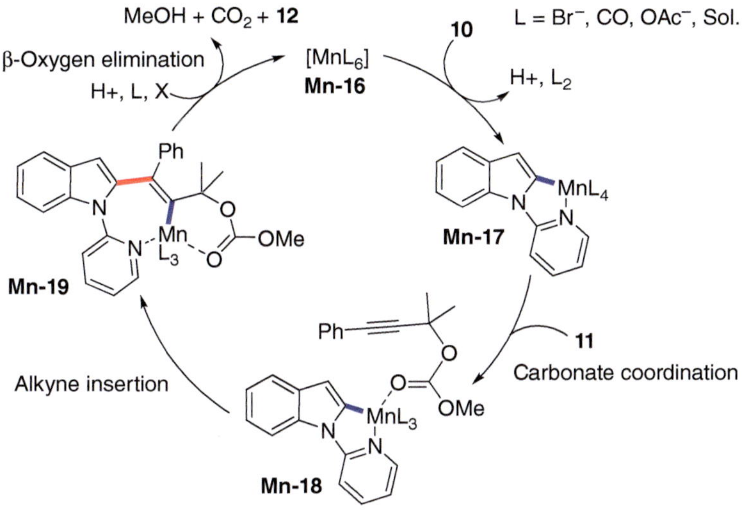

Scheme 7.16 Proposed mechanism of manganese-catalyzed C–H allenylation with propargylic carbonates.

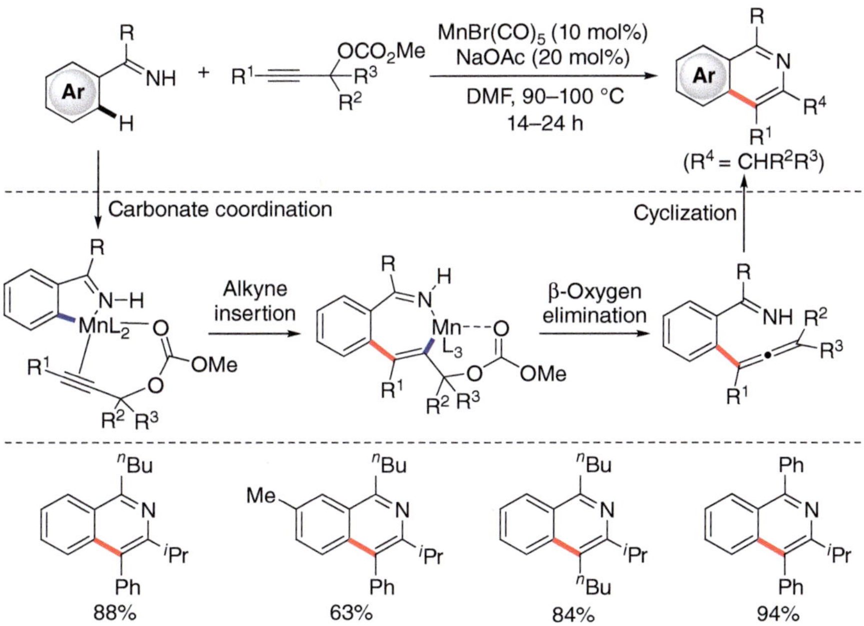

Scheme 7.17 Manganese-catalyzed C–H annulations of propargyl carbonates with *N*–H imines. Source: Redrawn from Lu et al. [29].

of alkynes with a leaving group in β-position led to a different cyclization process from the Wang's report [21]. Herein, the authors proposed that the reaction commenced with the formation of the five-membered manganacycle via base-assisted cyclomanganation of imine, which was followed by a sequence of coordination of the carbonyl oxygen atom, regioselective alkynes insertion, and selective β-oxygen elimination to deliver an allenylated intermediate. Finally, a fast intramolecular cyclization would give the isoquinoline product as a single regioisomer.

In a subsequent report, Glorius and coworkers demonstrated a manganese-catalyzed C—H (2-indolyl)methylation of heteroarenes employing ethynyl benzoxazinanones to access diheteroarylmethanes (Scheme 7.18) [30]. Analogously, the authors proposed that the decarboxylative β-oxygen elimination furnished an allene intermediate, which finally underwent regioselective allene insertion and protonation successively or protonation first and then intramolecular cyclization to deliver the desired product.

Very recently, Li's [31] and Ackermann's [32] groups independently devised a manganese-catalyzed C—H activation/Diels–Alder/retro-Diels–Alder domino annulation sequence between various heteroarenes and propargyl carbonates providing fused carbo/heterocycle derivatives. Under the optimized reaction conditions of Li's work, not only *N*-pyridyl-2-pyridone and analogous isoquinolones proved amenable to this protocol, but also *N*-pyridylindoles were successfully employed in this cascade reaction (Scheme 7.19) [31]. Additionally, detailed mechanistic studies were performed, which led to the proposed catalytic cycle depicted briefly in Scheme 7.19. After a fast and reversible C—H manganation and subsequent migratory insertion of alkyne, a seven-membered manganacycle

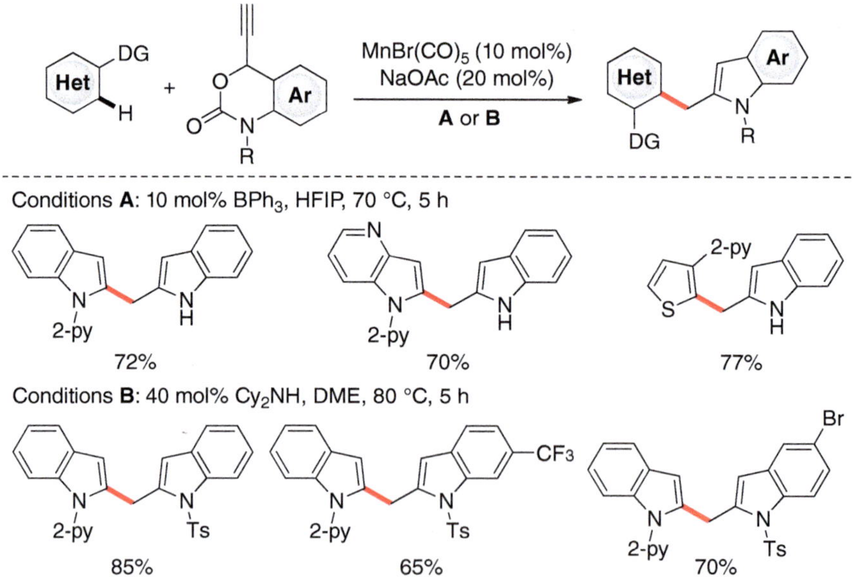

Scheme 7.18 Manganese-catalyzed C—H (2-indolyl)methylation of heteroarenes. Source: Redrawn from Lu et al. [30].

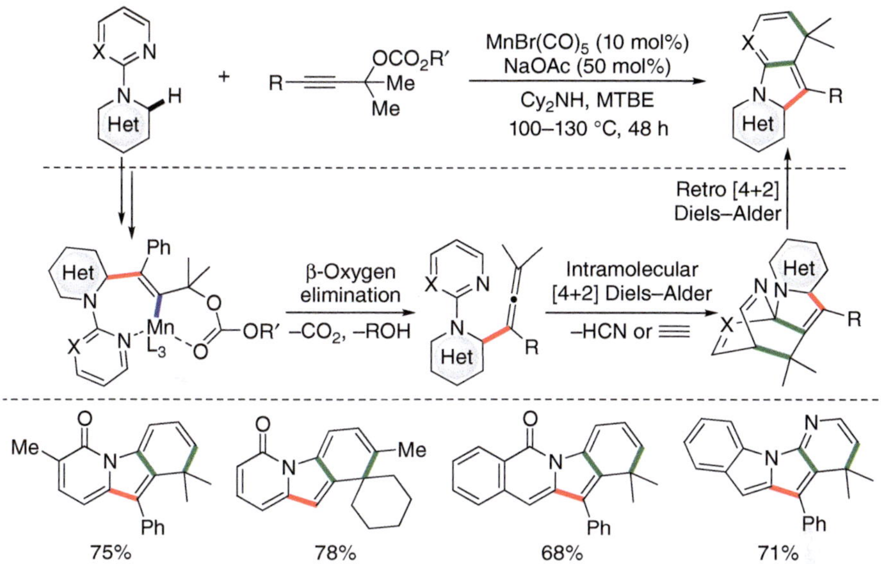

Scheme 7.19 Manganese-catalyzed C–H activation/Diels–Alder/retro-Diels–Alder domino annulation sequence. Source: Redrawn from Zheng et al. [31].

was formed. Then, a β-oxygen elimination gave an allene intermediate, which underwent intramolecular [4+2] Diels–Alder reaction, along with cyanide extrusion via retro-Diels–Alder reaction providing the final product. As a particular highlight, pyridine and pyrimidine that represented typical "permanent" DG were used as a transformable one in this protocol. Concurrently, Ackermann and coworkers reported the same manganese-catalyzed tandem annulation reactions of *N*-pyridyl-2-pyridone with propargylic carbonates in the presence of BPh_3 to access indolone alkaloid derivatives [32]. Experimental and DFT studies set the stage for a plausible catalytic cycle involving a sequential of acetate-assisted organometallic C—H activation, β-oxygen-elimination, BPh_3-mediated Diels–Alder reaction, and retro-Diels–Alder reaction, which showed similarity to that of Li's work [31].

To further access fused carbo/heterocycle derivatives with propargyl carbonates, the group of Li reported manganese-catalyzed synthesis of fused eight- and four-membered carbocycles through redox-neutral coupling of 3-alkenyl- and 3-allylindoles with propargylic carbonates, respectively (Scheme 7.20) [33]. This protocol proceeded with broad substrate scopes and good tolerance of functional groups. Mechanistically, after the C—H cyclomanganation/alkyne coordination/migratory insertion sequence, the β-oxygen elimination and ensuing decarboxylation selectively furnished the allenylation intermediate, which then underwent pericyclic reactions to deliver the final products.

In sharp contrast, the group of Ackermann described a synergistic Brønsted acid/manganese-catalyzed C—H functionalization with propargylic carbonates to solely afford allylic carbonates without the β-oxygen elimination (Scheme 7.21) [34]. Experimental and computational mechanistic studies shed light on the

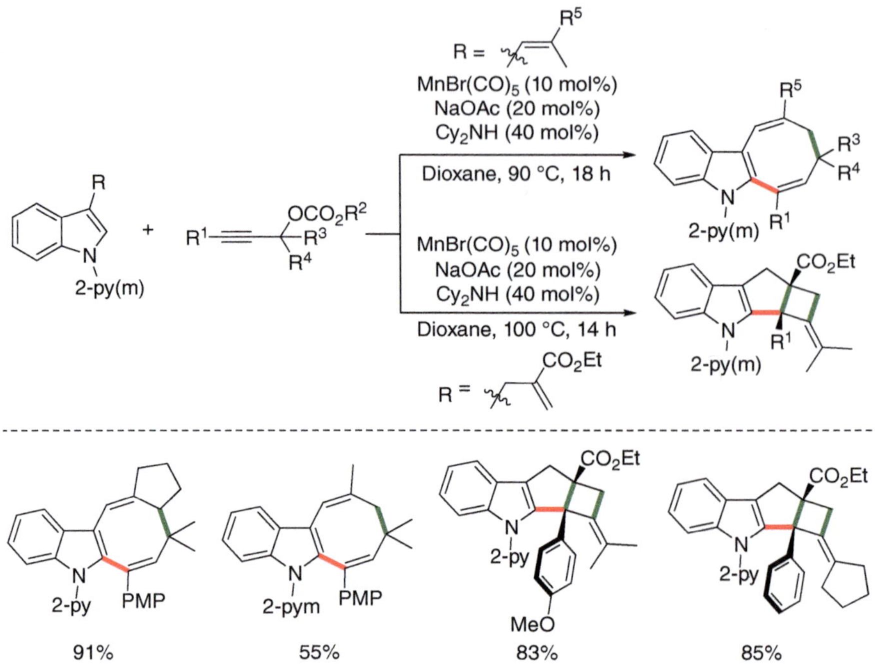

Scheme 7.20 Manganese-catalyzed synthesis of fused eight- and four-membered carbocycles with propargylic carbonates. Source: Redrawn from Xu et al. [33].

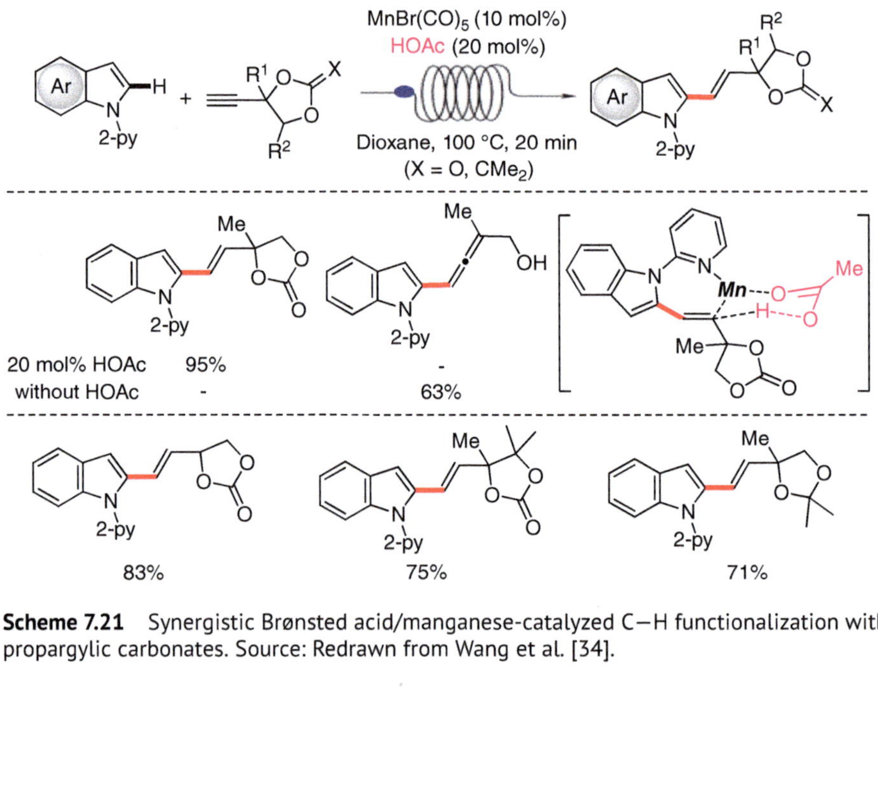

Scheme 7.21 Synergistic Brønsted acid/manganese-catalyzed C–H functionalization with propargylic carbonates. Source: Redrawn from Wang et al. [34].

essential role of the Brønsted acid for the unique chemoselectivity of this approach. The additive of 20 mol% of HOAc set the stage for the alkenylated product in 95% yield through an intramolecular proton transfer process, while the allenylation process occurred via undesired β-oxygen-elimination in the absence of carboxylic acids. Notably, a continuous flow system was exploited for the first time in manganese/base-catalyzed C—H activation reactions, thus improving reaction efficiency obviously.

For manganese-catalyzed C—H activation of arenes with alkynes, the key elementary step of the C≡C bond insertion into Mn—C_{aryl} bond of the five-membered manganacycle could form a seven-membered manganacycle, an alkenyl manganese species, which were usually transformed by direct protonation or H-transfer delivering the C—H alkenylation products, dehydrogenative [4+2] annulation affording diverse isoquinolines, or β-elimination resulting in C—H alkynylation or C—H allenylation of arenes. Recently, Li and coworkers demonstrated that such alkenyl manganese species might also undergo a subsequent tandem intramolecular Michael addition due to the high nucleophilicity of the Mn—C bond (Scheme 7.22) [35]. The optimized catalytic system consisted $Mn_2(CO)_{10}$ as the catalytic precursor and BPh_3 as the key additive. Under the optimal reaction conditions, a broad scope of indoles underwent C—H insertion/Michael addition sequence smoothly to yield an *exo*-cyclized olefin attached to a tetrahydrofuran skeleton. Interestingly, fused phenols could be obtained as the major product by judicious choice of $Zn(OAc)_2$/PivOH as additives.

In addition, the authors performed mechanistic studies and showed the proposed pathways in Scheme 7.23. After a fast and reversible C—H activation of indole giving

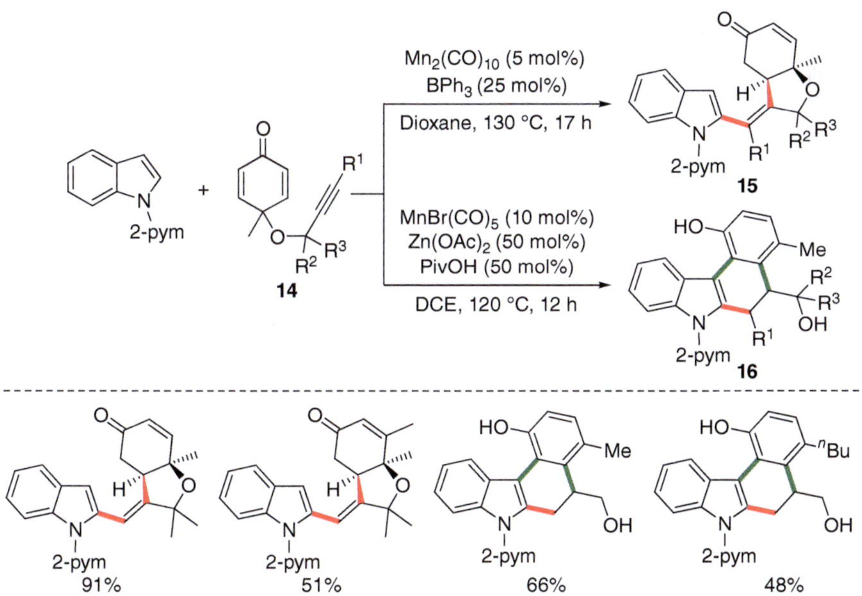

Scheme 7.22 Manganese-catalyzed divergent annulations of indoles with 1,6-enynes. Source: Redrawn from Liu et al. [35]

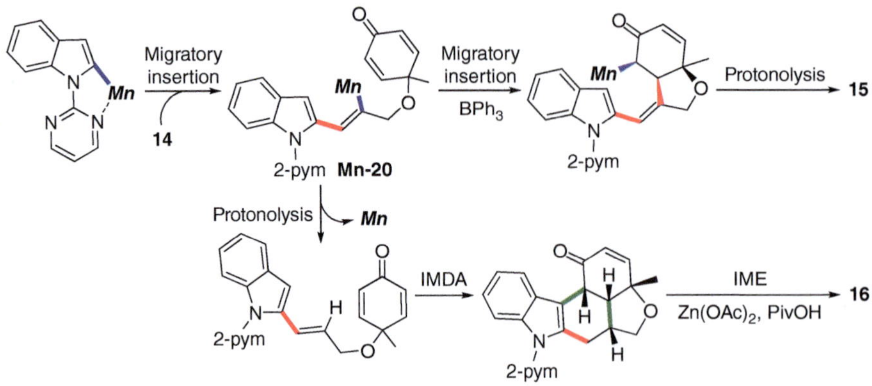

Scheme 7.23 Proposed pathways of the manganese-catalyzed cascade reactions.

the five-membered manganacycle, a regioselective migratory insertion of C≡C bond into the Mn—C bond generates an alkenyl intermediate **Mn-20**. Depending on the reaction additives, either a conjugate addition/protonation furnished product **15** promoted by the BPh_3 additive or a subsequent protonolysis/intramolecular Diels–Alder addition (IMDA)/intramolecular elimination (IME) cascade yielded the aromatization product **16** facilitated by $Zn(OAc)_2$ and PivOH additives.

Likewise, the group of Lin independently disclosed a related manganese-catalyzed tandem process of 2-arylpyridines and 1,6-enynes in the presence of a catalytic amount of Cy_2NH under milder conditions (Scheme 7.24) [36]. This reaction proceeded smoothly with high chemoselectivity and atomic economy. Mechanistically, the authors proposed that the C—H metalation process was probably involved in the turnover-limiting step, which differentiated from that of Li's work on indoles [35]. Thereafter, regioselective syn-insertion of the Mn—C bond into the C≡C bond delivered the key alkenyl manganese intermediate, which underwent Michael addition and protonation to give the desired product.

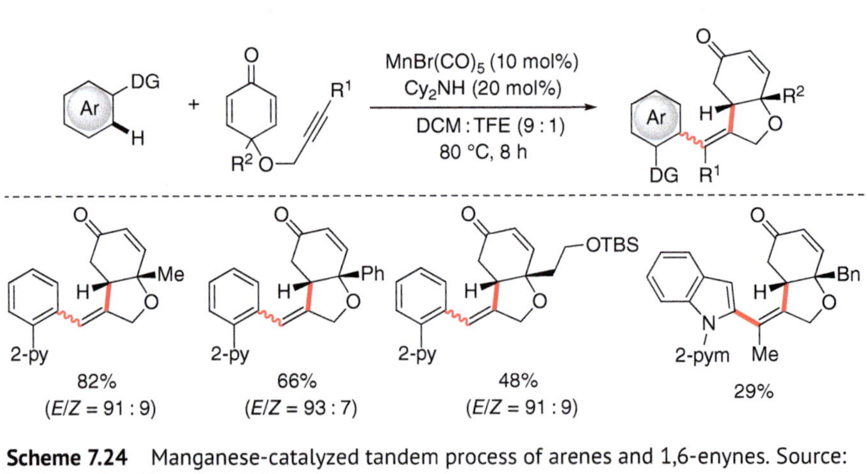

Scheme 7.24 Manganese-catalyzed tandem process of arenes and 1,6-enynes. Source: Redrawn from Tan et al. [36].

In 2018, Chen and coworkers reported a manganese-catalyzed sequential annulation reaction involving an oxidative dehydrogenation process [37]. The annulation proceeded well between indoles and various terminal and internal 1,6-diynes to provide numbers of fused carbazole derivatives in a step-economical way (Scheme 7.25). Remarkably, three C—C bonds, even six C—C bonds in one case, were formed in one pot, albeit in relatively low and moderate yields. Based on the mechanistic studies, the authors put forward a plausible mechanism as shown in Scheme 7.26. As it was previously unrevealed for manganese-catalyzed C—H activation of arenes, the five-membered manganacycle **Mn-17** was formed initially by the assistance of pyridine DG. Afterwards, alkyne insertion into the Mn—C

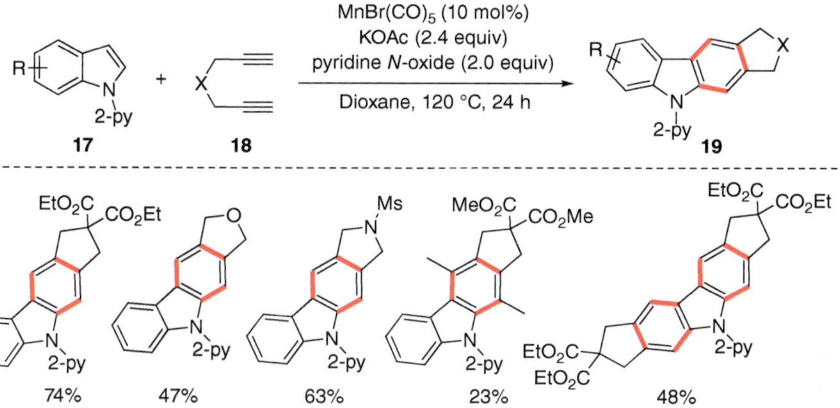

Scheme 7.25 Manganese-catalyzed sequential annulation reaction between indoles and 1,6-diynes.

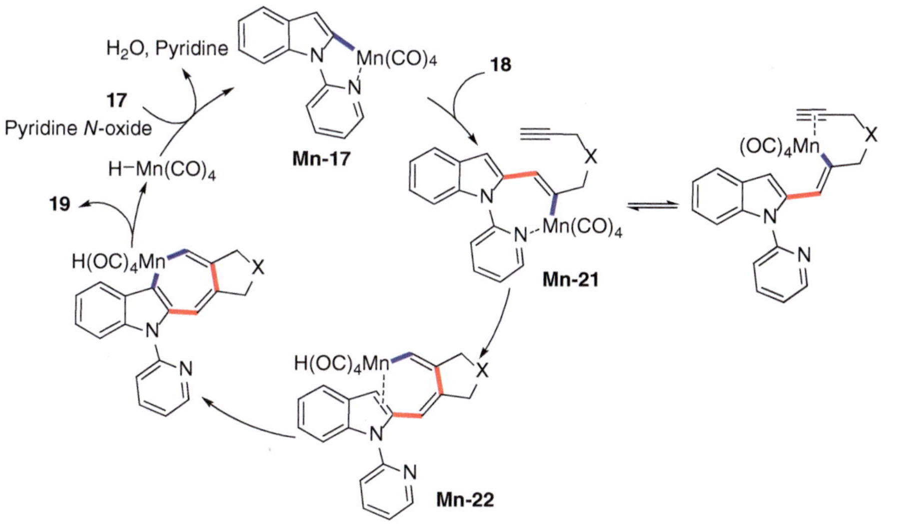

Scheme 7.26 Proposed mechanism for sequential annulations involving oxidative dehydrogenation.

bond furnished the alkenyl manganese species **Mn-21**, which was transformed to intermediated **Mn-22** via a secondary alkyne insertion. Further presumed C—H oxidative addition followed by reductive elimination afforded the carbazole product along with $MnH(CO)_4$. Then the Mn–H species reacted with indole substrate in the presence of pyridine *N*-oxide as the oxidant to regenerate **Mn-17** and close the catalytic cycle. Of note, the previously mentioned cyclization mechanism may need further explorations.

7.3.2 Insertion of C=C Bonds into Manganacycles

With the success of the first manganese-base co-catalytic system in C—H activation reactions with alkyne derivatives [11a], and studies on stoichiometric manganese-mediated C—H bond insertion into C=C double bond [13c, 13d, 15c], a seminal manganese-base synergistically co-catalyzed aromatic C—H alkylation with olefin substrates was first reported by Wang and coworkers in 2014 (Scheme 7.27) [11b]. It is worth noting that excellent monoselectivity was achieved with a wide range of substitution patterns on 2-arylpyridines. Based on experimental and computational studies, a mechanistic proposal that showed similarities to the manganese-catalyzed C—H alkenylation [11a] was put forward. The reaction commenced with a Cy_2NH-assisted C—H manganation to generate the five-membered manganacycle. Thereafter, migratory insertion of the C=C double bond to Mn—C bond occurred, followed by a proto-demetalation with the conjugated acid of Cy_2NH, which contrasted to the previously proposed LLHT regime [11a] (Scheme 7.7). Similarly, Song, Gong, and coworkers showed the alkylation of indoles with electron-deficient maleimides via additive-free manganese-catalyzed C—H activation to produce synthetically meaningful 3-substituted succinimide derivatives in moderate to excellent yields (Scheme 7.28) [38].

In 2015, Ackermann and coworkers reported an access to *cis*-β-amino acid esters through manganese-catalyzed C—H annulations of ketimines with α,β-unsaturated esters (Scheme 7.29a) [39]. The reaction proceeded with excellent regio- and chemoselectivity, delivering the corresponding β-amino acid derivatives with high

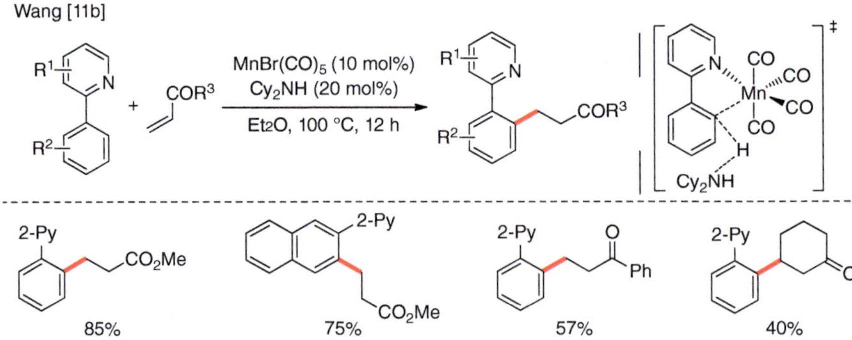

Scheme 7.27 Manganese-catalyzed C—H alkylation with activated olefins. Source: Redrawn from Zhou et al. [11b].

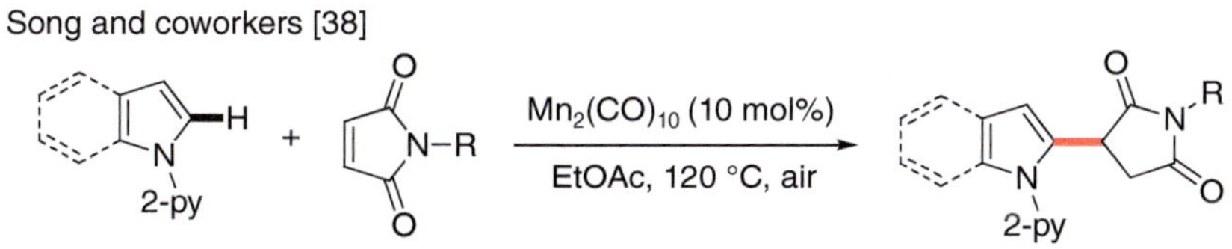

Scheme 7.28 Manganese-catalyzed C–H alkylation of indoles with maleimides. Source: Redrawn from Liu et al. [38].

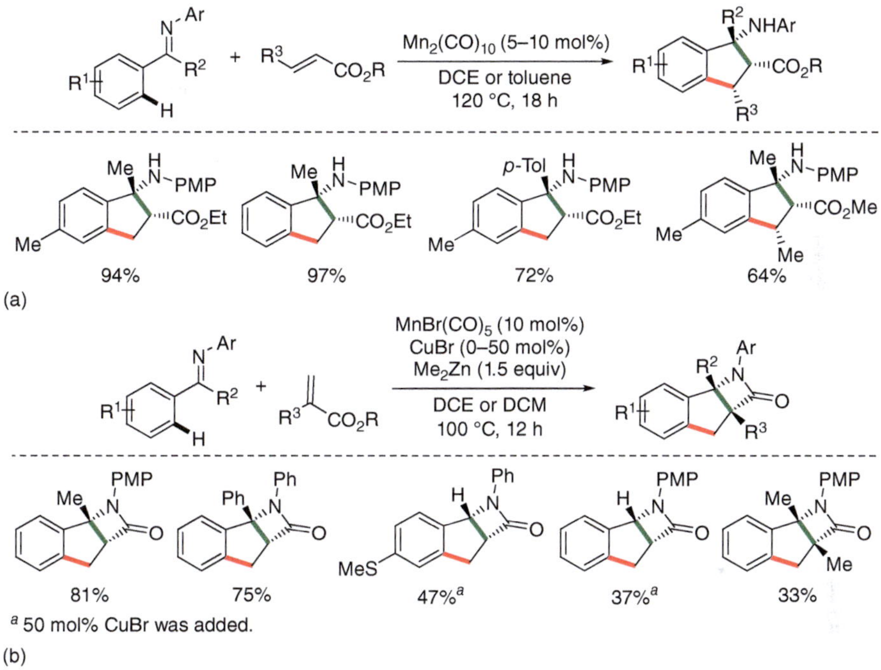

Scheme 7.29 Manganese-catalyzed C–H annulations of imines with acrylates. (a) Ackermann and coworkers [39] and (b) Wang and coworkers [40]. Sources: (a) Redrawn from Liu et al. [39]; (b) Redrawn from Hu et al. [40].

efficiency, good functional tolerance, and sole *cis*-diastereoselectivity. A related process was also reported by the group of Wang [40]. The authors described a manganese-catalyzed bicyclic annulation of imines with α,β-unsaturated esters, which provided access to fused β-lactams in the presence of dimethylzinc (Scheme 7.29b). Notably, this transformation was applied to both ketimines and previously untouched aldimines to form *cis*-β-lactams with excellent chemo- and diastereoselectivity in a step-economical way.

An efficient method for the synthesis of tetra- and pentacyclic aniline derivatives was described by Ackermann and coworkers [41], which involved a manganese-catalyzed annulation of methylenecyclopropanes (MCPs) with ketimines and $ZnCl_2$-promoted Povarov cycloaddition [42] cascade (Scheme 7.30). This reaction nicely merged C—H activation with C—C bond cleavage delivering

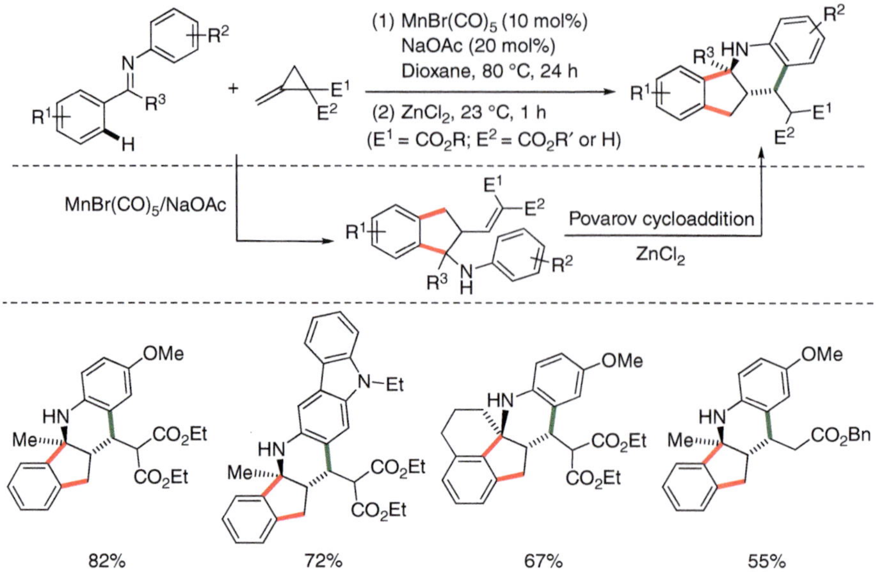

Scheme 7.30 Manganese-catalyzed annulations of imines with MCPs via C–H/C–C bond cleavage.

the complex products in moderate to high yields and with good compatibility of functional groups.

On the way to extend the scope of manganese catalysis, Ackermann and coworkers demonstrated the base-assisted manganese-catalyzed C—H allylation of imines with allylic carbonates providing the allylated ketones upon acidic workup (Scheme 7.31) [43]. The optimized catalytic system was highly tolerant of valuable electrophilic functional groups such as amino, halo, and cyano substituents. The authors proposed that a catalytic cycle invoked a mode of C—H activation via base-assisted internal electrophilic-type substitution (BIES) in the manganese-carboxylate complex **Mn-23**, followed by a sequence of migratory insertion of the C=C double bond and the β-oxygen elimination.

As mentioned earlier, after a migratory insertion of the C=C bond of allylating reagents into the five-membered manganacycle, the seven-membered manganacycle bearing heteroatom at the β-position to the metal center could be formed, which is usually prone to undergo elimination reactions. Based on this mechanistic scenario, a plethora of manganese-catalyzed C—H allylation reactions had been developed, which proceeded with concurrent β-X (X = O, N, F, Br, etc.) elimination.

Two independent reports from the group of Glorius [44] and Ackermann [45] in 2017 further extended the scope of allylating reagents for manganese-catalyzed C—H allylation of arenes. In both cases, products of allylic alcohols were achieved via sequential C—H activation and β-oxygen elimination process by using dioxolanone as allylating reagents and sodium acetate as key catalytic additive. The group of Glorius employed 2-pyri(mi)dinyl-indole, -benzene, and -thiophene derivatives as the C—H substrates and dimethyl ether as solvent or neat to furnish

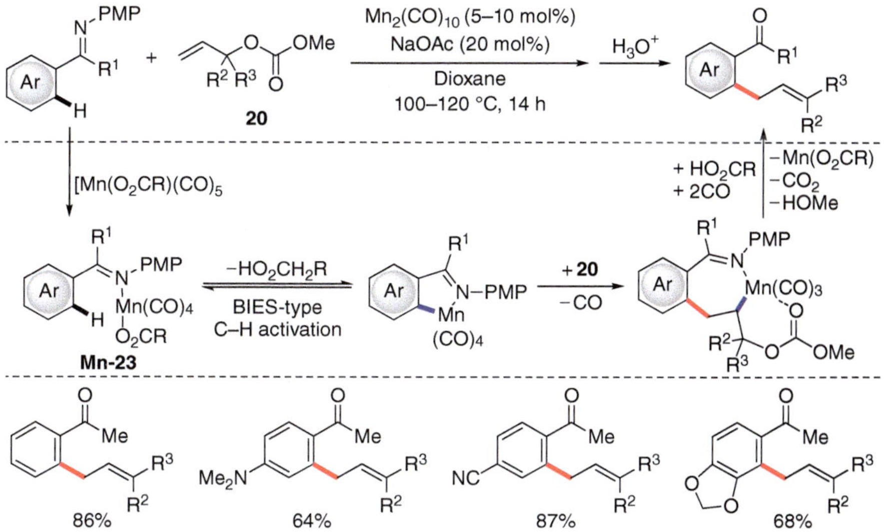

Scheme 7.31 Manganese-catalyzed C–H allylation of imines with allyl carbonates. Source: Redrawn from Liu et al. [43].

the corresponding products in moderate to excellent yields (Scheme 7.32a). Moreover, when 2-(prop-1-en-2-yl)pyridine was used as the coupling partner, the olefinic C—H allylation product could be achieved in 75% yield under the optimized reaction conditions. In Ackermann's report, a variety of DG-containing indoles and benzenes as well as ketimines were accomplished to afford the desired allylic alcohols in moderate to excellent yield under a robust manganese catalytic system, which was tolerant of water and air in some cases (Scheme 7.32b).

Furthermore, the strategy described by Glorius and coworkers was applied to the synthesis of aryl- and aminoalkyl cyclopentenes **22** from 2-pyridinylindoles, -arenes, or -olefins with diazabicycle **21** through sequential C_{sp^2}–H activation and β-nitrogen elimination (Scheme 7.33) [44].

Interestingly, Zhang and coworkers disclosed a novel manganese-catalyzed strategy for the synthesis of 3,3-difluoroallyl 2-pyridones and indoles by using 3-bromo-3,3-difluoropropene as the allylating reagent (Scheme 7.34) [46]. Mechanistically, this methodology involved a facile C—H activation, alkene insertion, and a β-bromo elimination process.

Concurrently, the group of Ackermann reported a related protocol of manganese-catalyzed allylative C—H functionalization of indoles, imines, and peptides with perfluoroalkyl alkenes via C—F bond cleavage (Scheme 7.35) [47]. Experimental and computational mechanism studies showed that organometallic C—H activation, slow migratory insertion, and β-fluoro elimination took place sequentially under the reaction conditions to give the allylated product. In principle, β-elimination could also occur if there is a proper leaving group at the terminal carbon of olefins when seven-membered manganacycle is formed after the olefin insertion. As such, Ackermann and coworkers elegantly developed

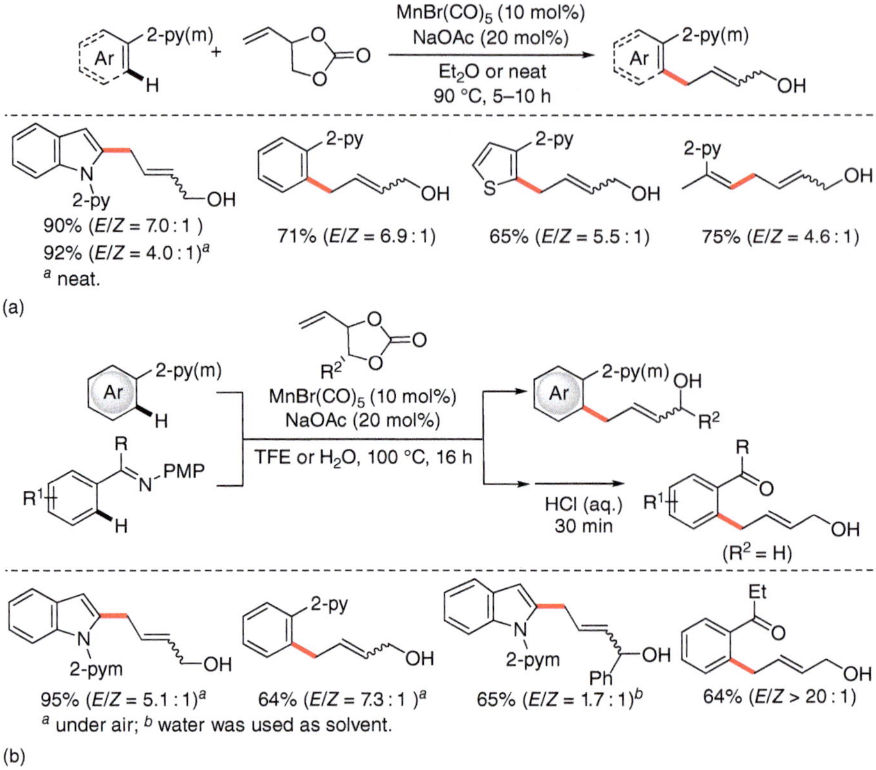

Scheme 7.32 Manganese-catalyzed C–H allylation of arenes with dioxolanones. (a) Glorius and coworkers [44] and (b) Ackermann and coworkers [45]. Source: (a) Redrawn from Lu et al. [44]; (b) Redrawn from Wang and coworkers [45].

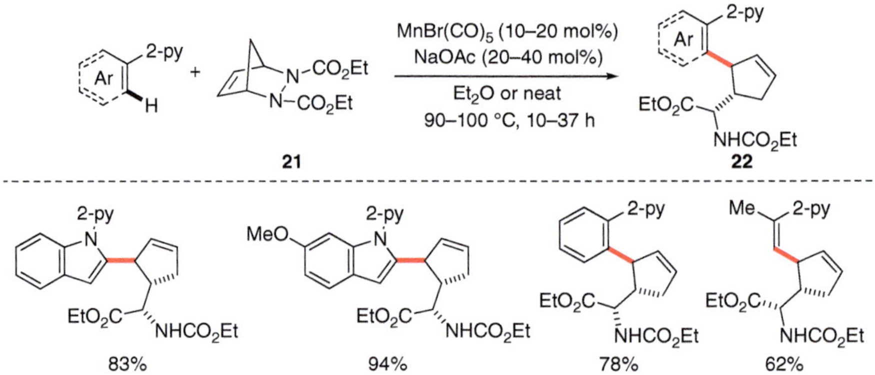

Scheme 7.33 Manganese-catalyzed C–H functionalization with diazabicycles. Source: Redrawn from Lu et al. [44].

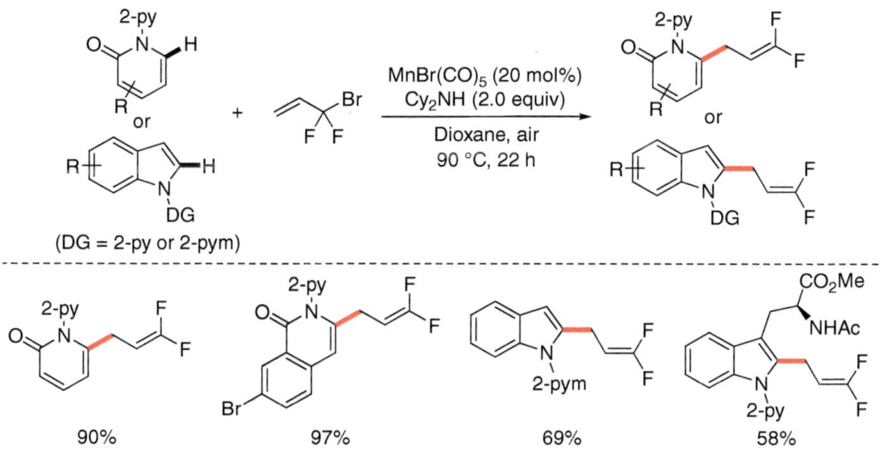

Scheme 7.34 Manganese-catalyzed C–H allylation with 3-bromo-3,3-difluoropropene. Source: Redrawn from Ni et al. [46].

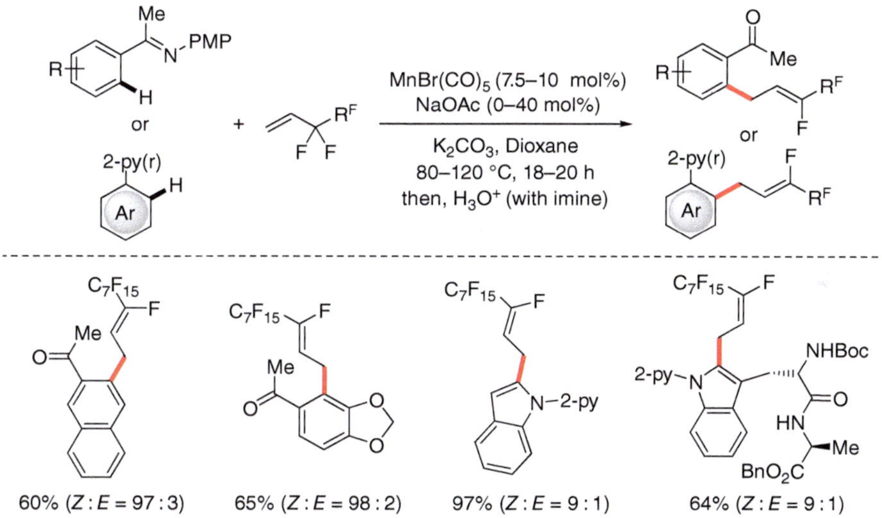

Scheme 7.35 Manganese-catalyzed C–H allylation with perfluoroalkyl alkenes. Source: Redrawn from Zell et al. [47].

manganese-catalyzed C—H activation reactions of indoles with *gem*-difluoro alkenes/perfluoroalkenes (Scheme 7.36a) [47]. The C—H fluoroalkenylated products were obtained with good diastereoselectivity through cleavage of both C—H and C—F bonds. An analogical method for the synthesis of mono-fluoroalkenylated indoles with *gem*-difluorostyrenes was described independently by Loh and coworkers, which involved a reaction sequence of manganese-catalyzed C—H activation and C—F bond cleavage (Scheme 7.36b) [48].

In 2018, the group of Wu exploited the manganese catalysis for the C2–H allylation of indoles, in which 2-(bromomethyl) acrylates were employed as the allylation

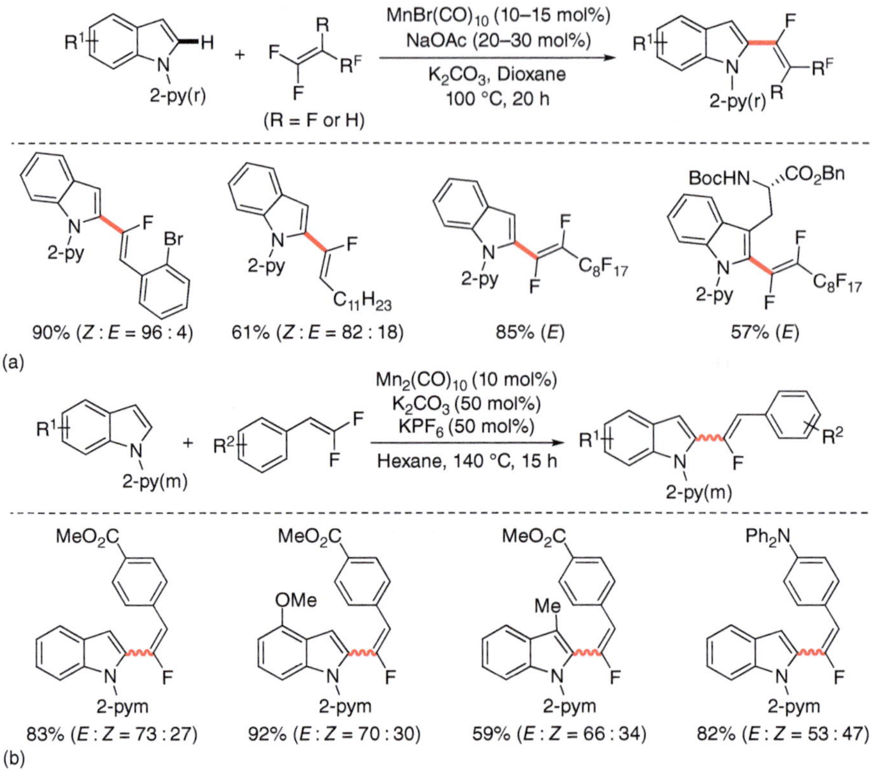

Scheme 7.36 Manganese-catalyzed C—H fluoroalkenylation of indoles with *gem*-difluoroalkenes. (a) Ackermann and coworkers [47] and (b) Loh and coworkers [48]. Source: (a) Redrawn from Zell et al. [47] and (b) Redrawn from Cai et al. [48].

partner (Scheme 7.37) [49]. A plausible catalytic cycle was suggested to involve successive base-assisted C—H cyclomanganation, alkene insertion, and β-bromo elimination.

Very recently, Wang and coworkers disclosed manganese-catalyzed aromatic C—H allylation of ketones through the use of allyl halides or allyl trifluoroacetates as the allylation reagents (Scheme 7.38) [50]. After an extensive optimization process, the desired products, *ortho*-C—H allylated ketones, were obtained in high yields under mild conditions with copper triflate as a Lewis acid and dimethylzinc as a key additive. This transformation was shown to be compatible with a broad scope of allylic electrophiles bearing substituents at α-, β-, or γ-positions, which yielded the mono-allylated ketones with good regio- and stereoselectivity. Furthermore, mechanistic studies highlighted the importance of the catalytic amount of copper triflate, which not only promoted the C—H activation step but also inhibited the aldol-type condensation of ketones.

As important applications of manganese-catalyzed C—H allylations of arenes with various allylating reagents, Ackermann and coworkers described the C—H allylation of decorated peptides, nucleotides, and drug molecules enabled by

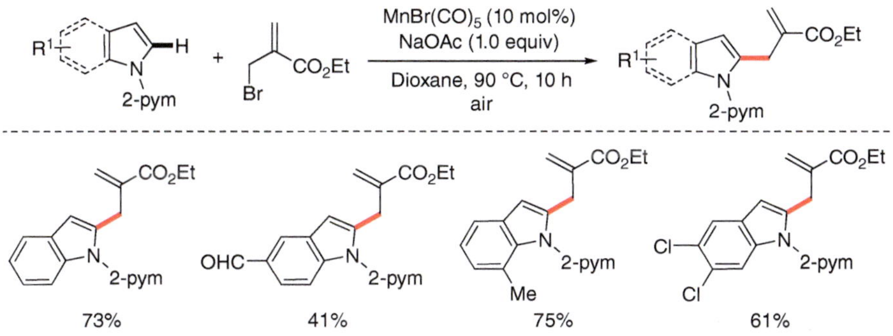

Scheme 7.37 Manganese-catalyzed C–H allylation of indoles with 2-(bromomethyl) acrylates. Source: Redrawn from Wu et al. [49].

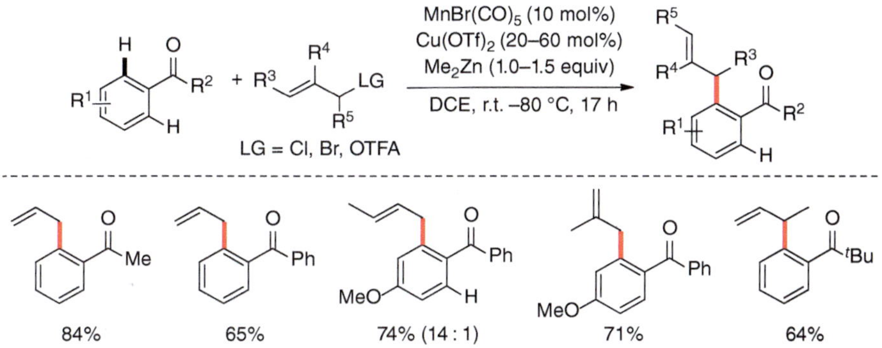

Scheme 7.38 Manganese-catalyzed aromatic C–H allylation of ketones. Source: Redrawn from Ali et al. [50].

manganese/base co-catalysis (Scheme 7.39a) [51]. This methodology exhibited good compatibility with a plethora of sensitive functional groups, delivering the corresponding products in moderate to excellent yields. Furthermore, the manganese–base co-catalysis regime proved applicable to the late-stage C—H glycosylation of amino acids and peptides bearing key indole skeletons with excellent chemo- and position-selectivity (Scheme 7.39b) [52].

Apart from the allylating reagents through β-heteroatom elimination to release the allylated products, activated vinylcyclopropanes could also enable the C—H allylation of arenes through successive C—H/C–C activation. In 2017, the group of Glorius reported an approach to the synthesis of allylated (hetero)arenes, in which $Mn_2(CO)_{10}$ was used as the catalyst and 2-vinylcyclopropane-1,1-dicarboxylate was employed as the coupling partner, delivering the desired product via a β-carbon elimination process. (Scheme 7.40a) [44]. Concurrently, the group of Ackermann accomplished manganese–base co-catalyzed similar transformations in better diastereoselectivity (Scheme 7.40b) [53]. As a result of β-carbon elimination,

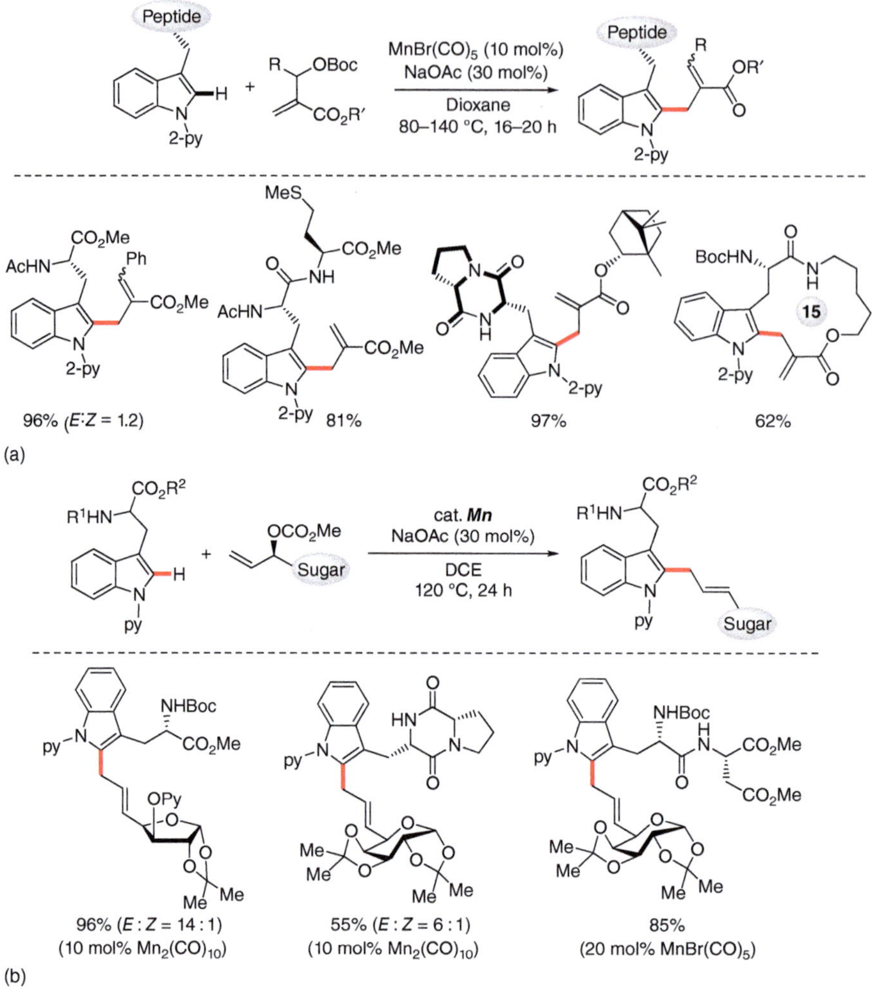

Scheme 7.39 Manganese-catalyzed late-stage C–H allylation of important compounds. (a) Ackermann and coworkers [51] and (b) Ackermann and coworkers [52]. Source: (a) Redrawn from Kaplaneris et al. [51]; (b) Redrawn from Wang et al [52].

the corresponding C2–H allylated indoles/pyrroles were formed in good yields from reactions of a range of *N*-pyrimidylindoles/pyrroles with vinylcyclopropanes bearing strong electron-deficient group substituents.

Inactivated olefins without a leaving group are very challenging reaction partners for manganese-catalyzed C—H activation reactions. The first example of manganese-catalyzed C—H olefination of ketones with inactivated alkenes was reported in 2018 by Wang and coworkers (Scheme 7.41) [12]. In the reaction, the carbonyl moiety of ketones acted not only as a DG to control the reaction regioselectivity but also as an intramolecular hydrogen acceptor to deliver the C—H olefinated benzylic alcohol products. Overall, this redox-neutral reaction proceeded

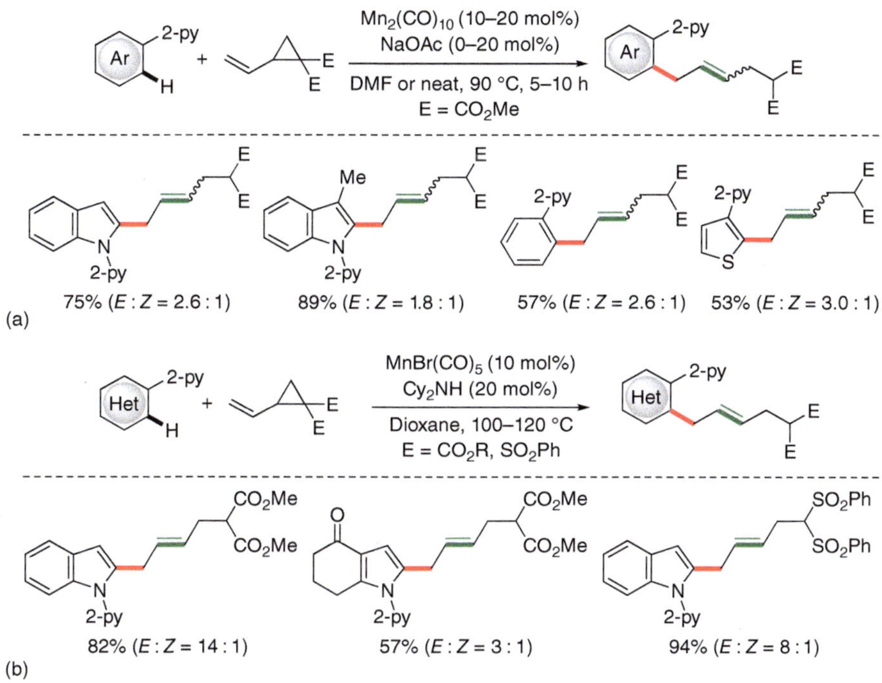

Scheme 7.40 Manganese-catalyzed C–H allylation with activated vinylcyclopropanes. (a) Glorius [44] and (b) Ackermann [53]. Source: (a) Redrawn from Lu et al. [44]; (b) Redrawn from Meyer et al. [53].

well under mild reaction conditions and with diverse functional group tolerance, thus providing the desired products in a step-economical way. Experimental and computational studies suggested the C—H bond cleavage via a Mn–Zn-enabled CBMD pathway. In the catalytic cycle, dimethylzinc functioned both as a transmetalation reagent and as a promotor for C—H activation through a Mn–Zn synergetic mode. Notably, after alkene insertion into the Mn—C bond, a subsequent direct intramolecular H-transfer rather than a traditional β-H elimination occurred to furnish the alkenylated-benzyl alcohol intermediate, which could further undergo transmetalation and hydrolysis to give the final product.

7.3.3 Insertion of Allene into Manganacycles

Allenes containing two cumulated C=C double bonds are important building blocks in organic synthesis. The application of allenes in manganese-catalyzed C—H allylation reactions was demonstrated by Wang and coworkers in 2017 (Scheme 7.42) [54]. Therein, 1,1-disubstituted allenes were employed as the reaction partners, and the reaction proceeded smoothly with various (hetero)arenes affording the desired products in an atom-efficient manner. The mechanism was proposed to involve organometallic C—H activation, regioselective migratory insertion of the terminal C=C bond of allenes, and protonation of the Mn—C bond with acid.

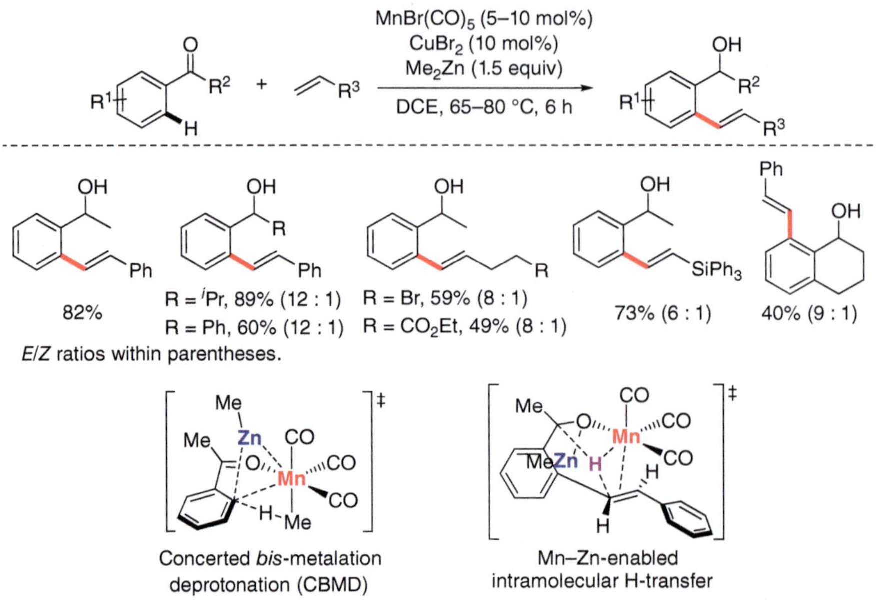

Scheme 7.41 Manganese-catalyzed redox-neutral C—H olefination of ketones with inactivated alkenes. Source: Redrawn from Hu et al. [12].

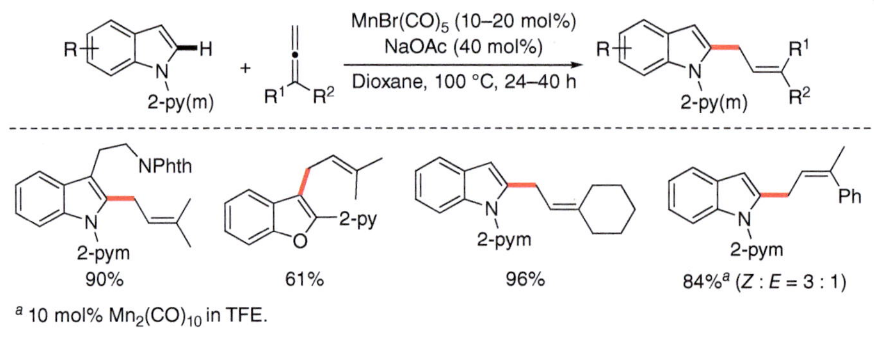

Scheme 7.42 Manganese-catalyzed C—H allylation with 1,1-disubstituted allenes. Source: Redrawn from Chen et al. [54].

The group of Wang also accomplished the synthesis of polycyclic compounds by employing 1,1-disubstituted allenes for the C—H activation reactions of imines (Scheme 7.43) [55]. This transformation proceeded through a sequence enabled by Mn/Ag relay catalysis, namely, manganese-catalyzed C—H allylation of imines with allenes and subsequent silver-promoted Povarov cyclization to deliver the tetracyclic products with high regioselectivity.

The same group further expanded the scope of substrates to tri- and tetra-substituted allenes (Scheme 7.44) [56]. In these cases, an efficient synthesis of diverse bicyclic or tricyclic compounds **23** bearing an exocyclic C=C double bond was achieved through sequential manganese-catalyzed C—H activation and Smiles

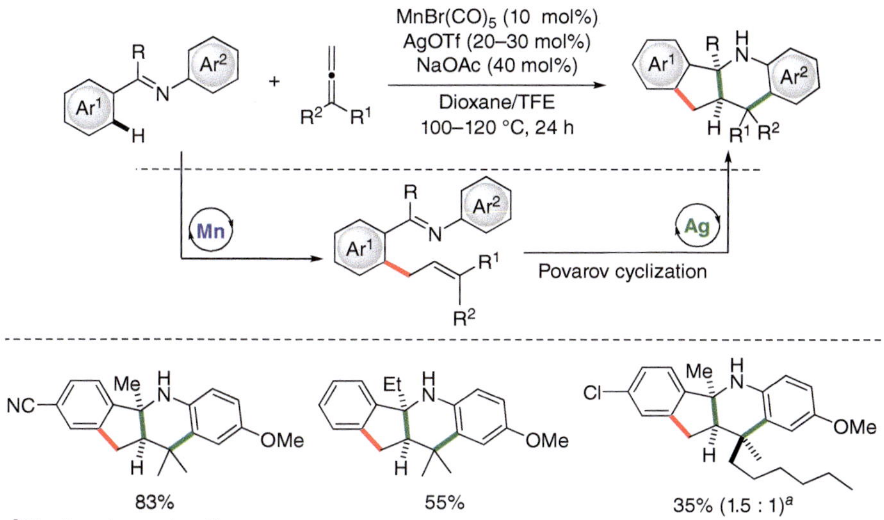

a Diastereoisomeric ratio.

Scheme 7.43 Polycyclization with 1,1-disubstituted allenes via Mn/Ag relay catalysis. Source: Redrawn from Chen et al. [55].

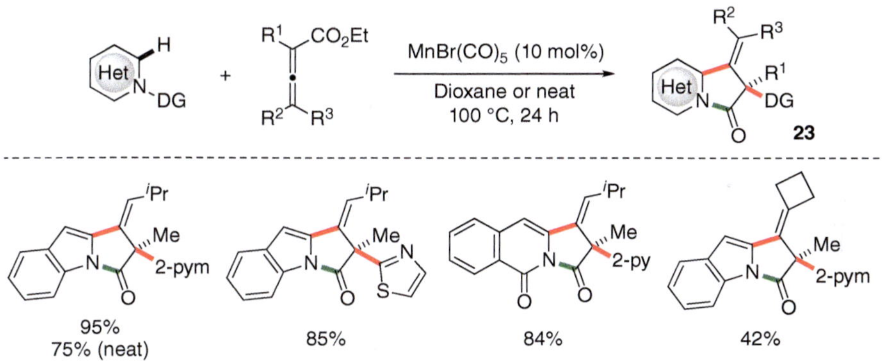

Scheme 7.44 Manganese-catalyzed C–H allylation and Smiles rearrangement cascade with tri- and tetra-substituted allenes. Source: Redrawn from Chen et al. [56].

rearrangement [57]. This reaction proceeded under additive-free and even neat conditions. It exhibited a broad scope of substrates and tolerated various DGs on the indoles. The proposed mechanism suggested that a facile and reversible C—H manganation occurred after coordination of the DG with manganese to furnish the five-membered manganacycle (Scheme 7.45), which underwent a regio- and stereo-selective migratory insertion of allene from the sterically less hindered face to deliver the allylated manganese species **Mn-24**. The control of reaction conditions provided either the alkenylated product **24** through protonation or the cyclic compounds via a subsequent N-to-C 1,4-migration of the DG, namely, Smiles rearrangement. It was supposed that the concomitant Smiles rearrangement

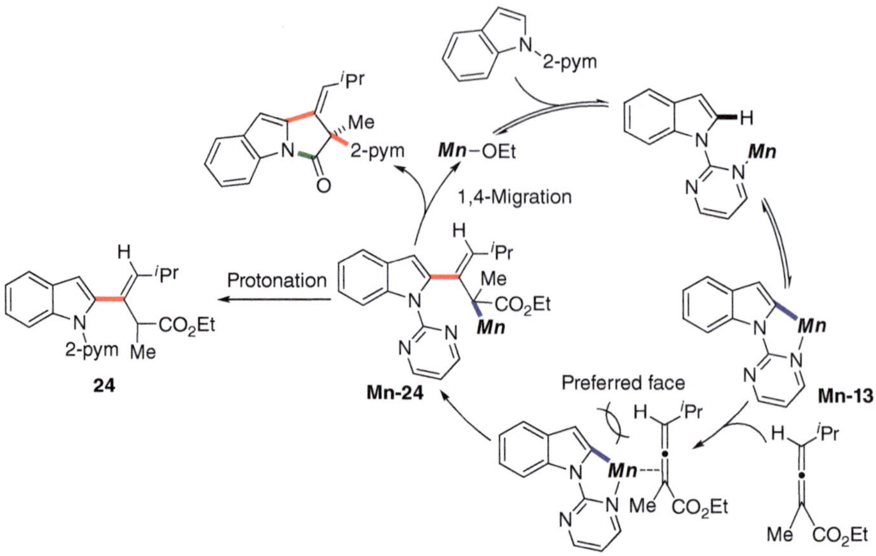

Scheme 7.45 Proposed mechanism of manganese-catalyzed C–H allylation and Smiles rearrangement cascade.

occurred most likely due to the strong nucleophilicity of the polarized Mn—C bond in the intermediate **Mn-24**.

At the same time, Rueping and coworkers independently developed a similar strategy for the synthesis of product **23** (Scheme 7.46) [58]. The reaction proceeded with a wide range of tri-substituted allenes in the presence of a catalytic amount of $MnBr(CO)_5$ and stoichiometric amount of NaOAc at lower temperature. When 1,3-disubstituted allenes were applied in the preceding reaction, the C2–H alkenylated products **24** were obtained in moderate to good yields with excellent regio- and stereoselectivity.

A remarkable application of allenes in manganese-catalyzed C—H propargylation of heterocycles was described by the group of Glorius in 2018 (Scheme 7.47) [59]. The optimal catalytic system, comprising of $MnBr(CO)_5$ as the catalytic precursor and triphenylboron as the cocatalyst, allowed for the C—H propargylation of various heteroarenes with bromoallenes. Mechanistically, the reaction proceeded through a selective attack of the Mn—C bond of the five-membered manganacycle at C1 rather than C3 position of allenes and β-bromo elimination of the seven-membered manganacycle giving the final propargylic heterocyclic compounds. Moreover, this protocol showed good functional group compatibility, which exhibited potentials for applications in late-stage diversification of complex molecules.

Further studies by Ding and coworkers demonstrated that the synthesis of 1-aminoindanes from ketimines with allenes could be achieved by using a catalytic $MnBr(CO)_5$/NaOAc system (Scheme 7.48) [60]. The proposed catalytic cycle invoked a C—H manganation assisted by acetate anion and migratory insertion at the electron-deficient central carbon of allene from the sterically less hindered face,

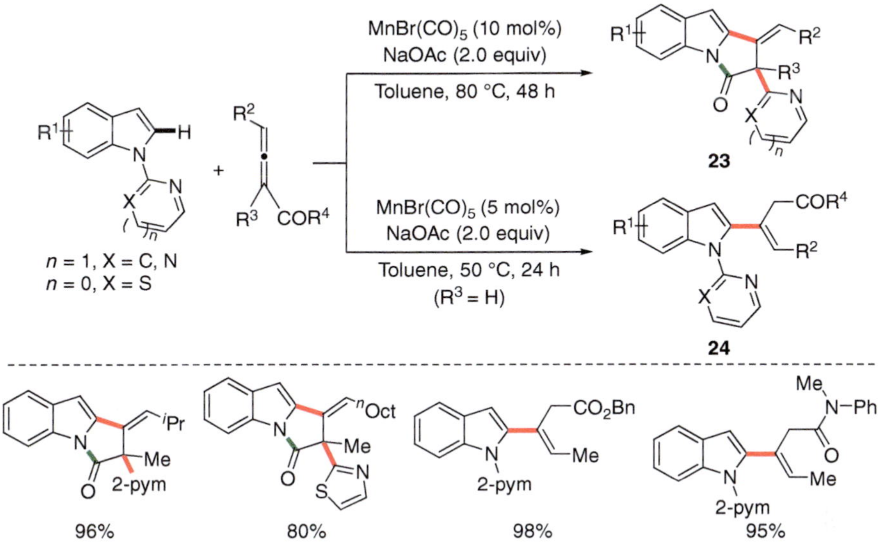

Scheme 7.46 Manganese-catalyzed C–H functionalization of indoles with allenes. Source: Redrawn from Wang et al. [58].

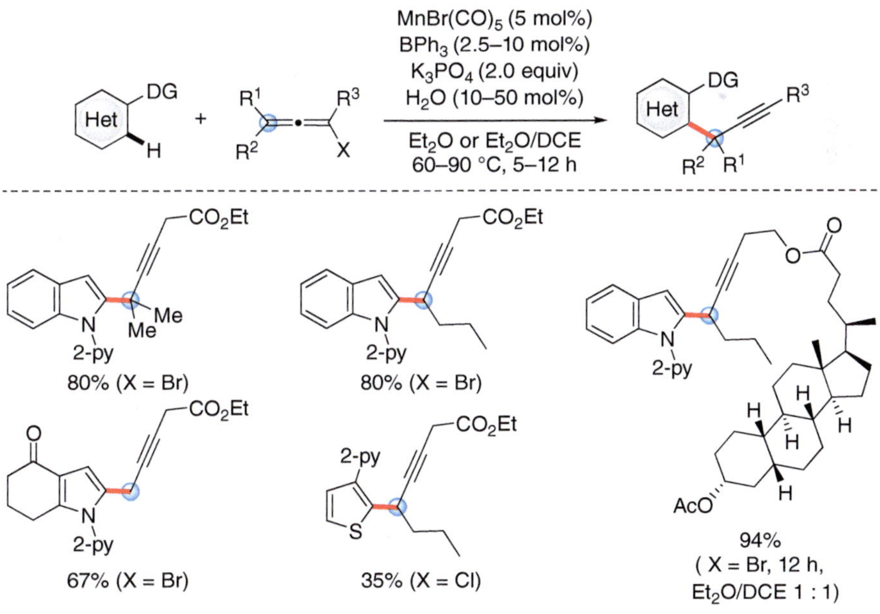

Scheme 7.47 Manganese-catalyzed C–H propargylation with bromoallenes. Source: Redrawn from Zhu et al. [59].

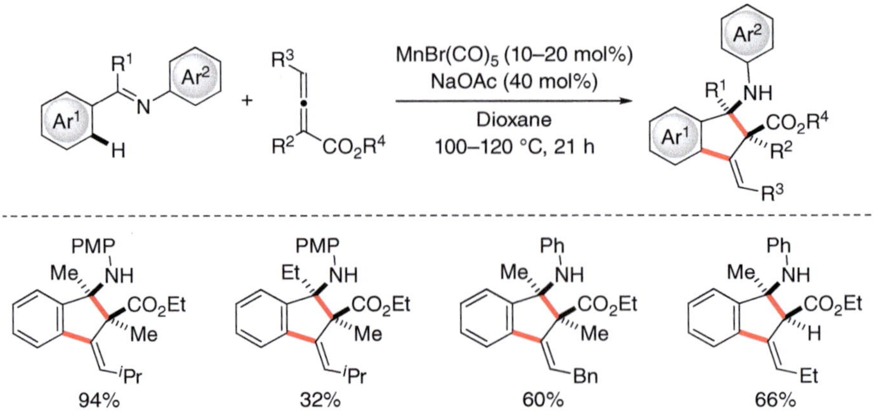

Scheme 7.48 Manganese-catalyzed synthesis of 1-aminoindanes from ketimines and electron-deficient allenes.

followed by protonation with HOAc. Therefore, high diastereoselectivity and *E*/*Z* selectivity were obtained in this protocol.

7.3.4 Insertion of C=X (X = O, N) Bond into Manganacycles

The first example of manganese-catalyzed addition of the C—H bond of 2-arylimidazoles to the C=O bond of aldehydes in the presence of stoichiometric quantities of silane was reported by Kuninobu, Takai, and coworkers in 2007 (Scheme 7.49a) [8a]. Notably, varied diastereomeric excess of the benzyl alcohol silyl ether products was achieved through employing enantiopure imidazolines as the DGs. In this seminal paper, it was also uncovered that the desired benzylic alcohol product was obtained from 2-phenylimidazole and benzaldehyde in 52% yield by employing a stoichiometric amount of $MnBr(CO)_5$ and treating the reaction mixture with TBAF (tetrabutylammonium fluoride). Meanwhile, benzylic alcohol silyl ether was formed in 58% yield by treatment of the stoichiometric reaction with an excess amount of triethylsilane (Scheme 7.49b). In addition, the catalytic reaction could give only trace amount of the desired benzyl alcohol in the absence of silane. Thus, the key success of this manganese-catalytic reaction lay in the addition of silanes. Mechanistically, the authors proposed an oxidative C—H addition to manganese(I) center to form manganese(III) species, which unfortunately awaited further experimental confirmation.

In 2015, the group of Wang explored a dual activation mode, that is, $MnBr(CO)_5$ as the catalytic precursor to cleave the aromatic C—H bond and $ZnBr_2$ as the Lewis acid to activate the C=O bond of aldehydes, which enabled the nucleophilic C—H bond addition to aldehydes to access benzylic alcohols directly (Scheme 7.50a) [61]. Remarkably, a broad range of aldehydes, including electron-deficient, electron-neutral, and electron-rich aromatic aldehydes as well as various aliphatic aldehydes were smoothly reacted with (hetero)arenes to give the desired products in moderate to excellent yields. It is noteworthy that this reaction also proceeded

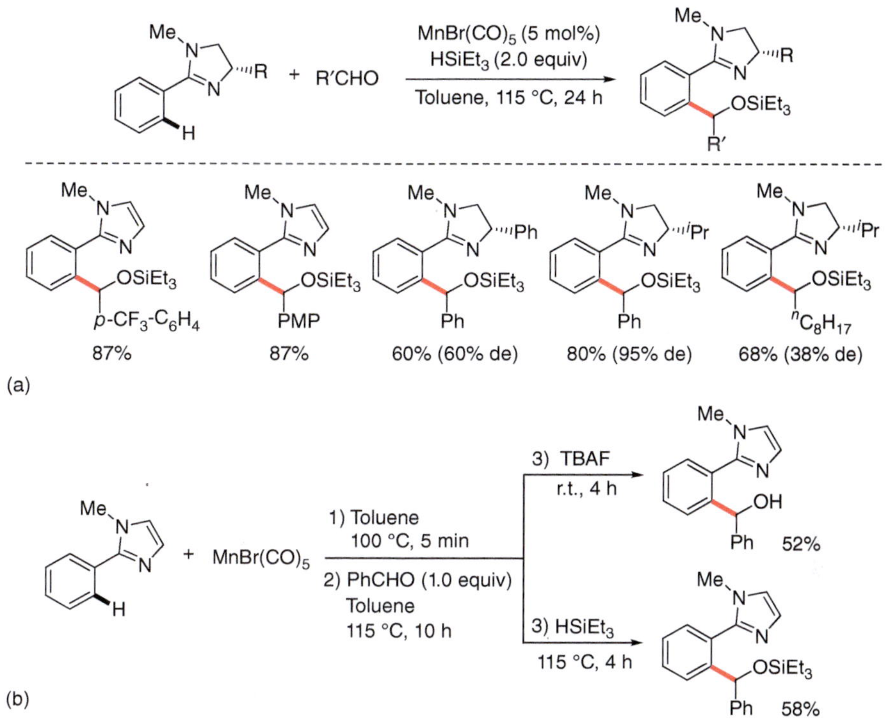

Scheme 7.49 Manganese-catalyzed insertion of aldehydes into aromatic C–H bonds in the presence of $HSiEt_3$. (a) Catalytic reactions and (b) stoichiometric reactions. Source: (a) Redrawn from Kuninobu et al. [8a].

well with olefinic C—H bonds to undergo addition to aldehydes affording the corresponding allylic alcohols as sole *Z*-stereoisomers. This methodology was further successfully applied to the directed C—H addition to nitriles [61]. Both aromatic and olefinic C—H bonds could react with aryl and alkyl nitriles to deliver the corresponding ketone products after acidic hydrolysis (Scheme 7.50b).

The authors performed a series of mechanistic experiments to shed light on the possible reaction mechanism (Scheme 7.51). Initially, $MnBr(CO)_5$ reacted with dimethylzinc to give the active catalytic precursor $MnMe(CO)_5$ (**Mn-25**), which was detected by NMR (nuclear magnetic resonance) analysis. Thereafter, a facile and irreversible C_{sp^2}–H cyclomanganation of the arene substrate with **Mn-25** furnished the five-membered manganacycle **Mn-1** and released methane (detected by GC-MS (gas chromatography-mass spectrometry)). Subsequently, migratory insertion of the C=O or C≡N bond activated by Lewis acid $ZnBr_2$ into the Mn—C bond afforded the seven-membered manganacycle **Mn-26**, which underwent a ligand metathesis with dimethylzinc to deliver **Mn-Zn-1** species. Then, a ligand exchange with another molecule of the arene substrate furnished **Mn-27** and **Zn-1** simultaneously. Finally, intramolecular C—H activation of **Mn-27** regenerated **Mn-1** with the evolution of methane, and hydrolysis of **Zn-1** delivered the desired products. It is of note that the five-membered manganacycle **Mn-1** was prepared independently through the

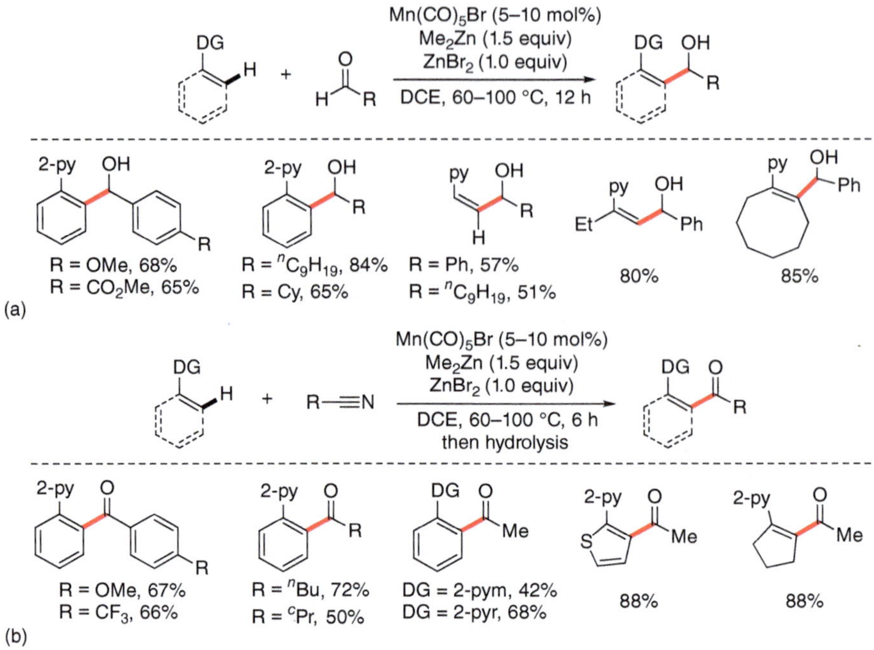

Scheme 7.50 Manganese-catalyzed insertion of aldehyde/nitriles into C_{sp^2} −H bonds. (a) Reactions with aldehydes and (b) reactions with nitriles. Source: (a) Redrawn from Zhou et al. [61].

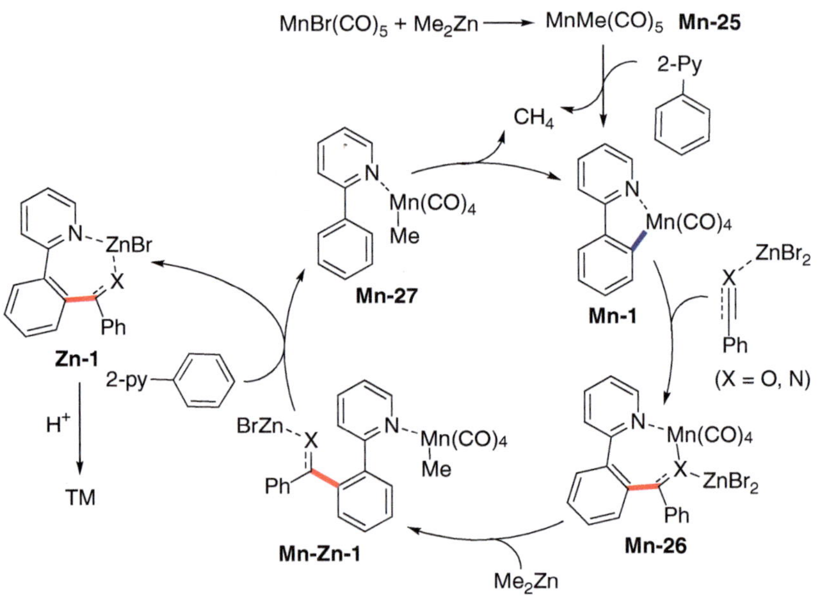

Scheme 7.51 A plausible reaction mechanism for manganese-catalyzed insertion of aldehyde/nitriles into C_{sp^2} −H bonds.

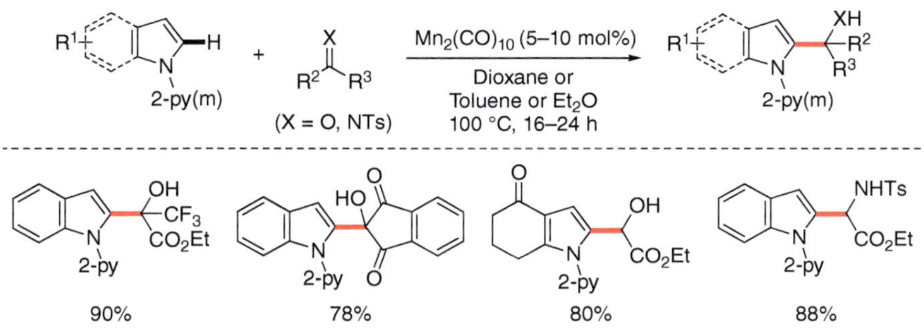

Scheme 7.52 Manganese-catalyzed insertion of electron-deficient aldehydes, ketones, and imines into C–H bonds. Source: Redrawn from Liang et al. [62].

stoichiometric reaction between 2-phenylpyridine and $MnBr(CO)_5$ in the presence of dimethylzinc, and not only **Mn-1** but also **Mn-25** could catalyze the reaction effectively.

In a subsequent report, Ackermann and coworkers described a manganese-catalyzed additive-free C—H addition of indole/pyrroles to electron-deficient aldehydes, ketones, and imines (Scheme 7.52) [62]. This reaction exhibited good compatibility with various functional groups, delivering the corresponding alcohols or amines in good yields with excellent C2-selectivity on indole/pyrroles. The mechanism is proposed to involve the formation of five-membered manganacycle, C=X (X = O, N) insertion into the Mn—C bond of the manganacycle, and a LLHT process with a second substrate molecule.

In 2016, the group of Kuninobu developed manganese-catalyzed synthesis of isobenzofuranones from aromatic esters and oxiranes (Scheme 7.53) [8b]. The authors proposed that this transformation proceeded through manganese-promoted C—H activation with the assistance of oxygen DG, insertion of aldehydes, which was isomerized from oxiranes under Lewis acidic conditions, into the Mn—C bond, and subsequent intramolecular nucleophilic cyclization. Though they surmised that the catalytic cycle involved an oxidative addition and reductive

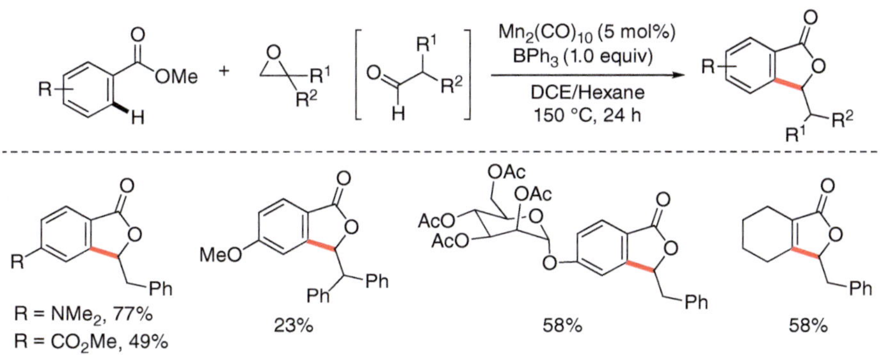

Scheme 7.53 Manganese-catalyzed C–H activation of aromatic esters with oxiranes. Source: Redrawn from Sueki et al. [8b].

elimination step, the possible reaction intermediates are still elusive. Moreover, the Lewis acid of BPh_3 was indispensable for this protocol, while its role remains ambiguous.

Another example of the C=O bond insertion into the Mn—C bond under manganese/base catalysis was described by Glorius and coworkers in 2018 [63]. In this protocol, paraformaldehyde was employed to react with decorated indoles, pyrroles, and other arenes delivering the hydroxymethyllation products in low to excellent yields (Scheme 7.54a). In addition, this strategy could be further applied to the one-pot C—H formylation of arenes by sequential manganese-catalyzed hydroxymethylation and iron-catalyzed aerobic oxidation (Scheme 7.54b).

Ketones are prevalent in a variety of natural products, pharmaceuticals, and material molecules. Meanwhile, they are also important building blocks in organic synthesis. In 2017, Wang and coworkers developed the manganese-catalyzed aromatic C—H addition of ketones to the C=N bond of imines by adopting the dual activation mode (Scheme 7.55a) [64]. Notably, the classic Mannich reaction was suppressed almost completely with the cooperative $MnBr(CO)_5/ZnBr_2/Me_2Zn$ regime. Under the optimized reaction conditions, the addition reaction of aryl alkyl ketones with sulfonyl imines proceeded successfully yielding *mono*-C—H addition products **27** at room temperature or 40 °C in moderate to good yields. Moreover, sole *exo*-olefinic isoindolines **28** could be obtained by tuning the reaction conditions slightly. Interestingly, a three-component reaction took place smoothly by using two equivalents of dimethylzinc at higher temperature, furnishing the methylated isoindolines **29** with high efficiency. As a particular highlight, a chiral ketone was shown to react with the imine in a racemization-free manner giving the major diastereoisomer of the desired C—H aminoalkylation products (Scheme 7.55b).

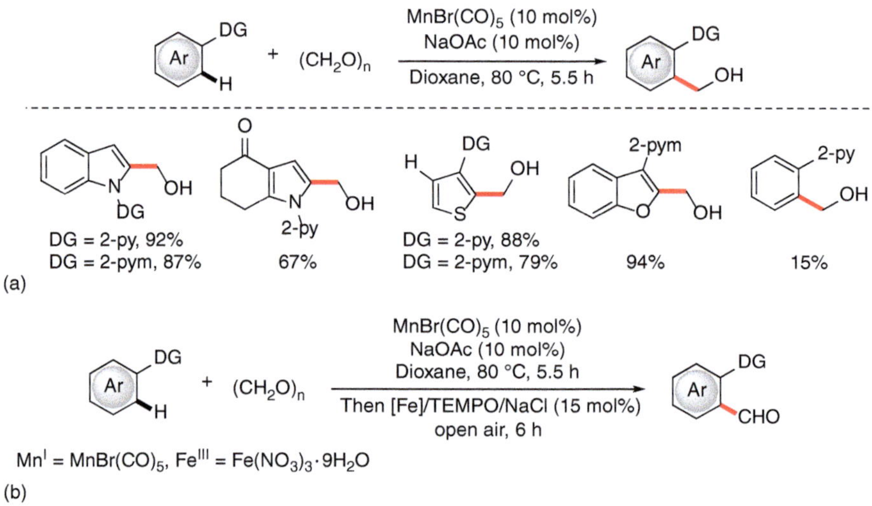

Scheme 7.54 Manganese-catalyzed C–H functionalization with paraformaldehydes. (a) C–H hydroxymethylation catalyzed by Mn^I and (b) relay catalyst of Mn^I and Fe^{III}. Mn_I = $MnBr(CO)_5$, Fe_{III} = $Fe(NO_3)_3 \cdot 9H_2O$.

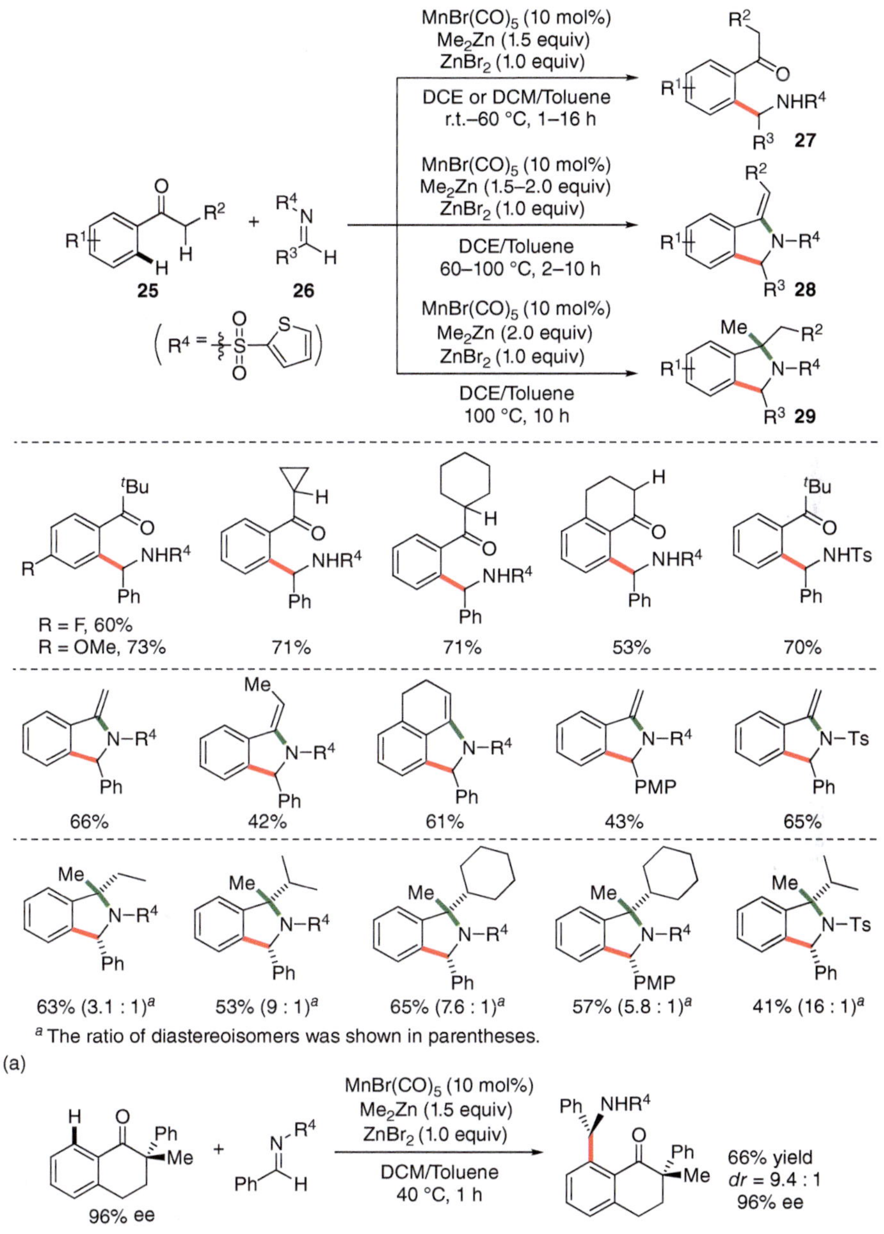

Scheme 7.55 Manganese-catalyzed aromatic C–H addition of ketones to imines. (a) Inert C–H addition of ketones to imines and (b) a reaction of a chiral ketone. Source: (a) Redrawn from Zhou et al. [64].

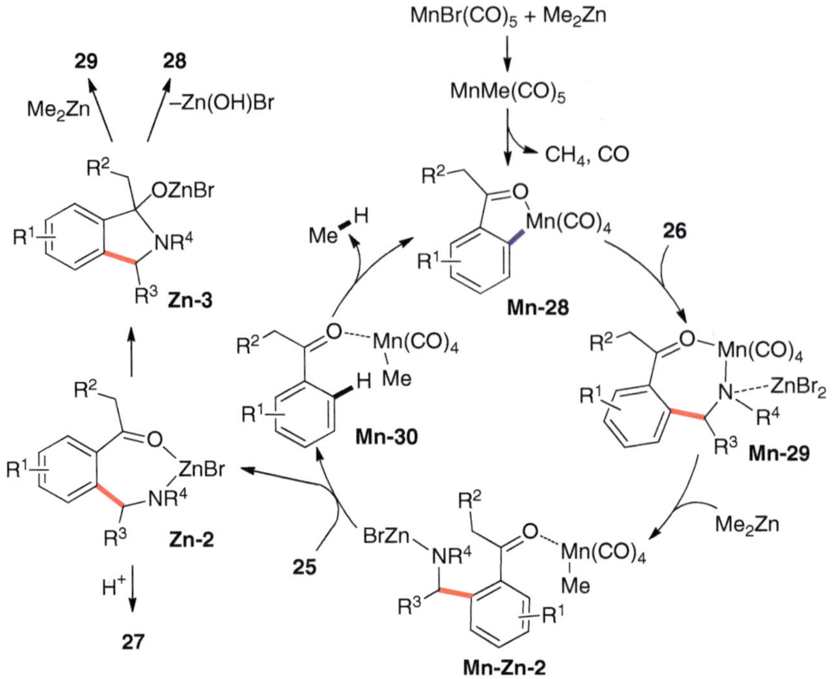

Scheme 7.56 A proposed reaction mechanism for the manganese-catalyzed C–H transformations of ketones with imines.

Similar to the C—H addition reactions of arenes to aldehydes and nitriles [61], the catalytic cycle of the previously mentioned reaction started with the formation of five-membered manganacycle of ketone **Mn-28** (Scheme 7.56), which underwent insertion of imine **26** to give seven-membered manganacycle **Mn-29**. Further reaction of **Mn-29** with Me_2Zn delivered the dimetallic intermediate **Mn-Zn-2**, which was converted into the methylmanganese species **Mn-30** and the zinc species **Zn-2** via ligand exchange with a second ketone. C—H activation occurred in **Mn-30** would regenerate **Mn-28** with concomitant release of methane. Under well-controlled reaction conditions, **Zn-2** could be either hydrolyzed to furnish the aminoalkylated product **27** or cyclize to give intermediate **Zn-3**, which further underwent elimination affording *exo*-olifinic isoindolines **28** or intermolecular nucleophilic substitution yielding methylated isoindolines **29**, respectively.

Later on, Ackermann and coworkers disclosed a manganese-catalyzed C—H activation reaction of indoles with aldimines under additive-free conditions at the temperature of 100 °C (Scheme 7.57) [65]. Indoles and pyrroles were suitable substrates for this protocol and a series of C2–H aminoalkylation products were obtained efficiently.

In 2015, Ackermann and coworkers showed that the combination of indoles/pyrroles and isocyanates under manganese catalysis gave the C2–H aminocarbonylated products (Scheme 7.58) [66]. The proposed mechanism suggested that the

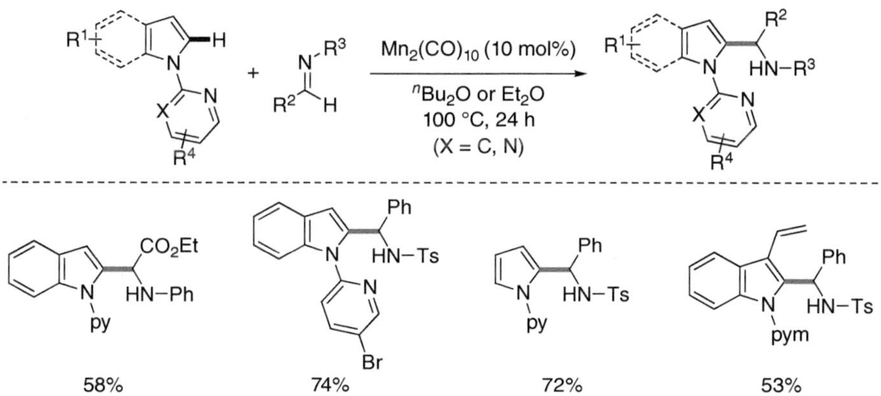

Scheme 7.57 Manganese-catalyzed C–H addition of indoles/pyrroles to imines. Source: Redrawn from Liang et al. [65].

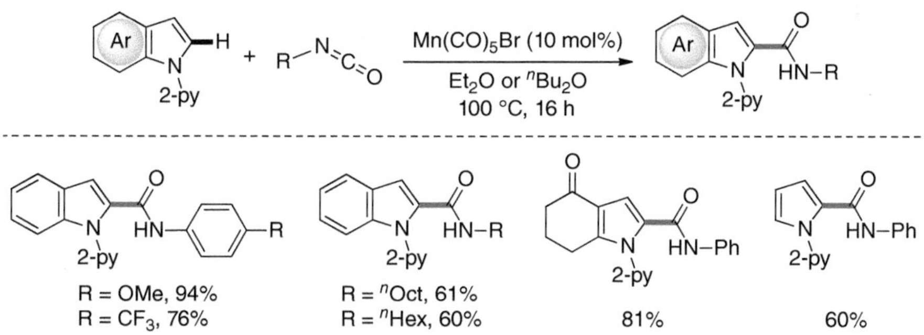

Scheme 7.58 Manganese-catalyzed C–H addition of indoles/pyrroles to isocyanates. Source: Redrawn from Liu et al. [66].

unsaturated C=N moiety of isocyanate underwent migratory insertion into the Mn—C bond of the five-membered manganacycle, leading to amides in excellent regioselectivity. Furthermore, this reaction was compatible with both aryl and alkyl isocyanates, providing the corresponding products in moderate to excellent yields.

Ketenimines were similarly employed as the reaction partner in a manganese-catalyzed C—H enaminylation reaction (Scheme 7.59), as reported by the group of Wang [67]. The reaction of 1-(pyrimidin-2-yl)-1*H*-indoles and triaryl ketenimines proceeded via sequential C—H activation/enaminylation/DG migration, giving access to a range of 1*H*-2-enaminylindoles derivatives. The authors found that 2-enaminylindoles could be obtained from ester-substituted ketenimines without DG-migration.

In contrast to Wang's previous report on manganese-catalyzed C—H addition to nitriles [61], Ackermann and coworkers disclosed manganese-catalyzed C—H cyanation of electron-rich heteroarenes with *N*-cyano-*N*-phenyl-*p*-toluenesulfonamide (NCTS) (Scheme 7.60) [68]. The optimal catalytic system comprised of $MnBr(CO)_5$

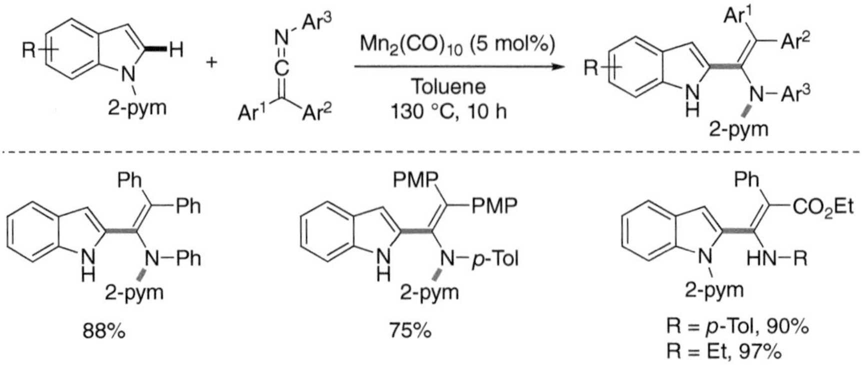

Scheme 7.59 Manganese-catalyzed C–H enaminylation of indoles with ketenimines. Source: Redrawn from Zhou et al. [67].

as a catalyst, zinc chloride as a crucial additive, and dicyclohexylamine as a base. Based on experimental and computational insights, the authors suggested that the rate-determining step involved the C—C bond formation process, and the stabilizing coordinative interaction between zinc chloride and manganacycle in the transition state led to energy decrease during the C—C bond formation.

Another example of manganese-catalyzed C—H cyanation under Lewis acid-free conditions was described by Bao and coworkers in 2018 (Scheme 7.61) [69]. In this reaction, various 2-arylpyridines underwent C—H cyanation with an electron-rich cyanating reagent, *N*-cyano-*N*-(4-methoxy)phenyl-*p*-toluenesulfonamide. The proposed catalytic cycle invoked a facile C—H activation via synergy of manganese and base, insertion of the C≡N bond into the Mn—C bond of the five-membered manganacycle, and elimination to release the desired product and regenerate active manganese species.

7.3.5 Miscellaneous Reactions Involving C–H Activation

Manganese carbonyl catalysts, especially $MnBr(CO)_5$ and $Mn_2(CO)_{10}$, are extensively used in C_{sp^2}—H bond activation reactions in the recent several years, which generally occur without the change of oxidation states of manganese throughout the entire catalytic cycles. Contrary to examples mentioned earlier, C—H transformations employing non-carbonyl manganese complexes that involve valence changes of manganese are emerging new trends in manganese-catalyzed organometallic C—H functionalization.

In 2017, the group of Ackermann described the manganese(II)-catalyzed C—H alkylations with challenging alkyl bromides (Scheme 7.62) [9]. With the assistance of the triazolyldimethylmethyl (TAM) amide DG, a variety of arenes and heteroarenes were monoselectively alkylated in good yields by using $MnCl_2$ as the catalytic precursor, a Grignard reagent and tetramethylethylenediamine (TMEDA) as the essential additives. Particularly, this reaction proceeded smoothly with a broad range of alkyl bromides without β-hydride elimination side reactions. The authors further

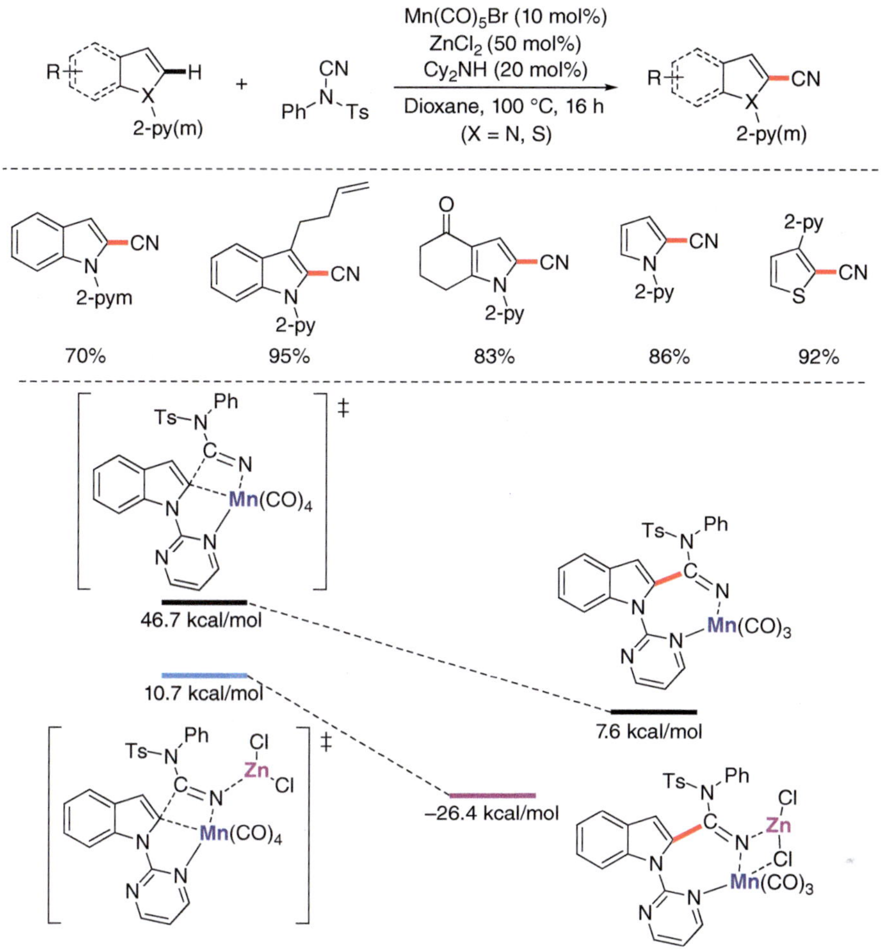

Scheme 7.60 Manganese-catalyzed C—H cyanation reactions of indole derivatives. Source: Redrawn from Liu et al. [68].

extended their manganese-catalyzed C—H alkylation methodology to an oxidative coupling reaction with MeMgBr, in which 2,3-dichlorobutane (DCB) was used as an oxidant. The proposed mechanism suggested that the C—H bond was cleaved via σ-bond metathesis as the rate-determining step to deliver the manganacycle **Mn-31** (Scheme 7.63). A subsequent one-electron oxidation of **Mn-31** by DCB occurred to give **Mn-32**, followed by a transmetalation step to furnish **Mn-33**. Finally, the desired product was released via a reductive elimination process of **Mn-33**, and the active species **Mn-34** was generated through a second one-electron oxidation step.

An analogous report on direct methylation of C_{sp^2}—H bonds with MeMgBr from the group of Ilies and Nakamura has utilized a similar Mn^{II} precatalyst, simple *N*-methyl amide motif as the DG, and 1-bromo-2-chloroethane (BCE) as the terminal oxidant (Scheme 7.64) [70]. This methodology was compatible with both (hetero)aryl and alkenyl carboxamides, providing the desired products in

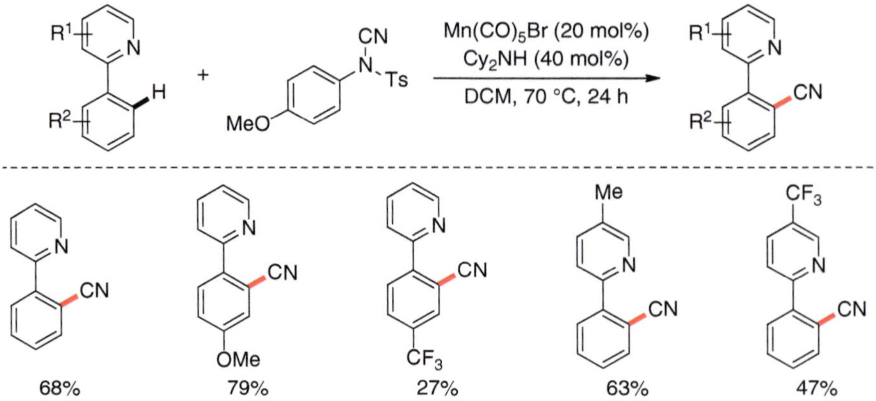

Scheme 7.61 Manganese-catalyzed C–H cyanation of 2-arylpyridines. Source: Redrawn from Yu et al. [69].

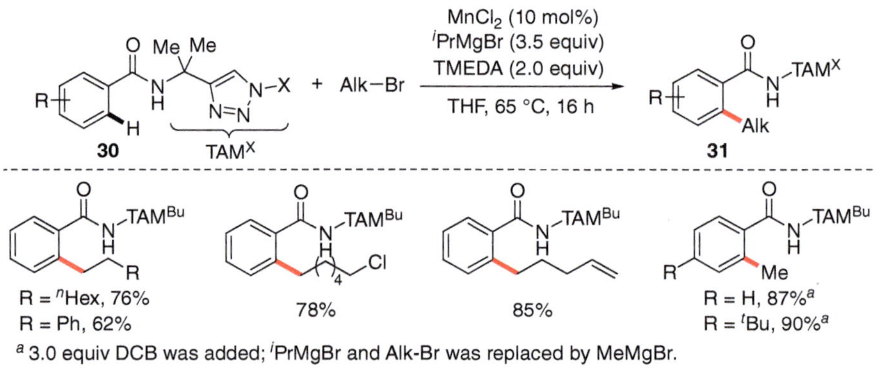

Scheme 7.62 Manganese(II)-catalyzed C–H alkylation with alkylbromides. Source: Redrawn from Liu et al. [9].

moderate to excellent yields. The ratios of mono- and dimethylated products varied in some cases. Furthermore, the authors put forward a catalytic cycle as shown in Scheme 7.65, which was a bit similar to that of Ackermann's proposal [9].

In 2018, Ackermann and coworkers further developed Mn^{II}-catalyzed C–H arylation of azines with Grignard reagents as the coupling partners through employing a sustainable flow technique (Scheme 7.66) [71]. The optimized reaction system required neocuproine as the ligand, 1,2-dichloro-2methylpropane (DCIB) as the terminal oxidant, and TMEDA for the deaggregation of the Grignard reagents. A Mn^{II}–Mn^{III}–Mn^{I} reaction pathway was invoked in the reaction mechanism, which was in agreement with their previous proposal for the C—H alkylation catalyzed by manganese [9].

Subsequently, the Ackermann group expanded the C—H cross-coupling of azines to that with challenging secondary as well as primary alkyl halides (Scheme 7.67)

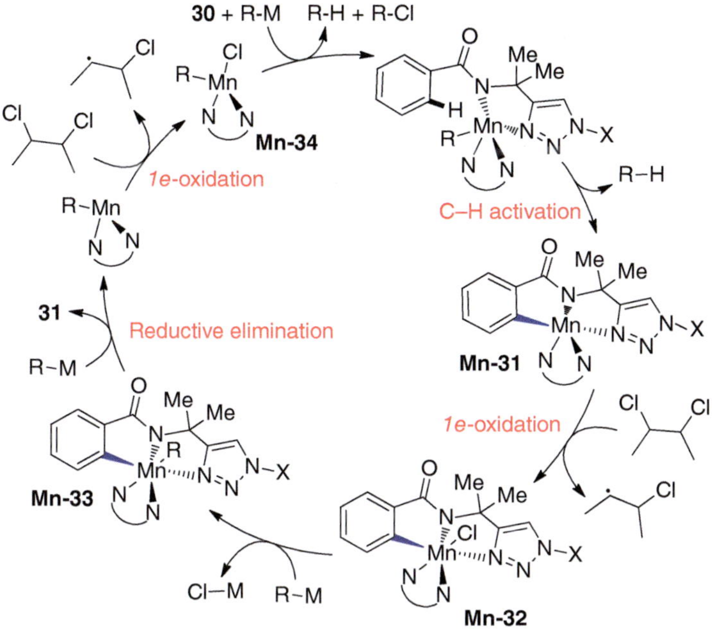

Scheme 7.63 Proposed mechanism for $MnCl_2$-catalyzed C–H methylation.

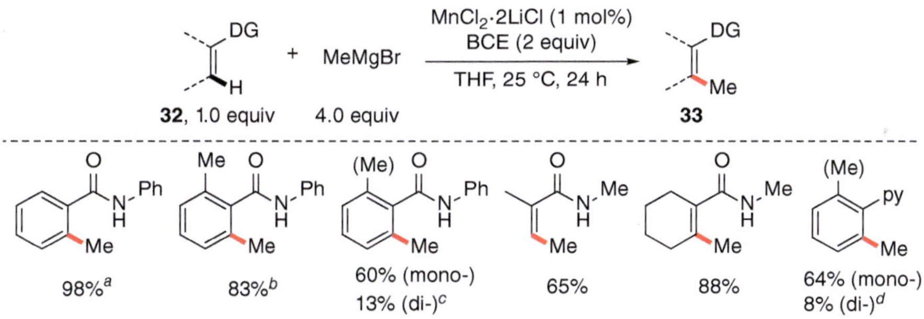

[a] 48 h; [b] 5.5 equiv MeMgBr, 3.0 equiv $BrCH_2CH_2Cl$, 48 h; [c] 3.1 equiv MeMgBr;
[d] 5 mol% $MnCl_2 \cdot 2LiCl$, 4.5 equiv MeMgBr, 3.0 equiv $BrCH_2CH_2Cl$, 48 h.

Scheme 7.64 Manganese(II)-catalyzed C–H methylation with MeMgBr. Source: Redrawn from Sato et al. [70].

[72]. This catalytic method provided access to a number of *o*-alkylated picolinamide derivatives with good chemoselectivity. Notably, unexpected double alkylated products **35** were obtained with bio-relevant nicotinamides as the coupling partners.

Diazo compounds had been employed as carbene precursors in a variety of transition metal-catalyzed transformations. In 2016, Pérez and coworkers disclosed the first manganese-catalyzed C—H alkylation of simple benzene

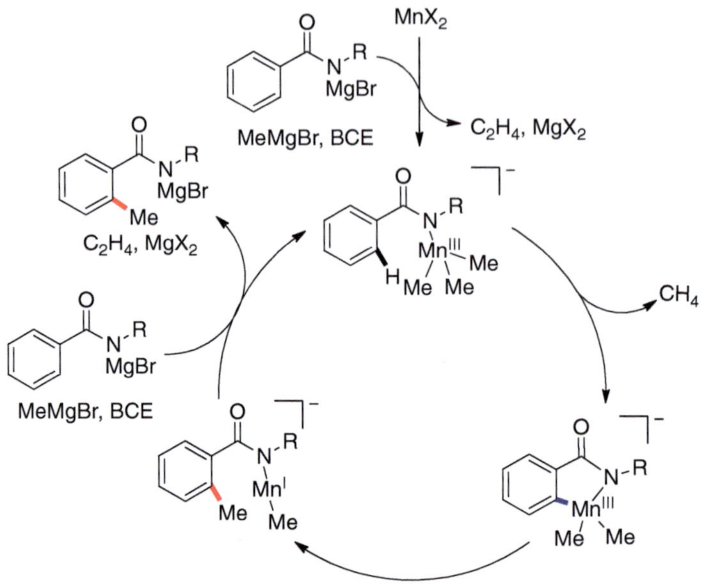

Scheme 7.65 A plausible catalytic cycle of manganese-catalyzed C–H methylation.

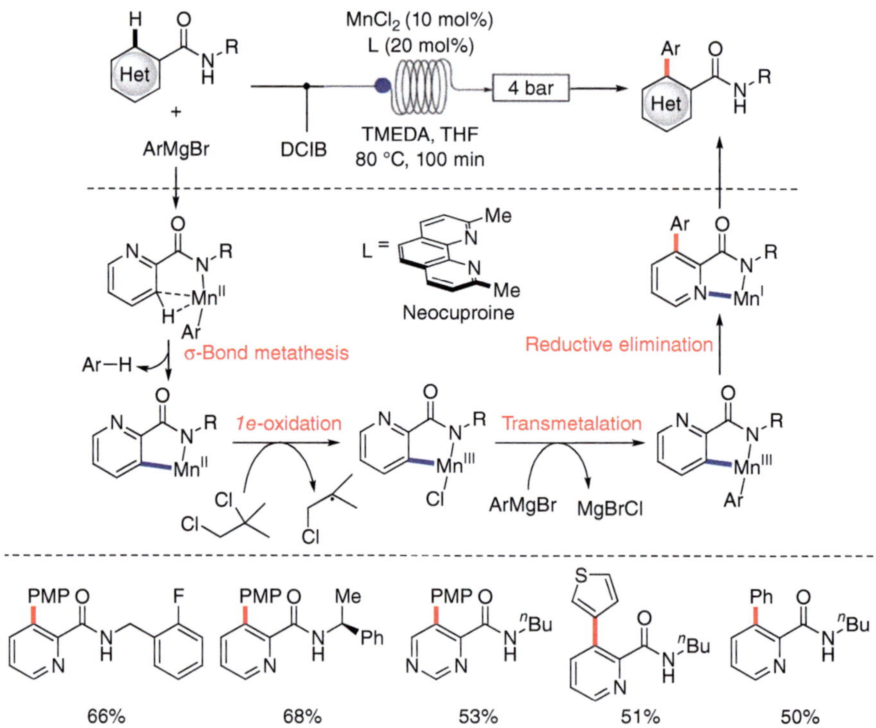

Scheme 7.66 Manganese(II)-catalyzed C–H arylation of azines in continuous flow. Source: Redrawn from Zhu et al. [71].

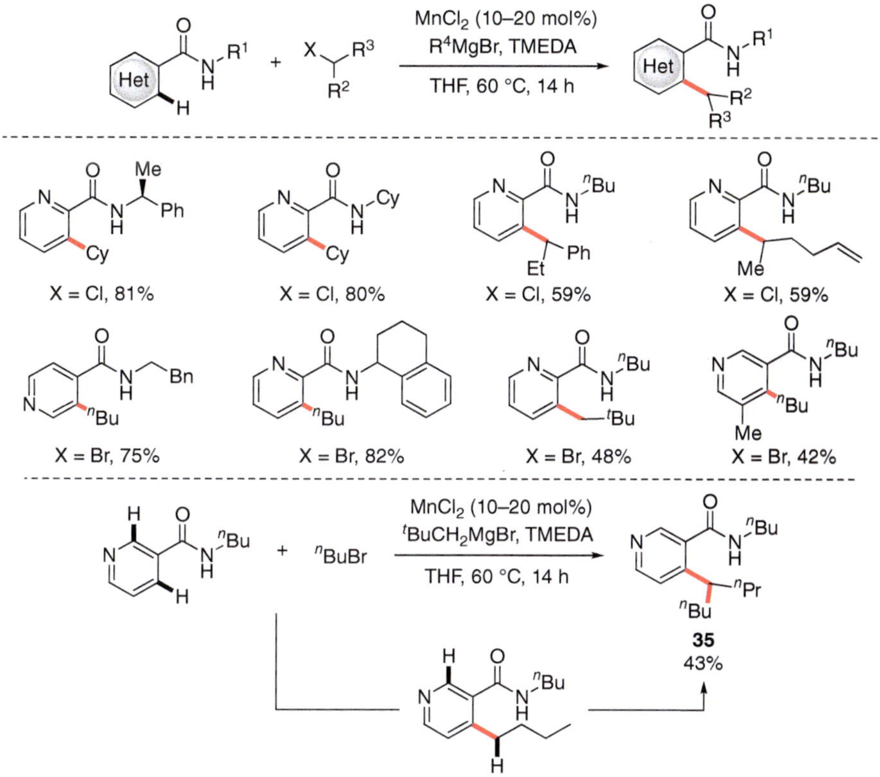

Scheme 7.67 Manganese(II)-catalyzed C–H alkylation of azines with alkyl halides. Source: Redrawn from Shen et al. [72].

derivatives with ethyl diazoacetate (Scheme 7.68) [73]. The reaction was enabled by a Mn^{II} precatalyst supported by a tetradentate pytacn ligand (pytacn = 1-(2-pyridylmethyl)-4,7-dimethyl-1,4,7-triazacyclononane) affording alkylated arenes in acceptable yields with excellent chemoselectivity. The authors proposed a possible mechanism that started with the formation of the dicationic manganese species **Mn-35** by exchange of the OTf anion to bulky $BAr^F{}_4$ anion. Then, metallocarbene species **Mn-37** was formed by releasing N_2 via intermediate **Mn-36**, which was generated from the coordination of diazo ester to the manganese center. Thereafter, an outer-sphere interaction between the carbene ligand in **Mn-37** and the arene led to the Wheland-type intermediate **Mn-38**, which could liberate the desired product via a subsequent 1,2-hydrogen shift.

Another example of manganese-catalyzed aryl C—H alkylation with diazo esters was achieved by Rueping and coworkers in 2018 (Scheme 7.69) [74]. By using the catalytic combination of $MnBr(CO)_5$ and NaOAc, the C—H alkylation of *N*-pyrimidinyl indoles proceeded with good functional group tolerance. Furthermore, the authors performed experimental and computational mechanistic studies to shed light on the possible reaction pathways (Scheme 7.69). The proposed mechanism suggested the formation of $(CO)_3Mn(\kappa^2\text{-OAc})$ complex as the

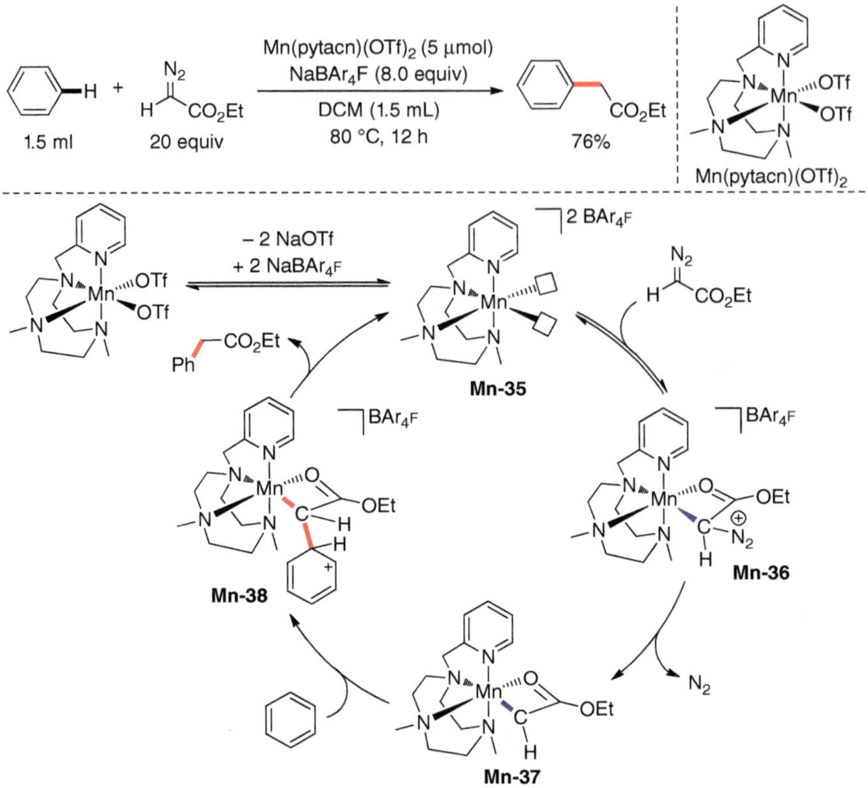

Scheme 7.68 Manganese-catalyzed C—H alkylation of arenes with ethyl diazoacetate. Source: Redrawn from Conde et al. [73].

catalytically active species, which coordinated with the DG of indoles and underwent rate-determining reversible C—H bond cleavage delivering the five-membered manganacycle **Mn-39**. It was transformed into manganese carbenoid intermediate **Mn-40** through coordination with the diazo ester. Subsequently, intramolecular carbene insertion into the Mn—C bond of **Mn-40** generated intermediate **Mn-41**, followed by protonation and release of the desired product.

Interestingly, the group of Sen applied the same catalytic system to the manganese-catalyzed cyclopropanation of *N*-acetyl indoles with diazo esters (Scheme 7.70) [75]. Control experiments and DFT calculations set the stage for a plausible mechanistic hypothesis of this transformation. The catalytic cycle was initiated by the formation of the manganese carbenoid intermediate **Mn-42** through the reaction of $(CO)_4Mn(\kappa^2\text{-OAc})$ complex and α-aryldiazoester. After coordination of the *N*-acetyl indole to the manganese center, cyclization occurred to generate species **Mn-44**, followed by the manganese release to afford the desired product. The authors highlighted the crucial role of the acetyl moiety, which facilitated the cyclization via the intermediate **Mn-43**.

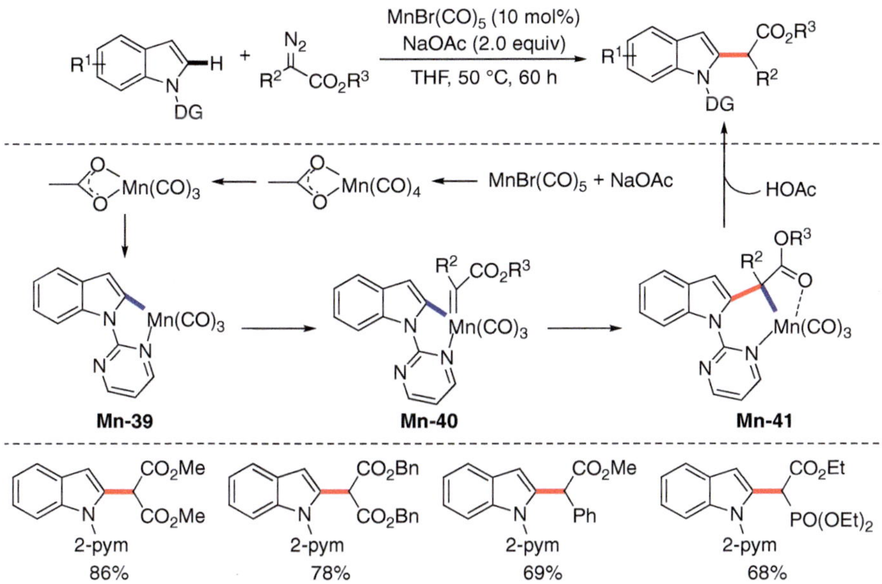

Scheme 7.69 Manganese-catalyzed C–H alkylation of indoles with diazoesters.

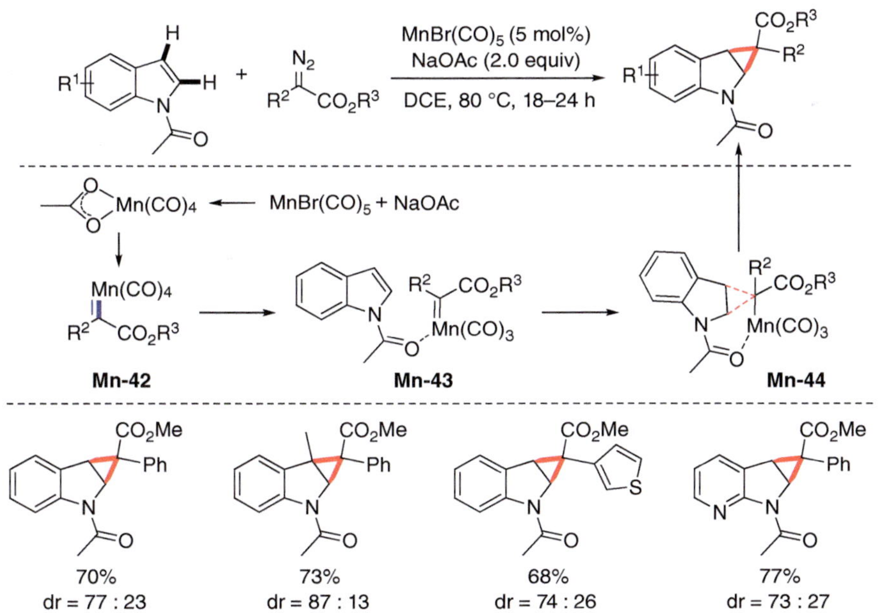

Scheme 7.70 Manganese-catalyzed cyclopropanation of indoles with diazoesters. Source: Redrawn from Dutta et al. [75].

7.4 Summary and Outlook

Due to its low toxicity, natural abundance, good biocompatibility, and unique catalytic performance, manganese-catalyzed inert C—H bond activation assisted by heteroatom DGs has aroused intensive interest over the past few years and emerged as a rapid and effective methodology to construct carbon–carbon bonds. A variety of C—H alkylation, alkenylation, and arylation reactions catalyzed by manganese have been achieved by the judicious choice of the DGs, the bases, the acids, other synergistic additives, and the reaction parameters. An overwhelming majority of these successful examples have been addition reactions of C—H bonds to unsaturated multiple bonds based on the nucleophilicity of Mn(I)—C bonds, involving base-assisted C—H cyclomanganation and migratory insertion of C=X (X = C, N, O) bonds without change of the oxidation state of manganese throughout the entire catalytic cycle. Furthermore, Mn(II)-based catalytic systems often proceed with single-electron transfer (SET) processes, during which the manganese intermediates exhibited variable oxidation states, setting the stage for developing novel redox manganese catalysis. In terms of the exact nature of the C—H bond activation step, two major categories of mechanisms have been invoked. One is redox-neutral C—H metalation based on manganese carbonyl complexes, which may involve base-assisted intermolecular deprotonation (BAID), CBMD, σ-bond metathesis, and so on. The other includes SET with manganese(II) catalysts and oxidative C—H addition though in controversial still.

Despite the remarkable advances that have been achieved in this rapidly evolving research area so far, further developments in the manganese-catalyzed organometallic C—H activation for C—C bond construction are still highly desirable. Future research in this field may be focused on the rational design of ligands, which is potentially the key factor to achieve more challenging C_{sp^3}—H bond functionalization, enantioselective C–H transformations, and DG-free C—H bond activation in future. The design of cooperative manganese catalysis with the assistance of base and/or acid would inspire further developments of manganese-catalyzed C—H functionalization with more diverse substrates, broader reaction types, and higher reaction efficiency. In addition, thorough mechanistic understandings of manganese-promoted catalytic cycles through detailed experimental and computational studies are also in high need in order to devise new manganese catalysis regimes.

References

1 (a) Huang, X., Zhuang, T., Kates, P.A. et al. (2017). *J. Am. Chem. Soc.* 139: 15407–15413. (b) Kong, X., Lin, L., and Xu, B. (2018). *Adv. Synth. Catal.* 360: 2801–2805. (c) Kong, X. and Xu, B. (2018). *Org. Lett.* 20: 4495–4498. (d) Paradine, S.M., Griffin, J.R., Zhao, J. et al. (2015). *Nat. Chem.* 7: 987–994. (e) Kim, J.Y., Cho, S.H., Joseph, J., and Chang, S. (2010). *Angew. Chem. Int. Ed.* 49: 9899–9903. (f) Huang, X., Bergsten, T.M., and Groves, J.T. (2015). *J. Am. Chem. Soc.* 137:

5300–5303. (g) Yu, X.Q., Huang, J.S., Zhou, X.G., and Che, C.M. (2000). *Org. Lett.* 2: 2233–2236. (h) Liang, J.-L., Huang, J.-S., Yu, X.-Q. et al. (2002). *Chem. Eur. J.* 8: 1563–1572.

2 (a) Wang, W., Xu, D., Sun, Q., and Sun, W. (2018). *Chem. Asian J.* 13: 2458–2464. (b) Shen, D., Miao, C., Wang, S. et al. (2014). *Org. Lett.* 16: 1108–1111. (c) Adams, A.M., Du Bois, J., and Malik, H.A. (2015). *Org. Lett.* 17: 6066–6069. (d) Ottenbacher, R.V., Samsonenko, D.G., Talsi, E.P., and Bryliakov, K.P. (2012). *Org. Lett.* 14: 4310–4313. (e) Breslow, R., Zhang, X., and Huang, Y. (1997). *J. Am. Chem. Soc.* 119: 4535–4536. (f) Das, S., Incarvito, C.D., Crabtree, R.H., and Brudvig, G.W. (2006). *Science* 312: 1941–1943.

3 (a) Liu, W., Huang, X., and Groves, J.T. (2013). *Nat. Protoc.* 8: 2348–2354. (b) Liu, W. and Groves, J.T. (2015). *Acc. Chem. Res.* 48: 1727–1735. (c) Liu, W. and Groves, J.T. (2013). *Angew. Chem. Int. Ed.* 52: 6024–6027. (d) Huang, X., Liu, W., Ren, H. et al. (2014). *J. Am. Chem. Soc.* 136: 6842–6845. (e) Liu, W. and Groves, J.T. (2010). *J. Am. Chem. Soc.* 132: 12847–12849. (f) Liu, W., Huang, X., Cheng, M.J. et al. (2012). *Science* 337: 1322–1325.

4 (a) Layfield, R.A. (2008). *Chem. Soc. Rev.* 37: 1098–1107. (b) Cahiez, G., Duplais, C., and Buendia, J. (2009). *Chem. Rev.* 109: 1434–1476. (c) Khusnutdinov, R.I., Bayguzina, A.R., and Dzhemilev, U.M. (2012). *Russ. J. Org. Chem.* 48: 309–348. (d) Saisaha, P., de Boer, J.W., and Browne, W.R. (2013). *Chem. Soc. Rev.* 42: 2059–2074. (e) Ottenbacher, R.V., Talsi, E.P., and Bryliakov, K.P. (2018). *Chem. Rec.* 18: 78–90. (f) Valyaev, D.A., Lavigne, G., and Lugan, N. (2016). *Coord. Chem. Rev.* 308: 191–235.

5 Bruce, M.I., Lqbal, M.Z., and Stone, F.G.A. (1970). *J. Chem. Soc. A*: 3204–3209.

6 Hartwig, J. (2009). *Organotransition Metal Chemistry: From Bonding to Catalysis.* University Science books.

7 Djukic, J.-P., Maisse, A., Pfeffer, M. et al. (1997). *Organometallics* 16: 657–667.

8 (a) Kuninobu, Y., Nishina, Y., Takeuchi, T., and Takai, K. (2007). *Angew. Chem. Int. Ed.* 46: 6518–6520. (b) Sueki, S., Wang, Z., and Kuninobu, Y. (2016). *Org. Lett.* 18: 304–307.

9 Liu, W., Cera, G., Oliveira, J.C.A. et al. (2017). *Chem. Eur. J.* 23: 11524–11528.

10 Young, K.M. and Wrighton, M.S. (1990). *J. Am. Chem. Soc.* 112: 157–166.

11 (a) Zhou, B., Chen, H., and Wang, C. (2013). *J. Am. Chem. Soc.* 135: 1264–1267. (b) Zhou, B., Ma, P., Chen, H., and Wang, C. (2014). *Chem. Commun.* 50: 14558–14561.

12 Hu, Y., Zhou, B., Chen, H., and Wang, C. (2018). *Angew. Chem. Int. Ed.* 57: 12071–12075.

13 (a) Tully, W., Main, L., and Nicholson, B.K. (1995). *J. Organomet. Chem.* 503: 75–92. (b) Robinson, N.P., Main, L., and Nicholson, B.K. (1989). *J. Organomet. Chem.* 364: C37–C39. (c) Depree, G.J., Main, L., and Nicholson, B.K. (1998). *J. Organomet. Chem.* 551: 281–291. (d) Gommans, L.H.P., Main, L., and Nicholson, B.K. (1987). *J. Chem. Soc., Chem. Commun.* 1987: 761–762.

14 Liebeskind, L.S., Gasdaska, J.R., McCallum, J.S., and Tremont, S.J. (1989). *J. Org. Chem.* 54: 669–677.

15 (a) Cambie, R.C., Metzler, M.R., Rutledge, P.S., and Woodgate, P.D. (1992). *J. Organomet. Chem.* 429: 41–57. (b) Cambie, R.C., Metzler, M.R., Rutledge, P.S., and Woodgate, P.D. (1990). *J. Organomet. Chem.* 398: C22–C24. (c) Cambie, R.C., Metzler, M.R., Rutledge, P.S., and Woodgate, P.D. (1990). *J. Organomet. Chem.* 381: C26–C30.
16 (a) Grigsby, W.J., Main, L., and Nicholson, B.K. (1990). *Bull. Chem. Soc. Jpn.* 63: 649–651. (b) Grigsby, W.J., Main, L., and Nicholson, B.K. (1993). *Organometallics* 12: 397–407.
17 Ackermann, L. (2011). *Chem. Rev.* 111: 1315–1345.
18 Suárez, A., Faraldo, F., Vila, J.M. et al. (2002). *J. Organomet. Chem.* 656: 270–273.
19 Yahaya, N.P., Appleby, K.M., Teh, M. et al. (2016). *Angew. Chem. Int. Ed.* 55: 12455–12459.
20 Bruce, M.I., Goodall, B.L., Iqbal, M.Z. et al. (1971). *J. Chem. Soc. D* 1971: 1595–1596.
21 He, R., Huang, Z.T., Zheng, Q.Y., and Wang, C. (2014). *Angew. Chem. Int. Ed.* 53: 4950–4953.
22 He, R., Huang, Z.-T., Zheng, Q.-Y., and Wang, C. (2014). *Tetrahedron Lett.* 55: 5705–5713.
23 Yang, Y., Zhang, Q., Shi, J., and Fu, Y. (2016). *Acta Chim. Sinica* 74: 422–428.
24 Shi, L., Zhong, X., She, H. et al. (2015). *Chem. Commun.* 51: 7136–7139.
25 Yang, X., Jin, X., and Wang, C. (2016). *Adv. Synth. Catal.* 358: 2436–2442.
26 Jia, T. and Wang, C. (2019). *ChemCatChem* 11: 5292–5295.
27 Ruan, Z., Sauermann, N., Manoni, E., and Ackermann, L. (2017). *Angew. Chem. Int. Ed.* 56: 3172–3176.
28 Lu, Q., Gressies, S., Klauck, F.J.R., and Glorius, F. (2017). *Angew. Chem. Int. Ed.* 56: 6660–6664.
29 Lu, Q., Gressies, S., Cembellin, S. et al. (2017). *Angew. Chem. Int. Ed.* 56: 12778–12782.
30 Lu, Q., Cembellin, S., Gressies, S. et al. (2018). *Angew. Chem. Int. Ed.* 57: 1399–1403.
31 Zheng, G., Sun, J., Xu, Y. et al. (2019). *Angew. Chem. Int. Ed.* 58: 5090–5094.
32 Zhu, C., Kuniyil, R., and Ackermann, L. (2019). *Angew. Chem. Int. Ed.* 58: 5338–5342.
33 Xu, Y., Zheng, G., Kong, L., and Li, X. (2019). *Org. Lett.* 21: 3402–3406.
34 Wang, H., Pesciaioli, F., Oliveira, J.C.A. et al. (2017). *Angew. Chem. Int. Ed.* 56: 15063–15067.
35 Liu, B., Li, J., Hu, P. et al. (2018). *ACS Catal.* 8: 9463–9470.
36 Tan, Y.X., Liu, X.Y., Zhao, Y.S. et al. (2019). *Org. Lett.* 21: 5–9.
37 Yi, X., Chen, K., and Chen, W. (2018). *Adv. Synth. Catal.* 360: 4497–4501.
38 Liu, S.L., Li, Y., Guo, J.R. et al. (2017). *Org. Lett.* 19: 4042–4045.
39 Liu, W., Zell, D., John, M., and Ackermann, L. (2015). *Angew. Chem. Int. Ed.* 54: 4092–4096.
40 Hu, Y. and Wang, C. (2016). *Sci. China Chem.* 59: 1301–1305.

41 Liang, Y.F., Muller, V., Liu, W. et al. (2017). *Angew. Chem. Int. Ed.* 56: 9415–9419.

42 Povarov, L. (1967). *Russ. Chem. Rev.* 36: 656–670.

43 Liu, W., Richter, S.C., Zhang, Y., and Ackermann, L. (2016). *Angew. Chem. Int. Ed.* 55: 7747–7750.

44 Lu, Q., Klauck, F.J.R., and Glorius, F. (2017). *Chem. Sci.* 8: 3379–3383.

45 Wang, H., Lorion, M.M., and Ackermann, L. (2017). *Angew. Chem. Int. Ed.* 56: 6339–6342.

46 Ni, J., Zhao, H., and Zhang, A. (2017). *Org. Lett.* 19: 3159–3162.

47 Zell, D., Dhawa, U., Müller, V. et al. (2017). *ACS Catal.* 7: 4209–4213.

48 Cai, S.H., Ye, L., Wang, D.X. et al. (2017). *Chem. Commun.* 53: 8731–8734.

49 Wu, S., Yang, Q., Hu, Q. et al. (2018). *Org. Chem. Front.* 5: 2852–2855.

50 Ali, S., Huo, J., and Wang, C. (2019). *Org. Lett.* 21: 6961–6965.

51 Kaplaneris, N., Rogge, T., Yin, R. et al. (2019). *Angew. Chem. Int. Ed.* 131: 3514–3518.

52 Wang, W., Subramanian, P., Martinazzoli, O. et al. (2019). *Chem. Eur. J.* 25: 10585–10589.

53 Meyer, T.H., Liu, W., Feldt, M. et al. (2017). *Chem. Eur. J.* 23: 5443–5447.

54 Chen, S.Y., Li, Q., and Wang, H. (2017). *J. Org. Chem.* 82: 11173–11181.

55 Chen, S.Y., Li, Q., Liu, X.G. et al. (2017). *ChemSusChem* 10: 2360–2364.

56 Chen, S.Y., Han, X.L., Wu, J.Q. et al. (2017). *Angew. Chem. Int. Ed.* 56: 9939–9943.

57 Levy, A.A., Rains, H.C., and Smiles, S. (1931). *J. Chem. Soc.* 1931: 3264–3269.

58 Wang, C., Wang, A., and Rueping, M. (2017). *Angew. Chem. Int. Ed.* 56: 9935–9938.

59 Zhu, C., Schwarz, J.L., Cembellin, S. et al. (2018). *Angew. Chem. Int. Ed.* 57: 437–441.

60 Lei, C., Peng, L., and Ding, K. (2018). *Adv. Synth. Catal.* 360: 2952–2958.

61 Zhou, B., Hu, Y., and Wang, C. (2015). *Angew. Chem. Int. Ed.* 54: 13659–13663.

62 Liang, Y.F., Massignan, L., Liu, W., and Ackermann, L. (2016). *Chem. Eur. J.* 22: 14856–14859.

63 Zhu, C., Pinkert, T., Greßies, S., and Glorius, F. (2018). *ACS Catal.* 8: 10036–10042.

64 Zhou, B., Hu, Y., Liu, T., and Wang, C. (2017). *Nat. Commun.* 8: 1169–1177.

65 Liang, Y.-F., Massignan, L., and Ackermann, L. (2018). *ChemCatChem* 10: 2768–2772.

66 Liu, W., Bang, J., Zhang, Y., and Ackermann, L. (2015). *Angew. Chem. Int. Ed.* 54: 14137–14140.

67 Zhou, X., Li, Z., Zhang, Z. et al. (2018). *Org. Lett.* 20: 1426–1429.

68 Liu, W., Richter, S.C., Mei, R. et al. (2016). *Chem. Eur. J.* 22: 17958–17961.

69 Yu, X., Tang, J., Jin, X. et al. (2018). *Asian J. Org. Chem.* 7: 550–553.

70 Sato, T., Yoshida, T., Al Mamari, H.H. et al. (2017). *Org. Lett.* 19: 5458–5461.

71 Zhu, C., Oliveira, J.C.A., Shen, Z. et al. (2018). *ACS Catal.* 8: 4402–4407.

72 Shen, Z., Huang, H., Zhu, C. et al. (2019). *Org. Lett.* 21: 571–574.

73 Conde, A., Sabenya, G., Rodriguez, M. et al. (2016). *Angew. Chem. Int. Ed.* 55: 6530–6534.

74 Wang, C., Maity, B., Cavallo, L., and Rueping, M. (2018). *Org. Lett.* 20: 3105–3108.

75 Dutta, P.K., Chauhan, J., Ravva, M.K., and Sen, S. (2019). *Org. Lett.* 21: 2025–2028.

8

Manganese-Catalyzed Cross-Coupling Processes

Xiaoping Liu, Florian Jaroschik, and Marc Taillefer

ICGM – Université de Montpellier – CNRS – ENSCM, 240 Avenue du Professeur Emile Jeanbrau, 34080 Montpellier, France

8.1 Introduction

The use of highly abundant, cheap, and nontoxic manganese salts in cross-coupling reactions has emerged over the last 25 years. In contrast to Pd- or Ni-catalyzed processes, the Mn-catalyzed reactions are less well understood from a mechanistic point of view, the scope is not fully exploited, and most of the reactions are based on simple Mn halide salts without any specific ligand. We herein describe recent applications of $MnCl_2$ in the catalytic C–C coupling of aryl, vinyl, or alkyl halides with various organometallic reagents. This first part includes Kumada and Stille-type cross-couplings as well as the homo- and heterocoupling of different types of organometallic species (Li, Mg, Sn) under oxidative conditions. In the second part, the recently emerging field of Mn-catalyzed arylation or alkylation of nitrogen-, sulfur-, oxygen-, and boron-based nucleophiles is presented. In both parts, selected mechanistic pathways as well as the possible influence of contamination or presence of other metals in small quantities are discussed. Coupling reactions involving hydrogen formation via C–H activation or hydrogen atom transfer (HAT) processes are not considered in this chapter. Several reviews have covered this research area [1–4], including a nice comprehensive review by Valyaev et al. and a review on the use of Grignard reagents in Mn-catalyzed cross-coupling reactions by Cahiez and Moyeux, both describing the literature until 2015 [3, 4]. This chapter aims to provide a non-exhaustive overview on the development of different Mn-catalyzed cross-coupling processes over the last 20 years with the literature leading up to the end of 2019.

Manganese Catalysis in Organic Synthesis, First Edition. Edited by Jean-Baptiste Sortais.

8.2 C—C Bond Formation via Mn-Catalyzed Cross-Coupling Reactions

8.2.1 Mn-Catalyzed Kumada Cross-Coupling Reactions

8.2.1.1 Aryl Halides with Grignard Reagents

In 1999, Cahiez reported that activated aryl halides, especially cheap and widely available aryl chlorides, could undergo a cross-coupling with different Grignard reagents with a simple manganese catalyst to form the corresponding biaryls (Scheme 8.1) [5]. The list of efficient activating groups involved nitriles, imines, or oxazolidines in the ortho position, whereas a stilbene moiety used as ortho substituent provided lower yields. A substrate bearing an imine group in the para position was also a suitable reagent, whereas for its isomer with the imine in the meta position, no reaction took place. In this initial study, aryl halides containing ester or amide groups could not be employed, mainly due to side reactions on the carbonyl functionality. In general, good to moderate results were obtained with several aryl, 1° and 2° alkyl, or alkenyl Grignard reagents, and the reaction was successful with chloride, bromide, fluoride, and methoxide as leaving group. Interestingly, with the aryl fluoride reagent, the reaction was slower than with the other halides, providing first evidence that a mechanism other than nucleophilic aromatic substitution might be at play. It is important to note that blank experiments in the absence of metal catalyst with different substrates showed no or very slow reactions, leading to the assumption that the metal-catalyzed reaction was the largely favored process compared to an S_NAr pathway.

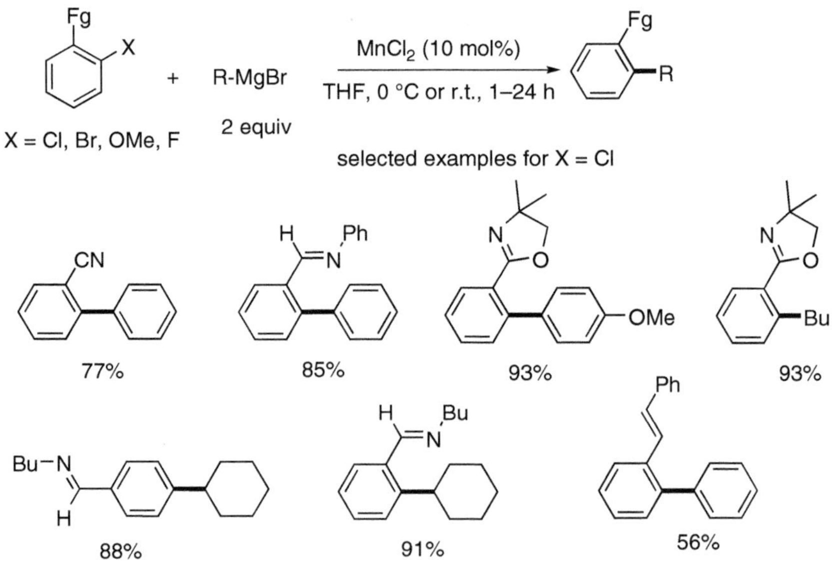

Scheme 8.1 First example of Mn-catalyzed Kumada cross-coupling between activated arylchlorides and various Grignard reagents reported by Cahiez et al. [5]. Source: Based on Cahiez et al. [5].

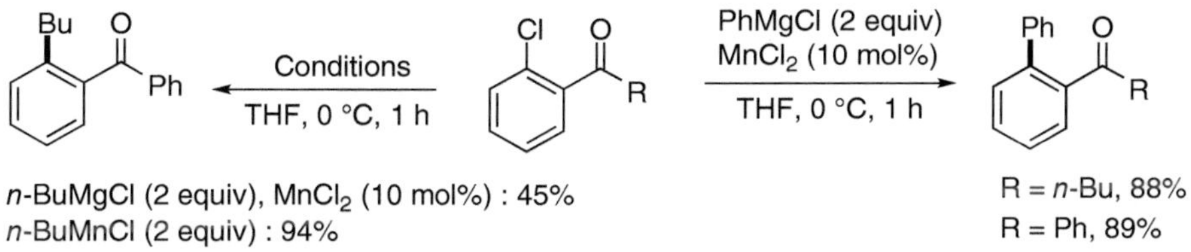

Scheme 8.2 Extension of Mn-catalyzed cross-coupling to *ortho*-chloroaryl ketones . Source: Based on Cahiez et al. [6].

In an attempt to extend this chemistry to *ortho*-chloroaryl ketones, for example, *o*-chlorovalerophenone or *o*-chlorobenzophenone, the limitation to aryl Grignard reagents became evident (Scheme 8.2) [6]. With PhMgCl, high yields of the cross-coupled biaryls were obtained, however, with a more reactive alkyl Grignard reagent, such as *n*-BuMgCl, only a low yield of 45% was observed, due to competitive reaction with the carbonyl group. This problem was overcome by replacing the Grignard reagent by stoichiometric or excess amounts of less reactive organomanganese reagents RMnCl in the coupling reaction. This way a large range of *ortho*-chloroaryl ketones could be coupled with different alkyl and arylmanganese compounds.

A further study indicated that *o*-halobenzoyl chlorides could be efficiently reacted with diorganyl magnesium complexes $R_2Mg{\cdot}LiCl$ under $MnCl_2$ catalysis to provide tandem cross-coupling products (Scheme 8.3) [7]. This reaction worked in general best with *o*-chlorobenzoyl chlorides in the case of dialkylmagnesium compounds,

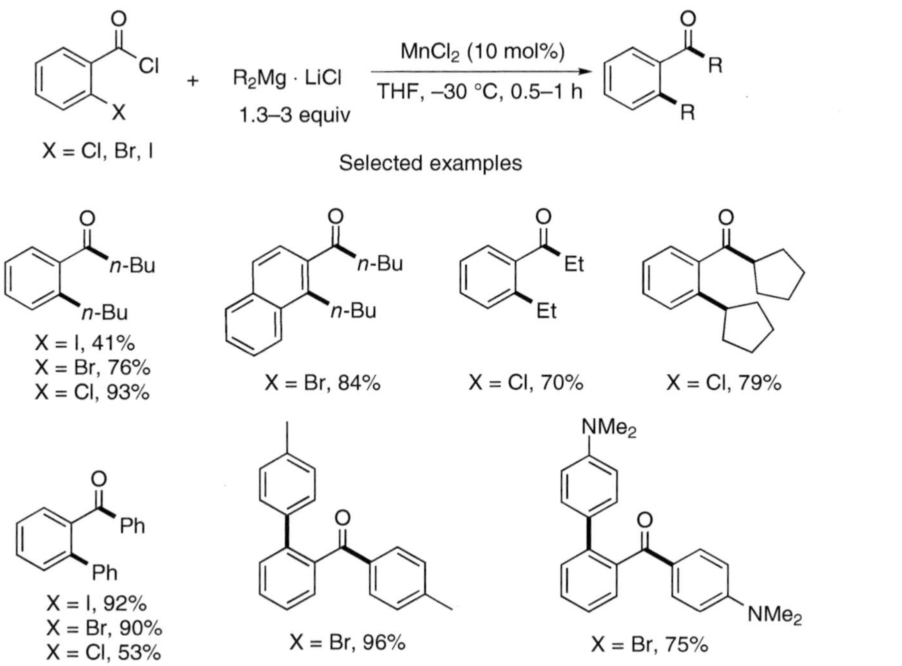

Scheme 8.3 Mn-catalyzed tandem cross-coupling of diorganomagnesium complexes with *o*-halobenzoyl chlorides. Source: Based on Zhang et al. [7].

whereas with diarylmagnesium compounds higher yields were observed for the bromo and iodobenzoyl chloride derivatives. No coupling reaction occurred in the case of *meta*- or *para*-halobenzoyl chlorides. The limitation of this process resides in the introduction of the same group on the ring and on the carbonyl group.

Nearly 20 years after the initial report by Cahiez, two contributions by Madsen and coworkers in 2017 provided some synthetic extensions and new mechanistic insights into the field of $MnCl_2$-catalyzed Kumada cross-coupling reactions [8, 9]. The coupling of *para*-halo benzonitrile reagents with the cyclohexyl Grignard complex was first investigated (Scheme 8.4) [8]. In this case, the chloride compound provided the highest yield with 94%, whereas with the bromide analogue only a moderate 45% yield was obtained, and no reaction was observed with iodide or fluoride substrates. Furthermore, Madsen showed that an ester group in the para position was a suitable activating group, providing 65% yield of coupling product, but an amide group was not tolerated similarly as observed by Cahiez et al. (Scheme 8.1) [5]. Also, *p*-CF_3 or *p*-NO_2 groups did not provide the desired cross-coupled product. The *p*-chlorobenzonitrile substrate was then used for testing the influence of solvent and temperature on the reaction outcome, showing that THF provided the best results and that the reaction was fast above 6 °C, whereas below 0 °C nearly no reaction took place. This reaction could be conveniently carried out at room temperature providing the cross-coupling of various aryl and alkyl Grignard reagents with *ortho*- or *para*-chlorobenzonitriles, while allyl Grignards gave no reliable results (Scheme 8.4).

A competition experiment using PhMgCl and cyclohexylMgCl in combination with *p*-chlorobenzonitrile provided by far the *p*-cyclohexylbenzonitrile as the major product, in agreement with the higher reactivity of alkyl vs. aryl Grignard reagents. No remarkable difference was observed in the reactivity of *p*-CN or *p*-CO_2Me aryl halide substrates with cyclohexylMgCl, as shown in another competition experiment, affording the coupling products in equal yields. The

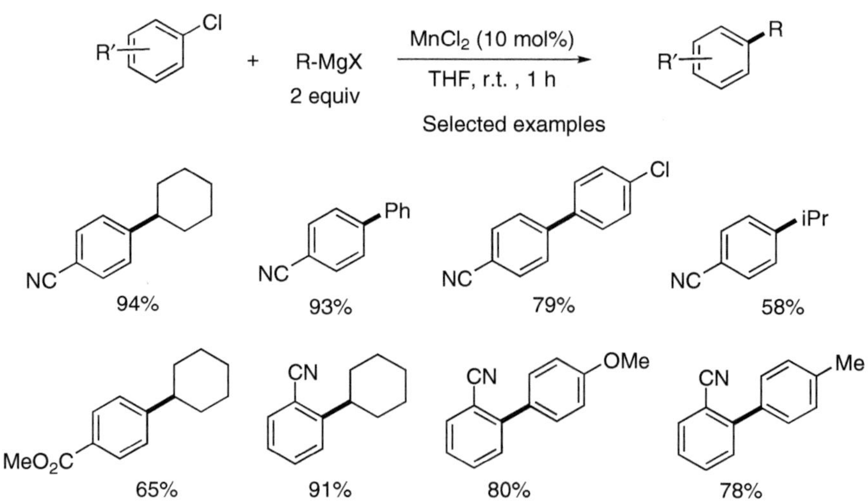

Scheme 8.4 Extending the scope of activated aryl halide substrates in $MnCl_2$-catalyzed cross- coupling reactions. Source: Based on Antonacci et al. [8].

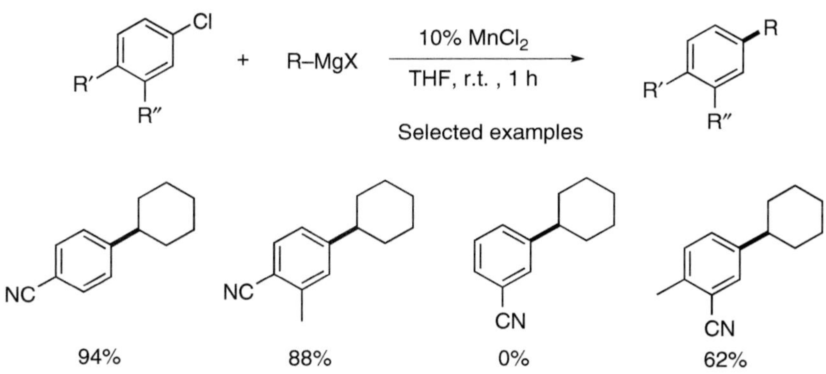

Scheme 8.5 Influence of methyl substituent in the cross-coupling reaction of chlorobenzonitriles. Source: Based on Antonacci et al. [8].

influence of an additional methyl group in the aromatic ring was then investigated. No remarkable difference was obtained in the case of the *p*-CN substrate having a methyl group in the meta position with respect to the substrate without methyl group (Scheme 8.5). However, an interesting result was the reaction of 1-chloro-3-cyano-4-methylbenzene with cyclohexylMgCl, providing the coupled product in 62% yield, whereas the similar substrate without the methyl group gave no conversion due to the cyano group in the meta position (Scheme 8.5).

Further variation of the reaction conditions, i.e. change of the solvent from THF to diethyl ether and using microwave heating in sealed vials with an increase in Grignard concentration (3 M), provided some new substrates in the $MnCl_2$-catalyzed Kumada cross-coupling reaction [9]. Non-activated substrates, such as different iodotoluenes could now be coupled with moderate to good yields with phenylmagnesium bromide; however, other electron–donating substituents (OMe, NMe_2) on the aryliodides provided lower yields (Scheme 8.6). Other halides (Br, Cl) were less

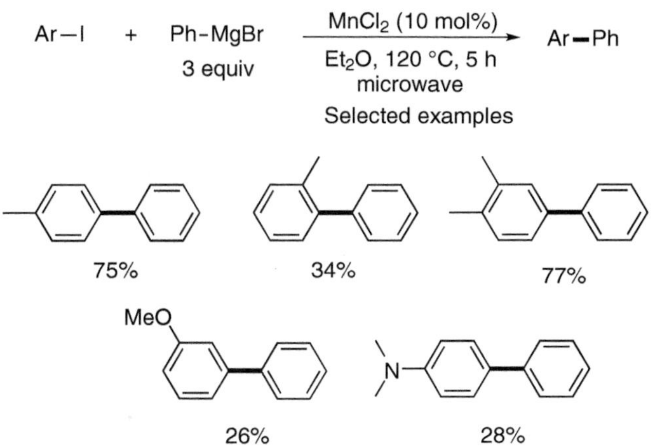

Scheme 8.6 $MnCl_2$-catalyzed cross-coupling with non-activated aryl halides. Source: Based on Santilli et al. [9].

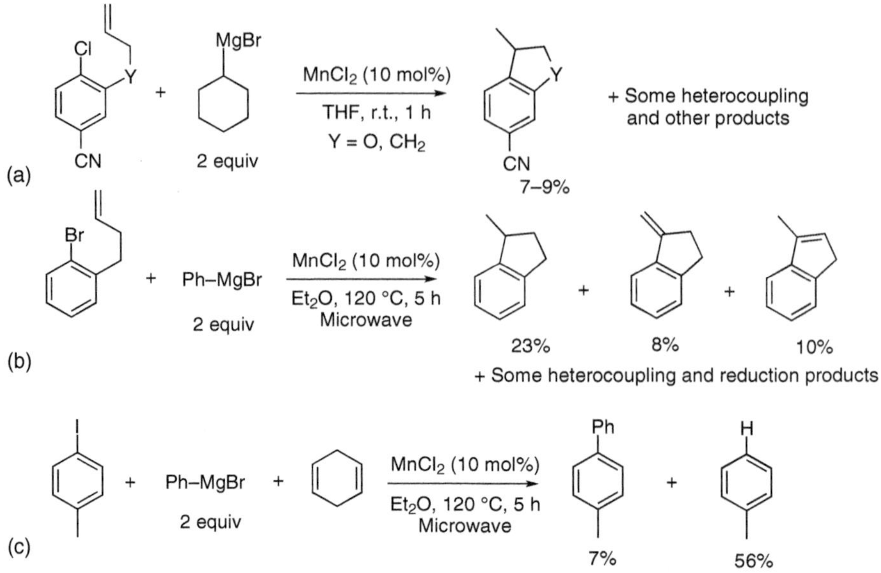

Scheme 8.7 Mechanistic investigations: radical clock experiments (a, b) and radical trapping with 1,4-cyclohexadiene (c). Source: Based on Antonacci et al. [8]; Based on Santilli et al. [9].

efficient and triflates did not work at all. The reaction could also be conducted in 2-methyl-THF and conventional heating was also possible, albeit with longer reaction times (18 hours at 120 °C).

Mechanistic studies on these $MnCl_2$-catalyzed Kumada cross-coupling reactions are scarce. Madsen has provided some important contributions to the understanding of these processes. Several radical clock experiments were conducted to obtain insights on the involvement of radical species in the reaction course [8, 9]. Substrates containing allyl ether or but-3-enyl moieties were reacted with Grignard reagents under different reaction conditions (solvent, temperature) to provide mixture of compounds from which the cyclized products were isolated as major compounds, indicative of radical intermediates (Scheme 8.7a,b). Only small amounts of cross-coupled products were observed. Furthermore, under microwave conditions, the reaction in the presence of 1,4-cyclohexadiene provided mainly the dehalogenation product, thus being another indication of radical formation (Scheme 8.7c).

From these experiments, Madsen deduced that aryl radical species might form in the reaction course, leading to the proposed reaction mechanism of an $S_{RN}1$ reaction, shown in Scheme 8.8. The first step would involve the reduction of the aryl halide by an *in situ* formed triorganomanganate complex **A**, to yield radical anion **B**, which would lose a halide anion to form the intermediate free aryl radical **C**. Radical processes triggered by triorganomanganate complexes have been studied since the 1990s by Oshima and coworkers [10]. Radical **C** would be attacked either by the nucleophilic Grignard reagent or by **A**. This rapid reaction would be followed by a single-electron transfer (SET) from the resulting radical

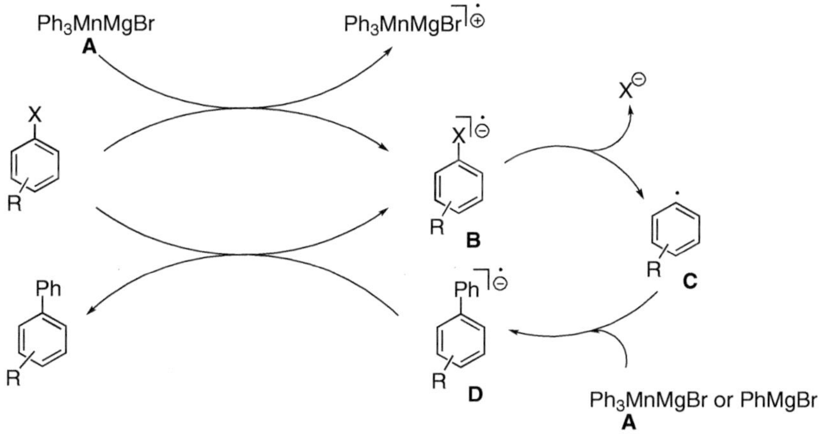

Scheme 8.8 Mechanistic proposal of an $S_{RN}1$ reaction involving free aryl radicals for Mn-catalyzed Kumada cross-coupling. Source: Based on Antonacci et al. [8]; Based on Santilli et al. [9].

anion **D** to another starting aryl halide. The limited scope of this transformation under the room temperature conditions in THF can be attributed to the necessity of having an electron withdrawing group (EWG) in the ortho or para position, responsible for stabilizing the radical anion and facilitating the dehalogenation process (see Scheme 8.4). On the other hand, in diethyl ether higher temperatures are presumably necessary as the electron-donating group (EDG) destabilizes the radical anion and also makes the aryl radical less electrophilic for the reaction with another Grignard or triorganomanganate complex (see Scheme 8.6).

In this context, it should be mentioned that some $MnCl_2$-catalyzed coupling reactions occur with C–O cleavage, for which radical mechanisms based on triorganomanganate complexes have been proposed: (i) in the $MnCl_2$-catalyzed coupling of 2-iodobenzofurane with Grignard reagents, the ring opening of the benzofuran moiety was proposed, based on the literature, to proceed via mixed manganate complex **E** obtained from iodo/manganese exchange between starting compound and an *in situ* formed triorganomanganate complex (Scheme 8.9a) [11] and (ii) for the $MnCl_2$-catalyzed three-component coupling of benzonitriles, Grignard reagent, and THF, there is experimental and theoretical evidence that the mechanism for the formation of the final product involves α-manganated THF **F**. Reductive coupling of two phenyl groups on the manganese(IV) center is followed by the transfer of the remaining phenyl group onto the THF moiety with concomitant ether cleavage to afford manganate(II) complex **G**, which is stabilized by another PhMgCl moiety. Complex **G** can then react with different nitriles to afford the three-component coupling product (Scheme 8.9b) [12].

8.2.1.2 Heteroaryl Halides with Grignard Reagents

In 2007, Rueping and Ieawsuwan reported on the efficient Mn-catalyzed cross-coupling of several nitrogen-containing heterocyclic chlorides with phenyl and alkylmagnesium chlorides under mild conditions [13]. Various

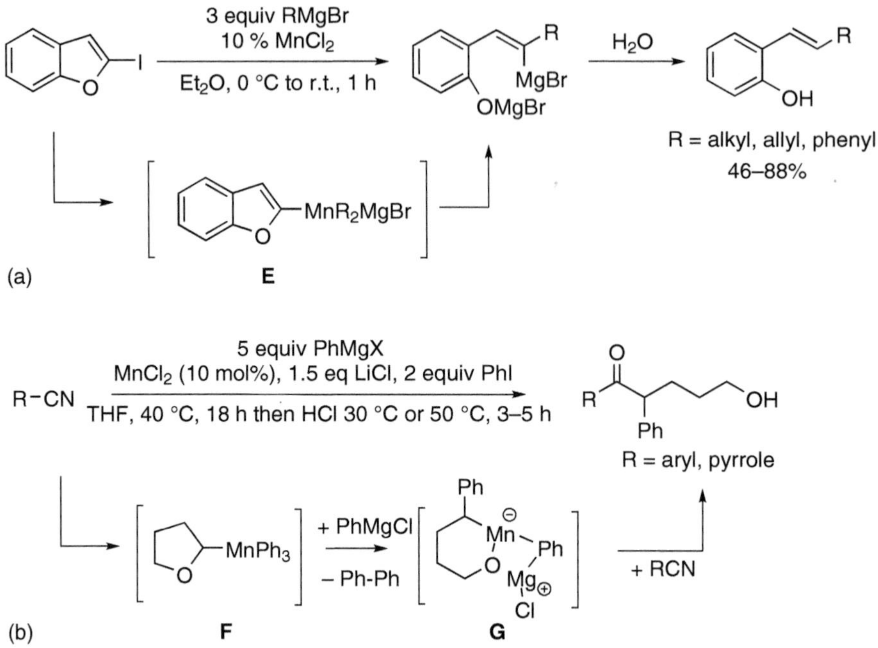

Scheme 8.9 C—O bond cleavage involving triorganomanganate-triggered radical processes in $MnCl_2$-catalyzed coupling reactions. (a) ring -opening of benzofurane derivatives. Source: Kakiya et al. [11]; (b) three component coupling involving THF ring opening. Source: He et al. [12].

2- or 4-chloroquinoline derivatives were successfully reacted with primary or secondary alkyl Grignard reagents as well as phenylmagnesium chloride in the presence of 1–5 mol% of $MnCl_2$ as catalyst between 0 °C and room temperature with short reaction times (often between 1.5 and 4 hours) (Scheme 8.10). $MnCl_2$ sources of different quality, ranging from 99.999% to 98% purity, provided the same yields. Depending on the substrate, a two- to fourfold excess of Grignard reagent was required to obtain good yields. Various other heterocyclic substrates such as quinazoline, benzothiazole, purine, pyrimidine, quinoxaline, and pyridine could also be successfully employed, but the yields were generally lower with less conjugated heterocycles.

This initial study was extended to the coupling of various pyridine and pyrazine derivatives with alkyl Grignard reagents by Matson and coworkers in 2018 [14]. Preliminary results concerning the coupling of 2-chloropyridine with *n*-butylmagnesium chloride (2.6 equiv) at room temperature revealed that among numerous first-row transition metal halides (3 mol%) tested as catalysts, only $MnCl_2$ gave efficient cross-coupling in 88% yield. Furthermore, it was shown that with lithium or zinc reagents, for example, *n*-BuLi, PhLi, or PhZnBr, lower yields or no cross-coupling product were obtained. In the absence of Mn catalyst, no reaction occurred. The coupling reaction was successful with several aliphatic Grignard reagents (Scheme 8.11), however the reaction furnished low yields with

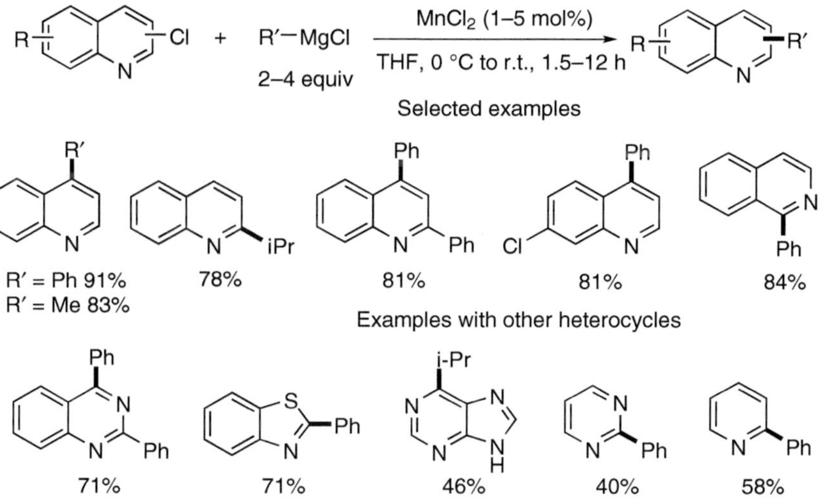

Scheme 8.10 Mn-catalyzed cross-coupling of N-heterocyclic chlorides with Grignard reagents. Source: Based on Rueping and Ieawsuwan [13].

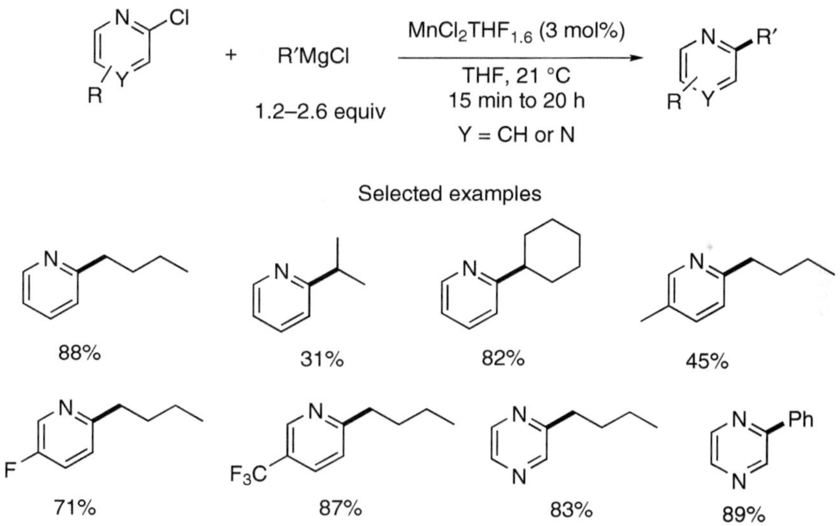

Scheme 8.11 Extension of Mn-catalyzed cross-coupling of pyridine and pyrazine derivatives with Grignard reagents. Source: Based on Petel et al. [14].

methylmagnesium chloride and no product with trimethylsilylmethylmagnesium chloride or tertiary Grignard reagents. An improved yield of 90% was obtained for the reaction of PhMgCl with 2-chloropyridine at higher temperature and longer reaction time (cf. 58% yield in Scheme 8.10). The fact that attempts with 3- or 4-halopyridines provided the cross-coupling products only in low yields led the authors to the assumption that the nitrogen atom in the pyridine ring might act as a directing group in these transformations. It was further observed that the influence of the halide in the 2-halopyridine on the reaction rate follows the often observed order Cl > Br ≫ F (see Section 2.1.1). The substrate scope was further extended to substituted pyridines, showing that electron-withdrawing groups, such as 5-F or 5-CF_3 groups, were well tolerated and electron-donating methyl groups to a lesser extent. In the case of the CF_3 group, the reaction provided high yields with shorter reaction times, and the amount of Grignard reagent could be reduced to 1.2 equiv. With the even more activated 2-chloropyrazine heterocycle, highly efficient cross-coupling was observed with primary and secondary alkyl Grignard reagents and even PhMgBr reacted at r.t. in only three hours to provide a 89% product yield.

8.2.1.3 Vinyl(Pseudo)Halides with Grignard Reagents

Following a preliminary report on the $MnCl_2$-catalyzed cross-coupling reaction of a vinyl iodide with phenylmagnesium chloride (Scheme 8.12a) by Cahiez et al. in 1976 [15], Oshima and coworkers showed in 1987 that vinyl triflates could undergo cross-coupling with an excess of Grignard reagent in the presence of Li_2MnCl_4 salt at 0 °C (Scheme 8.12b) [16]. Depending on the nature of the Grignard reagent, moderate to high yields were obtained, often accompanied by some reduction or homocoupling products. The best results were observed with phenyl, benzyl, and allyl magnesium halides. A postulated mechanism was based on the existing literature, involving a Mn/triflate exchange by reaction of a triphenylmanganate complex with the iodoalkene to afford intermediate **H**, followed by reductive elimination.

In 1998, Cahiez and coworkers reported on the efficient and highly stereoselective coupling of chloroenynes with an excess of primary, secondary, or tertiary aliphatic

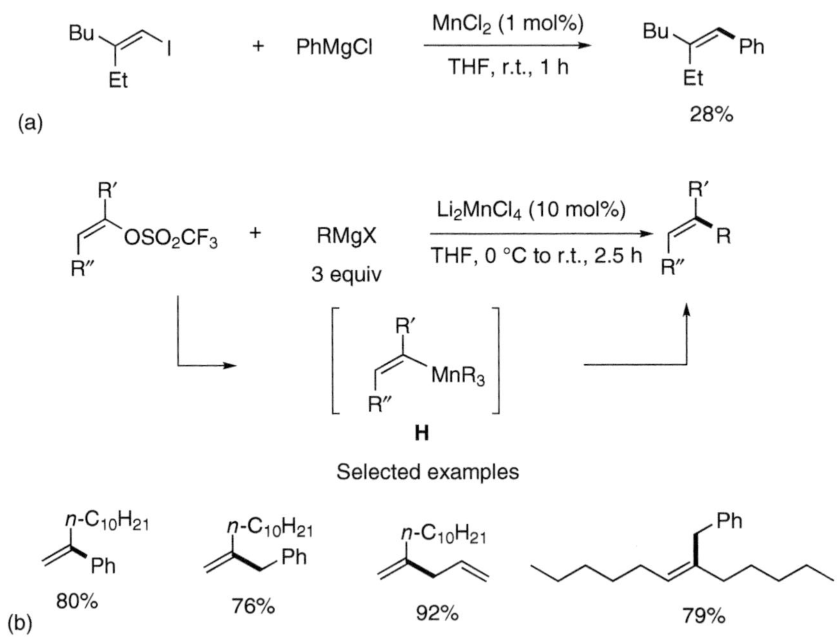

Scheme 8.12 Early work by Cahiez/Normant (a) and Oshima (b) on the Mn-catalyzed cross-coupling of vinylhalides or triflates with Grignard reagents. Source: Cahiez et al. [15]; Fugami et al. [16].

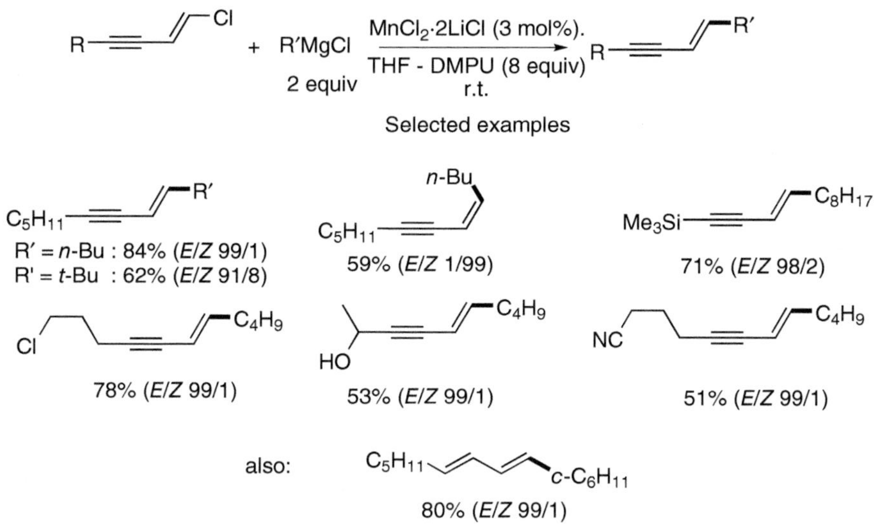

Scheme 8.13 Mn-catalyzed cross-coupling of chloroenynes or chlorodiene with Grignard reagents. Source: Based on Alami et al. [17].

Grignard reagents (Scheme 8.13) [17]. The use of an excess of tetrasubstituted urea derivatives, such as *N*,*N*-dimethylpropylene urea (DMPU), turned out to be crucial for high product yields. Under these conditions, phenylmagnesium chloride gave only a low 25% yield and non-conjugated vinyl chlorides did not react. The reaction showed good chemoselectivity, e.g. aliphatic chloride or nitrile groups and free alcohols were tolerated. The procedure could also be extended to a chlorodiene substrate to afford the coupled product in high 80% yield. Interestingly, $PdCl_2$ or $NiCl_2$-based catalysts provided much lower yields in the coupling of chloroenyne with *n*-butylmagnesium chloride than the $MnCl_2$/DMPU system.

An extension toward the stereoselective coupling of simple, non-activated alkenyl halides with various aryl Grignard reagents was reported in 2008 by Cahiez et al. (Scheme 8.14) [18]. Higher catalyst loadings (10 mol% of $MnCl_2$) and sometimes temperatures up to 50 °C were required, especially with β,β-disubstituted alkenyl iodides. Aryl Grignard reagents bearing an electron-withdrawing group (*p*-fluoro) or sterically more hindered aryl groups (*o*-tolyl) provided lower yields, whereas electron-donating groups (*p*-methoxy, *p*-NMe_2) did not significantly influence the outcome. In these reactions, alkenyl bromides and iodides provided good results, whereas chlorides were much less reactive. The vast majority of substrates provided a highly stereospecific reaction outcome; however, the (*Z*)-pyridinyl-alkenyl iodide led to complete inversion of the stereochemistry, providing the cross-coupled (*E*)-alkene in 71% yield. It should further be noted that in contrast to many $MnCl_2$-catalyzed reactions, which frequently employ more soluble lithium manganate Li_2MnCl_4 (see Scheme 8.12), addition of LiCl slowed down the reaction significantly from 3 to 18 hours.

Some alkyl Grignard reagents, such as methyl or cyclohexyl magnesium halides, could also be successfully coupled with styryl bromide (Scheme 8.15). However,

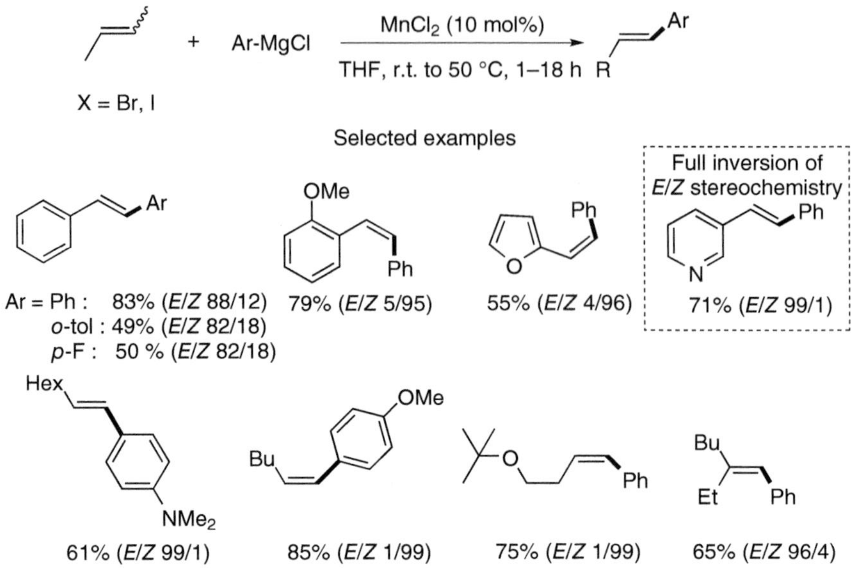

Scheme 8.14 Mn-catalyzed stereospecific cross-coupling of non-activated alkenyl halides with aryl Grignard reagents. Source: Based on Cahiez et al. [18].

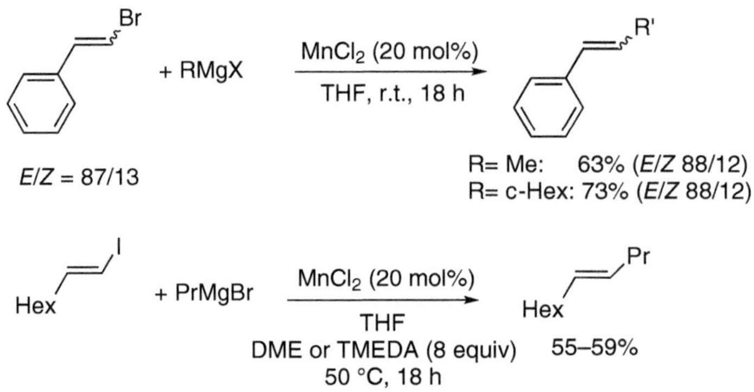

Scheme 8.15 Mn-catalyzed cross-coupling of non-activated alkenyl halides with alkyl Grignard reagents. Source: Based on Cahiez et al. [18].

with propylmagnesium bromide only reduction and homocoupling of the alkenyl derivatives was observed due to the ready β-hydrogen elimination. A solution to this problem was found by the addition of 8 equiv of dimethoxyethane (DME) or tetramethylethylenediamine (TMEDA) to the reaction mixture. Under these conditions, the reaction of 1-iodo-1-octene with PrMgBr the cross-coupled product was obtained in 55–59% yield after heating for 18 hours (Scheme 8.15) [18].

In addition to alkenyl halides and triflates, the groups of Comasseto and coworkers have shown that alkenyl tellurides could also be employed in the cross-coupling reaction with aryl Grignard reagents under a combined Mn/Cu catalysis [19].

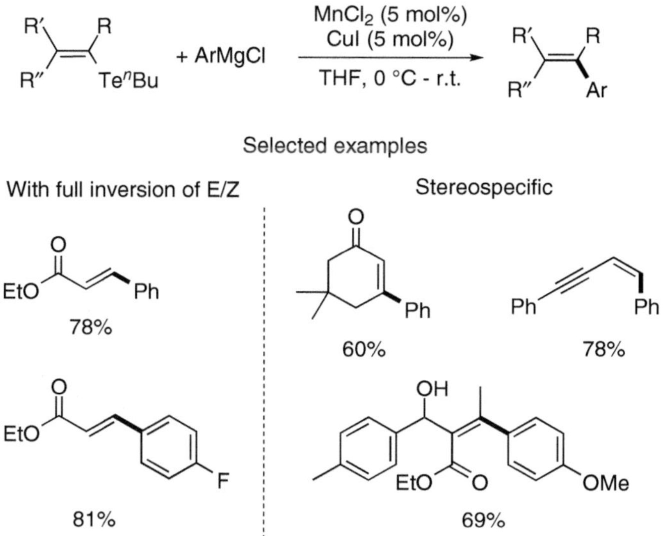

Scheme 8.16 Mn/Cu-catalyzed cross-coupling of alkenyl tellurides with aryl Grignard reagents. Source: Based on Silva et al. [19].

The optimization of reaction conditions has shown that a 1 : 1 mixture of $MnCl_2$ with CuI (5 mol% each) provided good to moderate yields in the coupling of various alkenyl tellurides with phenylmagnesium chloride under mild conditions (Scheme 8.16). In contrast, using only $MnCl_2$ or only CuI provided very low yields. The majority of substrates concerned conjugated alkenyl tellurides based on styrene, enyne, or α,β-unsaturated carbonyl compounds. Even tetrasubstituted alkenyl telluride substrates could be employed, as well as *p*-substituted aryl Grignard reagents with electron-withdrawing or electron-donating substituents. Apart from the disubstituted ester substrate, which undergoes full inversion of the double bond configuration, all the other substrates reacted with a retention of the double bond configuration. The mechanism of this transformation still needs to be inquired.

8.2.1.4 Conclusion on Mn-Catalyzed Kumada Cross-Coupling Reactions

Mn-catalyzed Kumada cross-coupling reactions of aryl, heteroaryl, and vinyl halides or pseudohalides with a variety of Grignard reagents have made considerable progress over the last 20 years, even though only a limited amount of heterocyclic halide compounds (only nitrogen heterocycles) has been investigated. Recent contributions on the mechanism in the case of aryl halides have opened up the way for further progress. The generality of the implication of radical species in these transformations still has to be explored. Further mechanistic studies, especially supported by theoretical studies, are required to elucidate the role of coordinating solvents (such as DMPU or TMEDA) in some of the transformations and to improve the potential of these easy-to-perform cross-coupling reactions.

8.2.2 Mn-Catalyzed Stille Cross-Coupling Reactions

In 1997, the first report on a Mn-catalyzed Stille coupling involving various organostannanes (vinyl, styryl, aryl, acetylenic) and styryl or aryl iodides showed the importance of the presence of one equivalent of NaCl or KCl to avoid homocoupling of stannanes (Scheme 8.17) [20]. Whereas $MnBr_2$ gave the best yields, $MnCl_2$ provided lower yields and MnI_2 did not yield any coupled product. High temperatures (90–120 °C) and slow addition of organostannanes and polar solvents such as N-methyl-2-pyrrolidone (NMP) or dimethylformamide (DMF) were necessary to obtain the cross-coupled products in high yields. Aryl bromides or triflates did not lead to the coupling product. In the same study it was shown that the results are analogous to a procedure involving catalytic CuI instead of $MnBr_2$.

In an attempt to repeat the cross-coupling reaction between *p*-tolyliodide and phenylstannane under exactly the same Mn-catalyzed conditions with extremely pure reagents and solvents, Madsen and coworkers in 2017 did not observe any product formation, despite several attempts (Scheme 8.18) [9]. In addition, experiments with CuI or $NiCl_2$ as catalysts were also unsuccessful. However, in the presence of very small amounts of $Pd(OAc)_2$ (down to 0.003 mol%), the cross-coupling product was obtained in moderate to good yields. This report casts therefore some doubt on the involvement of Mn catalysis in the original report, and some Pd contamination may be the actual catalyst source.

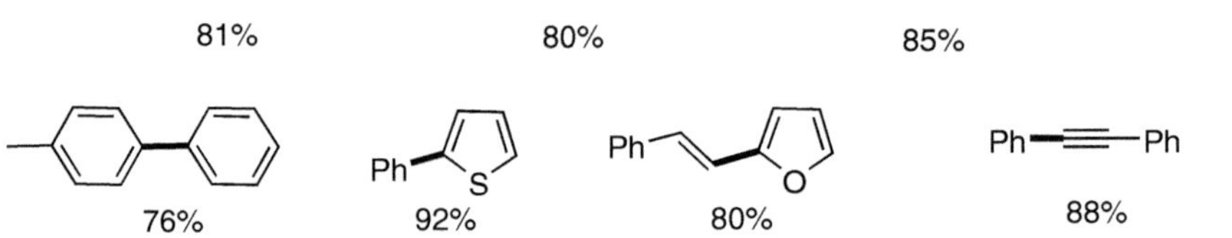

Scheme 8.17 First reported Mn-catalyzed Stille cross-coupling reaction by Kang et al. [20]. Source: Based on Kang et al. [20].

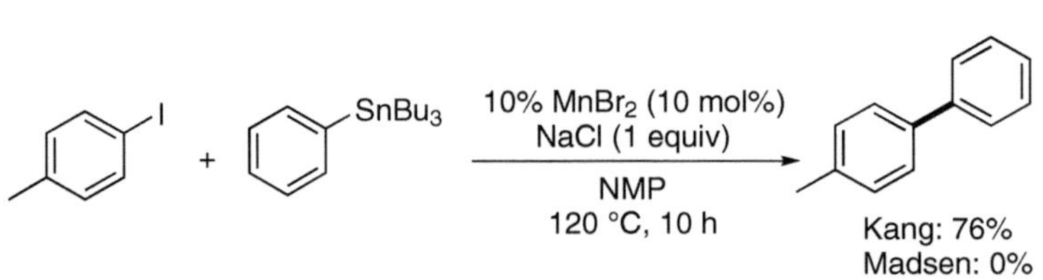

Scheme 8.18 Mn- or Pd-catalyzed cross-coupling reaction: Kang and coworkers vs. Madsen and coworkers [9, 20]. Source: Santilli et al. [9]; Kang et al. [20].

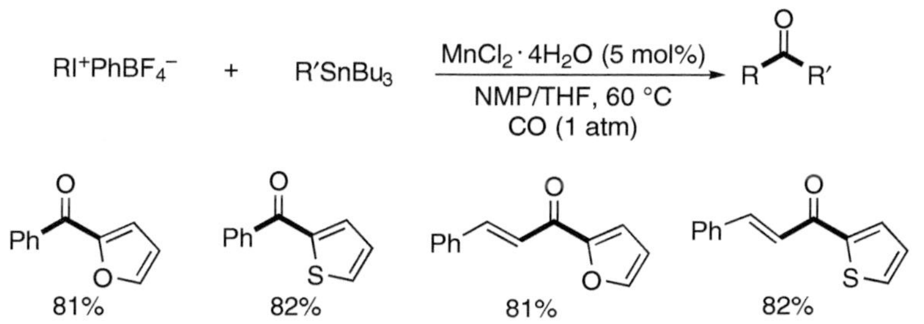

Scheme 8.19 Mn-catalyzed carbonylative cross-coupling with iodonium salts. Source: Based on Kang et al. [21].

One year after the initial report, Kang et al. reported in 1998 on the Mn-catalyzed cross-coupling and on the carbonylative cross-coupling of various organostannanes with phenyl or styryl iodonium salts (Scheme 8.19) [21]. Moderate to high yields were obtained with $MnCl_2$ (5 mol%) after refluxing the organostannane with iodonium salt under a CO atmosphere in an NMP/THF mixture. In this reaction $MnBr_2$ was not active and the yields were significantly lower when only NMP or THF was employed.

Twenty years later, Falck and Dakarapuand disclosed a report on the Mn-catalyzed stereospecific Stille cross-coupling in the reaction of protected α-alkoxy(tri-*n*-butyl)stannanes with organic electrophiles (Scheme 8.20) [22]. The $MnCl_2$/(*S*)-$BnCH_2$PyBOX system, employed in 50 mol%, mediated the coupling of various aryl iodides bearing electron-rich or electron-withdrawing groups and even heteroaryl iodides with the organostannanes (Scheme 8.20a). The reaction was rather slow (two to four days), even when performed at 70 °C. Interestingly, no Newman–Kwart rearrangement of the thiocarbamate occurred, which is often observed in copper(I) thiophene-2-carboxylate (CuTC)-mediated cross-couplings. Importantly, with enantiomerically enriched substrates, the reaction proceeded with a variety of electrophiles with >98% retention of the configuration (Scheme 8.20b). Furthermore, an adjacent asymmetric center as in the acetonide substrate had no influence on the stereofidelity of the reaction (Scheme 8.20c). A large variety of electrophiles (allylic, aryl, heteroaryl, and olefinic) could be employed, and the reaction conditions were compatible with many functionalities, such as nitro, cyano, acetyl, methoxy, CF_3, Br, or I groups. The stereochemistry of olefinic iodides was also maintained in the product.

In order to improve the reaction kinetics, it was found that a mixture of $MnCl_2$ (15 mol%) with $Cu(OTf)_2$ (20 mol%) and Jackphos (15 mol%) as ligand was superior to the Mn/Pybox-only system. Addition of KF and K_3PO_4 provided cleaner products under overall significantly faster and milder conditions (6–24 hours, room temperature) (Scheme 8.21) [22].

The high stereospecificity of the reaction led the authors to claim the involvement of an oxidative addition of an intermediate alkyl manganese complex **K** with the organic halide R–X to form complex **L**, followed by reductive elimination as key steps in the catalytic cycle (Scheme 8.22). As the reaction outcome was not altered in the

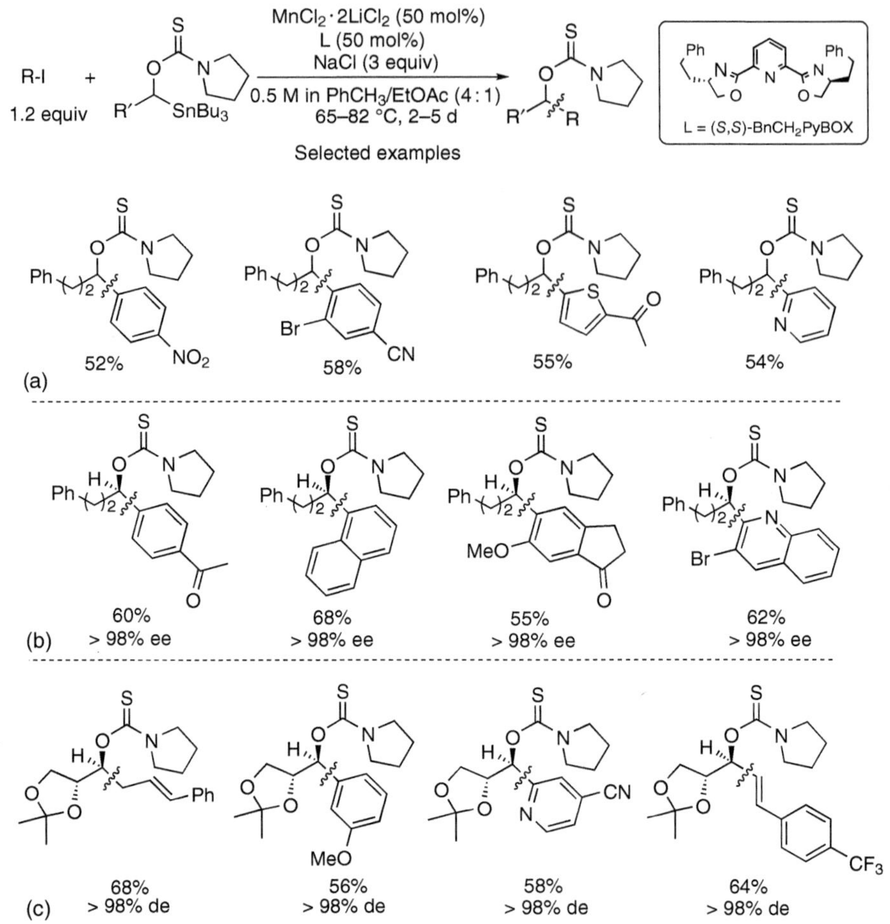

Scheme 8.20 Scope of stereospecific Mn-mediated cross-coupling with alkylstannanes. (a) racemic alkylstannane as starting material; (b) starting from (S)-enantiomer of alkylstannane; (c) starting from (R),(S) diastereomer of alkylstannane. Source: Dakarapu and Falck [22].

presence of TEMPO, the formation of radicals was excluded by the authors. It should however be noted that such a direct oxidative addition has never been observed for a Mn(II) complex. Very recently, the first example of an oxidative addition involving a Mn(I) complex was reported [23]. The role of Cu in the bimetallic system would be to accelerate the formation of a reactive organometallic species **J** from the stannane via transmetallation (complex **I**), which would then transmetallate with MnX_2 to afford complex **K**.

To date, only very few examples of Mn-catalyzed Stille coupling reactions have been reported, which have led, in one case, to some controversy about the catalytic metal species involved in the reaction process. More in-depth mechanistic studies would be required to understand the role of the Mn catalyst in these reactions and to reduce the catalytic charge.

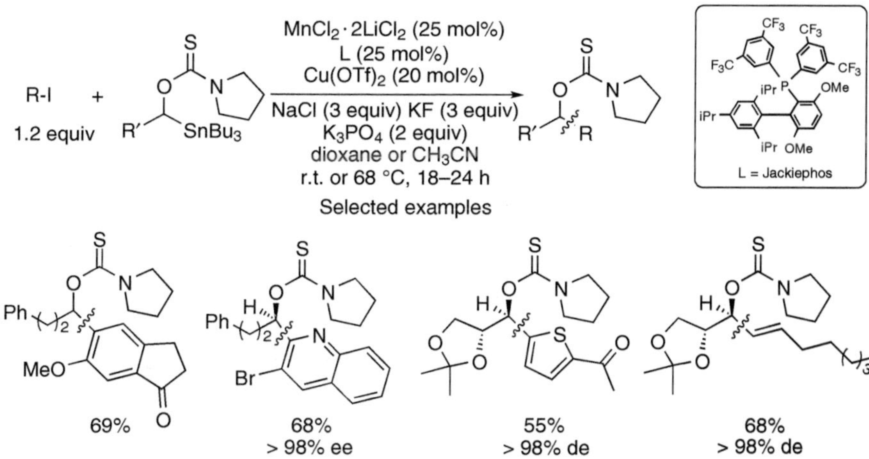

Scheme 8.21 Scope of stereospecific Mn/Cu-catalyzed cross-coupling with alkylstannanes. Source: Dakarapu and Falck [22].

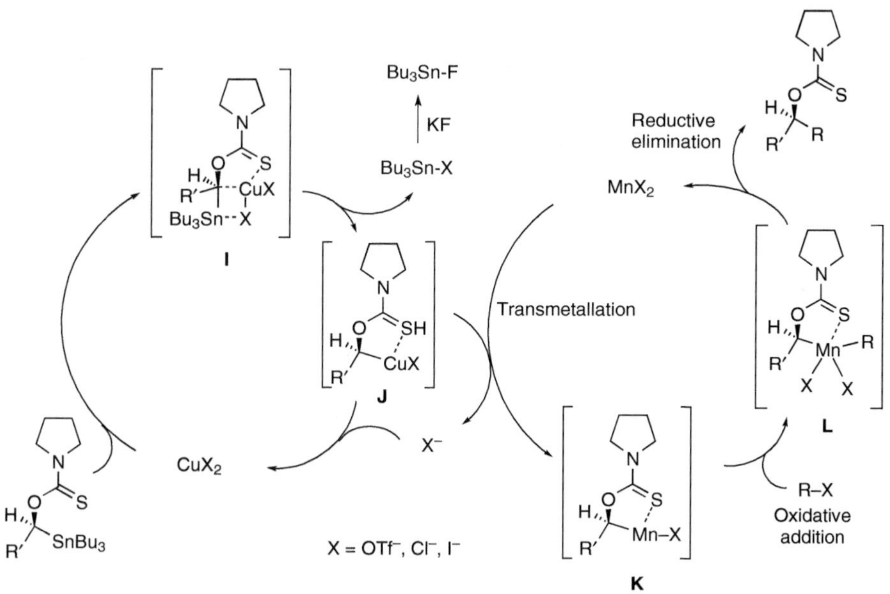

Scheme 8.22 Proposed mechanism for Mn/Cu-catalyzed Stille coupling. Source: Dakarapu and Falck [22].

8.2.3 Mn-Catalyzed Coupling Reactions of Organometallic Reagents Under Oxidative Conditions

The Mn-catalyzed oxidative homo- or heterocoupling of two organometallic species is an intriguing transformation with high synthetic potential. The general reaction scheme is the mixing of 2 equiv of the same or of different organometallic complexes (R–Li, R–MgX, R–SnR'_3) in the presence of a catalytic amount of $MnCl_2$ or

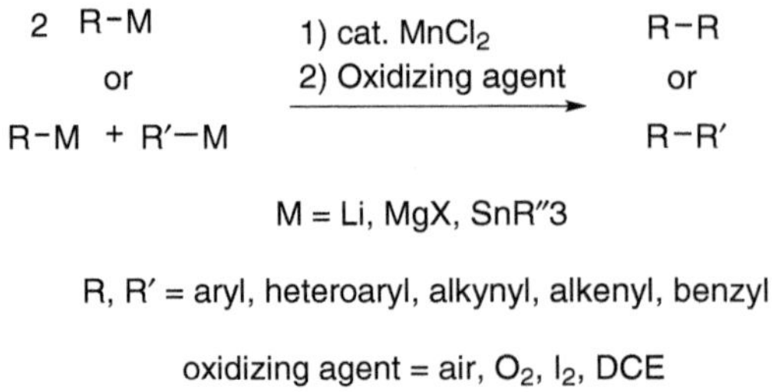

Scheme 8.23 General scheme for Mn-catalyzed oxidative homo- or heterocoupling of organometallics.

$MnCl_2$·2Li (often between 10 and 20 mol%) at low temperature followed by addition of an oxidizing reagent, such as iodine, dichloroethane (DCE), or dry air, to afford the corresponding homo- or heterocoupled products (Scheme 8.23). This reaction has been best explored for the synthesis of biaryls, but other coupling processes have also been successfully conducted, involving, for example, alkynyl, alkenyl, or benzyl organometallics.

8.2.3.1 Homocoupling Reactions

The first oxidative homocoupling of *in situ* generated vinyllithium reagents, catalyzed by small amounts of $MnCl_2$, was reported in 1975 by Cahiez et al. (Scheme 8.24) [24]. In this case, the iodobutane, produced during the halogen–lithium exchange between vinyliodide and *n*-BuLi, has been proposed to act as the oxidizing agent. Interestingly, in THF only the heterocoupling product was obtained, whereas from the reaction in diethylether, the dienes were isolated in high yields.

A more classical oxidizing agent, i.e. elemental iodine, was used by Kang et al. in 1999, for the homocoupling of various organotin reagents under manganese catalysis. Elevated temperatures were necessary (NMP, 100 °C) to obtain the coupled products in good yields (Scheme 8.25) [25].

Ten years later, in 2007, Cahiez et al. showed in a seminal contribution that atmospheric oxygen could be employed in combination with Grignard reagents to

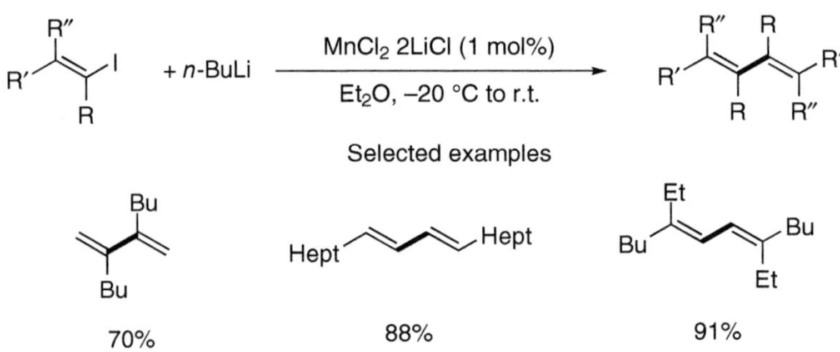

Scheme 8.24 Mn-catalyzed homocoupling of *in situ* generated vinyllithium reagents. Based on Cahiez et al. [24].

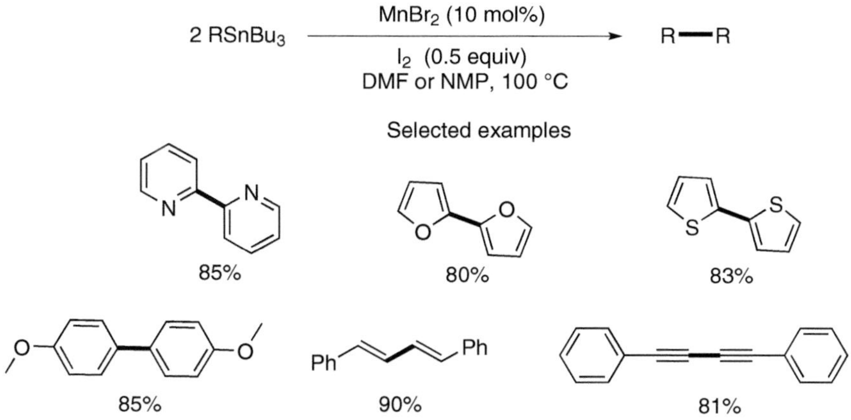

Scheme 8.25 Mn-catalyzed homocoupling of organostannanes under oxidative conditions. Source: Based on Kang et al. [25].

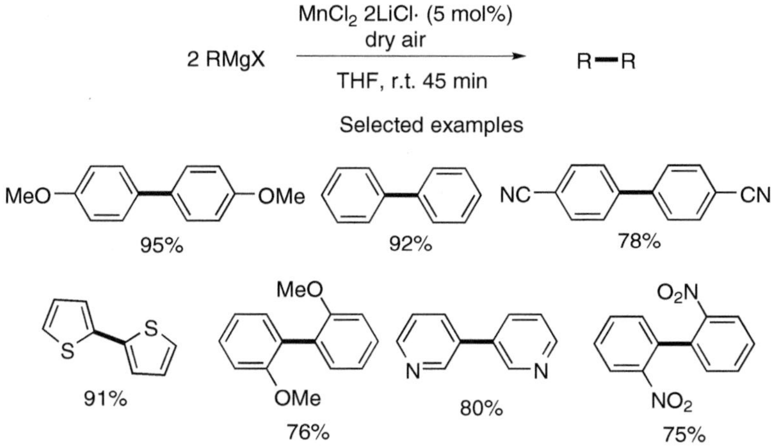

Scheme 8.26 Scope of Mn-catalyzed homocoupling of Grignard reagents using air as oxidant. Source: Based on Cahiez et al. [26].

access a wide range of homocoupled products in very high yields (Scheme 8.26) [26]. Among the limitations were sterically hindered arenes, e.g. mesityl groups, or very electron-poor arenes, e.g. C_6F_5, which did not provide the homocoupling products.

In the following year, the group of Zhou and Xue showed that DCE was also a suitable oxidizing reagent for the homocoupling of various aryl magnesium chlorides, even though the yields were generally lower than with dry air [27].

The combination of Mg/Zn superbases with several arenes or heteroarenes to generate *in situ* Grignard reagents provided the homocoupling products in the presence of $MnCl_2$ and oxygen as shown by Daugulis and coworkers in 2010 (Scheme 8.27) [28]. Even an ester group was well tolerated under these conditions. It should be

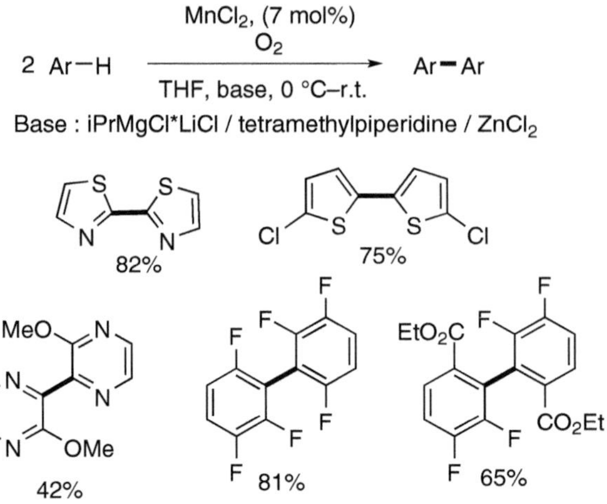

Scheme 8.27 Mn-catalyzed homocoupling of *in situ* generated Grignard reagents from arenes using air as oxidant. Source: Truong et al. [28].

noted that these superbases could also be employed in combination with $NiCl_2$ or $CoCl_2$ and oxygen to furnish efficient homocoupling reactions.

In the following years, the use of organic quinone-type oxidizing agents was reported in homocoupling reactions. In 2014, Peng et al. showed the homocoupling of *in situ* generated organomanganese halides with 3,3,5,5-tetra-*tert*-butyl-[1,1-bi(cyclohexylidene)]-2,2,5,5-tetraene-4,4-dione [29]. In 2015, Knochel and coworkers reported the stoichiometric *in situ* preparation of diaryl and di(heteroaryl) manganese complexes that could oxidatively be coupled to the diaryl compounds in the presence of chloranil (Scheme 8.28) [30]. A good functional group tolerance was observed under these conditions. Interestingly, air or DCE as oxidizing agents did not provide the homocoupling products under these conditions.

In 2019, the combination of dry air and *in situ* generated organolithiums (via Li/halide exchange) was reported for the successful Mn-catalyzed homocoupling reaction of aryl, heteroaryl, or benzyl halides (Scheme 8.29) [31]. The best yields were obtained with aryliodides or bromides; however, in certain cases chlorides also furnished satisfactory yields. *tert*-Butyllithium provided better results than *sec*- or *n*-butyl lithium.

In addition, substrates, such as alkynes, certain arenes (anisole, trifluorotoluene), or ferrocene, could be employed after deprotonation with the organolithium base (Scheme 8.30) [31].

8.2.3.2 Heterocoupling Reactions

With respect to heterocoupling reactions, Cahiez et al. demonstrated for the first time that two Grignard reagents with different electronic or steric properties could be coupled using $MnCl_2$ (20 mol% with respect to the excess reagent) and dry air at low temperature (0 °C) (Scheme 8.31) [32]. The combination of alkynyl/aryl or alkynyl/alkenyl Grignards and two different alkynyl or alkenyl

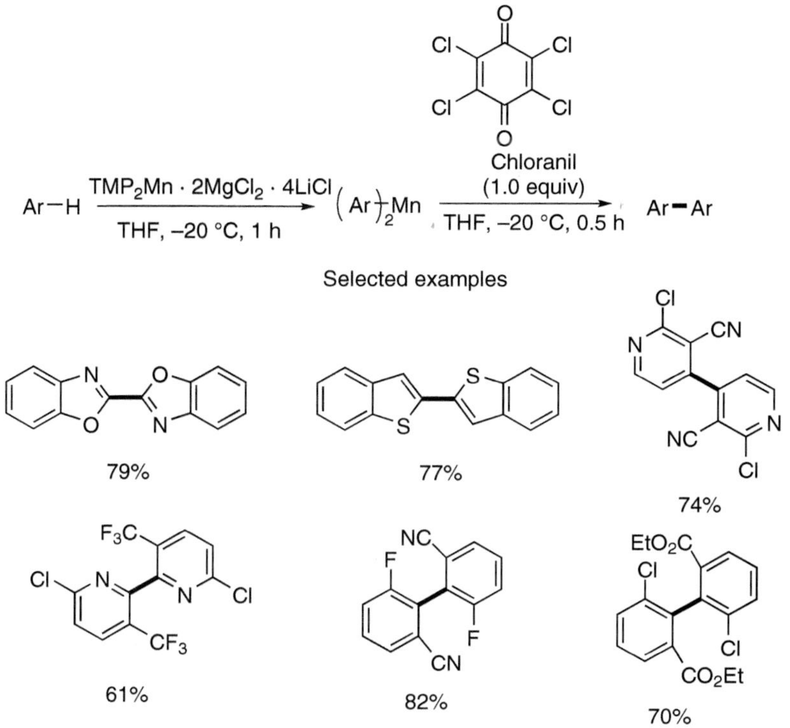

Scheme 8.28 Homocoupling of *in situ* generated diarylmanganese complexes with chloranil as oxidizing agent. Source: Based on Haas et al. [30].

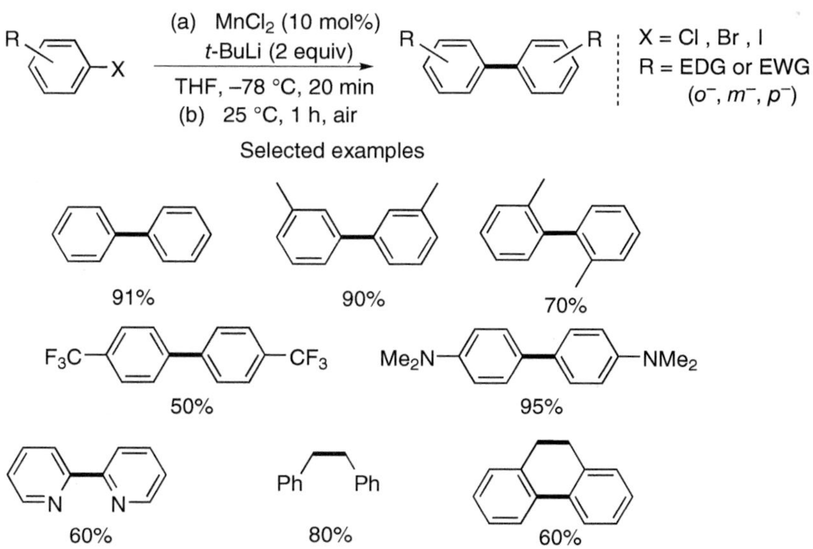

Scheme 8.29 Mn-catalyzed homocoupling of *in situ* generated lithium reagents by Li/halide exchange using air as oxidant. Source: Liu et al. [31].

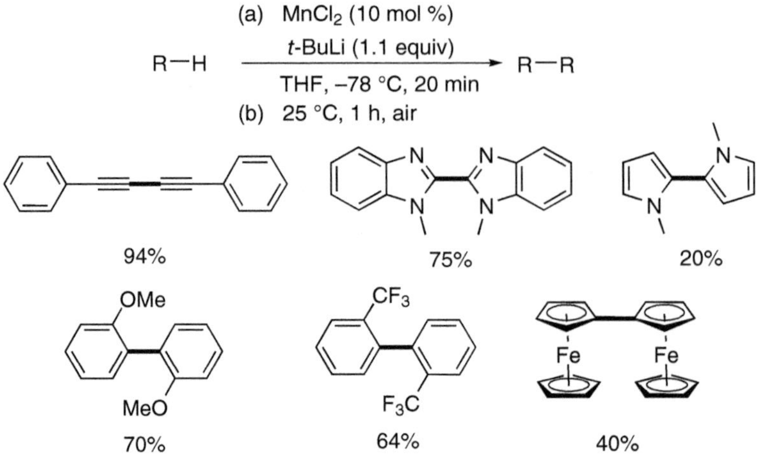

Scheme 8.30 Mn-catalyzed homocoupling of *in situ* generated lithium reagents by deprotonation using air as oxidant. Source: Liu et al. [31].

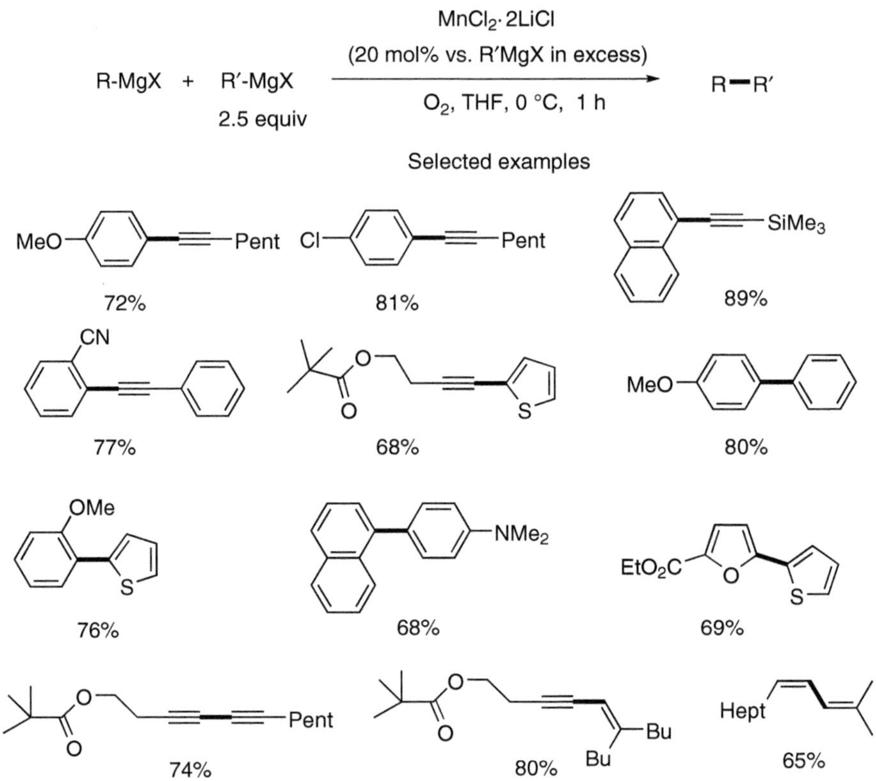

Scheme 8.31 Mn-mediated oxidative heterocoupling of Grignard reagents with air. Source: Based on Cahiez et al. [32].

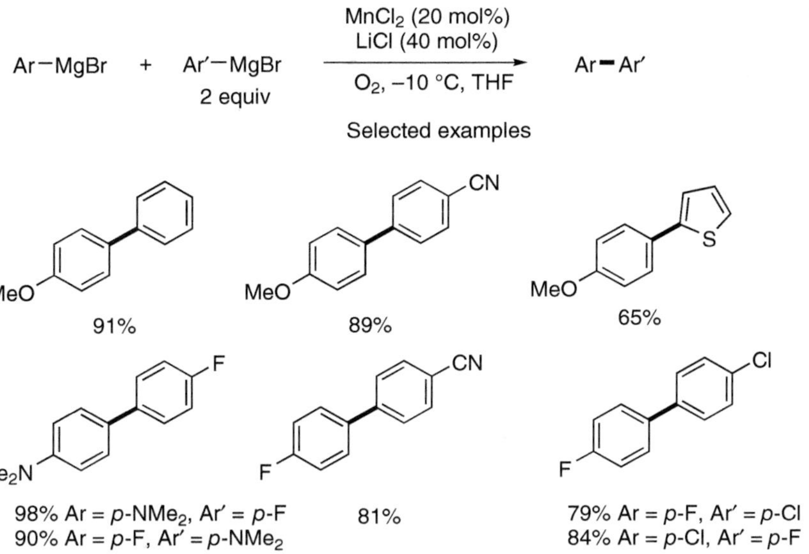

Scheme 8.32 Extension of Mn-catalyzed oxidative heterocoupling of Grignard reagents with air. Source: Ghaleshahi et al. [33].

Grignards in a 2.5 : 1 ratio gave good results. In the case of two different aryl Grignard reagents one of the coupling partners needed to be an electron-rich arene.

In 2017, Madsen and coworkers further examined this heterocoupling reaction of aryl Grignard reagents by lowering the catalyst loading (20 mol% with respect to the default reagent), using a 2 : 1 ratio between the coupling partners and decreasing the temperature to −10 °C [33]. Under these conditions, Grignard reagents having *p*-OMe, *p*-NMe_2, *p*-F, or *p*-Cl substituted aryl groups afforded no or only very small amounts of homocoupling by-product. Hence, these reagents turned out to be excellent coupling partners with various other aryl Grignard reagents. In this case, even two Grignard reagents with electron-withdrawing substituents provided good yields, whereas heterocyclic reagents gave only moderate outcomes (Scheme 8.32).

8.2.4 Mechanistic Insights

From a mechanistic point of view, some recent theoretical and mass-spectrometric studies have been conducted concerning the homocoupling reaction of organolithium and organomagnesium compounds to provide support for the generally proposed mechanism shown in Scheme 8.33 [31, 34]. Initially, a diarylmanganese complex **M** is formed via the transmetallation from the organometallic reagent onto $MnCl_2$. The addition of oxygen affords a monomeric diarylmanganese(IV) peroxo complex **N** that rapidly undergoes reductive elimination to yield the coupled biaryl product and the concomitant formation of a Mn(II)-peroxo species **O**. The latter reacts with two equivalents of the organometallic reagent to provide

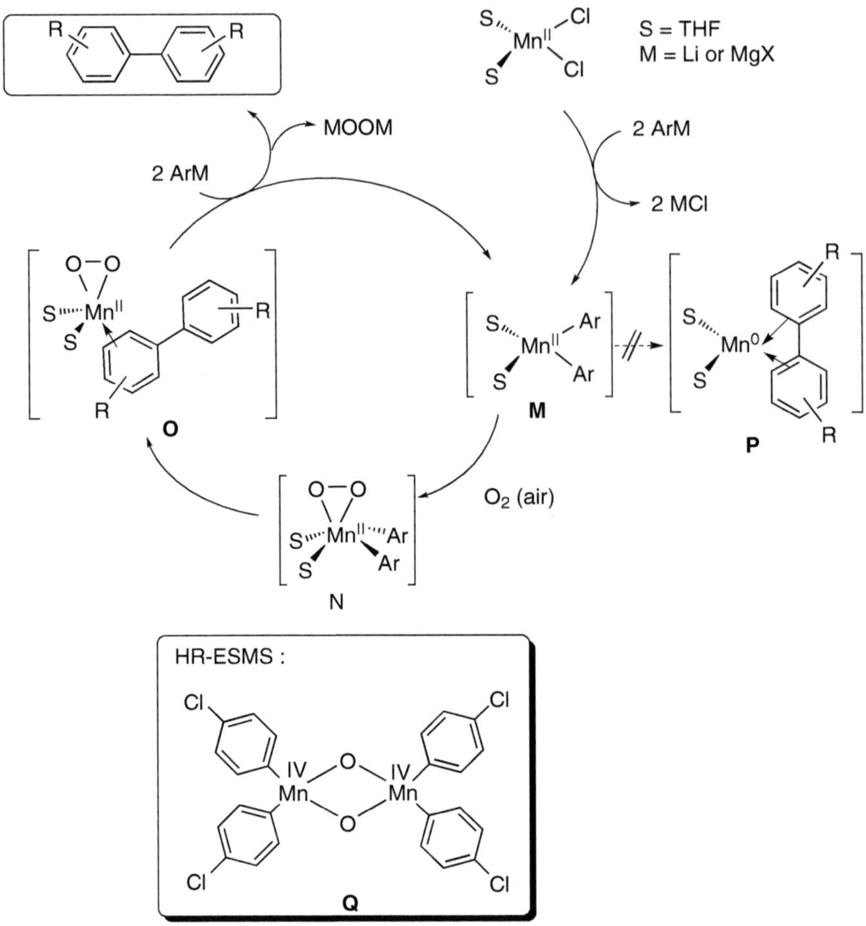

Scheme 8.33 Generally proposed mechanism for Mn-catalyzed coupling of organometallics under air. Source: Liu et al. [31]; Bottoni et al. [34].

again a diarylmanganese species **M**. Experimentally it has been observed that after the addition of air, the coupling reaction proceeds very quickly, in often less than 10 minutes, so that side reactions between the oxidizing agent and the organometallic reagents are very limited and rarely observed.

From a theoretical point of view, the formation of species **M**, **N,** and **O** could be confirmed for Grignard and organolithium reagents with the transmetallation and oxygen coordination steps being two strongly exothermic processes [31, 34]. An attempt to calculate the formation of the homocoupling product **P** directly from diarylmanganese complex **M** without oxygen revealed a very high energy barrier for the transition state (c. 35 kcal/mol) [31]. On the other hand, the reductive elimination from the diarylmanganese(IV) peroxo complex **N** has a reasonable energy barrier (c. 10 kcal/mol), and the process is exergonic by −13 kcal/mol. In an attempt to observe an intermediate Mn–oxo complex, such as **N**, experimentally,

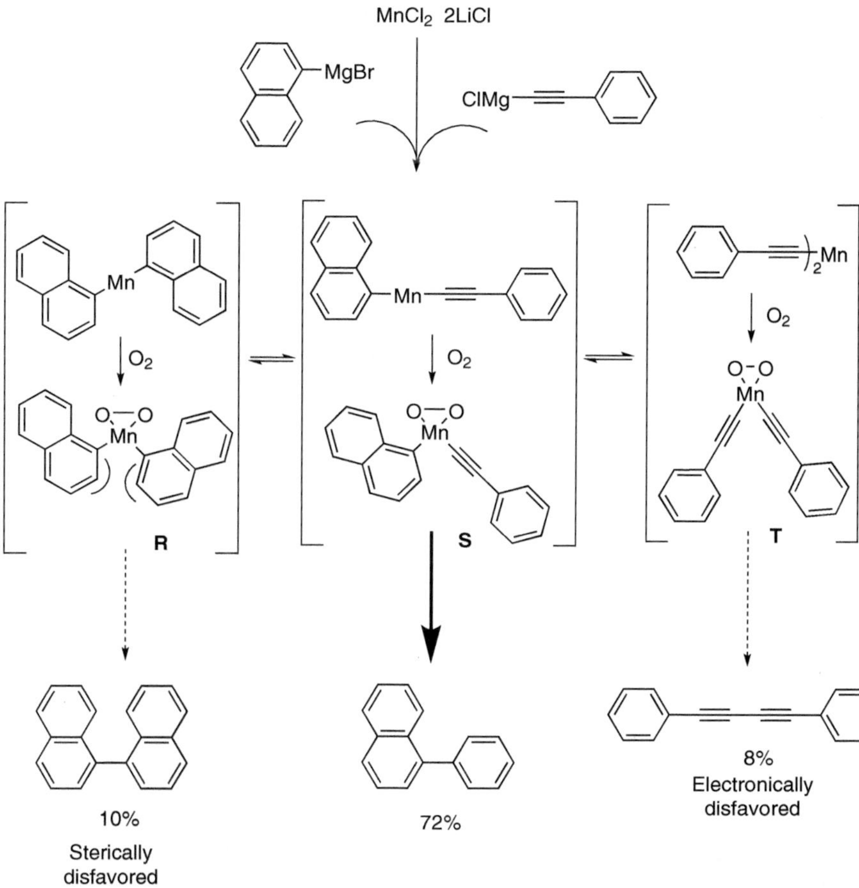

Scheme 8.34 Proposed mechanism for Mn-catalyzed heterocoupling reaction of two different Grignard reagents. Source: Based on Cahiez et al. [32].

the solution of a diarylmanganese complex after addition of air was studied by electrospray mass spectrometry (ESMS) [31]. The proposed peroxo species **N** could not be detected; however, a bimetallic oxygen-containing complex **Q** was clearly identified by isotope distribution and high resolution MS. According to DFT calculations, this complex cannot provide the homocoupled product [31]. Further theoretical studies confirmed that for mesityl or C_6F_5 groups, the energy barrier for the homocoupling product was too high (over 20 kcal/mol) to provide such products under the given reaction conditions [34].

For the heterocoupling reaction of an alkynyl Grignard with a bulky aryl group (mesityl or 1-naphthyl), it has been proposed that a rapid equilibrium between three organometallic manganese complexes **R, S, T** operates, which will preferentially lead to the heterocoupling product when the steric or electronic properties disfavor the homocoupling reaction shown in Scheme 8.34 [32].

8.2.5 Conclusion on Mn-Catalyzed Coupling Reactions of Organometallic Reagents Under Oxidative Conditions

The results presented earlier clearly show the potential of Mn-catalyzed oxidative homo- and heterocoupling reactions: (i) the products can be obtained from readily available organic halide precursors (mainly Br, I) or arenes/heteroarenes/alkynes via *in situ* deprotonation; (ii) the reaction conditions are mild (−10 °C to room temperature); (iii) atmospheric oxygen can be used as widely available, green oxidizing agent; (iv) and a wide scope of C—C bond formation reactions is possible. Despite the recent progress in understanding the reaction mechanism, further studies on the reaction intermediates are still necessary to get the full picture of these transformations.

8.3 Mn-Catalyzed Carbon–Heteroatom Bond Formation

The formation of carbon–heteroatom bonds C—Y (Y = N, O, S, B) starting from readily available organic halide precursors under metal-catalyzed (Pd, Ni, Cu) or metal-free conditions has been widely studied. However, until about 10 years ago, no Mn-catalyzed reaction was reported in this flourishing field. This section will describe some recent advances in the formation of C—Y bonds employing catalytic Mn or Mn/Cu systems with a critical analysis on the role of the manganese catalyst and possible trace metal contaminations.

8.3.1 C–N Bond Formation

Several Mn-catalyzed C—N bond formation reactions using aryl or heteroaryliodides have been described. In 2009, Teo et al. reported a high-temperature protocol for the C–N cross-coupling reaction of nitrogen nucleophiles and aryl iodides, which employed $MnCl_2{\cdot}4H_2O$ as catalyst, *trans*-1,2-diaminocyclohexane as ligand, K_3PO_4 as base, and water as solvent (Schemes 8.35) [35].

A wide range of *N*-arylation products were successfully synthesized according to this procedure. Limitations concerned *ortho*-substituted aryl iodides which afforded products in poor yields. On the other hand, the electronic character of the aryl iodides did not significantly influence the reactivity, for example, *p*-CF_3 or *p*-OMe substituents provided the coupling products with pyrazole in equally good yields. Concerning the nitrogen nucleophiles, pyrazole, 7-azaindole, indazole, 3-methylpyrazole, and indole derivatives all gave good results, whereas the reaction did not proceed with pyrrole, imidazole, triazole, benzylamine, pyrrolidin-2-one, benzamide, or aniline. It is worth noting that control experiments using pyrazole and iodobenzene as parent reagent under optimized reaction conditions were carried out to test the catalytic activity of high purity $MnCl_2{\cdot}4H_2O$ (99.99%, Sigma Aldrich), providing the final product in 64% yield [35]. Nevertheless, in an attempt to gain more insights into the reaction mechanism, especially the intermediacy of a radical species, in 2017, Madsen and coworkers repeated the reaction of

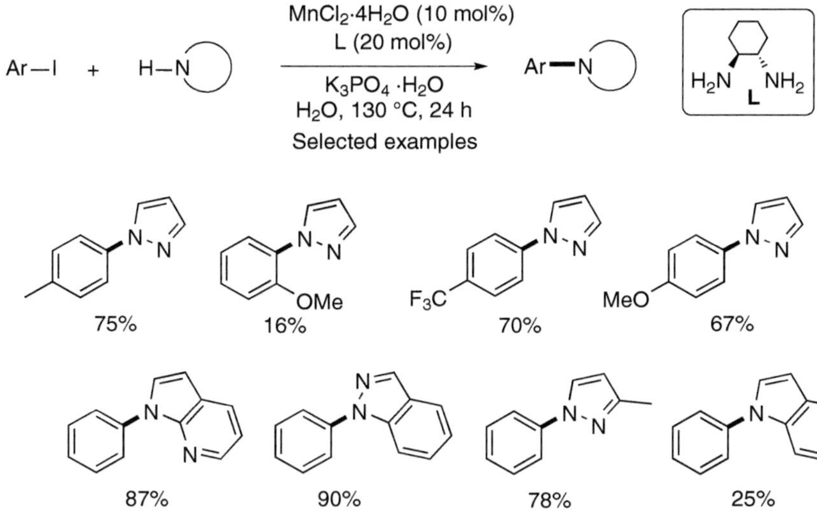

Scheme 8.35 Mn-catalyzed C–N cross-coupling of aromatic nitrogen nucleophiles and aryl halides in water. Source: Based on Teo et al. [35].

m-iodotoluene and pyrazole with freshly distilled starting materials and deionized water, obtaining the corresponding 1-(*m*-tolyl)-1H-pyrazole in an only very low yield of 6–8% (instead of the 75% yield reported by Teo) (Scheme 8.36) [9]. In addition, under the same conditions, but in the absence of Mn source, the yield was quite similar (around 2–3%). However, when deionized water was replaced with undistilled tap water, the yield of the product increased to 45%. As no indication on the water purity was given in the original paper by Teo, Madsen suggested that a possible metal contamination was responsible for the observed catalytic reaction. Moreover, the fact that the reaction only works well in water and not in organic solvents such as toluene, THF, DMF, DMSO, or acetonitrile might be a further indication toward the interplay of another metal in water. It should be noted that Madsen further showed that the above C–N coupling reaction works already with very small amounts of Cu catalyst (0.01 mol% of $CuCl_2{\cdot}H_2O$) under the exact same conditions, to provide the coupled product in 78% yield [9], in agreement with the literature [36].

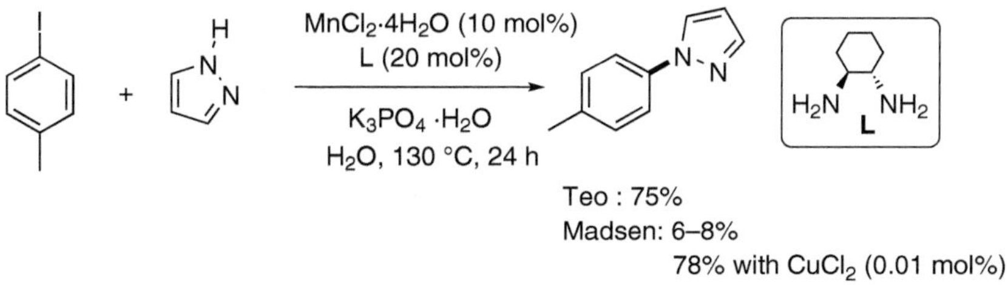

Scheme 8.36 Mn- or Cu-catalyzed arylation of pyrazole: Madsen and coworkers vs. Teo et al. [9, 35]. Source: Santilli et al. [9]; Teo et al. [35].

Teo and Yong's initial work was extended in 2012 to the coupling of indole, 7-azaindole, and indazole derivatives with pyridine and thiophene iodides in water under quite similar conditions (Scheme 8.37) [37]. The use of MnF_2 and Cs_2CO_3 provided the best results for several C–N coupling products. It should be noted that a blank test without Mn gave only 6% of the expected product but the purity of the MnF_2 was only 98%.

In order to extend the scope of N-nucleophiles, a catalytic system based on $MnCl_2·4H_2O$ associated with L-proline as a ligand was presented, for the N-arylation of aliphatic amines using aryl halides (Scheme 8.38) [38].

This protocol provided good to moderate yields for a range of aliphatic amines, such as morpholine and several primary and secondary amines. For some meta-

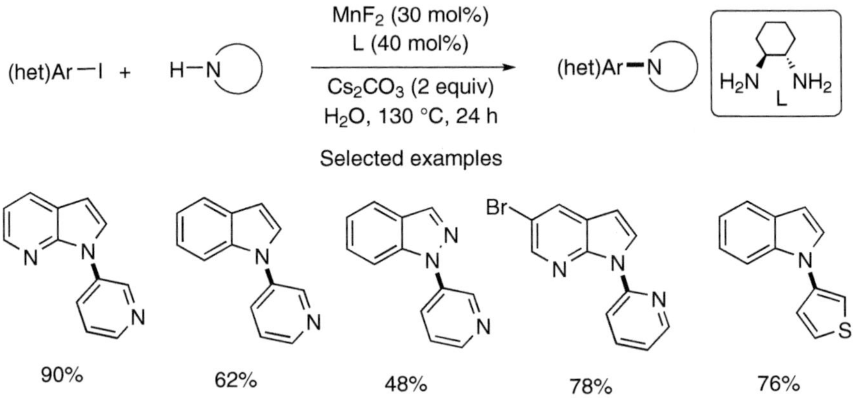

Scheme 8.37 Mn-mediated N-heteroarylation of indoles and indazoles in water. Source: Based on Yong and Teo [37].

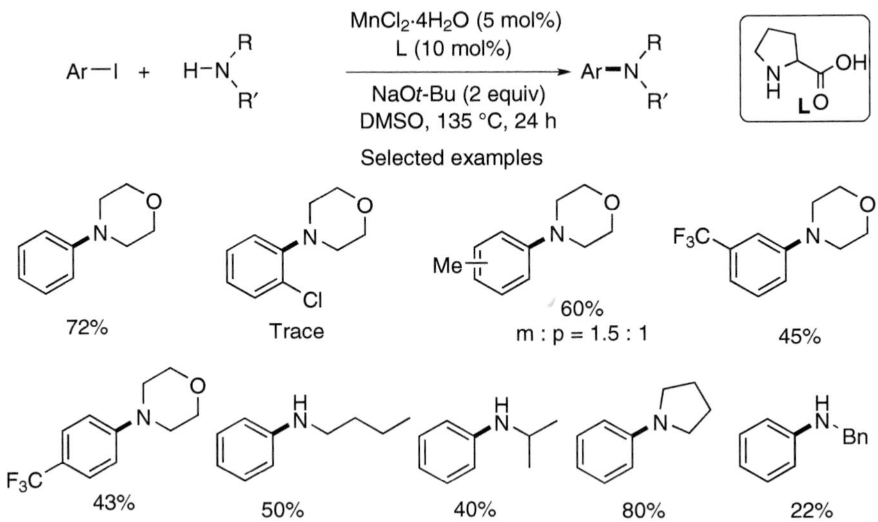

Scheme 8.38 Mn-mediated N-arylation of aliphatic amines with aryl halides. Source: Based on Yong and Teo [38].

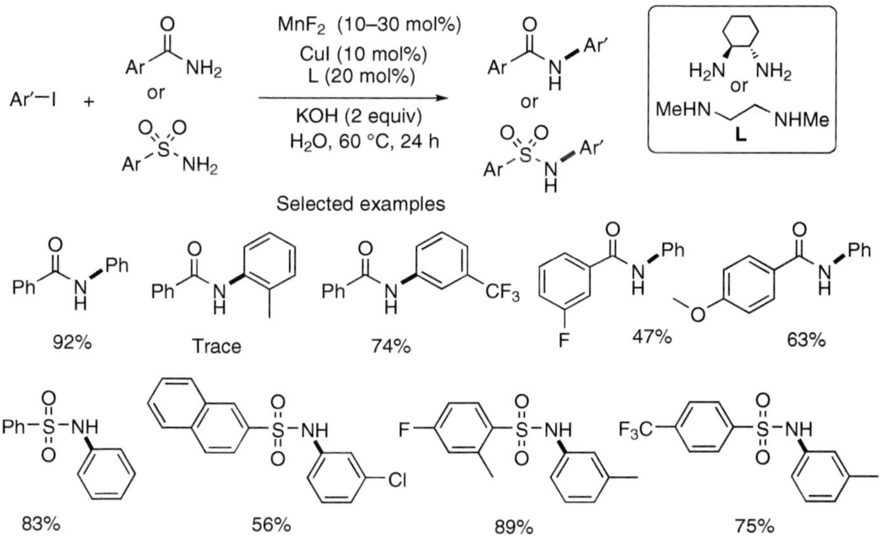

Scheme 8.39 Mn/Cu co-catalyzed N-arylation of benzamides and sulfonamides with various substituted aryl iodides in water. Source: Based on Teo, et al. [40].

or para-substituted aryliodides, two regioisomers were observed in the product mixture pointing toward possible aryne intermediates. The best yield was obtained in the case of pyrrolidine (80%), whereas benzylamine (22%) gave the lowest result. It should further be noted that an intriguing blank test with morpholine and iodobenzene under Mn-free condition revealed that the ligand L-proline with NaO*t*Bu and DMSO provided 50% of the corresponding *N*-arylation product, whereas in the presence of Mn the yield was 72%. The influence of the Mn salt as catalyst seems therefore rather limited in this process, which is further corroborated by a report on a transition metal-free process using the KO*t*Bu/DMSO system, which appeared some years earlier [39].

In order to lower the reaction temperatures in the previously presented reactions and to further broaden the substrate scope, a bimetallic system MnF_2/CuI was subsequently developed (Scheme 8.39) [40]. Under these conditions many C–N coupling reactions proceeded at 60 °C. Furthermore, this catalytic system also enabled the coupling of benzamide and sulfonamide derivatives with various aryl halides. In all cases, electron-withdrawing groups and ortho-substituents on the aryliodide led to lower yields. Experiments with different concentrations of Cu and Mn catalysts and different bases showed that a bimetallic cooperation may occur. Decreasing the amount of Mn salt to 5% led to significant drop in yield; however, no blank experiment without Mn catalyst was reported.

8.3.2 C–O Bond Formation

To date, there is only one report on the synthesis of biaryl ethers from phenols and aryl halides employing Mn catalysis. In 2019, Anilkumar and coworkers

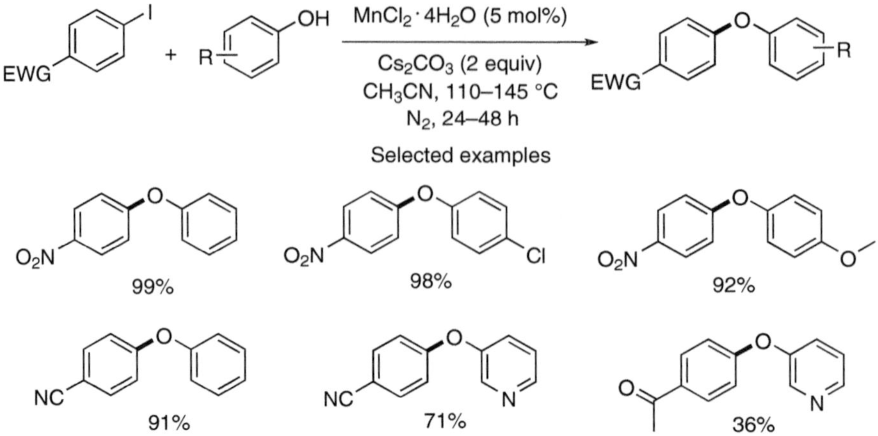

Scheme 8.40 Manganese-catalyzed C–O coupling for the O-arylation of phenols. Source: Rohit et al. [41].

provided a protocol that employed $MnCl_2{\cdot}4H_2O$ as catalyst at elevated temperatures in a Cs_2CO_3–CH_3CN system under ligand-free conditions (Scheme 8.40) [41]. The coupling was limited to electron-deficient aryl iodides derivatives, i.e. 4-iodonitrobenzene, 4-iodobenzonitrile, and 4-iodoacetophenone, but was successful with a variety of phenols.

Blank experiments without Mn catalyst revealed a strong decrease in yield from 99% to 29%, whereas an experiment with Mn but without base gave only trace amounts of product. No detailed information of the purity of Mn catalysts was provided. As an experiment with TEMPO did not influence the reaction outcome, a radical mechanism was ruled out by the authors. A postulated mechanism for this catalytic system is depicted in Scheme 8.41. The Mn(II) complex **U** obtained from salt metathesis reaction between $MnCl_2$ with phenoxide ions activates the aryl iodide leading to intermediate **V**, which evolves to intermediate **W** via C—O bond formation. The latter releases the biaryl ether product and the catalytically active species. As indicated earlier, the reaction proceeds also partially in the absence of a Mn source. As a quite similar reaction has already been described without the presence of any transition metal precatalyst [42], the exact role of the metal or even the necessity of a metal needs to be clarified. Nucleophilic aromatic substitution (S_NAr), typically involving electron-poor substituents, could also explain the reaction.

8.3.3 C–S Bond Formation

A report in 2010 on the cross-coupling of thiophenols with aryl or vinyl halides using a high-temperature $MnCl_2{\cdot}4H_2O$/DMSO/KOH system provided first hints into the potential of Mn catalysis in C—S bond formation reactions (Scheme 8.42) [43].

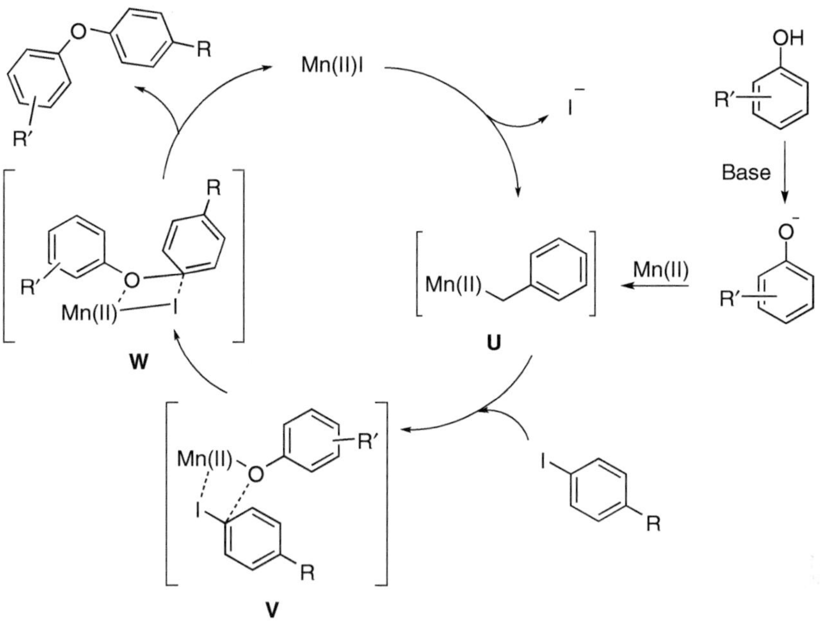

Scheme 8.41 Proposed mechanism for Mn-catalyzed O-arylation of phenols. Source: Rohit et al. [41].

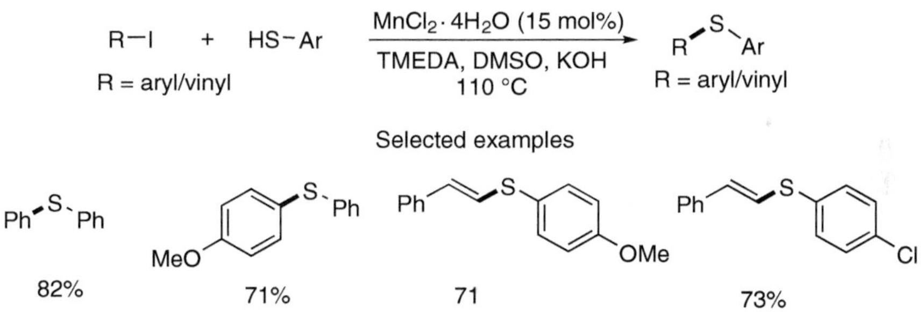

Scheme 8.42 Synthesis of aryl and vinyl sulfides in $MnCl_2 \cdot 4H_2O$/DMSO/KOH system. Source: Based on Bandaru et al. [43].

However, some years later it was shown by Lee that using this system with high purity $MnCl_2$ and KOH (99.99%) led to a significant drop in yield, for example, from 82% to 19% in the coupling of thiophenol with iodobenzene, indicating that other metal impurities might be involved in the catalytic process [44]. Consequently, a new Mn-catalytic system was devised by Lee using 1,10-phenanthroline derivatives as ligands in toluene or dioxane with Cs_2CO_3 as a base with high purity reagents (Scheme 8.43) [44]. These conditions enabled the coupling of a wide array of mainly electron-rich aryl halides, even sterically encumbered ortho-substituted substrates, with a range of thiophenols and thioalkanes in good yields. A mechanistic proposal has not been provided.

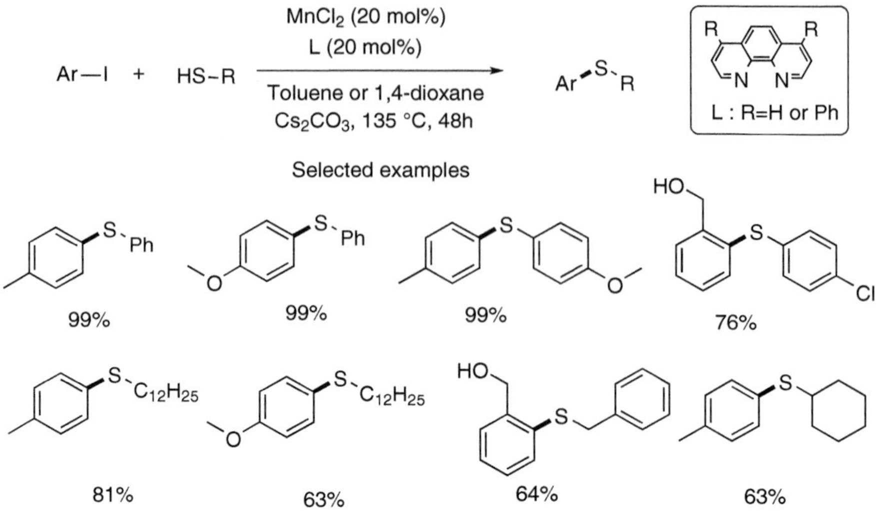

Scheme 8.43 Mn-catalyzed S-arylation of aryl iodides with aryl thiols. Source: Based on Liu et al. [44].

8.3.4 C—B Bond Formation

The Mn-catalyzed cross-coupling of alkyl halides with bis(pinacolato)-diboron provided an efficient access to a wide range of primary, secondary, and tertiary boronic esters [45]. Linear, branched, and cyclic alkyl chlorides were employed successfully in such reactions, which proceeded in the presence of small amounts of $MnBr_2$/TMEDA as catalyst system (0.1–5 mol%) in DME at room temperature (Scheme 8.44).

Even though no detailed mechanism was provided, first insights were obtained from radical clock experiments and reactions with different halide precursors. The involvement of a radical process was suggested from the isolation of a racemic product from an enantiomerically enriched secondary alkyl halide as well as from the faster reaction of tertiary over secondary over primary alkyl halides. Furthermore, the reaction of 6-bromo-1-hexene under standard conditions provided a 3 : 1 mixture of products derived from direct borylation vs. a ring-closing process. By varying the temperature, it was shown that at 0 °C only the borylation product was obtained with 79% yield, whereas at 65 °C the ring-closing product was dominant (Scheme 8.45a). This unique temperature-controlled switch in selectivity was further exploited in the first synthesis of cyclopropylmethylboronate from the corresponding alkyl bromide, in a 1 : 2.5 mixture with the ring-opened product (Scheme 8.45b).

From these results, the reaction was proposed to proceed via a stepwise, radical oxidative addition process during which the metal has a close association to the intermediate radical, as the radical recombination with Mn occurred faster than the 5-*exo*-dig cyclization or could at least compete with the ring opening of the cyclopropyl ring. Finally, the higher reactivity of primary alkyl bromides over alkyl chlorides was demonstrated in the selective borylation of 1-bromo-6-chlorohexane.

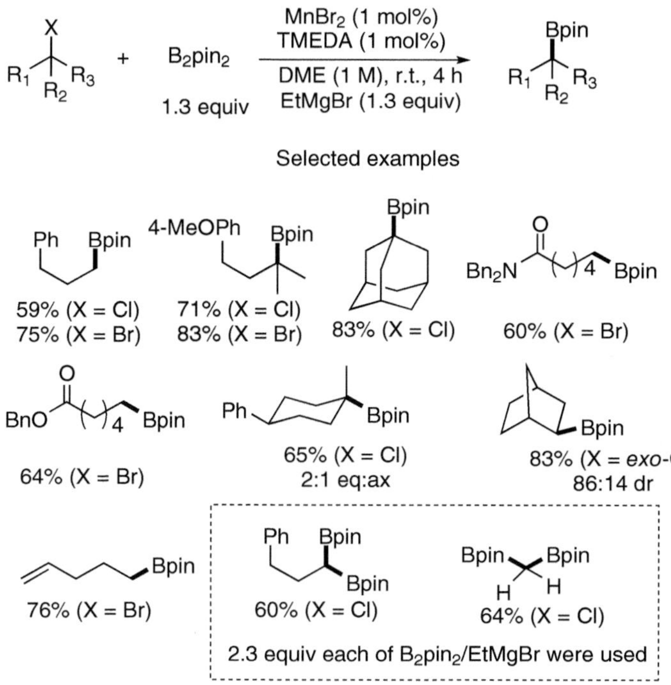

Scheme 8.44 Mn-catalyzed C—B bond formation cross-coupling reaction. Source: Based on Atack et al. [45].

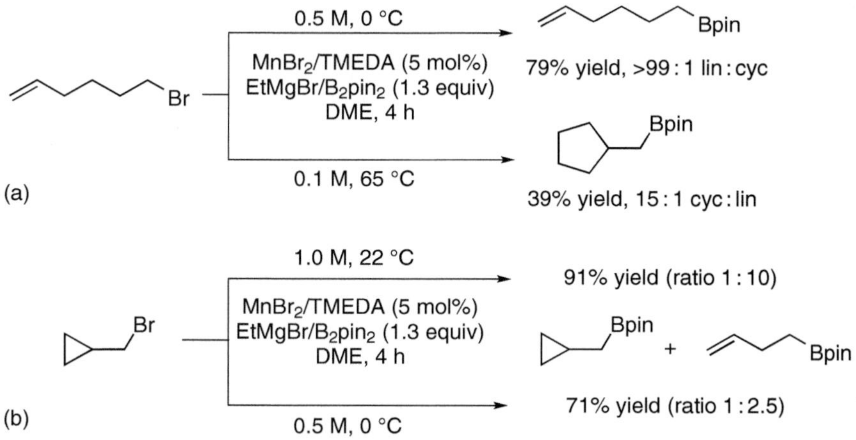

Scheme 8.45 Radical clock experiments: (a) temperature-controlled linear vs. cyclized reaction. (b) Temperature and concentration controlled substitution vs. ring-opening reaction. Source: Based on Atack et al. [45].

8.3.5 Conclusion on Mn-Catalyzed C—Y Bond Formation

This part has provided an overview on the currently known C—Y bond formation processes (Y = N; O, S, B) in the presence of Mn compounds. Most reactions involve high temperatures and strong bases rendering the determination of the catalytic species complicated. This still young research area also lacks mechanistic studies to determine the exact role of the Mn salt in certain processes. It further shows the difficulty to determine whether the catalytic activity is due to the Mn species or to the presence of other transition metals in small quantities. A rigorous series of blank experiments involving not only the metal but also the ligand, the base, and the solvents or additives should be conducted to avoid any misinterpretations of results.

8.4 Summary and Conclusions

The recent progress made in the field of Mn-catalyzed cross-coupling reactions has revealed the high potential of this transition metal for the development of more sustainable processes to access various classes of organic compounds, such as substituted alkenes, biaryls, heteroaryls, and other heteroatom containing compounds (Scheme 8.46).

As shown all along the chapter, most reactions work well with simple manganese salts, especially $MnCl_2$, providing cost-efficient processes due to its high availability and reduced waste production. In addition, the absence of sophisticated organic ligands requires less synthetic efforts and makes these coupling reactions readily available to the wider chemistry community. Some reactions have also been shown to be compatible with green solvents such as 2-methyl-THF or water.

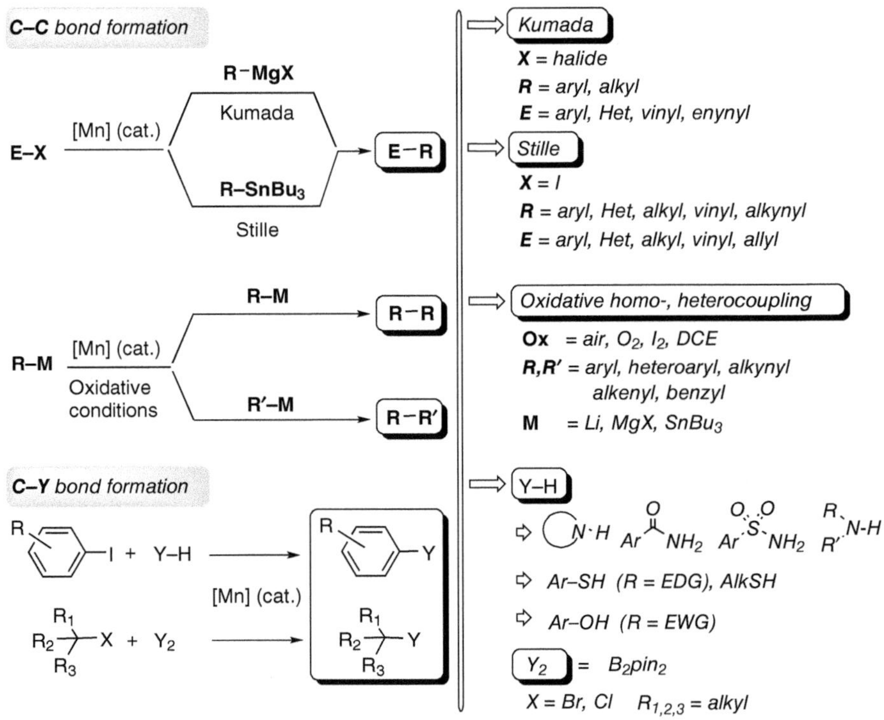

Scheme 8.46 Overview on Mn-catalyzed C—C and C—heteroatom bond formation processes described in this chapter.

Compared with other transition metals, cross-coupling reactions with Mn are still much less developed and future directions may go toward the development of specific ligands to reduce the amount of catalyst required or explore photocatalytic processes. The area is still dominated by the use of Grignard reagents as coupling partners, so many other types of organometallic species and organic substrates need to be investigated. In addition, in-depth mechanistic studies, combining experimental, analytical, and theoretical approaches, are required to gain further knowledge on these non-trivial processes to further explore their potential for organic synthesis. The difficulty related to Mn chemistry lies in the paramagnetism of many Mn complexes making the identification of reaction intermediates by conventional means, especially NMR spectroscopy, difficult. The combination of other analytical methods such as EPR or ESI-MS, monitoring the reactions by UV/Vis or *in situ* IR techniques, and the isolation and characterization of reaction intermediates by X-ray diffraction studies may help to advance this area in combination with theoretical studies.

Finally, special care has to be taken on the identification of the true catalytic species, and rigorous methodologies are required to exclude any metal contaminations, which may induce erroneous conclusions.

Acknowledgment

We thank the China Scholarship Council (CSC) for a PhD grant to Mr. Xiaoping LIU and the Centre National de la Recherche Scientifique (CNRS) for financial support.

References

1 Cahiez, G., Duplais, C., and Buendia, J. (2009). *Chem. Rev.* 109: 1434–1476.
2 Carney, J.R., Dillon, B.R., and Thomas, S.P. (2016). *Eur. J. Org. Chem.*: 3912–3929.
3 Valyaev, D.A., Lavigne, G., and Lugan, N. (2016). *Coord. Chem. Rev.* 308: 191–235.
4 Cahiez, G. and Moyeux, A. (2016). *Grignard Reagents and Transition Metal Catalysts* (ed. J. Cossy), 210–243. Berlin, Germany: de Gruyter.
5 Cahiez, G., Lepifre, F., and Ramiandrasoa, P. (1999). *Synthesis*: 2138–2144.
6 Cahiez, G., Luart, D., and Lecomte, F. (2004). *Org. Lett.* 6: 4395–4398.
7 Zhang, F., Shi, Z., Chen, F., and Yuan, Y. (2010). *Appl. Organomet. Chem.* 24: 57–63.
8 Antonacci, G., Ahlburg, A., Fristrup, P. et al. (2017). *Eur. J. Org. Chem.*: 4758–4764.
9 Santilli, C., Beigbaghlou, S.S., Ahlburg, A. et al. (2017). *Eur. J. Org. Chem.*: 5269–5274.
10 Inoue, R., Shinokubo, H., and Oshima, K. (1998). *J. Org. Chem.* 63: 910–911.
11 Kakiya, H., Inoue, R., Shinokubo, H., and Oshima, K. (2000). *Tetrahedron* 56: 2131–2137.
12 He, R., Jin, X., Chen, H. et al. (2014). *J. Am. Chem. Soc.* 136: 6558–6561.

13 Rueping, M. and Ieawsuwan, W. (2007). *Synlett*. 2007: 247–250.
14 Petel, B.E., Purak, M., and Matson, E.M. (2018). *Synlett* 29: 1700–1706.
15 Cahiez, G., Bernard, D., and Normant, J.F. (1976). *J. Organomet. Chem.* 113: 107–113.
16 Fugami, K., Oshima, K., and Utimoto, K. (1987). *Chem. Lett.* 16: 2203–2206.
17 Alami, V., Ramiandrasoa, P., and Cahiez, G. (1998). *Synlett*. 1998: 325–327.
18 Cahiez, G., Gager, O., and Lecomte, F. (2008). *Org. Lett.* 10: 5255–5256.
19 Silva, M.S., Ferrarini, R.S., Sousa, B.A. et al. (2012). *Tetrahedron Lett.* 53: 3556–3559.
20 Kang, S.-K., Kim, J.–.S., and Choi, S.–.C. (1997). *J. Org. Chem.* 62: 4208–4209.
21 Kang, S.–.K., Kim, W.-Y., Lee, Y.–.T. et al. (1998). *Tetrahedron Lett.* 39: 2131–2132.
22 Dakarapu, R. and Falck, J.R. (2018). *J. Org. Chem.* 83: 1241–1251.
23 Sarbajna, A., He, Y.-T., Dinh, M.H. et al. (2019). *Organometallics* 38: 4409–4419.
24 Cahiez, G., Bernard, D., and Normant, J.F. (1976). *J. Organomet. Chem.* 113: 99–106.
25 Kang, S.–.K., Baik, T.–.G., Jiao, X.H., and Lee, Y.–.T. (1999). *Tetrahedron Lett.* 40: 2383–2384.
26 Cahiez, G., Moyeux, A., Buendia, J., and Duplais, C. (2007). *J. Am. Chem. Soc.* 129: 13788–13789.
27 Zhou, Z. and Xue, W. (2009). *J. Organomet. Chem.* 694: 599–603.
28 Truong, T., Alvarado, J., Tran, L.D., and Daugulis, O. (2010). *Org. Lett.* 12: 1200–1203.
29 Peng, Z., Li, N., Sun, X. et al. (2014). *Org. Biomol. Chem.* 12: 7800–7809.
30 Haas, D., Hammann, J.M., Moyeux, A. et al. (2015). *Synlett* 26: 1515–1519.
31 Liu, Y., Bergès, J., Zaid, Y. et al. (2019). *J. Org. Chem.* 84: 4413–4420.
32 Cahiez, G., Duplais, C., and Buendia, J. (2009). *Angew. Chem. Int. Ed.* 48: 6731–6734.
33 Ghaleshahi, H.G., Antonacci, G., and Madsen, R. (2017). *Eur. J. Org. Chem.*: 1331–1336.
34 Bottoni, A., Cahiez, G., Calvaresi, M. et al. (2016). *J. Organomet. Chem.* 814: 25–34.
35 Teo, Y.–.C., Yong, F.–.F., Poh, C.–.Y. et al. (2009). *Chem. Commun.*: 6258–6260.
36 Monnier, F. and Taillefer, M. (2009). *Angew. Chem. Int. Ed.* 48: 6954–6971.
37 Yong, F.–.F. and Teo, Y.–.C. (2012). *Synlett* 23: 2106–2110.
38 Yong, F.–.F. and Teo, Y.–.C. (2010). *Tetrahedron Lett.* 51: 3910–3912.
39 Shi, L., Wang, M., Fan, C.-A. et al. (2003). *Org. Lett.* 5: 3515–3517.
40 Teo, Y.–.C., Yong, F.–.F., Ithnin, I.K. et al. (2013). *Eur. J. Org. Chem.*: 515–524.
41 Rohit, K.R., Saranya, S., Harry, N.A., and Anilkumar, G. (2019). *ChemistrySelect* 4: 5150–5154.
42 Pichette Drapeau, M., Ollevier, T., and Taillefer, M. (2014). *Chem. Eur. J.* 20: 5231–5236.
43 Bandaru, M., Sabbavarpu, N.M., Katla, R., and Yadavalli, V.D.N. (2010). *Chem. Lett.* 39: 1149–1151.
44 Liu, T.–.J., Yi, C.–.L., Chan, C.–.C., and Lee, C.–.F. (2013). *Chem. Asian J.* 8: 1029–1034.
45 Atack, T.C. and Cook, S.P. (2016). *J. Am. Chem. Soc.* 138: 6139–6142.

9

Manganese(III) Acetate-Mediated Cyclizations

Barry B. Snider

Brandeis University, Department of Chemistry MS 015, Waltham, MA 02453-2728, USA

9.1 Introduction

The oxidative addition of acetic acid to an alkene to give a γ-lactone using 2 equiv of $Mn(OAc)_3$ in acetic acid at reflux was first reported in 1968. In the past 50 years, the use of $Mn(OAc)_3$ to oxidatively initiate free radical reactions of mono- and 1,3-dicarbonyl compounds has been extensively developed and widely reviewed. The author surveyed oxidative cyclization reactions in 1996 [1]. More recent reviews comprehensively cover oxidative cyclization, oxidative addition, and functionalization reactions inclusively [2–4] the formation of 1,2-dioxanes [5], the formation of carbon–heteroatom bonds [6], and mechanistic aspects [7]. This chapter will use the same organization as the 1996 review [1], covering only oxidative cyclizations in which the ketone or 1,3-dicarbonyl compound that is oxidized is tethered to the alkene. Intermolecular additions and annulations, in which an intermolecular addition is followed by a cyclization, are not covered. The focus will be on those examples published since 1996, although sufficient mechanistic details will be provided so that this chapter can be understood without reading previous reviews. The examples chosen are largely selected from natural product syntheses since applications in highly functionalized, complex ring systems most effectively illustrate the scope and limitations of a synthetic method.

9.2 Mechanistic Considerations

β-Keto esters are the most widely used oxidizable group and react with $Mn(OAc)_3$ at or near room temperature. Other 1,3-dicarbonyl compounds including malonic esters, 1,3-diketones, and β-keto acids, amides, sulfoxides, and sulfones can also be used. Oxidative cyclization of simple unsaturated ketones can be successfully carried out at 80 °C if the ketone enolization is regiospecific and the product ketone cannot enolize, thus preventing further oxidation of the product.

Manganese Catalysis in Organic Synthesis, First Edition. Edited by Jean-Baptiste Sortais.

$Mn(OAc)_3{\cdot}2H_2O$, which is commercially available and most often used, is actually an oxo-centered trimer of Mn(III) with bridging acetates. Anhydrous $Mn(OAc)_3$ is slightly more reactive leading to shorter reaction times, but not necessarily higher product yields. Acetic acid is the usual solvent, but DMSO, ethanol, trifluoroethanol, methanol, dioxane, and acetonitrile have advantages in some reactions. The use of ethanol as a solvent is advantageous when a reductive termination is desired as with the alkenyl radicals formed by addition to alkynes since the solvent can act as a hydrogen atom donor reducing the alkenyl radical to an alkene and giving the α-hydroxyethyl radical, which is oxidized to acetaldehyde by Mn(III). The use of trifluoroacetic acid as a co-solvent usually increases the reaction rate, but often the yields of products are lower. Addition of acetate salts may facilitate enolization and act as a buffer.

Manganese is not inherently expensive, but $Mn(OAc)_3{\cdot}2H_2O$ is much more expensive than $Mn(OAc)_2$ so it would be highly desirable if Mn(III) can be used catalytically and generated initially by oxidation of Mn(II). Electrochemical regeneration of Mn(III) has been extensively investigated with limited success since yields are often lower than with stoichiometric oxidant [1, 8]. However, Baran and coworkers recently reported that comparable yields are obtained electrochemically with 0.5 equiv of $Mn(OAc)_2$ and 0.02 equiv of Cu(II) 3,5-diisopropylsalicylate in AcOH, EtOAc, and KOAc using a divided cell [9]. Sodium periodate has been used successfully as a stoichiometric oxidant with catalytic $Mn(OAc)_3$ in DMSO [8, 10].

With 1,3-dicarbonyl compounds, the mechanism is variable depending on the acidity of the α-proton and the oxidation potential of the resulting enolate [7, 11]. With an α-alkyl β-keto ester such as **1**, reaction with $Mn(OAc)_3$ to give Mn(III) enolate **2** is the alkene-independent rate-determining step (Scheme 9.1). The electron-donating alkyl group decreases the acidity by 1–2 orders of magnitude and therefore the enolization rate. The alkyl group also decreases the oxidation potential by 0.25–0.4 V so that enolate **2** is oxidized readily, losing Mn(II) to form free radical **3**, which is no longer complexed to manganese. Comparable regio- and stereochemical results were obtained in several cases in which free radical **3** was generated oxidatively or by atom transfer reaction from an α-halo β-keto ester with hexamethylditin, suggesting that the same free radical **3** is involved in both reactions [12].

On the other hand, more acidic α-unsubstituted β-keto esters such as **4** undergo rapid and reversible formation of enolate **5** (Scheme 9.2). Loss of Mn(II) from enolate **5** to give free radical **6** is very slow. The rate-determining step is the reaction of the alkene with the Mn(III) enolate of **5** to give cyclic radical **7** [7]. Thus, the

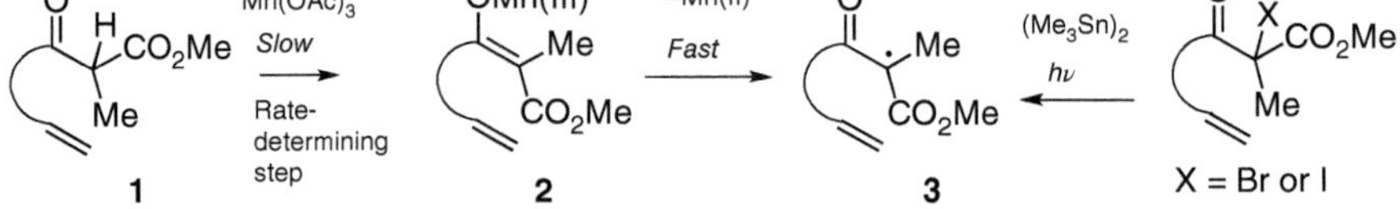

Scheme 9.1 Rate-determining step for oxidation of α-substituted β-keto esters.

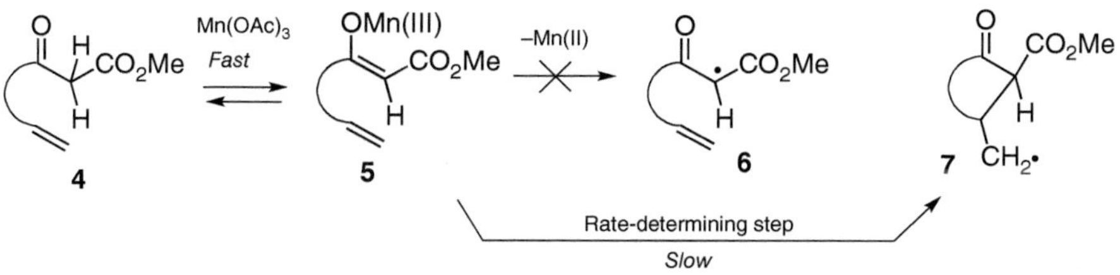

Scheme 9.2 Rate-determining step for oxidation of α-unsubstituted β-keto esters.

nature of the tether affects the rate of these reactions. Unfortunately, the direct conversion of manganese enolate **5** to cyclic radical **7** without the intermediacy of free radical **6** does not fit well with arrow-pushing models of radical reactions. The mechanistic details with other classes of 1,3-dicarbonyl compounds also depend on the acidity of the α-proton and the oxidation potential of the enolate. For instance, enolization is the rate-determining step in the oxidative cyclization of α-substituted dialkylmalonates, whereas α-substituted Meldrum's acids are much more acidic, so cyclization of the Meldrum's acid enolate is the rate-determining step leading to very different product mixtures [13].

Although Mn(III) is a strong oxidant, it is only effective in oxidizing tertiary radicals to carbocations. Isolated secondary and primary radicals (such as **7** or **8**) are not oxidized rapidly by Mn(III) so that hydrogen abstraction is the primary reaction if no co-oxidant is used (Scheme 9.3). In the 1960s, Kochi showed that Cu(II) reacts rapidly with secondary radicals such as **8** to give reactive alkyl Cu(III) intermediate **9** [14]. The formation of an alkene is the major pathway with copper acetate and other carboxylates. The elimination probably occurs by an intramolecular abstraction of a proton by the acetate oxygen with loss of CuOAc and HOAc. As with other syn-eliminations (see copper(III) complex **10**), there is a strong preference for the formation of the less substituted double bond (Hofmann product) and (*E*)-alkene **11**. On the other hand, ligand transfer to give halide or pseudohalide **13** occurs with cupric halides and thiocyanate, and oxidation to carbocation **14** occurs with copper

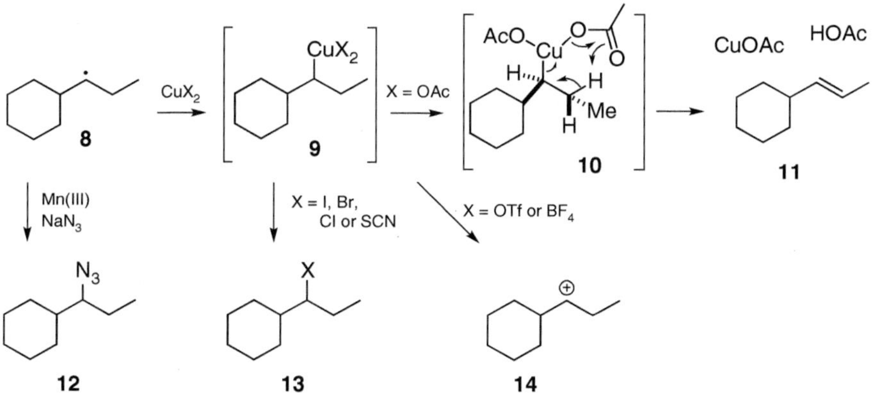

Scheme 9.3 Oxidation of secondary radicals by cupric salts.

triflate or tetrafluoroborate [15]. The copper(I) product is immediately re-oxidized to Cu(II) by Mn(III), so these reactions still require 2 equiv of $Mn(OAc)_3$ and are catalytic in Cu(II) although a full equivalent is usually used. Radicals such as **8** will also react with LiCl and Mn(III) to give alkyl chlorides and with NaN_3 and Mn(III) to give alkyl azides **12** [16].

9.3 Monocyclizations

9.3.1 Radicals Derived from β-Keto Esters and β-Diketones that Lead to Cycloalkanones

We found that the oxidative cyclization of unsaturated β-keto ester **15** with 2 equiv of $Mn(OAc)_3$ and 1 equiv of $Cu(OAc)_2$ gives an equilibratable mixture of tautomers, ketone **16**, and enol **17** (71%) (Scheme 9.4) [17]. On the other hand, the analogous oxidation of β-keto ester **18** affords cyclopentanone **19** in only 36% yield. Both products **16/17** and **19** are enolizable β-keto esters that can be oxidized further. However, this appears to be faster for cyclopentanone **19**, since dienone **20** (10%) is also isolated. Many other early examples have been reviewed [1].

Newhouse used an oxidative cyclization to construct the D-ring of protoaustinoid D and berkeleyone A [18]. The treatment of β-keto ester **21** with 3 equiv of $Mn(OAc)_3$ in AcOH at 60 °C for two hours forms tetracycle **22** as a mixture of isomers in 44% yield for a multi-step sequence (Scheme 9.5). Pattenden explored transannular cyclizations of large ring β-keto esters [19]. Oxidative cyclization of β-keto ester **23** provides allylic acetate **26** (40%). The transannular 6-*exo* cyclization gives radical **24**, which undergoes a 1,5-hydrogen atom shift to give the more stable allylic radical **25** that is oxidized to **26**. On the other hand, β-keto ester **27** undergoes a 6-*endo* cyclization to provide cyclohexenone **28** (44%). Nicolaou used the oxidative cyclization of β-keto ester **29** to tricycle **30** (76%) to prepare a key tricyclic intermediate in the synthesis of vannusal A [20]. In a model study for the synthesis of macquarimicin C, Gössinger converted allylic keto lactone **31** to

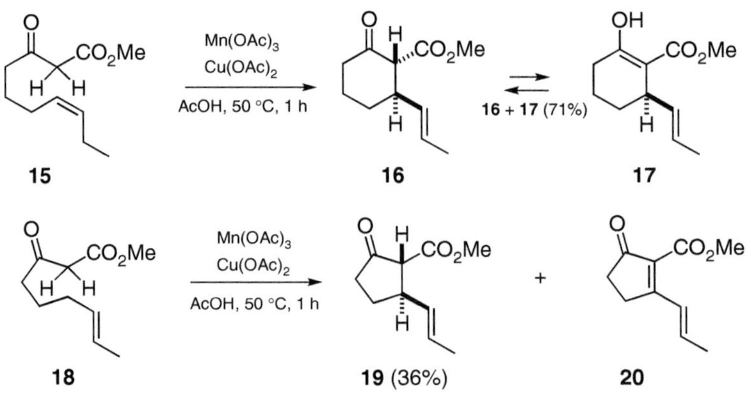

Scheme 9.4

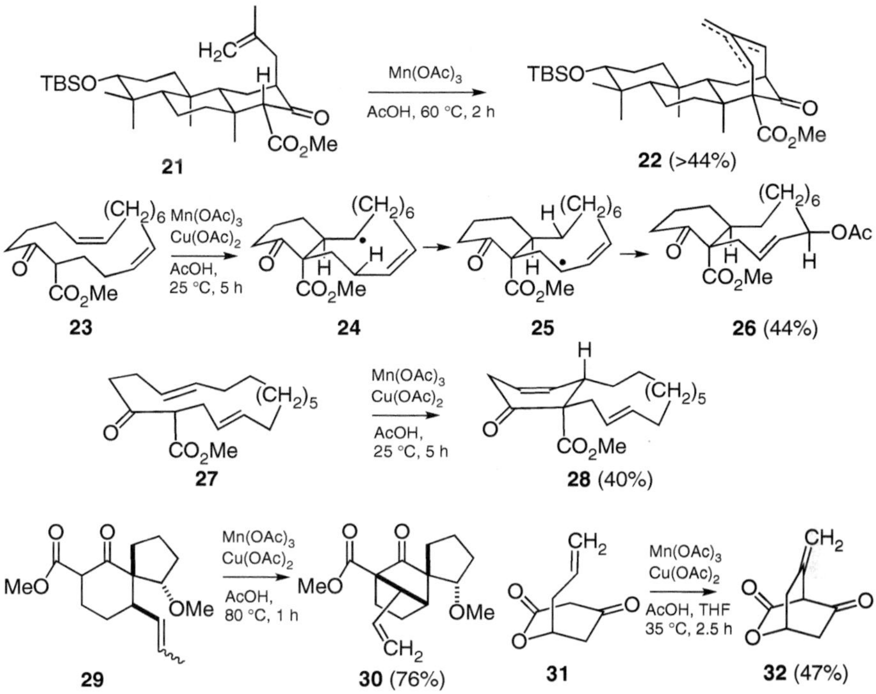

Scheme 9.5

bicyclic lactone **32** (47%) [21]. Unfortunately, the oxidative cyclization fails in the more highly functionalized system needed for the natural product.

We developed a general synthesis of salicylate esters by oxidative cyclization and aromatization of 3-oxohept-6-enoate esters [22]. Treatment of **33** (methyl 3-oxohept-6-enoate) with 4 equiv of $Mn(OAc)_3$ and 1 equiv of $Cu(OAc)_2$ in AcOH for 24 hours at 25 °C gives **36** (methyl salicylate, 78%) (Scheme 9.6). Oxidative cyclization with a carbonyl group in the tether occurs preferentially 6-*endo* to give secondary radical **34**, which is oxidized by $Cu(OAc)_2$ to give a mixture of alkenes **35**, which is further oxidized by $Mn(OAc)_3$ to give **36**. Alvarez-Manzaneda used this reaction for the preparation of phenol abietane diterpenes [23]. Treatment of β-keto ester **37** with 4 equiv of $Mn(OAc)_3$ and 3 equiv of LiCl in Ac_2O at 120 °C for 12 hours affords salicylate **38** (74%), which was elaborated to sugikurojin A and 19-hydroxyferruginol. Koo reported that oxidative cyclization of δ-acetoxy-β-keto ester **39** with $Mn(OAc)_3$ provides tertiary radical **40**, which abstracts a hydrogen atom to give decalin **41** [24]. Treatment of **41** with DBU in THF gives enone **42** (65% overall) as the major product. Presumably the electron-withdrawing acetate or steric hindrance retards oxidation of the tertiary radical of **40** by Mn(III) so that hydrogen abstraction becomes the major pathway. If $Cu(OAc)_2$ is used as a co-oxidant, radical **40** is oxidized to a carbocation, which loses a proton to form an acetoxyalkene. DBU elimination of the acetate gives methyl 5,6,7,8-tetrahydro-2-hydroxynaphthalene-1-carboxylate as the major product [24].

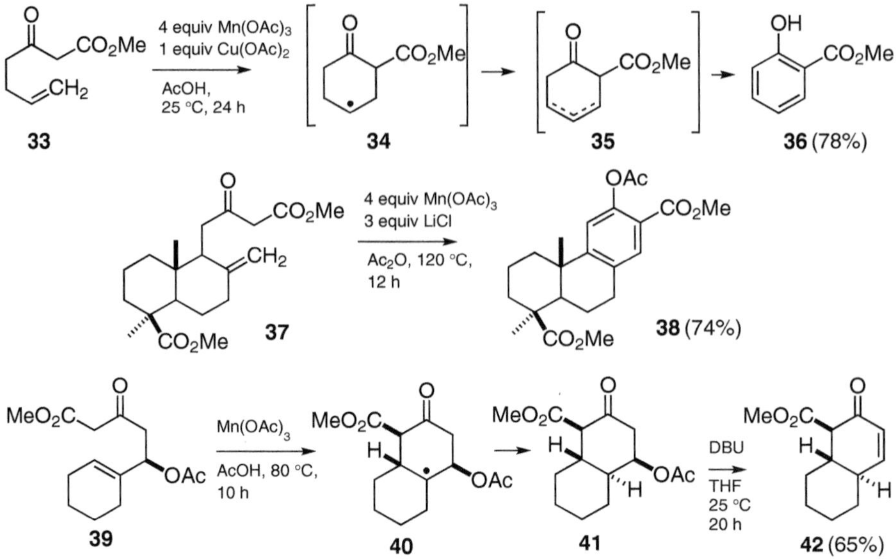

Scheme 9.6

9.3.2 Radicals Derived from β-Keto Esters, β-Diketones, or Malonate Esters that Lead to Cycloalkanes

The tether can also be attached to the α-position of the β-dicarbonyl compound leading to a cycloalkane product [1]. Taber recently oxidatively cyclized β-keto ester **43** with $Mn(OAc)_3$ and $Cu(OAc)_2$ in AcOH at 80 °C for two hours to give isopropenylbicyclo[3.3.1]octane **44** (38%) and a mixture of hemiketal **45** and other isomers (Scheme 9.7) [25]. Treatment of this mixture with 0.05 equiv of TsOH in toluene at reflux for one hour gives additional **44** (37%, 75% overall). Further heating for four hours isomerizes the double bond of **44** to give isopropylidenebicyclo[3.3.0]octane **46** (82%), which can be elaborated to cameroon-7α-ol. Heating the mixture of **44** and **45** with 1.05 equiv of TsOH in toluene at reflux affords lactone **47** (87%). Oxidative cyclization of **43** with $Mn(OAc)_3$ and $Cu(OAc)_2$ in acetonitrile at 25 °C affords **44** (59%) and 30% of a complex mixture of isomers. Wallace used an oxidative cyclization to prepare the polycyclic ring system of the sesquiterpene phenol natural product 15-oxopuupehenol [26]. The reaction of β-keto ester **48** with 2 equiv of $Mn(OAc)_3$ in AcOH at 80 °C for 12 hours affords pentacyclic lactone **49** (58%). A similar reaction of β-keto ester **50** with 2 equiv of $Mn(OAc)_3$ in AcOH at 65 °C for three hours provides pentacyclic hemiketal **51** (31%) from one of the two diastereomers of **50**. In all three of these cases, the cyclic radical is tertiary and is therefore oxidized by Mn(III) to a carbocation that reacts further.

Oxidative cyclization of malonate **52** with $Mn(OAc)_3$ affords the primary cyclopentylmethyl radical (Scheme 9.8). With $Cu(OAc)_2$ in acetic acid, this is oxidized to lactone **53** (46%) and methylenecyclopentane **54** (20%) [27]. With $Cu(OAc)_2$ in DMSO, methylenecyclopentane **54** (53%) is formed as the major product, while in EtOH without a Cu(II) salt, methylcyclopentane **55** (40%) is

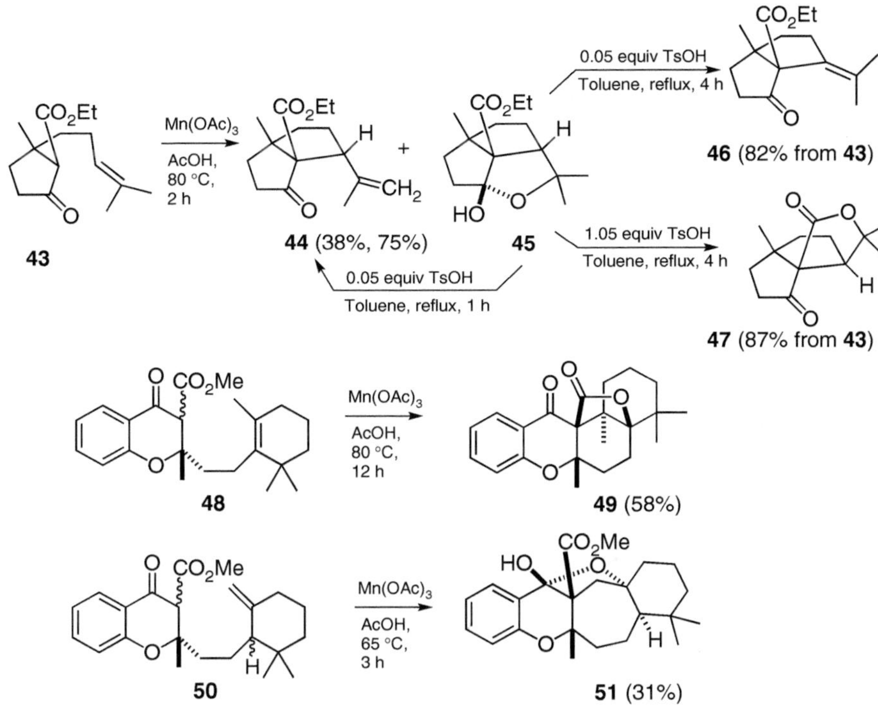

Scheme 9.7

formed by abstraction of a hydrogen from the solvent [27]. Burton found that use of $Cu(OTf)_2$ instead of $Cu(OAc)_2$ leads selectively to lactone **53** (92%) via the carbocation as discussed previously and shown in Scheme 9.3 [28]. Excellent stereocontrol was achieved with substituents on the tether. For instance, oxidative cyclization of malonate **56** affords lactone **57** (75–88%, >13 : 1 dr), which was used for the synthesis of 7,11-cyclobotryococca-5,12,26-triene [29], and oxidative cyclization of dienyl malonate **58** provides lactone **59** (89%, 6 : 1 dr), which was used for the synthesis of aphanamol I [30]. Oxidative cyclization of malonate **60a** proceeds analogously to give lactone **61** (84%, 4.1 : 1 dr) [31]. Oxidative cyclization with the tetrasubstituted alkene of malonate **60b** affords lactone **62** (83%, 8 : 1 dr) with control of two vicinal all-carbon stereocenters and the opposite stereochemistry at the benzylic center [31].

Burton found that the best conditions for the oxidative cyclization of hydroxyalkenylmalonates **63a–63d** depend on the length of the tether (Scheme 9.9) [15]. Tetrahydrofurans **64a** and **64b** are formed in highest yield without a copper additive. Tetrahydropyrans **64c** and **64d** are formed most effectively with $Cu(OTf)_2$ and $Cu(SbF_6)_2$, respectively [15]. Several tricyclic lactones **66** (71–86%) are formed by oxidative cyclization of hydroxyalkyl cycloalkenyl malonates **65** with 2 equiv of $Mn(OAc)_3$ and 1 equiv of $Cu(OTf)_2$ or $Cu(BF_4)_2$ in CH_3CN at reflux [32]. Oxidative cyclization of allenyl malonate esters gives a mixture of allylic radicals, which dimerize or are oxidized, and alkenyl radicals, which are reduced to alkenes or

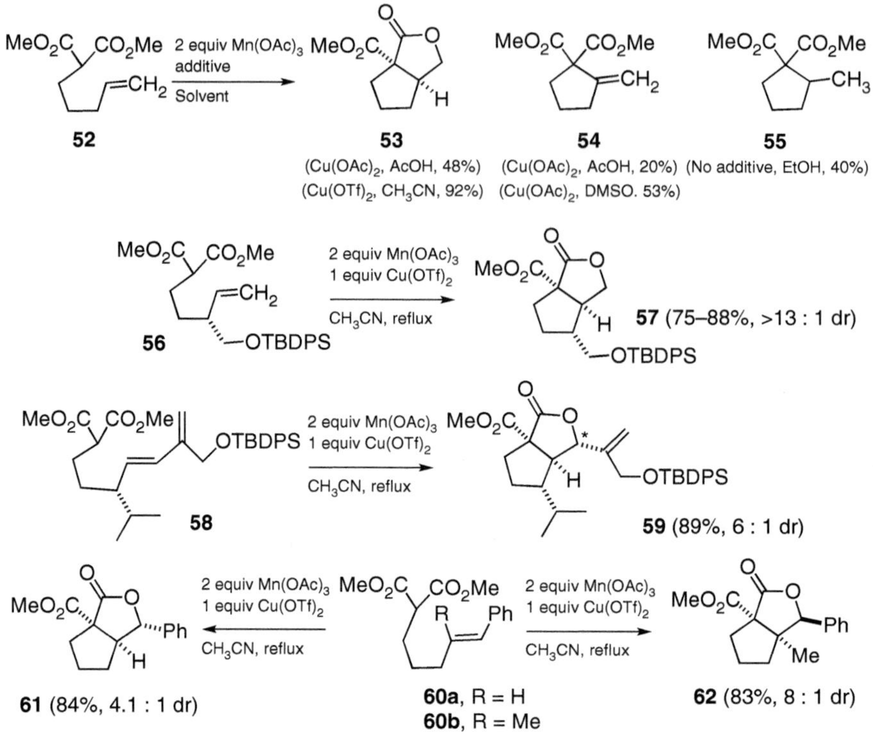

Scheme 9.8

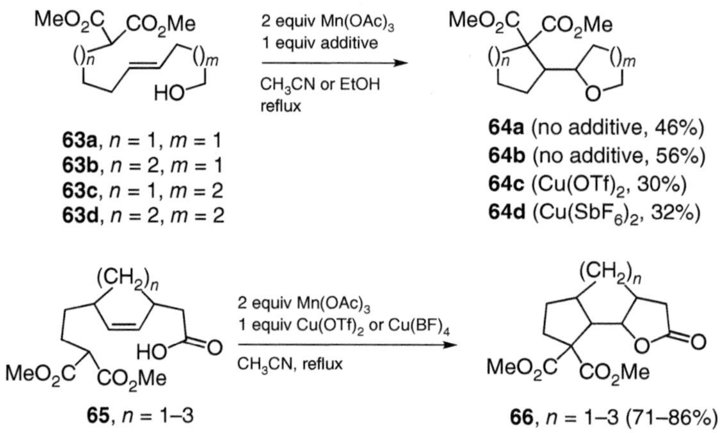

Scheme 9.9

oxidized to enol acetates [33]. The ratio of products depends on tether length and allene substitution.

9.3.3 Formation of Lactones

The oxidative cyclization of allylic and homoallylic β-keto esters and malonic esters leads to γ- and δ-lactones, respectively [1]. DFT calculations and mechanistic studies have been reported on the oxidative cyclizations of substituted allyl α-methyl β-keto esters [34]. Oxidative cyclization of allyl β-keto esters gives cyclopropanes as in the synthesis of *endo,exo*-furofuranone intermediate **69** (67–74%) from allylic β-keto ester **67** via primary radical **68** reported by Brown (Scheme 9.10) [35]. Burton reported that tetrahydrofuran lactones **71a** and **71b** and bislactones **71c** and **71d** can be prepared by oxidative cyclization of α-alkenyloxy malonates **70a** and **70b** and α-alkenoyloxy malonates **70c** and **70d**, respectively, using $Cu(OTf)_2$ as an additive in CH_3CN [36]. Unfortunately, the analogous oxidation of α-alkenoyloxy malonate **72** with R = alkyl gives only 24% of the bislactone **75** [37]. The major product (40%) is the unsaturated monocyclic lactone formed by oxidative elimination of the monocyclic radical. Fortunately, oxidation of **72** with $Mn(OAc)_3$ and KI in CH_3CN gives iodo malonate **73**, which undergoes an atom transfer cyclization to give iodoalkyl lactone **74**, which then undergoes thermal lactonization to give bislactone **75**, which was used for the synthesis of several members of the avenaciolide family [37].

Deprés and Greene found that oxidative cyclization of β-keto ester **76** with $Mn(OAc)_3$ in EtOH at 20 °C affords methylene lactone **78** (68%) (Scheme 9.11) [38].

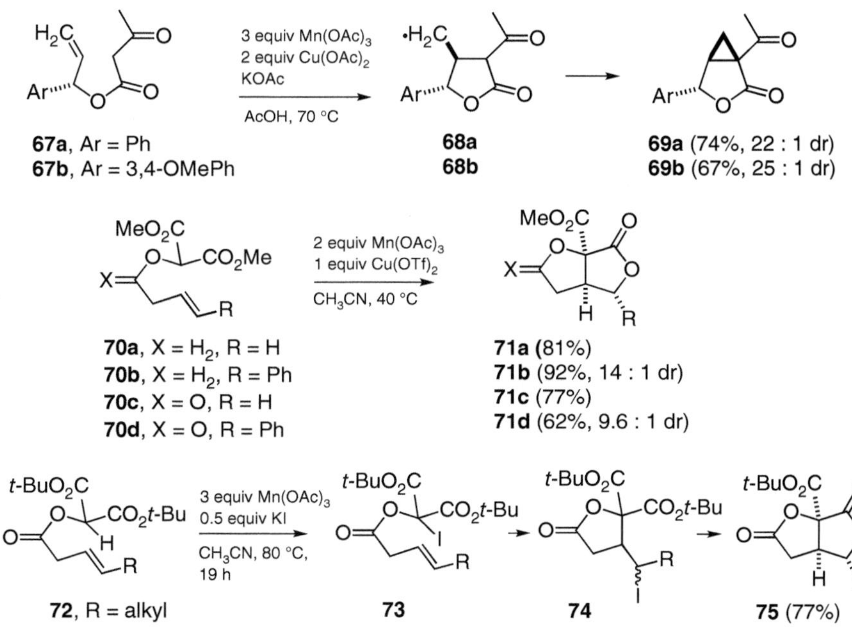

Scheme 9.10

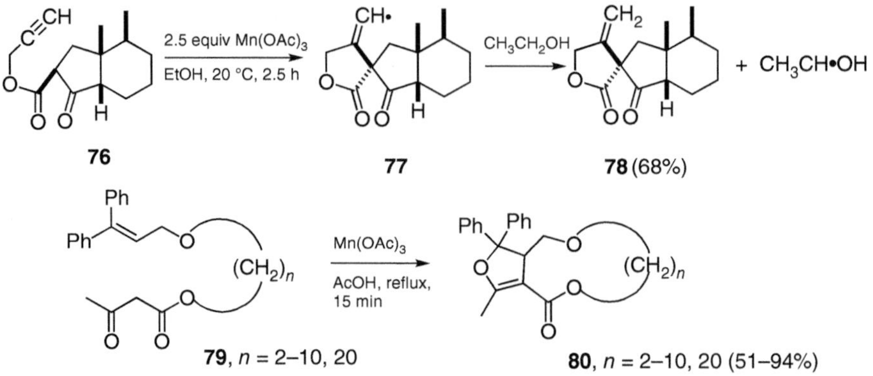

Scheme 9.11

The initial cyclization gives alkenyl radical **77**, which abstracts an α-hydrogen from the solvent ethanol to give **78** and $CH_3CH{\cdot}OH$, which is oxidized to acetaldehyde by $Mn(OAc)_3$. Lactone **78** was elaborated to 9-acetoxyfukinanolide [38] and later to several bakkenolides by Scheidt [39]. Nishino reported the oxidative cyclizations of diphenylalkenyl acetoacetates **79**, $n = 2$–10, 20 with $Mn(OAc)_3$ in AcOH at reflux to give lactones **80** (51–94%) with 8- to 26-membered rings [40a]. Macrolides with a variety of other tethers were also prepared [40].

9.3.4 Formation of Lactams

The oxidative cyclization of allylic and homoallylic β-keto amides and malonamides leads to γ- and δ-lactams, respectively [1, 41]. Cossy reported numerous examples, such as the oxidative cyclization of β-keto amide **81** with $Mn(OAc)_3$ and $Cu(OAc)_2$ in EtOH at reflux that affords unsaturated lactam **82** (58%) resulting from Cu(II) oxidation of the primary radical (Scheme 9.12) [41b]. In the absence of $Cu(OAc)_2$, the radical abstracts a hydrogen from EtOH to give saturated lactams **83** (56%) as a 7 : 1 mixture of diastereomers. Oxidative cyclization of the corresponding enamine **84** affords the same saturated lactams **83** (32%), but as a 1 : 19 mixture of diastereomers. Chuang found that oxidative cyclization of malonamide **85a** (C=C) under air provides quinolinones **86** (61–89%) [42]. The 6-*exo* cyclization gives a benzylic radical that reacts with oxygen to give a hydroperoxide that fragments to give **86** and ArCHO. Oxidative cyclization of malonamide **85b** (C≡C) yields ketone **87** (51–92%) [43]. The 6-*exo*-cyclization gives an alkenyl radical that reacts with oxygen to give a hydroperoxide that loses water to form the ketone.

Lactams can also be formed by oxidative cyclizations of enamides such as **88**, which gives β-lactam **89** (73%) by 4-*exo* cyclization (Scheme 9.13) [44]. Parsons found that oxidative cyclization of enamides such as **90** with 4 equiv of $Mn(OAc)_3$ provides bicyclic lactam **91** (52%) by a 5-*endo* cyclization and an additional oxidation [45]. Ishibashi reported the analogous cyclizations of methylthioacetamides, leading to the erythrinane skeleton [46]. Oxidative cyclization of enamide **92** with 10 equiv of

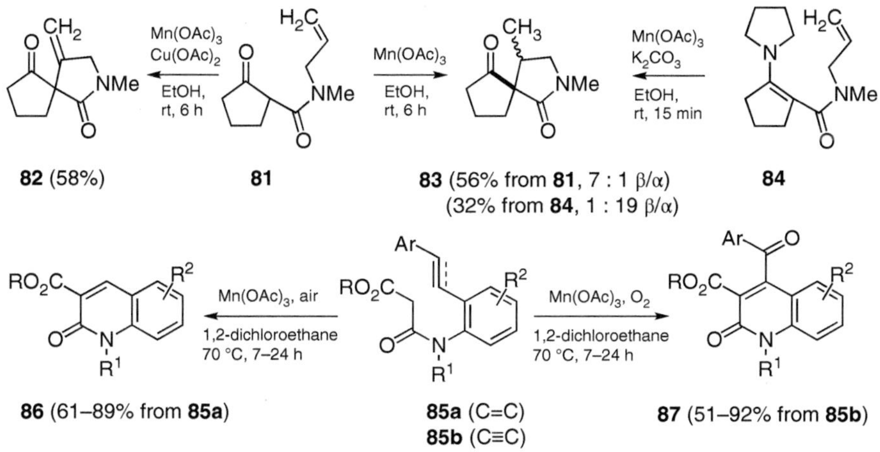

Scheme 9.12

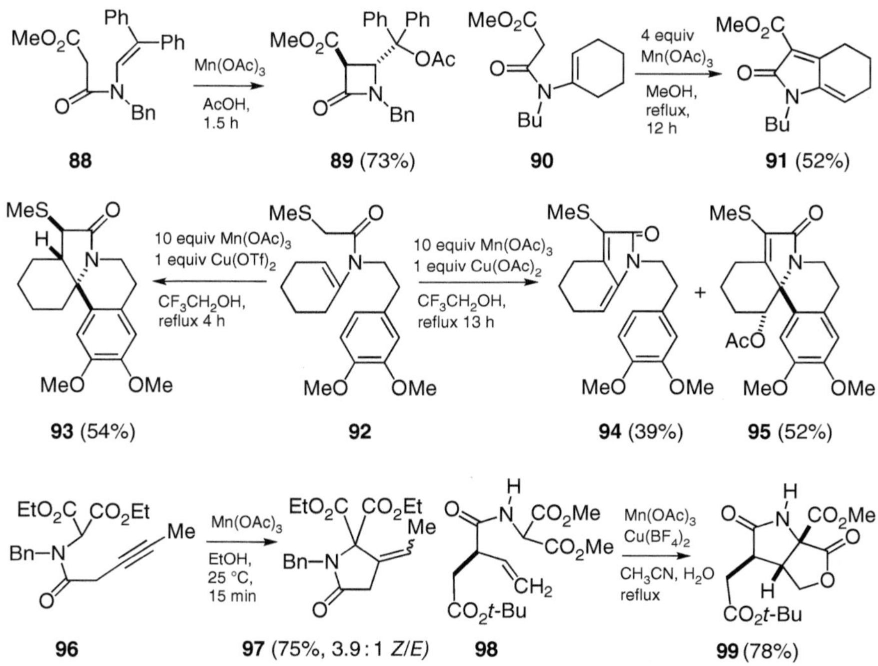

Scheme 9.13

$Mn(OAc)_3$ and 1 equiv of $Cu(OAc)_2$ affords dienyl lactam **94** (39%) and tetracyclic lactam **95** (52%) [46]. Resubjection of **94** to the reaction conditions provides additional **95**. Surprisingly, changing the copper salt to $Cu(OTf)_2$ results in the formation of tetracyclic lactam **93** (54%), presumably because the reaction medium is more acidic so the initially formed carbocation cyclizes to give **93** instead of losing a proton to give the alkene precursor to diene **94** [46]. This reaction was used for a formal synthesis of the natural product 3-demethoxyerythratidinone. Burton found that oxidative cyclization of amido malonates such as **96** with $Mn(OAc)_3$ in EtOH affords lactam **97** (75%) with the (*Z*)-isomer predominating [47]. Burton used the oxidative cyclization of amido malonate **98** with $Mn(OAc)_3$ and $Cu(BF_4)_2$ in wet acetonitrile at reflux to prepare lactam lactone **99** (78%), which was used as intermediate for the synthesis of salinosporamide A [48].

9.3.5 Cyclizations to Aromatic Rings

Electrophilic β-dicarbonyl radicals formed oxidatively will cyclize to aromatic rings, especially electron-rich ones [1]. Kerr reported that indoline malonates such as **100** are first oxidized by $Mn(OAc)_3$ to indole **101** and then to the malonyl radical, which cyclizes to give tricycle **102** (67%) (Scheme 9.14) [49]. The direct oxidative

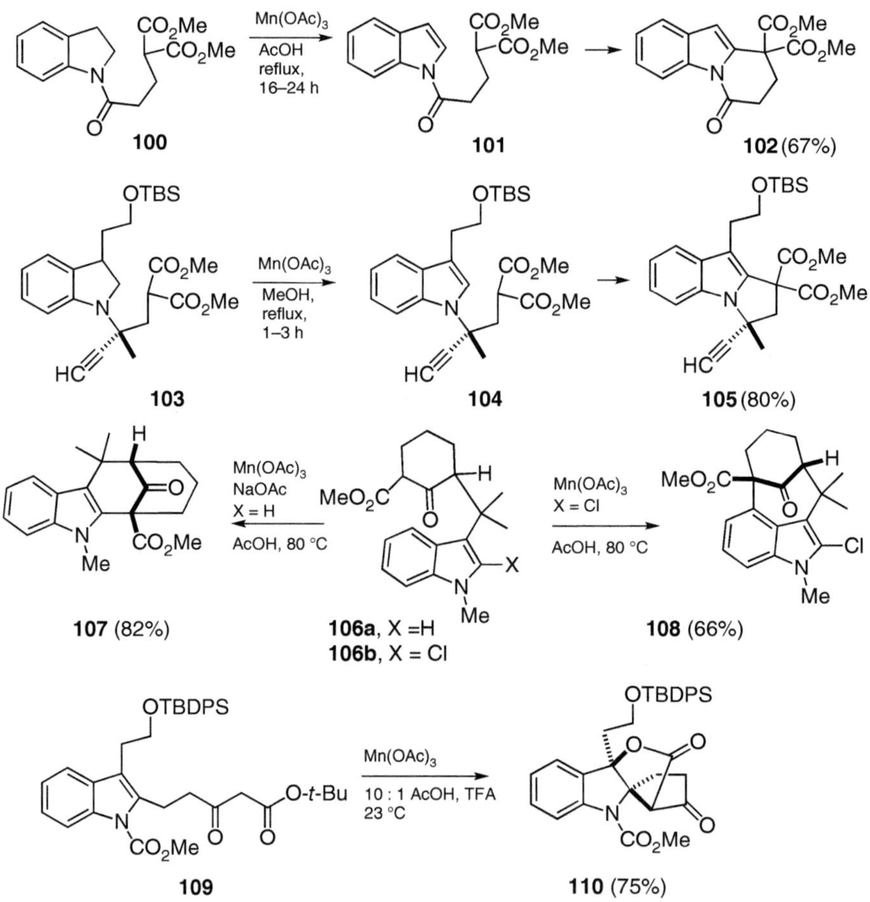

Scheme 9.14

cyclization of indole malonate **101** gives **102** (74%) in slightly higher yield. This cyclization proceeds in high yield with a variety of pyrroles and indoles attached to the nitrogen by either an alkyl group or an acyl group. Kerr used the reaction in an approach to mersicarpine [50] and to the flinderole skeleton [51]. Oxidative cyclization of indoline **103** first gives indole **104**; subsequent malonate oxidation and cyclization provides tricycle **105** (80%) [51]. Ding showed that the oxidative cyclizations analogous to that of **100–102** with a 3-substituted indoline can be carried out using $NaIO_4$ as the stoichiometric oxidant and catalytic $Mn(OAc)_2$ or $Mn(OAc)_3$ [52]. Rawal found that β-keto esters attached to the 3-position of the indoles such as **106a** oxidatively cyclize at the 2-position to give tetracycle **107** (82%) [53]. However, if a chlorine is present to block cyclization to the 2-position as in indole **106b**, oxidative cyclization occurs at the 4-position to give tetracycle **108** (66%), thus providing efficient access to the bicyclo[4.3.1]decane core of the welwitindolinones [53]. Oisaki and Kanai reported numerous examples in which malonates attached to either the 1- or 3-positions of indoles oxidatively cyclize to the 2-position using 10 mol% $Mn(acac)_3$ in xylene at 130 °C under air with a $CaCl_2$ drying tube [54]. Ma reported numerous examples of the oxidative cyclization of *N*-carbomethoxy 3-substituted indole β-keto esters to give spirocyclic lactones. For instance, oxidative cyclization of indole **109** affords indoline lactone **110** (75%), which was elaborated to lapidilectine B [55].

Myers reported a general route to 2-hydroxy-1-naphthoic esters by oxidative cyclization of (*Z*)-unsaturated β-keto esters such as **111**, which provides naphthoic ester **112** (93%) (Scheme 9.15) [56]. Nishino reported that oxidative cyclization of unsubstituted anilide malonate **113a** in AcOH affords lactam **114** (97%) [57]. On the other hand, the electrophilic radical cyclizes *ipso* to the electron-rich aromatic ring of anilide malonate **113b** to give cyclohexadiene **115** (85%). Oxidative cyclization of a variety of *N*-aryl-*N*-methyl-α-methylacetoacetamides provides a general route to 3-acetyl-1,3-dimethylindolin-2-ones [57].

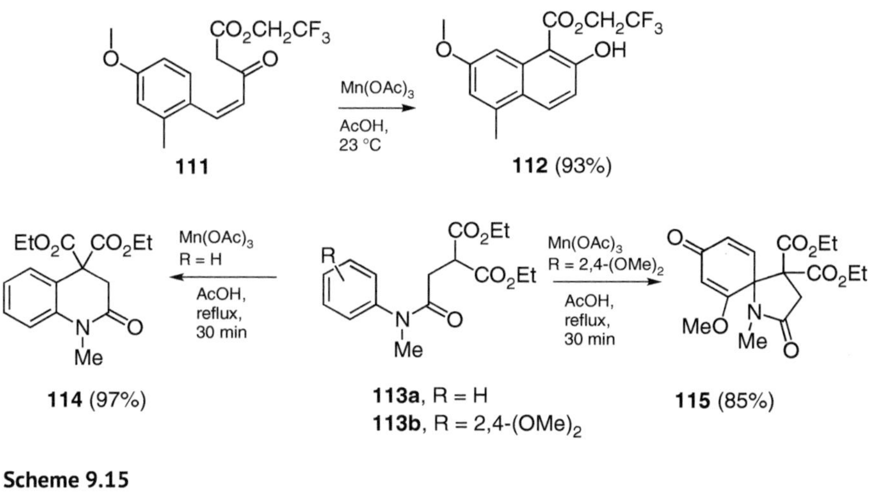

Scheme 9.15

9.4 Tandem, Triple, and Tetra Cyclizations

9.4.1 Addition to a Double Bond and Then an Aromatic Ring

In 1985 we found that oxidative cyclization of β-keto ester **116a** with 2 equiv of $Mn(OAc)_3$ affords tricycle **117a** (50%), which was converted to *O*-methylpodocarpic acid (Scheme 9.16) [58]. Enolization and oxidation gives an acyclic radical that preferentially cyclizes through a chair transition state with an axial ester group. A similar oxidative cyclization of β-keto ester **118** affords tricycle **119** (70%) with an equatorial ester group since this position is equilibratable. This reaction has been widely used for the synthesis of tricyclic terpenes [1]. Yamashita recently converted β-keto ester **116b** to tricycle **117b** (71%), which was used for the synthesis of the zoanthenol ABC ring system [59], and we found that oxidative cyclization of β-keto ester **120** with $Mn(OAc)_3$ and KOAc in MeOH affords methoxydihydrofuran **121** (59%), which loses MeOH on washing with HCl to generate a furan, which was elaborated to 15-acetoxypallescensin-A [60]. Attempts to prepare analogues with a cis-ring fusion by oxidative cyclization of analogues of β-keto esters **116** or **118** with a *Z*-double bond gave mixtures of cis- and trans-ring-fused isomers [61–62].

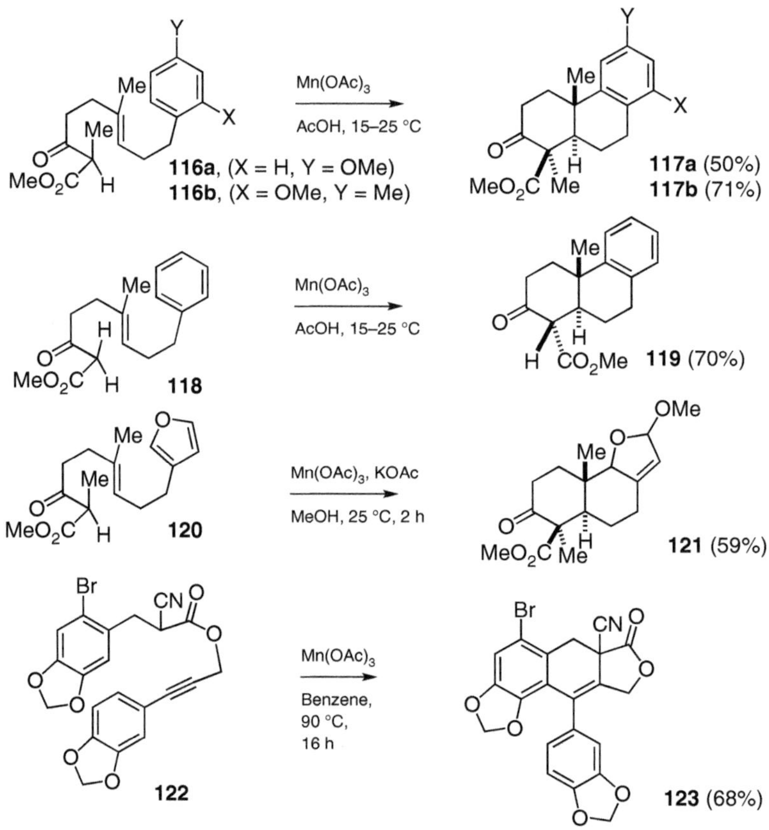

Scheme 9.16

Shia reported a general route to the cyclopenta[*b*]naphthalene ring system by oxidative cyclization of aryl 1-cyanoalk-5-ynyl ketones [63a]. For instance, oxidative cyclization of cyano ester **122** gives tricycle **123** (68%), which was elaborated to helioxanthin [63b].

9.4.2 Addition to Two Double Bonds

Tandem cyclizations to two double bonds followed by oxidation of the bicyclic radical with Cu(II) or Mn(III) provide an effective route to form bridged rings systems such as **125** and **131** or fused ring systems such as **127**, **129**, and **133** (Scheme 9.17) [1]. We found that oxidative cyclization of β-keto ester **124** with $Mn(OAc)_3$ and $Cu(OAc)_2$ affords bicyclo[3.2.1]octane **125** (86%) and β-keto ester **126** provides cis-fused hydrindane **127** (46%) [64]. Similarly, oxidative cyclization of β-keto ester **128a** with $Mn(OAc)_3$ without $Cu(OAc)_2$ provides decalin **129a** (46%) with a small amount of the endocyclic alkene isomer [64a]. Li prepared **129a** (48%) as an intermediate in the synthesis of the oridamycins [65]. The indole can also be attached before the oxidative cyclization in that β-keto ester **128d** undergoes oxidative cyclization analogously to form decalin **129d** (52%). Trotta oxidatively cyclized β-keto ester **128b** to give decalin **129b** (50%), which was then used as an intermediate in the synthesis of the oridamycins [66]. Zoretic oxidatively cyclized

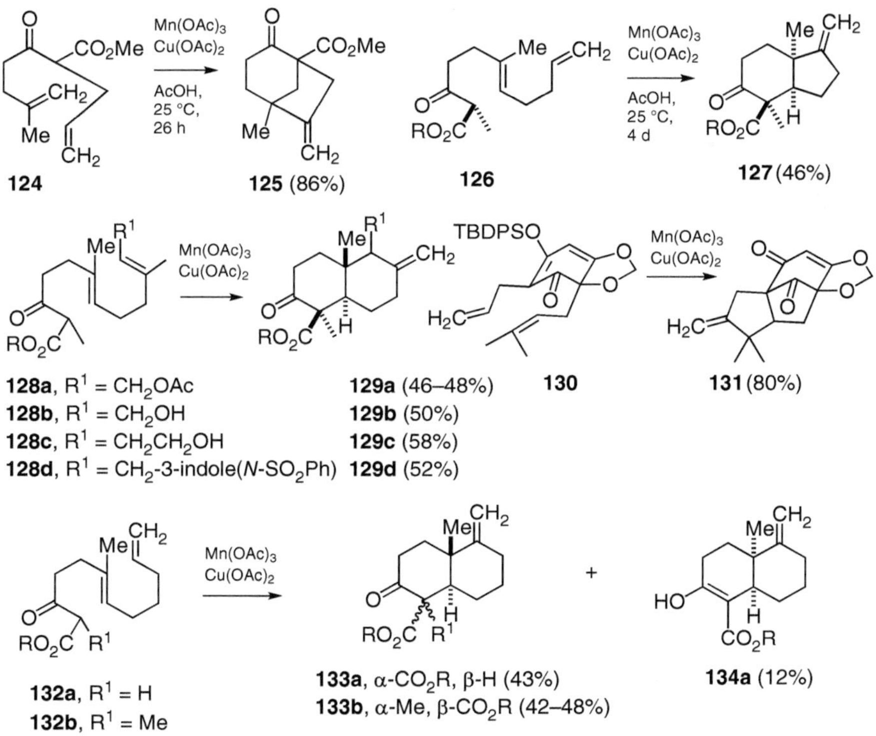

Scheme 9.17

β-keto ester **128c** to decalin **129c** (58%) [67]. Complex mixtures were obtained without $Cu(OAc)_2$. Barrero used decalin **129c** as an intermediate in the synthesis of the proposed structure of wentilactone B [68]. Danishefsky used an oxidative cyclization to convert cyclohexadienone **130** to tricycle **131** (80%), which was used for the synthesis of 3-oxytricycloillicinone [69]. The silyl enol of **130** is cleaved to give a β-diketone that is oxidized to the radical by Mn(III). Zoretic found that oxidative cyclization of β-keto ester **132a** gives a mixture of *trans*-decalin **133a** (43%) and *cis*-decalin **134a** (12%) [70]. Donaldson used **133a** as an intermediate in a route to the proposed structure of heteroscyphic acid [71]. Zoretic found that β-keto ester **132b** with an α-methyl group undergoes cyclization to give exclusively *trans*-decalin **133b** (48%) [72]. Baran developed conditions to carry out this reaction electrochemically with catalytic $Mn(OAc)_3$, giving **133b** (42%), which was used as an intermediate in the synthesis of the subglutinols and higginsianin A [9].

Chen extensively explored the oxidative cyclization of highly functionalized β-keto esters as a route to oroidin dimers [73]. Oxidative cyclization of β-keto ester **135** with $Mn(OAc)_3$ in AcOH at 60 °C affords tricycle **136** (63%) (Scheme 9.18) [70]. A similar oxidative cyclization of β-keto ester **137** provides tricycles **138** (18–25%) and **139** (9%) that were both elaborated to ageliferin [73]. Lee used the tandem oxidative cyclization of β-keto ester **140** to give guaianolide **141** (65%) as a key step in the synthesis of cladantholide [74]. The initial 5-*exo* cyclization forms a γ-lactone radical that undergoes a 7-*endo* cyclization to give a tertiary radical that is oxidized to **141** by Cu(II).

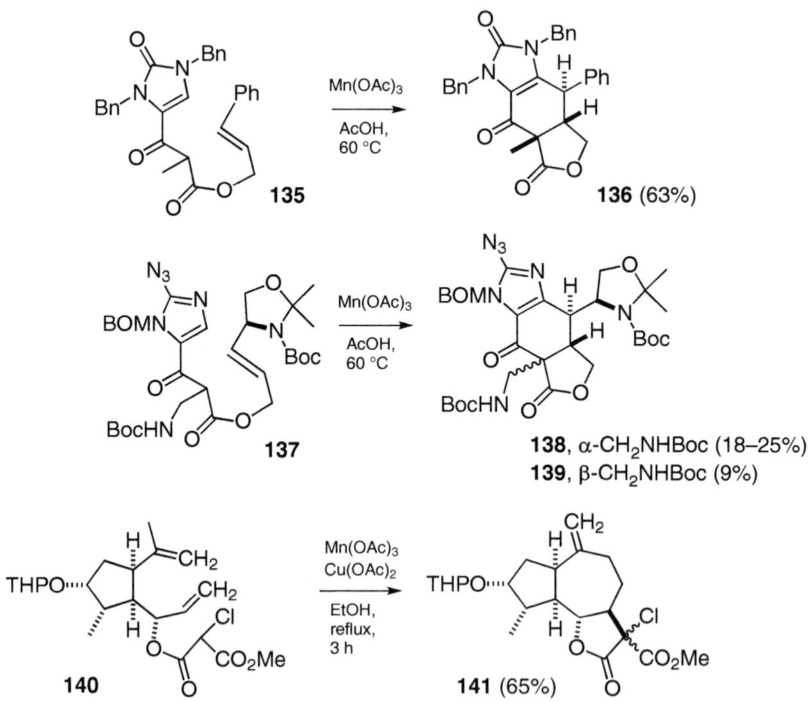

Scheme 9.18

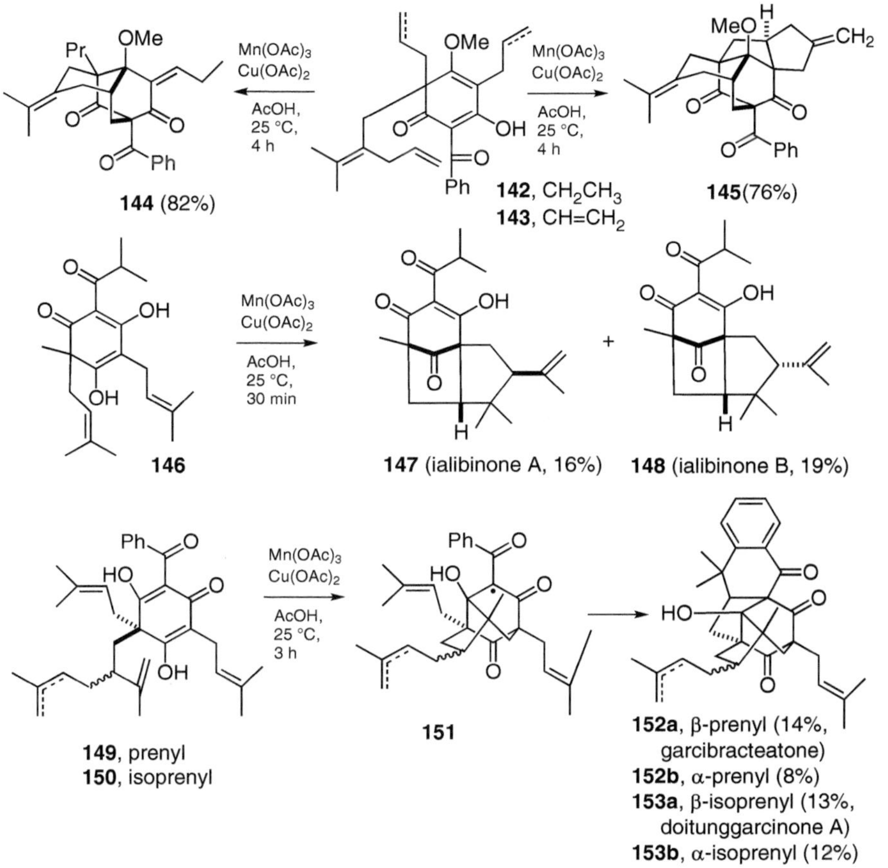

Scheme 9.19

Oxidative mono, tandem, triple, and tetra cyclizations have been extensively used for the synthesis of polycyclic polyprenylated acylphloroglucinol natural products. Porco found that oxidative cyclization of β-diketone **142** with two propyl side chains gives tricycle **144** (82%) by a tandem cyclization (Scheme 9.19) [75]. The initially formed radical undergoes a 7-*endo* cyclization to form a radical that then undergoes a 6-*endo* cyclization to give a radical that is oxidized by Cu(II) to introduce the propylidene substituent. A similar reaction of β-diketone **143** with two allyl side chains gives pentacycle **145** by a tetracyclization with the first two cyclizations the same as in the formation of **144** and the last two to the two allyl groups being 5-*exo* [75]. Simpkins reported the tandem oxidative cyclization of trione **146** to give ialibinones A (**147**, 16%) and B (**148**, 19%) [76]. George prepared garcibracteatone (**152a**) by an oxidative tetra cyclization [77]. Oxidation of trione **149** as a mixture of diastereomers gives radical **151** by a tandem cyclization, which undergoes a 5-*exo* cyclization to give a tertiary radical that cyclizes onto the aromatic ring to give garcibracteatone (**152a**, 19%) and the diastereomer **152b** (16%). A similar oxidative cyclization of β-diketone **150** as a mixture of isomers with an isoprenyl side chain provides doitunggarcinone A (**153a**, 13%) and the diastereomer **153b** (12%) [77].

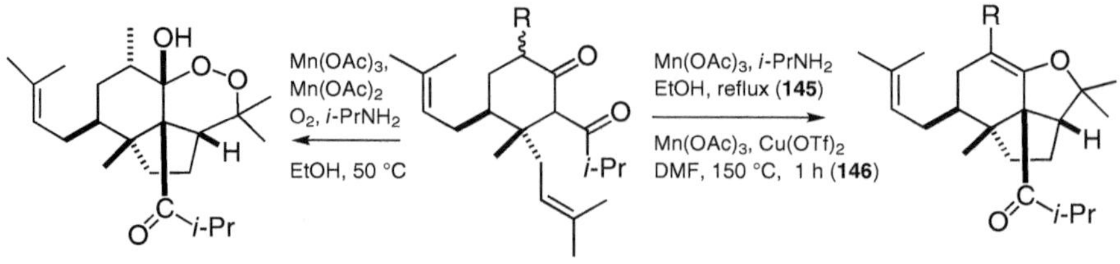

156 (15%)

154, R = α-Me
155, R = β-prenyl

157, R = Me (52%, yezo'otogirin C)
158, R = prenyl (29%, yezo'otogirin A)

Scheme 9.20

Lee reported two approaches to prepare yezo'otogirin C (**157**) using oxidative cyclization (Scheme 9.20) [78]. Oxidative cyclization of β-diketone **154** with $Mn(OAc)_3$ and $Mn(OAc)_2$ under an oxygen atmosphere (Nishino conditions) gives the 5-*exo* cyclization product that reacts with oxygen to give peroxyhemiketal **156** (15%) after reduction of the peroxy radical to a hydroperoxide and hemiketal formation. The reaction proceeds in 55% yield with the methyl ester instead of the isopropyl ketone. Reduction of peroxide **156** with thiourea in methanol at reflux affords **157** (85%). Alternatively, oxidatively cyclization of β-diketone **154** with $Mn(OAc)_3$ and isopropylamine in degassed ethanol at reflux gives **157** (52%) in a single step [78]. George reported an analogous synthesis of yezo'otogirin A (**158**, 29%) by oxidative cyclization of β-diketone **155** with $Mn(OAc)_3$ and $Cu(OTf)_2$ in DMF at 150 °C [79].

Gao synthesized **159** (equisetin) and converted it to **161** (fusarisetin A) by oxidative cyclization of **159** with $Mn(OAc)_3$ under an oxygen atmosphere (Nishino conditions) (Scheme 9.21) [80]. The initially formed radical reacts with oxygen to give a mixture of peroxyhemiketals **160** (60%) that can be reduced with Zn or $P(OMe)_3$ to give **161** (fusarisetin A) and the C-5 diastereomer. Theodorakis carried out a similar cyclization of **159** with $Mn(OAc)_3$ and TEMPO, which trapped the radical at C-5 [81]. Reduction gave exclusively the diastereomer of **161**. The oxidation of **159** with ceric ammonium nitrate under an oxygen atmosphere also provides **160** [81].

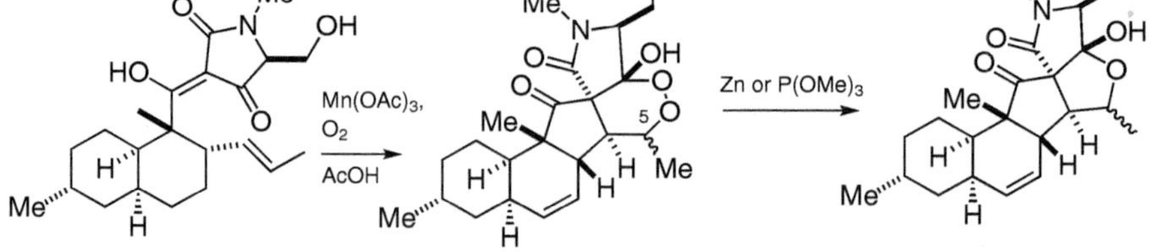

159 (equisetin)

160 (60%, 1.1 : 1 dr)

161 (fusarisetin A, β-Me)

Scheme 9.21

9.4.3 Triple and Tetra Cyclizations

9.4.3.1 Triple Cyclizations

A few examples of triple and tetra cyclizations were reported in the previous review in 1996 [1]. Zoretic has reported many additional examples. Oxidative cyclization of β-keto ester **162** provides a mixture of the all-trans-fused tricycle **164a** (24%), which was elaborated to isospongiadiol, and tricycle **164b** (11%) (Scheme 9.22) [82]. The products are formed by three successive 6-*endo* cyclizations followed by Hofmann selective elimination with Cu(II). A similar oxidative cyclization of β-keto ester **163** affords tricycles **165a** (47%) and **165b** (11%), establishing that nitriles can be used to provide functionality for further elaboration [83]. Tricyclizations of β-keto esters **166** and **167** to give tricycles **168** (43%) and **169** (60%), respectively, proceed by two 6-*endo* and then one 6-*exo* cyclizations to give a primary radical that is oxidized by Cu(II) to the alkene [83, 84].

González converted β-keto ester **170** to tetracycle **171** (40%) by termination of a triple cyclization by addition to a pyridine ring (Scheme 9.23) [85]. Shoji explored the oxidative cyclization of α-chloro β-keto ester **172** that gives diastereomeric tricycles **173** (34%) and **174** (29%) analogously to the formation of **169** [86]. Yamashita carried out the oxidative cyclization of dienynyl β-keto ester **175** with $Mn(OAc)_3$ in EtOH to give tricycle **176** (64%), which was elaborated to limonin [87]. The third cyclization, which forms a five-membered ring, gives a mixture of stereoisomers in which the cis-isomer predominates. The alkenyl radical formed by addition to the alkyne abstracts a hydrogen atom from EtOH to give the alkene.

9.4.3.2 Tetracyclizations

Zoretic has also reported examples of tetracyclizations. Oxidative cyclization of β-keto ester **177** affords fused tetracycle **178** (38–45%) as the only product (Scheme 9.24) [88]. Oxidative cyclization of β-keto ester **179** provides a tertiary

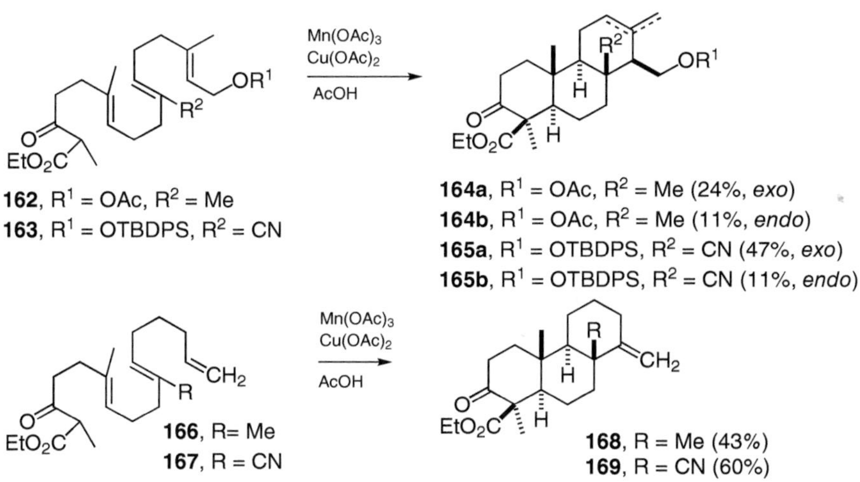

162, R^1 = OAc, R^2 = Me
163, R^1 = OTBDPS, R^2 = CN

164a, R^1 = OAc, R^2 = Me (24%, *exo*)
164b, R^1 = OAc, R^2 = Me (11%, *endo*)
165a, R^1 = OTBDPS, R^2 = CN (47%, *exo*)
165b, R^1 = OTBDPS, R^2 = CN (11%, *endo*)

166, R= Me
167, R = CN

168, R = Me (43%)
169, R = CN (60%)

Scheme 9.22

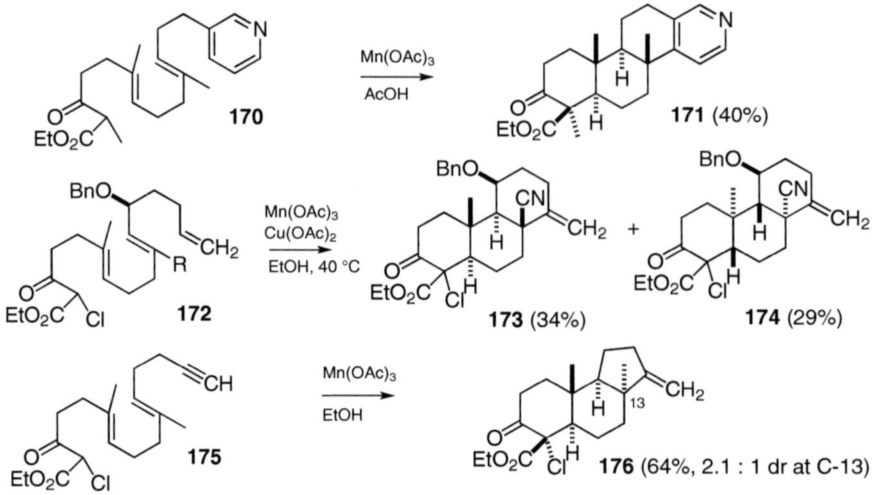

Scheme 9.23

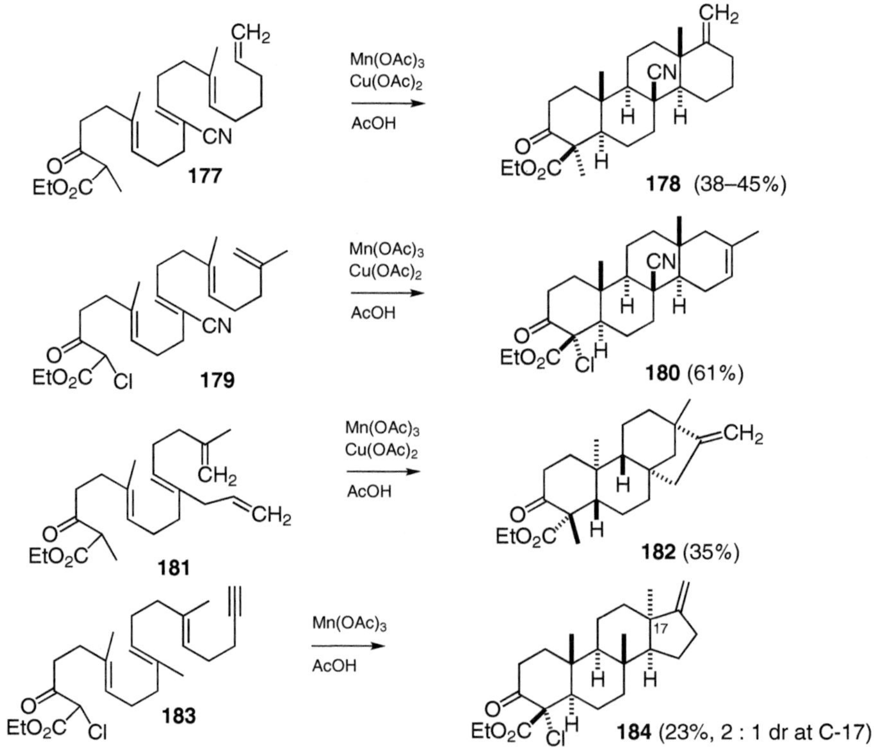

Scheme 9.24

radical that is oxidized to a mixture of tetracyclic alkenes and the tertiary acetate (71%) in which isomer **180** predominates [89]. We found that oxidative cyclization of β-keto ester **181** gives bridged tetracycle **182** (35%), which was elaborated to isosteviol [90]. Yamashita reported the oxidative cyclization of β-keto ester **183** to give tetracycle **184** (23%) as a mixture of isomers at C-17 in which the *cis*-C–D ring-fused isomer predominates as in the formation of **176** [91]. It is not clear whether the yield of **184** is lower than that of **176** because of the formation of four rather than three rings or because AcOH, which is less effective at reducing alkenyl radicals, was used as solvent.

9.5 Asymmetric Induction

We found that high levels of asymmetric induction were achieved with β-keto sulfoxides and 8-phenylmenthyl β-keto esters [1]. For instance, oxidative cyclization of (−)-8-phenylmenthyl ester **185** affords tricycle **186** (86%, 82% de), which was elaborated to *O*-methylpodocarpic acid (Scheme 9.25) [92]. Yang obtained similar results with α-chloro-(−)-8-phenylmenthyl ester **187a**, which gives tricycle **188a** (74%, 67% de) [93, 94]. On the other hand, the α-unsubstituted β-keto ester **187b** with $Mn(OAc)_3$ in AcOH provides the enantiomeric tricycle **188b** (60%, 50% de). Higher yields and much better selectivity for **188b** (77%, 95% de) were obtained by adding $Yb(OTf)_3$ and using trifluoroethanol as the solvent [94]. Triptolide, triptonide, and triptophenolide were prepared from **188b**. The stereoselectivity of the cyclizations is consistent with the following models. α-Unsubstituted β-keto esters in the presence of $Yb(OT)_3$ cyclize through radical **191** with the alkene adding to the top face [94]. On the other hand, β-keto esters with small α-substituents cyclize through radical **190**, whereas β-keto esters with large α-substituents cyclize through radical **189** [92, 94].

Although both isomers of 8-phenylmenthol are available, the starting material for the (+)-isomer, (*S*)-pulegone, is much more expensive than (*R*)-pulegone. Yang tried to prepare chiral alcohols that would give both diastereomers selectively from (*R*)-pulegone and to improve the selectivity by varying the aryl group [95]. The best results were obtained with the 8-(2-naphthyl)menthyl ester **192a** that gives tricycle **193a** (67%, >98% de), an improvement on the 95% de obtained with the 8-phenylmenthyl ester (Scheme 9.26). 8-Arylneomenthyl esters give the opposite enantiomer. The best results were obtained with 8-(3,5-diisopropoxyphenyl)neomenthyl ester **192b** that provides tricycle **193b** (55%, 92%), which was converted to (+)-triptocallol [91]. The same approach was used to prepare both enantiomers of wilforonide [96]. Oxidative cyclization of β-keto ester **194** with $Mn(OAc)_3$ affords wilforonide intermediate **195** (24%, 98% de), alkene position isomers **196** (27%), and some cis-fused ring isomers [96]. The yield of these products is low if $Yb(OTf)_3$ is added.

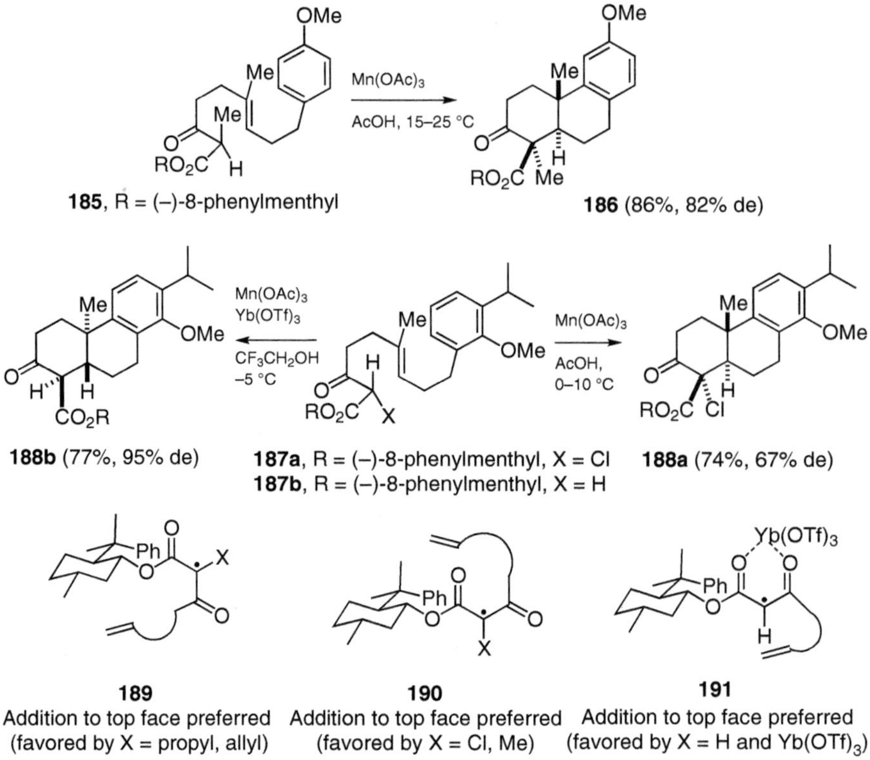

185, R = (−)-8-phenylmenthyl

186 (86%, 82% de)

188b (77%, 95% de)

187a, R = (−)-8-phenylmenthyl, X = Cl
187b, R = (−)-8-phenylmenthyl, X = H

188a (74%, 67% de)

189
Addition to top face preferred
(favored by X = propyl, allyl)

190
Addition to top face preferred
(favored by X = Cl, Me)

191
Addition to top face preferred
(favored by X = H and $Yb(OTf)_3$)

Scheme 9.25

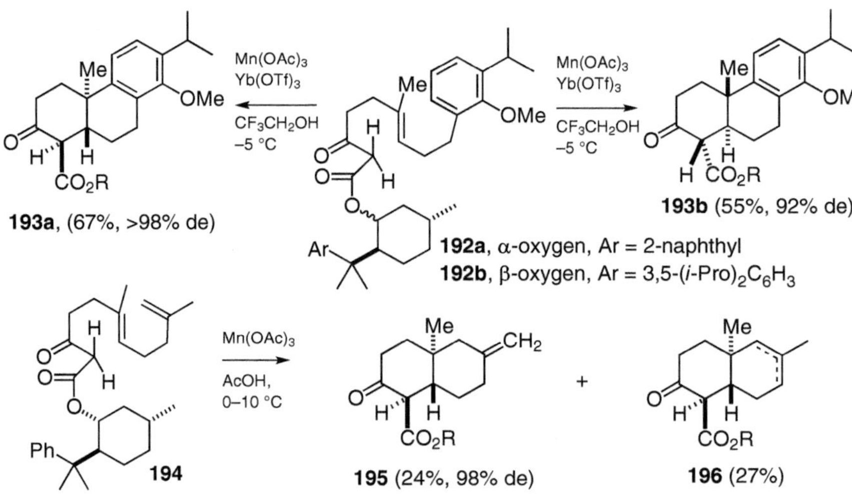

193a, (67%, >98% de)

193b (55%, 92% de)

192a, α-oxygen, Ar = 2-naphthyl
192b, β-oxygen, Ar = 3,5-(*i*-Pro)$_2$C$_6$H$_3$

194

195 (24%, 98% de)

196 (27%)

Scheme 9.26

9.6 Oxidations of Ketones

The oxidative cyclizations discussed earlier have all been initiated by oxidation of a β-dicarbonyl compound that occurs at or below room temperature. Oxidative cyclization of unsaturated ketones at 80–100 °C with $Mn(OAc)_3$ and $Cu(OAc)_2$ can also be a synthetically useful procedure with broad applicability if certain conditions are met [97]. These reactions proceed in synthetically useful yield if the starting ketone enolizes only to one side of the carbonyl group and the ketone product cannot enolize or enolizes much more slowly than the starting ketone. Cyclohexanone **197a** can only enolize to one side and the α-keto radical undergoes mainly 6-*endo* cyclization to give a radical that is oxidized to a mixture of bicycles **198a** (66%) and **199a** (7%) (Scheme 9.27) [97]. 5-*exo* cyclization and oxidation to give bicycle **200a** (4%) is a minor pathway. Kraus carried out the related oxidative cyclization of cyclohexanone **197b** to give bicycle **198b** [98]. In these cases, one side of the ketone is fully substituted and enolization of the product cannot occur because it would give a strained bridgehead double bond. Garcia Ruano found that keto sulfone **201a** reacts similarly, giving bicycle **202a** (95%) [99]. On the other hand, the radical formed from keto sulfide **201b** adds preferentially to the electron-rich aromatic ring to give tricycle **202b** (80%).

The analogous oxidative cyclizations of alkynyl ketones in 9–19 : 1 EtOH/AcOH at 80–90 °C with $Mn(OAc)_3$ affords alkenyl radicals that are reduced by the solvent ethanol to give alkenes [100]. For instance, trimethylsilylalkynyl ketone **203** gives alkenyl radical **204** that abstracts an α-proton from ethanol to give tricycle **205** (62%) as a mixture of isomers (Scheme 9.28) [100]. Protodesilylation and reduction of the ketone gives gymnomitral. Oxidative cyclization of cyclohexenones containing an unsaturated side chain with $Mn(OAc)_3$ and $Cu(OAc)_2$ proceeds efficiently in benzene as solvent. For instance, the radical formed from cyclohexenone **206** undergoes a 6-*exo* cyclization to give a radical that is oxidized to bicycle **207**

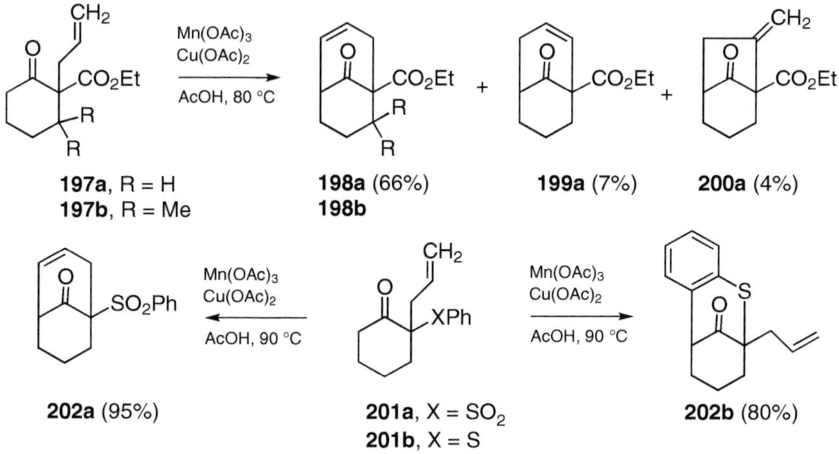

Scheme 9.27

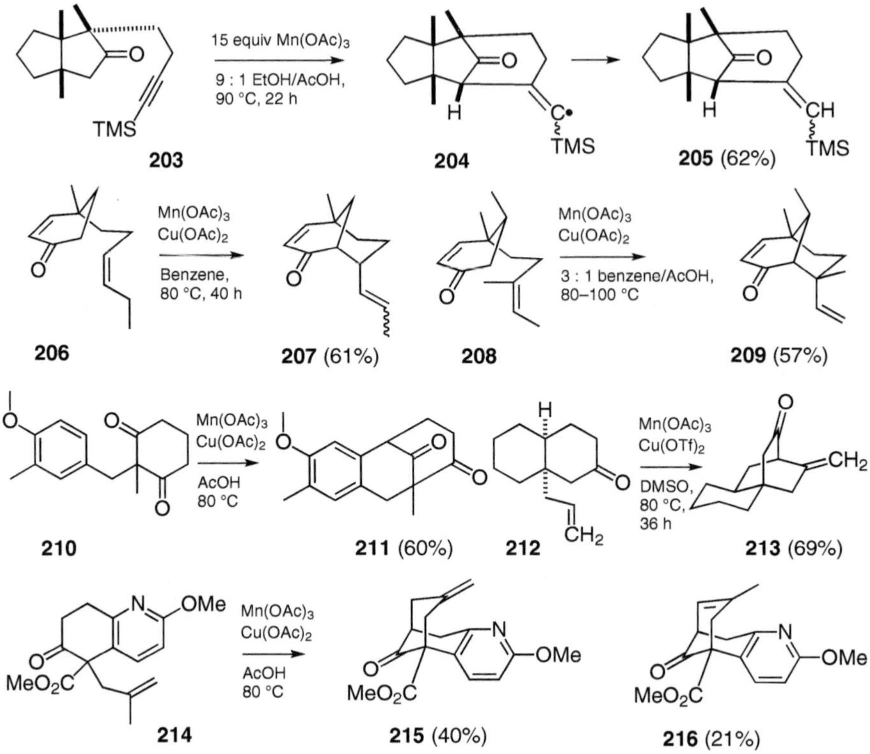

Scheme 9.28

(61%) [101]. Sorensen carried out a similar oxidation of cyclohexenone **208** in 3 : 1 benzene/AcOH to give bicycle **209** (57%) containing the bicyclo[3.3.1]nonane core of the proposed structure of edaxadiene [102]. The analogous oxidative cyclization of the (*E*)-isomer of **208** gives a 1.4 : 1 mixture of diastereomers. Venkateswaran reported the oxidative cyclization of cyclohexanedione **210** to give tricycle **211** (60%), which was converted to parvifolene [103]. The ketone of product **211** on the one-carbon bridge cannot enolize. The ketone on the 3-carbon bridge of **211** could enolize but must do so more slowly than either ketone of **210**. Vanderwal reported that oxidative cyclization of decalone **212** affords tricycle **213** (69%) [104]. The high yield of **213** indicates that enolization of **212** occurs regiospecifically and enolization of the product **213** is much slower than enolization of **212**. Lee reported the oxidative cyclization of several allylic 6-oxotetrahydroquinone carboxylate esters such as **214**, which affords a mixture of bridged tricycles **215** (40%) and **216** (21%) [105]. Triflic acid isomerizes **215** to **216**, which was used for the synthesis of huperzine A.

9.7 Summary and Conclusions

Over the past 35 years $Mn(OAc)_3$-based oxidative free radical cyclizations have been extensively developed into a broadly useful method for the synthesis of cycloalkanes, lactams, and lactones by oxidative radical cyclization that starts with simple starting materials such as unsaturated ketones and β-dicarbonyl compounds and leads to products that are highly functionalized and useful synthetic intermediates. Mono, tandem, triple, and tetra cyclizations can be carried out successfully with predictable regio- and stereochemistry. During the past 20 years, these reactions have been increasingly used in the synthesis of complex natural products. The successful synthetic applications that form the bulk of this review demonstrate the power of the reaction to construct rings with stereo- and regiocontrol in highly functionalized molecules.

9.8 Addendum

Yang, Cheng, and Lee recently carried out the oxidative cyclization of **217** with $Mn(OAc)_3$ and $Cu(OAc)_2$ in EtOH on a 20-g scale to provide picrotoxane intermediates **219a** (55%) and **219b** (22%) in combined 77% yield [106] (Scheme 9.29). Oxidative cyclization affords primary radical **218**, which cyclizes to the carbonyl oxygen to give a carbon radical that is oxidized to a cation that reacts with the solvent ethanol or water to give **219a** and **219b**, respectively. Lan, Huang, and Yang reported that the oxidative cyclization of norascyronone C (**220a**) to give

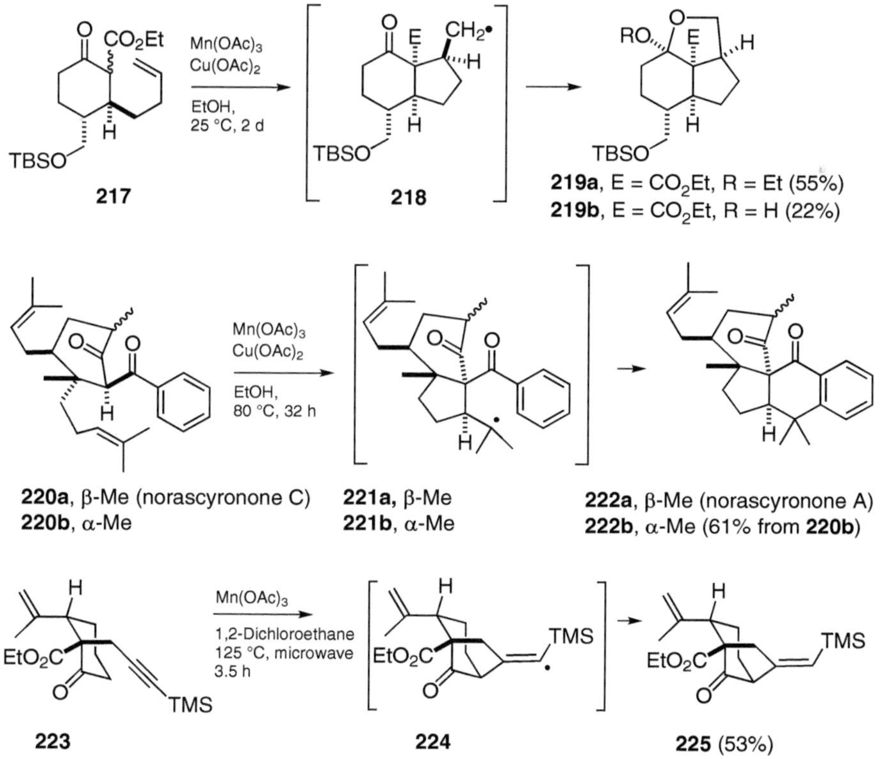

Scheme 9.29

norascyronone A (**222a**) was unsuccessful. Fortunately, oxidative cyclization of the methyl epimer **220b** with $Mn(OAc)_3$ and $Cu(OAc)_2$ in EtOH at 80 °C provides the tertiary radical **221b**, which cyclizes to the aromatic ring to give **222b** in 61% yield [107]. Enolization of **222b** and kinetic protonation affords norascyronone A (**222a**) in 68% yield. Jia used an oxidative cyclization to construct a bicyclic precursor to glaucocalyxin A. Microwave heating of **223** with $Mn(OAc)_3$ in 1,2-dichloroethane at 125 °C for 3.5 hours gives the vinyl radical **224**, which abstracts a hydrogen atom to give **225** in 53% yield as a mixture of isomers [108]. The yields are lower in the standard solvents for this reaction.

References

1 Snider, B.B. (1996). *Chem. Rev.* 96: 339–363.
2 Melikyan, G.G. (1997). *Org. React.* 49: 427–625.
3 Demir, A.S. and Emrullahoglu, M. (2007). *Curr. Org. Synth.* 4: 321–350.
4 Mondal, M. and Bora, U. (2013). *RSC Adv.* 3: 18716–18754.
5 Nishino, H. (2006). *Top. Heterocycl. Chem.* 6: 39–76.
6 Pan, X.-Q., Zhou, J.-P., and Zhang, W. (2009). *Mol. Diversity* 13: 421–438.
7 Snider, B.B. (2009). *Tetrahedron* 65: 10738–10744.
8 Snider, B.B. and McCarthy, B.A. (1994). *Benign by Design. Alternative Synthetic Design for Pollution Prevention*, American Chemical Society Symposium Series, vol. 577 (eds. P.T. Anastas and C.A. Ferris), 84–97. Washington, D C: American Chemical Society.
9 Merchant, R.R., Oberg, K.M., Lin, Y. et al. (2018). *J. Am. Chem. Soc.* 140: 7462–7465. For a recent example of electrochemical manganese(III)-assisted intermolecular carbon–carbon bond formation, see: Strehl, J. and Hilt, G. (2019). *Org. Lett.* 21: 5259–5263.
10 Hattori, K., Ziadi, A., Itami, K. et al. (2014). *Chem. Commun.* 50: 4105–4107.
11 Snider, B.B., Patricia, J.J., and Kates, S.A. (1988). *J. Org. Chem.* 53: 2137–2143.
12 Curran, D.P., Morgan, T.M., Schwartz, C.E. et al. (1991). *J. Am. Chem. Soc.* 113: 6607–6617.
13 Snider, B.B. and Smith, R.B. (2002). *Tetrahedron* 58: 25–34.
14 Kochi, J.K. (1973). *Free Radicals*, vol. 1 (ed. J.K. Kochi), 591–683. New York: Wiley.
15 Hulcoop, D.G., Sheldrake, H.M., and Burton, J.R. (2004). *Org. Biomol. Chem.* 2: 965–967.
16 Snider, B.B. and Duvall, J.R. (2004). *Org. Lett.* 6: 1265–1268.
17 Kates, S.A., Dombroski, M.A., and Snider, B.B. (1990). *J. Org. Chem.* 55: 2427–2436.
18 Elkin, M., Szewczyk, S.M., Scruse, A.C. et al. (2017). *J. Am. Chem. Soc.* 139: 1790–1793.
19 Jones, P. and Pattenden, G. (1997). *Synlett*: 398–400.

20 (a) Nicolaou, K.C., Jennings, M.P., and Dagneau, P. (2002). *Chem. Commun.*: 2480–2481. (b) Nicolaou, K.C., Ortiz, A., Zhang, H. et al. (2010). *J. Am. Chem. Soc.* 132: 7138–7152.

21 Chobrok, A., Gössinger, E., Grünerger, K. et al. (2007). *Tetrahedron* 60: 8336–8350.

22 Snider, B.B. and Patricia, J.J. (1989). *J. Org. Chem.* 54: 38–46.

23 Alvarez-Manzaneda, E., Chahboun, R., Cabrera, E. et al. (2007). *Synlett*: 2425–2429.

24 Lee, M., Yang, J.-D., Kim, M.-S. et al. (2002). *Bull. Korean Chem. Soc.* 23: 736–740.

25 Taber, D.C. and Nelson, C.G. (2011). *J. Org. Chem.* 76: 1874–1882. For a related cyclization leading to the ABC ring system of hexacyclinic acid see: Toueg, J. and Prunet, J. (2008). *Org. Lett.* 10: 45–48.

26 Crombie, B.S., Smith, C., Varnavas, C.Z. et al. (2001). *J. Chem. Soc., Perkin Trans. 1*: 206–215.

27 Snider, B.B. and McCarthy, B.A. (1993). *J. Org. Chem.* 58: 6217–6223.

28 Powell, L.H., Docherty, P.H., Hulcoop, D.G. et al. (2008). *Chem. Commun.*: 2559–2561.

29 Davies, J.J., Krulle, T.M., and Burton, J.W. (2010). *Org. Lett.* 12: 2738–2741.

30 Ferrara, S.J. and Burton, J.W. (2016). *Chem. Eur. J.* 22: 11597–11600.

31 Logan, A.W.J., Parker, J.S., Hallside, M.S. et al. (2012). *Org. Lett.* 14: 2940–2943.

32 Hulcoop, D.G. and Burton, J.W. (2005). *Chem. Commun.*: 4687–4689.

33 Curry, L., Hallside, M.S., Powell, L.H. et al. (2009). *Tetrahedron* 65: 10882–10892.

34 Sung, K. and Wang, Y.Y. (2003). *J. Org. Chem.* 68: 2771–2778.

35 (a) Swain, N.A., Brown, R.C.D., and Bruton, G. (2002). *Chem. Commun.*: 2042–2043. (b) Swain, N.A., Brown, R.C.D., and Bruton, G. (2004). *J. Org. Chem.* 69: 122–129. For the analogous synthesis of 3-azabicyclo[3.1.0]hexan-2-ones see: Galeazzi, R., Germania, S., Mobbili, G. et al. (1996). *Tetrahedron: Asymmetry* 7: 3573–3584.

36 Chany, A.-C., Marx, L.B., and Burton, J.W. (2015). *Org. Biomol. Chem.* 13: 9190–9193.

37 Martinez, S.A., Gillard, M., Chany, A.-C. et al. (2018). *Tetrahedron* 74: 5012–5021. For another $Mn(OAc)_3$-based synthesis of avenaciolide using different bond connections see: Snider, B.B. and McCarthy, B.A. (1993). *Tetrahedron* 49: 9447–9452.

38 (a) Hamelin, O., Deprés, J.-P., Greene, A.E. et al. (1996). *J. Am. Chem. Soc.* 118: 9992–9993. (b) Brocksom, T.J., Coelho, F., Deprés, J.-P. et al. (2002). *J. Am. Chem. Soc.* 124: 15313–15325.

39 Phillips, E.M., Roberts, J.M., and Scheidt, K.A. (2010). *Org. Lett.* 12: 2830–2833.

40 (a) Ito, Y., Jogo, S., Fukuda, N. et al. (2011). *Synthesis*: 1365–1374. (b) Matsumoto, C., Yasutake, K.-j., and Nishino, H. (2016). *Tetrahedron* 72: 6963–6971.

41 (a) Galeazzi, R., Mobbili, G., and Orena, M. (1996). *Tetrahedron* 52: 1069–1084. (b) Cossy, J., Bouzide, A., and Leblanc, C. (2000). *J. Org. Chem.* 65: 7257–7265.

42 Lin, G.-A. and Chuang, C.-P. (2015). *Tetrahedron* 71: 4795–4800.

43 Chuang, C.-P. and Chen, Y.-J. (2016). *Tetrahedron* 72: 1911–1918.

44 (a) D'Annibale, A.D., Pesce, A., Resta, S. et al. (1997). *Tetrahedron* 53: 13129–13138. (b) Atteni, B., Cerreti, A., D'Annibale, A. et al. (1998). *Tetrahedron* 34: 12029–12038. (c) D'Annibale, A., Nanni, D., Trogolo, C. et al. (2000). *Org. Lett.* 2: 401–402.

45 (a) Davies, D.T., Kapur, N., and Parsons, A.F. (1998). *Tetrahedron Lett.* 39: 4397–4400. (b) Davies, D.T., Kapur, N., and Parsons, A.F. (2000). *Tetrahedron* 56: 3941–3949.

46 (a) Ishibashi, H., Toyao, A., and Takeda, Y. (1999). *Synlett*: 1468–1470. (b) Toyao, A., Chikaoka, S., Takeda, Y. et al. (2001). *Tetrahedron Lett.* 42: 1729–1732. (c) Chikaoka, S., Toyao, A., Ogasawara, M. et al. (2003). *J. Org. Chem.* 68: 312–318.

47 Keane, H.A., Hess, W., and Burton, J.W. (2012). *Chem. Commun.* 48: 6496–6498.

48 Marx, L.B. and Burton, J.W. (2018). *Chem. Eur. J.* 24: 6747–6754.

49 Magolan, J. and Kerr, M.A. (2006). *Org. Lett.* 8: 4561–4564.

50 Magolan, J., Carson, C.A., and Kerr, M.A. (2008). *Org. Lett.* 10: 1437–1440.

51 Curiel Tejeda, J.E., Landschoot, B.K., and Kerr, M.A. (2016). *Org. Lett.* 18: 2142–2145.

52 Zhu, C., Liu, Z., Chen, G. et al. (2015). *Angew. Chem. Int. Ed.* 54: 879–882.

53 Bhat, V., MacKay, J.A., and Rawal, V.H. (2011). *Org. Lett.* 13: 3214–3217.

54 Oisaki, K., Abe, J., and Kanai, M. (2013). *Org. Biomol. Chem.* 11: 4569–4572.

55 Gao, Y., Fan, M., Geng, Q. et al. (2018). *Chem. Eur. J.* 24: 6547–6550.

56 Ji, N., Rosen, B.M., and Myers, A.G. (2004). *Org. Lett.* 6: 4551–4553.

57 (a) Kikue, N., Takahashi, T., and Nishino, H. (2015). *Heterocycles* 90: 540–547. (b) Katayama, S. and Nishino, H. (2019). *Synthesis* 51: 3277–3286.

58 Snider, B.B., Mohan, R., and Kates, S.A. (1985). *J. Org. Chem.* 50: 3661–3663.

59 Yamashita, S., Skuda, N., Hayashi, Y. et al. (2013). *Tetrahedron Lett.* 54: 1389–1391.

60 Snider, B.B. and O'Hare, S.M. (2001). *Synth. Commun.* 31: 3753–3758.

61 Snider, B.B. and Kiselgof, J.Y. (1998). *Tetrahedron* 54: 10641–10648.

62 Suzuki, E., Ueda, M., Ohba, S. et al. (2013). *Tetrahedron Lett.* 54: 1589–1592.

63 (a) Wong, Y.-C., Kao, T.-T., Huang, J.-K. et al. (2014). *Adv. Synth. Catal.* 356: 3025–3038. (b) Kao, T.-T., Lin, C.-C., and Shia, K.-S. (2015). *J. Org. Chem.* 80: 6708–6714.

64 (a) Snider, B.B., Mohan, R., and Kates, S.A. (1987). *Tetrahedron Lett.* 28: 841–844. (b) Dombroski, M.A., Kates, S.A., and Snider, B.B. (1990). *J. Am. Chem. Soc.* 112: 2759–2767.

65 Meng, Z., Yu, H., Li, L. et al. (2015). *Nat. Commun.* 6: 6096.

66 (a) Trotta, A.H. (2015). *Org. Lett.* 17: 3358–3361. (b) Trotta, A.H. (2017). *J. Org. Chem.* 82: 13500–13516.

67 Zoretic, P.A., Fang, H., and Riberio, A.A. (1998). *J. Org. Chem.* 63: 4779–4785.

68 Barrero, A.F., Herrador, M.M., Quílez del Moral, J.F. et al. (2002). *Org. Lett.* 4: 1379–1382.

69 Pettus, T.R.R., Inoue, M., Chen, X.-T. et al. (2000). *J. Am. Chem. Soc.* 122: 6160–6168.

70 Zoretic, P.A., Ramchandani, M., and Caspar, M.L. (1991). *Synth. Commun.* 21: 915–922.

71 Chadhury, S., Li, S., and Donaldson, W.A. (2006). *Chem. Commun.*: 2069–2070.

72 Zoretic, P.A. and Wang, M. (1996). *Synth. Commun.* 26: 2783–2796.

73 (a) Tan, X. and Chen, C. (2006). *Angew. Chem. Int. Ed.* 45: 4345–4348. (b) Wang, X., Ma, Z., Lu, J. et al. (2011). *J. Am. Chem. Soc.* 133: 15350–15333. (c) Wang, X., Wang, X., Tan, X. et al. (2012). *J. Am. Chem. Soc.* 134: 18834–18842.

74 Lee, E., Lim, J.W., Yoon, C.H. et al. (1997). *J. Am. Chem. Soc.* 119: 8391–8392.

75 Mitasev, B. and Porco, J.A. Jr., (2009). *Org. Lett.* 11: 2285–2288.

76 Simpkins, N.S. and Weller, M.D. (2010). *Tetrahedron Lett.* 51: 4823–4826.

77 (a) Pepper, H.P., Lam, H.C., Bloch, W.M. et al. (2012). *Org. Lett.* 14: 5162–5164. (b) Pepper, H.P., Tulip, S.J., Nakano, Y. et al. (2014). *J. Org. Chem.* 79: 2564–2573.

78 (a) He, S., Yang, W., Zhu, L. et al. (2014). *Org. Lett.* 16: 496–499. (b) Yang, W., Cao, J., Zhang, M. et al. (2015). *J. Org. Chem.* 80: 836–846.

79 Lam, H.C., Kuan, K.K.W., and George, J.H. (2014). *Org. Biomol. Chem.* 12: 2519–2522.

80 (a) Yin, J., Wang, C., Kong, L. et al. (2012). *Angew. Chem. Int. Ed.* 51: 7786–7789. (b) Yin, J., Kong, L., Wang, C. et al. (2013). *Chem. Eur. J.* 19: 13040–13046. (c) Yin, J. and Gao, S. (2014). *Synlett* 25: 1–7.

81 (a) Xu, J., Caro-Diaz, E.J.E., Trzoss, L. et al. (2012). *J. Am. Chem. Soc.* 134: 5072–5075. (b) Xu, J., Caro-Diaz, E.J.E., Lacoske, M.H. et al. (2012). *Chem. Sci.* 3: 3378–3386.

82 Zoretic, P.A., Wang, M., Zhang, Y. et al. (1996). *J. Org. Chem.* 61: 1806–1813.

83 Zoretic, P.A., Zhang, Y., Fang, H. et al. (1998). *J. Org. Chem.* 63: 1162–1167.

84 Zoretic, P.A., Zhang, Y., and Riberio, A.A. (1996). *Tetrahedron Lett.* 37: 1751–1754.

85 González, M.A. and Molina-Navarro, S. (2007). *J. Org. Chem.* 72: 7462–7465.

86 (a) Furuta, M., Hanaya, K., Sugai, T. et al. (2014). *Tetrahedron Lett.* 55: 3189–3191. (b) Furuta, M., Hanaya, K., Sugai, T. et al. (2017). *Tetrahedron* 73: 2316–2322.

87 Yamashita, S., Naruko, A., Nakazawa, Y. et al. (2015). *Angew. Chem. Int. Ed.* 54: 8538–8541.

88 Zoretic, P.A., Chen, Z., Zhang, Y. et al. (1996). *Tetrahedron Lett.* 37: 7909–7912.

89 Zoretic, P.A., Fang, H., and Riberio, A.A. (1998). *J. Org. Chem.* 63: 7213–7217.

90 Snider, B.B., Kiselgof, J.Y., and Foxman, B.M. (1998). *J. Org. Chem.* 63: 7945–7952.

91 Yamashita, S., Naruko, A., Yamada, T. et al. (2013). *Chem. Lett.* 42: 220–221.

92 Zhang, Q., Mohan, R.M., Cook, L. et al. (1993). *J. Org. Chem.* 58: 7640–7651.

93 (a) Yang, D., Ye, X.-Y., Xu, M. et al. (1998). *J. Org. Chem.* 63: 6446–6447. (b) Yang, D., Ye, X.-Y., Xu, M. et al. (2000). *J. Am. Chem. Soc.* 122: 1658–1663.

94 (a) Yang, D., Yang, X., Gu, S. et al. (1999). *J. Am. Chem. Soc.* 121: 5579–5580. (b) Yang, D., Ye, X.-Y., and Xu, M. (2000). *J. Org. Chem.* 65: 2208–2217.

95 Yang, D., Xu, M., and Bian, M.-Y. (2001). *Org. Lett.* 3: 111–114.

96 Yang, D. and Xu, M. (2001). *Org. Lett.* 3: 1785–1788.

97 McCarthy Cole, B.A., Han, L., and Snider, B.B. (1996). *J. Org. Chem.* 61: 7832–7847.

98 (a) Kraus, G.A., Nguyen, T.H., and Jeon, I. (2003). *Tetrahedron Lett.* 44: 959–961. (b) Kraus, G.A., Dneprovskaia, E., Nguyen, T.H. et al. (2203). *Tetrahedron* 59: 8975–8978.

99 García Ruano, J.L. and Rumbero, A. (1999). *Tetrahedron: Asymmetry* 10: 4427–4436.

100 O'Neil, S.V., Quickley, C.A., and Snider, B.B. (1997). *J. Org. Chem.* 62: 1970–1975.

101 Snider, B.B. and Kiselgof, E.Y. (1996). *Tetrahedron* 52: 6073–6084.

102 Spangler, J.E., Carson, C.A., and Sorensen, E.J. (2010). *Chem. Sci.* 1: 202–205.

103 Bhowmik, D.R. and Venkateswaran, R.V. (1999). *Tetrahedron Lett.* 40: 7431–7433.

104 Mai, D., Uchenik, D., and Vanderwal, C.D. (2017). *Synlett* 28: 1758–1762.

105 Lee, I.Y.C., Jung, M.H., Lee, H.W. et al. (2002). *Tetrahedron Lett.* 43: 2407–2409.

106 Cao, J., Thor, W., Yang, S. et al. (2019). *Org. Lett.* 21: 4896–4899.

107 Cao, T., Zhu, L., Lan, Y. et al. (2020). *Org. Lett.* 22: 2517–2521.

108 Guo, J., Li, B., Ma, W. et al. (2020). *Angew. Chem. Int. Ed.* 59: 15195–15198.

10

Manganese-Catalyzed Dihydroxylation and Epoxidation of Olefins

Niek N.H.M. Eisink, and Wesley R. Browne

University of Groningen, Stratingh Institute for Chemistry, Faculty of Science and Engineering, Molecular Inorganic Chemistry, Nijenborgh 4 9747 AG Groningen, Netherlands

10.1 Introduction

Epoxides and 1,2-diols are versatile intermediates in organic chemistry [1]. The inherent polarity and strain of epoxides makes them readily undergo stereospecific ring-opening reactions with nucleophiles to form 1,2-difunctionalized compounds [2]. 1,2-Diols are useful intermediates to a range of medicinal compounds; are a common motif frequently found in natural products, such as carbohydrates, polyketides, and alkaloids; and can serve as an intermediate toward $NaIO_4$ alkene cleavage [3–5]. The installation of two possible new stereocenters on a non-functionalized alkene allows for the simple diversification of readily available starting materials. Among the possible transition metal catalyst used for epoxidation and dihydroxylation, manganese complexes play an increasingly important role due to their negligible toxicity and environmental impact and their wide range of stable and chemically accessible oxidation states (II–V and VII). Indeed, manganese-catalyzed oxidations have been the subject of intense research over recent decades as effective oxygen atom transfer (OAT) catalysts [6, 7].

The field of manganese-catalyzed (ep)oxidation reactions was perhaps initiated in efforts to models for the iron-dependent enzyme cytochrome P-450 [8]. These models were typically based on porphyrin ligands and showed reasonable activity and selectivity for epoxidation in comparison to enzymes [9, 10]. Manganese-based epoxidations impacted the field of chemical synthesis with the development of the Jacobsen–Katsuiki enantioselective epoxidation [11–13]. Chiral salen ligand-based manganese complexes achieved already impressive enantioselectivities with useful yields. Due to their widespread use in synthesis, the mechanisms by which these manganese, salen, and porphyrin catalysts engage in oxidation have already received extensive attention. For both ligand systems, it is widely accepted that the oxidation proceeds via the formation of Mn^V–oxo species, which then relays the oxygen to the substrate, Scheme 10.1 [14, 15].

While the formation of the manganese(V)–oxo species has received wide acceptance, the exact mode of oxygen transfer remains a point of debate. The reaction

Manganese Catalysis in Organic Synthesis, First Edition. Edited by Jean-Baptiste Sortais.

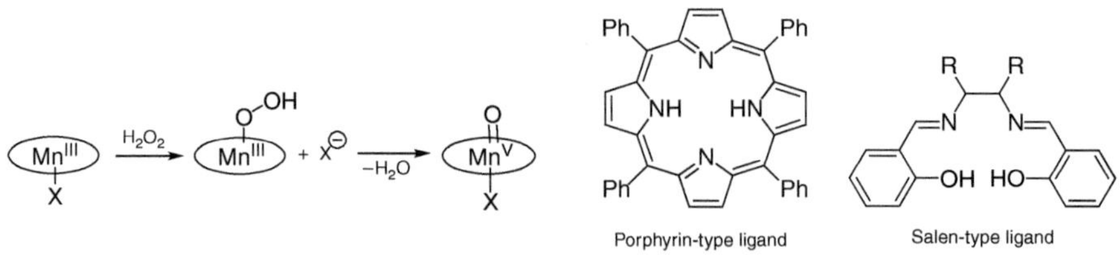

Scheme 10.1 Formation of manganese(V)–oxo species from Mn(III) complexes and an oxidation. Source: Méou et al. [14].

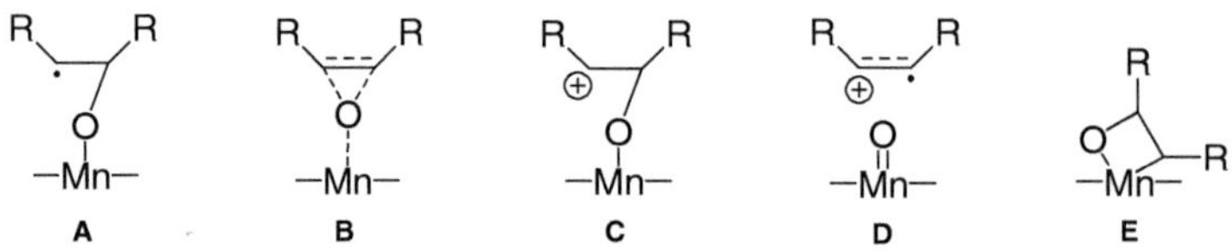

Scheme 10.2 Proposed intermediates for oxygen atom transfer. (A) a carbon radical, (B) a concerted transition state, (C) a carbocation, (D) a π–radical cation, (E) a manganaoxetane Source: Redrawn from Ostovic and Bruice [16].

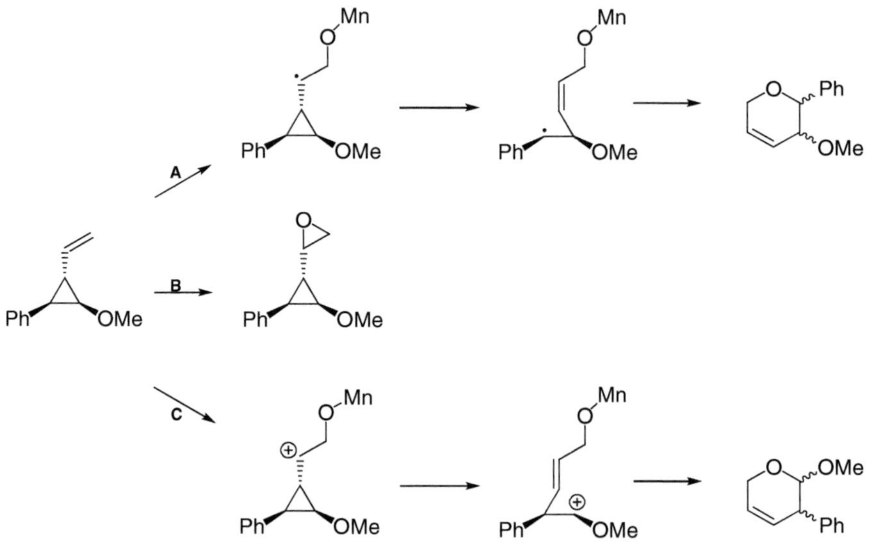

Scheme 10.3 A: Radical mechanism, B: concerted mechanism, and C: cationic mechanism.

could proceed either via (A) a carbon radical, (B) a concerted transition state, (C) a carbocation, (D) a π-radical cation, or (E) a manganaoxetane, Scheme 10.2 [16].

Olefins such as those in Scheme 10.3 were applied to investigate the mode of OAT [17]. Depending on the mode of oxygen transfer, the reaction can yield any of three different products. In path A, the OAT proceeds via the formation of carbon radical. This radical is stabilized most when at the benzylic position following cyclopropane opening. Ring closure yields 3-OMe-2-Ph-dihydropyran. In path C, the reaction proceeds via a carbocation intermediate, which is stabilized best when next to the methoxy group after cyclopropane opening. Ring closure results in the formation of 2-OMe-3-Ph-dihydropyran. While if the reaction is concerted (path B), the epoxide is formed directly retaining the stereo-information. These types of olefins give insight into the mode of oxygen transfer, through product formation, and therefore a definitive conclusion cannot be drawn from them in regard to the nature of the OAT species. Furthermore, due to the instability of the epoxide formed by the concerted reaction (path B), often results can be misinterpreted due to formation of side products (e.g. Lewis acid ring opening of the epoxide would yield similar, if not the same, products as via path C).

In recent years, improvements to known catalytic systems have accompanied the development of novel catalysts. Highly efficient systems employing catalyst loadings as low as 0.01 mol% and/or enantioselective systems with excellent enantiomeric excess achievable for selected substrate classes have been reported. Depending on the system (catalyst, solvent, and oxidant) and substrate, the oxidation of alkene moieties can result in the formation of epoxides, diols, alcohols (allylic oxidation), and even hydroxy ketones (over oxidation of the diol) or dicarbonyl derivatives (oxidative cleavage), Scheme 10.4 [18].

Although catalytic procedures are developed to focus on either epoxidation or dihydroxylation, often an undesired oxidation product is obtained as a minor by-product. It is notable that many systems employ essentially similar ligands and conditions (e.g. N4-ligands); however, the mechanisms proposed can vary significantly. In this chapter, attention will be given to the mechanism proposed for these reactions and their study and basis. One of the major challenges in studying these reactions is that Mn^{II} ions and complexes are essentially spectroscopically "white" or invisible, or their paramagnetism interferes with analysis of reactions by NMR spectroscopies, making detection, let alone quantification, difficult. EPR spectroscopy is the exception and is useful for detecting manganese complexes in many of its oxidation states including Mn^{II}. Both Mn^{III} and Mn^{IV} show intense UV absorption bands assigned to ligand to metal charge transfer transitions as well as characteristic EPR spectra but often form multinuclear complexes in which the paramagnetism is suppressed by coupling. With regard to mechanistic studies, however, perhaps two of the most powerful tools in manganese oxidation catalysis, and indeed in most areas of catalysis, are kinetics and isotopic labeling both for kinetic isotope effect determination and, more importantly, for atom tracking of oxygen. In this chapter we will not discuss all systems by class but instead focus on mechanistic aspects, such as oxidant, additives, the activation of H_2O_2 by high valent species, and enantioselectivity, to showcase the diversity this field already presents.

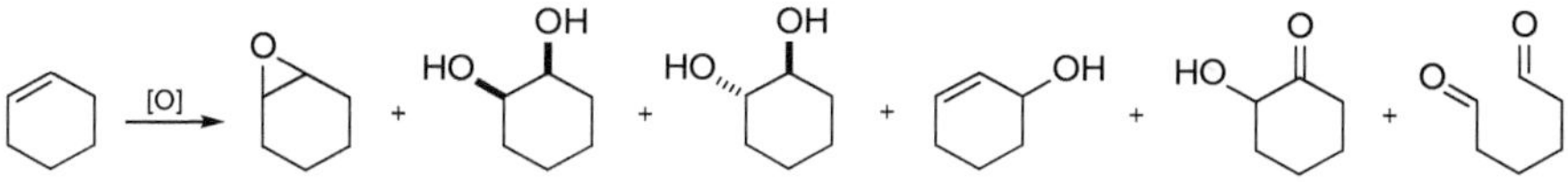

Scheme 10.4 Products obtained in olefin oxidation.

10.2 Oxidant

A wide range of oxidants are applicable in manganese-catalyzed (ep)oxidations, facilitating the tuning of the reactions' conditions and even in some cases influencing selectivity. The choice of oxidant is furthermore influenced by price, active oxygen content, availability, stability, selectivity, and the nature of by-product formed. On this basis oxygen and hydrogen peroxide are the obvious choice since they have high active oxygen contents and produce only water as a by-product, Table 10.1. However, the decomposition of H_2O_2 by disproportionation, inactivation of manganese catalysts, and formation of hydroxyl radicals make either NaOCl or iodosylbenzene the oxidant of choice. Recently, better understanding of catalyst properties have resulted in the application of oxidants such as H_2O_2 and organic peroxides. Molecular oxygen, although in many ways ideal, is used directly much less due to overoxidation (further oxidation of initial products) and side reactions as well as risk of explosions. Reductants such as aldehydes can be added to temper the reactivity of oxygen and allow for milder reaction conditions (temperature, O_2 pressure) to be used, for example, 1 atm. of O_2 in combination with 2 equiv of aldehyde, often isobutyraldehyde [19, 20]. However, its main advantage, that of atom economy, is lost in this way.

10.3 Use of Additives

Manganese salen/porphyrin complexes are perhaps the most widely investigated classes. Manganese-based catalysts are especially well known for the impact of additives (additional components added to improve catalyst performance such as imidazole) on catalyst stability, activity, and enantioselectivity [21–24]. In most cases their impact can be ascribed to them acting as axial substituents; however,

Table 10.1 Active oxygen content per oxidant.

Oxidant	Active oxygen (wt%)	By-product
O_2	50.0	H_2O
H_2O_2 [a)]	47.0	H_2O
NaOCl	21.6	NaCl
CH_3CO_3H	21.1	CH_3CO_2H/CO_2
t-BuOOH	17.8	*t*-BuOH
$KHSO_5$	10.5	$KHSO_4$
MCPBA	9.3	*m*-Cl-$C_6H_4CO_2H$
$NaIO_4$	7.5	$NaIO_3$ (NaI)
PhIO	7.3	PhI

a) Based on 100% H_2O_2.

it is increasingly the case that in non-salen/porphyrin-based catalytic systems, a diverse range of additives can affect catalyst performance. Typically encountered additives are carboxylic acids or their carboxylates. For example, Du et al. proposed that the specific role of carboxylic additives in an N4 ligated manganese-catalyzed reaction was proposed to be that the carboxylic acid residue aids in the formation of manganese–oxo species, Scheme 10.5 [25].

For this to be the case, the manganese complex has to have two free cis-binding sites available. One to bind to peroxide and one for the carboxylic acid. Indeed comparison of the reactivity of manganese complexes based on N4- and N5-type ligands, Scheme 10.6, shows that, for N5 ligands that leave only one coordination site available, carboxylates do not bind to the metal center.

Whereas the N4-complex showed little to moderate conversion and with low enantioselectivity with the peroxides (hydrogen peroxide, *tert*-butyl hydroperoxide,

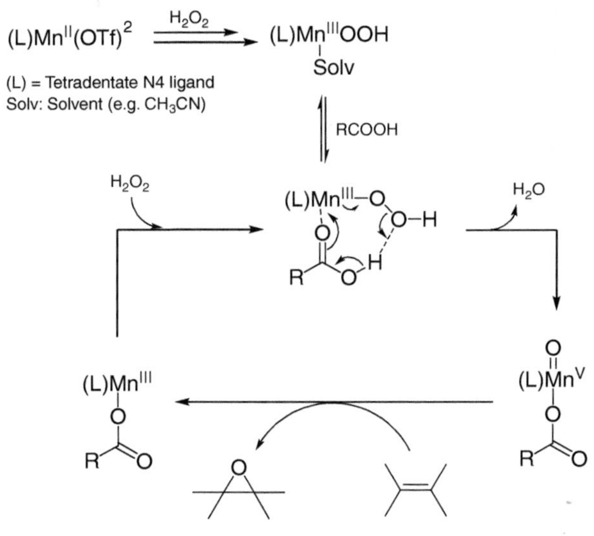

Scheme 10.5 Proposed carboxylic acid-assisted Mn–OOH cleavage. OTf$^-$: $CF_3SO_3^-$. Source: Redrawn from Du et al. [25].

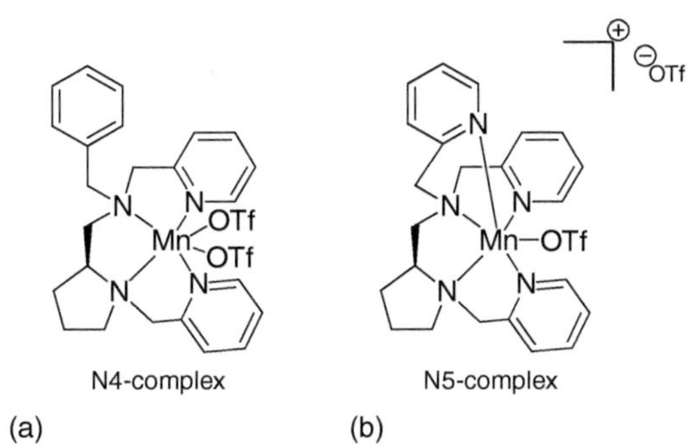

Scheme 10.6 Structure of (a) N4 and (b) N5 complexes.

cumyl hydroperoxide) alone, in the presence of 5 equiv of 2-ethylhexanoic acid (EHA) with respect to the complex, the catalyst provided full conversion with excellent yield. Furthermore, the enantiomeric excess increased to 90% in all three cases. In contrast, with the N5-complex, addition of EHA did not improve conversion with any of the three peroxides, consistent with the formation of the reactive manganese–oxo species requiring a cis-coordinated carboxylate ligand. For the N4-complex, the EHA facilitates O—O bond cleavage to yield the manganese–oxo species, as depicted in Scheme 10.5, while for the N5-complex this does not occur. In the case of iodosylbenzene, which is a single oxygen atom donor and therefore does not require the O—O bond cleavage step, the N4 and the N5 complexes show similar activity in the absence of a carboxylic acid. Interestingly, when the reactions were carried out in the presence of EHA, not only the product yields increased but also the enantiomeric excesses, which the authors ascribed to hydrogen bonding interactions with the manganese–oxo species. The increased conversion was ascribed to the increased solubility of the iodosylbenzene by the formation of phenyliodo dicarboxylate. Indeed with phenylidodo diacetate alone, the reaction outcome was comparable to that of iodosylbenzene in combination with EHA. Furthermore, Brønsted acids such as H_2SO_4 are also effective additives to promote O—O bond cleavage in the Mn-peroxide species [26]. In summary, the carboxylate additives seem to aid in O—O bond cleavage to form the reactive Mn=O species. In the case of peroxy oxidants, the effect of carboxylate additives is clear. However, in the case of single oxygen atom donors such as iodsylbenzene, the addition of carboxylates does not directly aid in the formation of the active species and instead acts as Bronsted acids.

Recently, non-redox active Lewis acids have proven effective in activating manganese-based oxygen transfer reactions, inspired in part by the role of the weak Lewis acid Ca^{2+} in the oxygen evolving complex of photosystem II [27]. In many cases the role of Lewis acids, and in particular Sc^{3+}, is proposed to be to bind to reactive metal–oxo complexes and increase their reactivity. In 2015, Chen et al. [28] demonstrated one of the first examples that non-redox metal ions promoting olefin epoxidation in combination with manganese–oxo species using the Mn(BPMEN)Cl_2 [BPMEN: *N,N′*-dimethyl-*N,N′*-bis(2-pyridylmethyl)-1,2-ethanediamine] [29] catalyst in the epoxidation of cyclooctene in combination with PhI$(OAc)_2$. The reaction proved sluggish and resulted in ~8% conversion after two hours without Lewis acids present. In contrast, in the presence of Lewis acids such as Ca^{2+}, Mg^{2+}, Sc^{3+}, Al^{3+}, Y^{3+}, and Yb^{3+}, full conversion within two hours was observed. A drawback of using Lewis acids is that they also catalyze the opening of the epoxide formed to their corresponding diol, reducing the effective yield of the desired epoxide. One notable exception to this is Sc^{3+}, which provided for full conversion to the epoxide (94% yield) and in the epoxidation of *cis*-stilbene, with Mn(BPMEN)Cl_2 alone a ~1 : 1 mixture of cis- and trans-epoxide was obtained, while in the presence of Sc^{3+} 52% yield of the cis-epoxide with only 3% yield of trans-epoxide with 90% conversion was obtained. EPR spectroscopy demonstrated that the manganese(II) complex Mn(BPMEN)Cl_2 was oxidized by PhI$(OAc)_2$ to provide the characteristic signal of a mixed valent di-μ-oxo-bridged dinuclear $Mn_2^{III,IV}$ core, Scheme 10.7, which was lost upon addition of Sc^{3+} with the appearance of a six-line signal of a mononuclear Mn^{II} ion. With these data, the authors confirmed that Sc^{3+} induces the dissociation of the μ-oxo-bridged dinuclear structure. However, a similar six-line EPR signal was observed after addition of Sc^{3+} directly to Mn(BPMEN)Cl_2. The absence of an EPR signal of a Mn^{IV} species resulting from the $Mn_2{}^{III,IV}$ core breaking up led to the conclusion that this type of species is most likely short-lived, i.e. highly reactive, and it was speculated that a Mn^{IV}=O species was formed, which was further activated by Sc^{3+}, Scheme 10.7. In the absence of Sc^{3+}, the reaction goes via stepwise oxygen transfer, with a carbon radical as a plausible intermediate, which allows for carbon–carbon bond rotation, leading to the observed mixture of cis/trans-epoxides. While in the case of addition of Sc^{3+}, the reaction forms the activated Mn^{IV}=O—Sc^{3+} intermediate that performs a concerted OAT, leading to the high cis-selectivity.

Continuing on the effect of Sc^{3+}, Nodzewska et al. reported its effect on the epoxidation catalyzed by $[Mn_2(\mu\text{-}O)_3\text{-}(TMTACN)_2](PF_6)_2$ [TMTACN: N,N′,N″-trimethyl-1,4,7-triazacyclononae] [30]. In this case, the amount of Sc(OTf)$_3$ applied also proved crucial for the selectivity. Increased amounts led to not only faster product formation but also faster epoxide conversion to the corresponding diol. Resulting in selective epoxidation of styrene within three minutes with c. 100 turnovers, 0.5 mol% in acetonitrile proved to be optimal. In this case, however, although the reactions were significantly accelerated, the cis/trans-selectivity was unaffected. Epoxidation of *cis*-stilbene led to a c. 1 : 1 mixture of cis/trans-epoxide. With the radical clock trans-1-phenyl-1(2-phenylcyclopropyl)-ethylene (similar to that in Scheme 10.3), initially, the epoxide was observed by ^{1}H-NMR spectroscopy in the first minute of the reaction, which was converted to corresponding dihydropyran, Scheme 10.8. The absence of products of cyclopropane ring opening indicates that a radical mechanism is not involved and taken together with the low

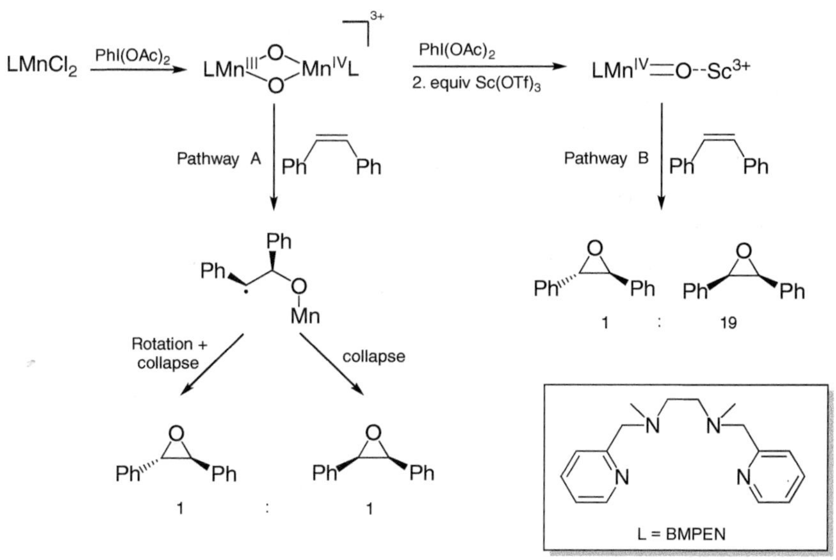

Scheme 10.7 Mechanistically changes upon addition of Sc(OTf)$_3$.

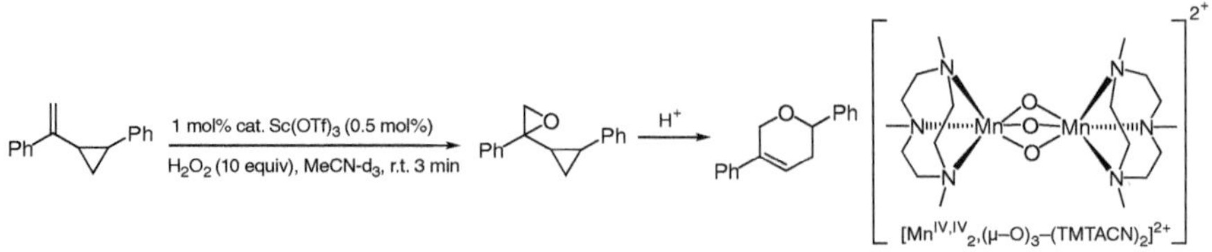

Scheme 10.8 Epoxidation of the radical clock trans-1-phenyl-1(2-phenylcyclopropyl)-ethylene.

stereoselectivity with *cis*-stilbene; it was concluded that the reaction proceeds via a two-step oxygen transfer involving a high valent Mn–oxo species and cationic intermediates.

A caveat to the mechanistic interpretations made in the role of Sc^{3+} and Lewis acids in general is that the soluble Lewis acids available are typically their triflate ions and in reactions where water is present, hydrolysis to yield, e.g. insoluble M(III)(hydr)oxides and carboxylates as well as the Bronsted acid CF_3SO_3H as demonstrated recently by Steen et al. [31] for reactions with the complex $[Mn_2(\mu\text{-}O)_3\text{-}(TMTACN)_2](PF_6)_2$. Indeed, the dependence of observed reactivity correlates well with the tendency of the specific Lewis acid to undergo hydrolysis and the strength of the Bronsted acid formed. As we will discuss in the following text, the key challenge in identifying the origin of effects of additives on catalytic reactivity is to recognize that these same additives often have interesting solution chemistry of their own that should be taken into account also.

10.4 *In Situ* Generated Catalysts

Catalytic oxidation methods based on safe and readily applicable "off the shelf" components (i.e. *in situ* preparation of the catalyst) are desirable for the oxidation for practical, economic, and environmental reasons. In this way the often extensive synthesis of catalyst can be circumvented.

Methods have been developed employing Mn oxides [32], $Mn(ClO_4)_2$ [33], and methods based on the electrochemical formation of H_2O_2 using Mn^{2+} salts [34].

A perhaps classic example is that reported by Lane et al. [35] in which Mn salts were shown to afford epoxidation with H_2O_2. Bicarbonate was proposed to play a crucial role in the epoxidation, and based on earlier reported work by Richardson et al. [36], it was concluded that sodium bicarbonate and hydrogen peroxide form peroxymonocarbonate *in situ*, Scheme 10.9.

Although peroxymonocarbonate can epoxidize alkenes without added catalyst, reaction times are long (one to two days), while with 1.0 mol% of $MnSO_4$, the reactions were complete within 16 hours. Additives were screened to increase conversion and efficiency in hydrogen peroxide, with carboxylic acids showing the best results. With NaOAc in ^{t}BuOH or salicylic acid in dimethylformamide (DMF), the hydrogen peroxide needed could be decreased from 10 to 2–5 equiv combined with an increase in yield. The substrate scope for this system is wide with good to excellent

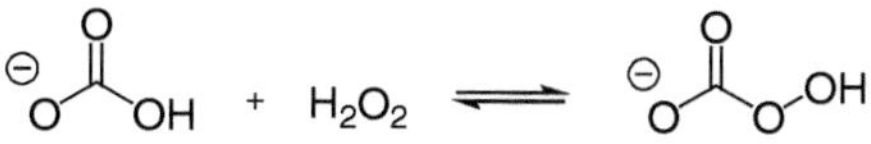

Scheme 10.9 Formation of peroxymonocarbonate.

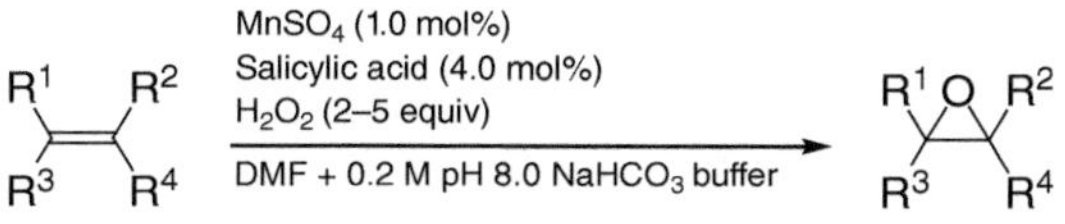

Scheme 10.10 Standard conditions for oxidations with H_2O_2 and $MnSO_4$.

yields, Scheme 10.10. Substrates such as benzylic, cyclic, tri-, and tetra-substituted double bonds all shown to be effectively converted to the desired epoxide. In the case of *cis*-oct-4-ene, the reaction proceeded with good conversion but the reaction was not stereospecific, an ~1 : 1 mixture of E/Z product was obtained in the end. The only substrate that proved problematic was 1-octene, a terminal mono-substituted alkene, which, surprisingly, did not show conversion.

It was proposed that the manganese(II) ions act as a Lewis acid and activate the peroxymonocarbonate to promote epoxidation. However, in EPR spectroscopy revealed that upon addition of hydrogen peroxide, the Mn^{II} signal disappeared and a signal corresponding to a Mn^{IV} appeared, which, together with the observation that *tert*-butyl peroxymonocarbonate is inactive, indicates that the Mn^{II} complex with peroxymonocarbonate, Scheme 10.11, collapses to a Mn^{IV}–oxo species. It is this latter species that acts as the active oxidant for epoxidation.

As discussed previously, although the formation of a high valent manganese–oxo species is reasonable, the exact mode of oxygen transfer to the alkene is not clear. Of the mechanistic pathways considered by the authors, stepwise oxygen delivery via carbon-centered radical intermediate would be most plausible due to the cis/trans-isomerization observed.

An *in situ* prepared catalyst that was reported more recently by Saisaha et al. shows the potential of these systems in terms of activity [37–39]. Complexes based on pyridyl ligands such TPEN [TPTN: *N,N,N′,N′*-tetrakis(2-pyridylmethyl)propane-1,3-diamine] and TPTN [TPEN: *N,N,N′,N′*-tetrakis(2-pyridylmethyl)ethane-1,2-diamine were already known to be effective in iron- and manganese-based alkene and alcohol oxidations [40]. However, these systems were hampered by extensive H_2O_2 disproportionation in the early stages of the reaction and prior to the onset of alkene oxidation. Therefore, typically 8–16 equiv of H_2O_2 were required to compensate for these losses. Intermediates in the synthesis of the TPEN ligands, the so-called aminals, were found to provide identical reactivity including the

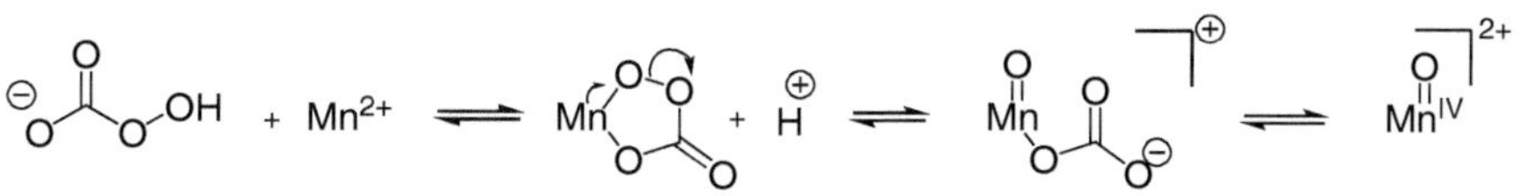

Scheme 10.11 Formation of Mn–oxo from peroxymonocarbonate.

wasteful loss of H_2O_2 and lag time of 20–60 minutes prior to the start of alkene oxidation. The addition of acid, e.g. acetic acid or trichloroacetic acid suppressed catalase-type activity completely, however in these cases conversion of the substrate or even consumption of the H_2O_2 was not observed. A notable observation was that although the lag time varied for the various ligands applied, reactivity and selectivity was invariant. The stability of the ligands under the reaction conditions was determined by ^{1}H-NMR spectroscopy, and in all cases the initial proton signals of the ligand were lost with signals of pyridine-2-carboylxic acid replacing them, Scheme 10.12 [41].

Tellingly, pyridine-2-carboxylic acid together with Mn^{II} salts afforded the same selectivity and conversion as any of the ligands in the oxidation of alkenes but with little or no wasteful decomposition of H_2O_2, and little if any lag period before the onset of alkene oxidation. Indeed whereas 4–8 equiv of H_2O_2 are required to achieve full conversion with the TPEN and aminal ligands, with pyridine-2-carboxylic acid, the efficiency in oxidant is substantially higher with only 1.5–2 equiv of H_2O_2 required for full conversion. These data demonstrated that in all cases picolinic acid (PCA) was being formed *in situ* during the lag period and was responsible for the reactivity observed.

Although, initially, the reaction was reported in acetone, safety concerns of using this solvent with the hydrogen peroxide prompted the search for other solvent systems. It became apparent that the presence of a carbonyl compound was essential to reactivity and that solvent mixtures in which only a small percentage of carbonyl functionality was present were sufficient for activity. Screening of ketones as additives rather than as solvent showed that $CF_3C(O)CH_3$ could be used sub-stoichiometrically and later butanedione was found as an optimal carbonyl additive, with the addition of 0.5 equiv with respect to substrate. In combination with butanedione, the solvent scope was essentially only limited to solvents that did not provide a homogeneous system with aqueous H_2O_2 (e.g. dichloromethane).

In this reaction, sodium acetate is used as an additive most likely to deprotonate the PCA. Other bases such as NaOH, $NaHCO_3$, or Na_2CO_3 all proved to be effective. After optimization of the reaction conditions, the catalyst loading could be reduced to 0.01 mol% in manganese with only 0.5 mol% of PCA, Scheme 10.13.

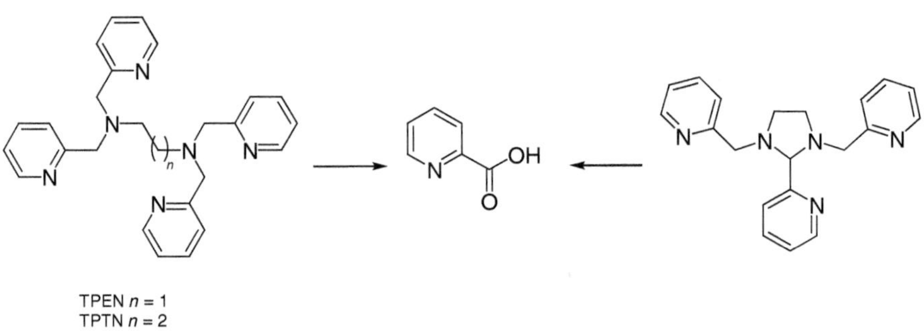

Scheme 10.12 TPEN/TPTN and aminal precursor to pyridine-2-carboylic acid (PCA). Source: Redrawn from Pijper et al. [41].

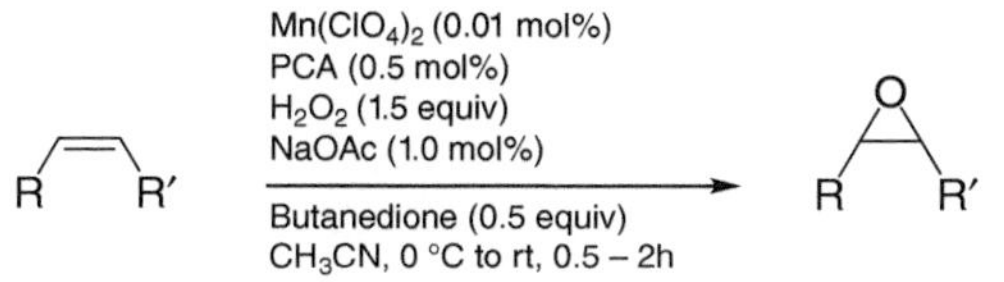

Scheme 10.13 Standard epoxidation with Mn/PCA system.

The system is capable of epoxidizing a wide range of alkenes including benzylic, cyclic, tri-, and tetra-substituted double bonds. Even terminal alkenes were converted with this system albeit with a lower isolated yield. The system shows excellent functional group tolerance due to the fact that the reaction conditions are essentially neutral. Furthermore, oxidation-sensitive moieties are also tolerated such as allylic alcohols and carbonyl groups.

This system showed surprising differences in reactivity dependent on the electronic nature of the alkene species. Whereas electron-rich alkenes show selectivity toward the epoxide product, electronic-deficient alkenes such as diethyl fumarate yielded in many cases exclusively the *syn*-diol product with almost complete conversion and selectivity, Scheme 10.14. In the case of alkenes substituted with only one electron-withdrawing moiety, e.g. α,β-unsaturated esters, low selectivity and a mixture of diol and epoxide was achieved.

UV/Vis absorption, ^{13}C-NMR, and Raman spectroscopy demonstrated that the ketone, either solvent (e.g. acetone) or additive (butanedione), forms a *gem*-hydroxyl-hydroperoxy species (Scheme 10.15) that is essential for the formation of the active catalyst. Although it was observed that the ketones undergo oxidation to their corresponding carboxylic acid, the deliberate addition of acetic acid did not affect the conversion or selectivity of the reaction. *In situ* formation of peracetic acid (PAA) from H_2O_2 and the ketone/acetic acid was excluded.

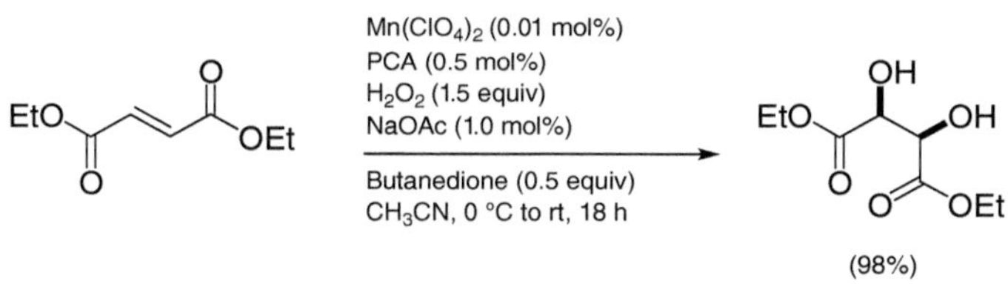

Scheme 10.14 Example of *syn*-dihydroxylation by Mn/PCA.

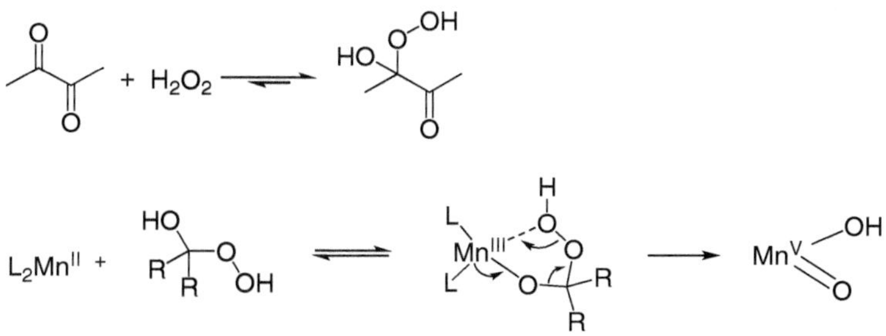

Scheme 10.15 Oxidation of Mn^{II} by 3-hydroxy-3-hydroperoxybutanaone.

While PAA is capable of the epoxidation of cyclooctene, diethylfumarate shows no reactivity. A lag period of c. 10 minutes was observed when low concentrations of Mn^{II} (~0.05 mM) were employed before substrate oxidation commenced. In contrast, UV/Vis absorption and Raman spectroscopy confirmed that upon addition of hydrogen peroxide, butanedione formed an adduct rapidly, and hence the lag period is not due to formation of this intermediate being delayed. Importantly, during the lag period, a part of the butanedione was converted to acetic acid. Indeed, addition of acetic acid at the start of the reaction eliminates the lag time with epoxidation beginning immediately, however, it is not yet clear what role(s) acetic acid plays. Perhaps, as discussed earlier, it aids the formation of the reactive manganese–oxo species. A proposed mechanism based on the formation of Mn(V) –oxo hydroxyl species, which oxidizes the substrate, by reaction of the Mn^{III} with the 3-hydroxy-3-hydroperoxybutanone is shown in Scheme 10.15. The absence of evidence for Mn(II) species by EPR spectroscopy is use to support the involvement of a Mn^{III}/Mn^{V} cycle.

A later report by Moretti et al. showed that a combination of $Mn(OTf)_2$, PCA, and PAA is also effective in epoxidations but with limited or no formation of *syn*-diol products [42]. In this case catalyst loading of 0.4 mol% Mn, 2.0 mol% PCA, and 1.1 equiv PAA (without the need for ketones) were sufficient to obtain high reactivity and selectivity including electron-deficient alkenes such as α,β-unsaturated esters, with full selectivity for the epoxide products. Although similar, the differences in substrate scope and reactivity demonstrate the latter two catalyst systems discussed in this section based on PCA and manganese salts are fundamentally distinct.

10.5 Activation of H_2O_2: Beyond High Valent Manganese Intermediates

Although high-valent manganese species are implied as oxygen transfer agents, it is also possible that the manganese catalyst activates oxidants as a Lewis acid, such as proposed for Mn^{V} imido/nitrido complexes [43, 44]. The complex $(TBP_8Cz)Mn^{V}(O)$ [TBP_8Cz = octakis(4-*tert*-butyl-phenyl)corrolazinato] catalyst (Scheme 10.16) was proposed by Leeladee et al. to engage in alkene oxidation

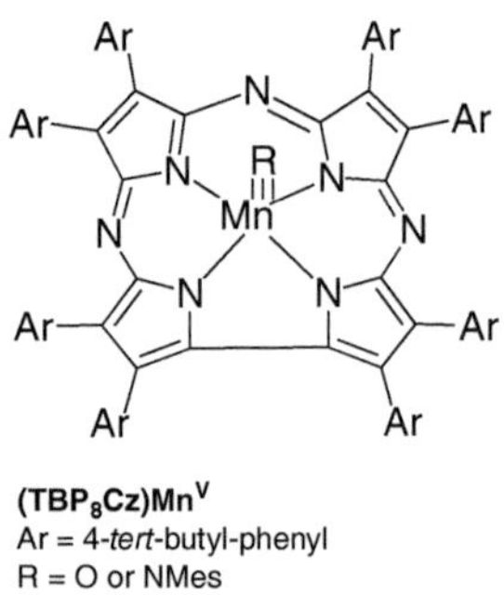

Scheme 10.16 Mn^{V}nitro complex with TBP_8Cz ligand.

by this pathway, in which oxygen transfer from the $Mn^{V}O$ species did not take place but instead via the formation of an activated $Mn^{V}(O)OIPh$. O^{18} labeling of the catalyst $(TBP_8Cz)Mn^{V}(O^{18})$ did not result in O^{18} being present in the epoxide product. However, ^{18}O labeling of PhIO and H_2O did lead to ^{18}O-labeled epoxide [45]. Further evidence in support of this pathway in which the Mn^{V}–oxo species is acting as a "spectator" oxo group was provided by NMes [Mes: $(CH_3)_3C_6H_{2-}$] nitrido variant of the catalyst, which provided the epoxide product while remaining intact.

The Mn^{V}nitro complex $(PPh_4)_2[Mn(N)(CN)_4]$ [43, 46] at one mol% catalyst loading in combination with 1–2 equiv of H_2O_2 provided epoxides from electron-rich olefins in high yields and from terminal alkenes with lower reaction rates and yields. The rate of alkene epoxidation was increased by addition of 1 equiv of acetic acid with less H_2O_2 (1.05 equiv) required for complete conversion of the alkene. Based on DFT calculation, the Mn^{V}nitrido complex is proposed to bind and activate hydrogen peroxide, with epoxidation taking place on the manganese peroxide, Scheme 10.17.

The complex $[Mn^{IV,IV}{}_2,(\mu\text{-}O)_3\text{-}(TMTACN)_2]^{2+}$ is an example of a high valent catalyst that shows an unusually high selectivity toward the *syn*-diol products. The selectivity of the system (epoxide/diol) in alkene oxidation depends heavily on the carboxylic additive added [47]. For example, in the case of oxidation of cyclooctene, 3.0 mol% 2,6-dichlorobenzoic acid shifts the selectivity toward the *syn*-diol product, while addition of 1 mol% of salicylic acid shifts selectivity almost fully to the epoxide product, Scheme 10.18. Such sensitivity of selectivity to the carboxylic acid used is much more marked than described for other catalytic systems (*vide supra*).

The origin of the dependence of selectivity on carboxylic acid used was revealed through a multi-technique spectroscopic study. Based on initial results, the authors already could conclude that the addition of the carboxylic acid resulted in formation of $[Mn^{IV,IV}{}_2,(\mu\text{-}O)(\mu\text{-}RCO_2)_2(TMTACN)_2]^{2+}$ prior to the onset of catalyzed oxidation of alkenes. This formation is also reflected in the time course of the reaction in which a lag time of ~1.5 hour is observed at the end

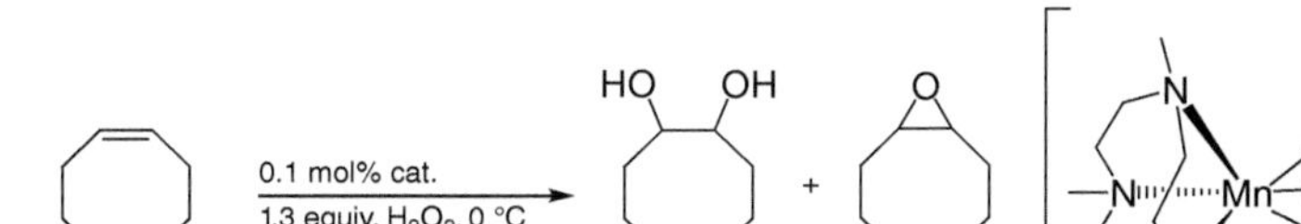

Scheme 10.17 Mechanistic proposal for epoxidation by Mn^{V}nitrido complex.

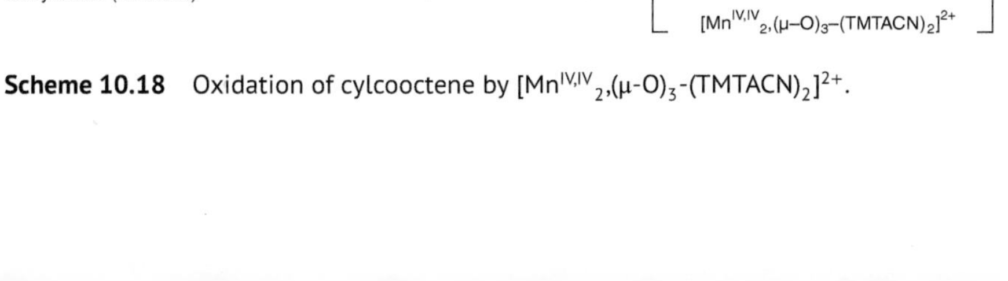

Scheme 10.18 Oxidation of cylcooctene by $[Mn^{IV,IV}{}_2,(\mu\text{-}O)_3\text{-}(TMTACN)_2]^{2+}$.

of which the initial complex is converted to the carboxylate bridged complex $[Mn^{IV,IV}{}_2,(\mu\text{-O})(\mu\text{-RCO}_2)_2(TMTACN)_2]^{2+}$ with several seconds through an autocatalytic process. Spectroscopic and electrochemical studies did not provide evidence for reactive mononuclear Mn^{IV} or Mn^{V} species. Instead the $Mn^{III,III}{}_2$ carboxylato bridged complexes formed, Scheme 10.19, underwent opening of the bridging oxo ligand as a rate-determining step, and therefore it was not possible to detect a reactive intermediate species. Indeed, all potential species (various oxidation states) prepared and characterized underwent rapid conversion to $[Mn^{IV,IV}{}_2,(\mu\text{-O})(\mu\text{-RCO}_2)_2(TMTACN)_2]^{2+}$ under reaction conditions, and it was speculated that after opening of the bridging oxo ligand, one of the hydroxyl groups exchanges with peroxide forming the active oxidant.

Based on O^{18} labeling, it became apparent that for the diol formation, one O originated from hydrogen peroxide and the other O from water. However, in the case of the epoxide product, incorporation depended on the carboxylic acid employed. Indeed there is a direct correlation between the extent of ^{18}O incorporation and the ratio of epoxide/*syn*-diol products. The ratio was also

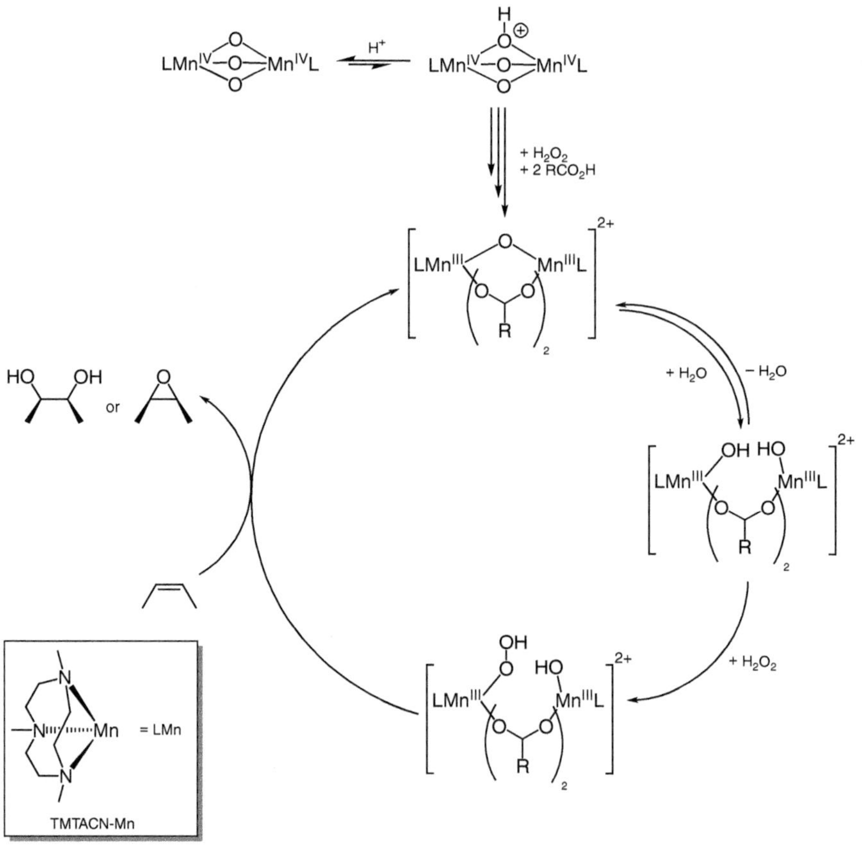

Scheme 10.19 Proposed mechanistic cycle for $[Mn^{IV,IV}{}_2,(\mu\text{-O})_3(TMTACN)_2]^{2+}$.

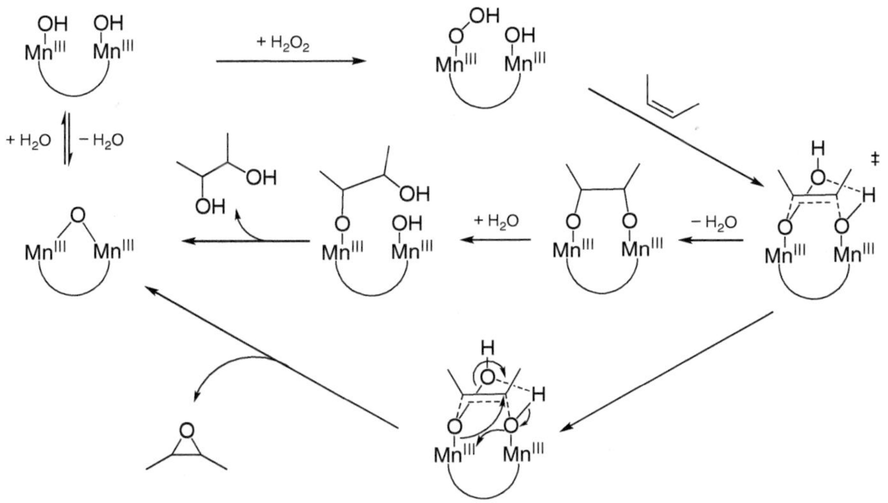

Scheme 10.20 Mode of epoxidation/*cis*-dihydroxylation.

correlated with the steric bulk of the carboxylate ligand with more sterically hindered acid favoring *syn*-dihydroxylation. The reactivity, however, depended on the electron-donating/withdrawing properties of the carboxylate. These data led to the proposal, Scheme 10.20, that following formation of the Mn–OH – Mn–OOH species, reaction with the alkene either to form the diol or the epoxide product takes place. In the case of the diol, the μ-oxo bridged complex has to be removed. After hydrolysis the active complex is recovered.

As shown in the previous example, oxygen-18 labeled compounds are employed to monitor the effect of oxygen incorporation and to elucidate oxidation mechanisms. Another example of such a study was reported by Yin et al. [48]. They reported the epoxidation of various olefins employing $Mn^{II}(Me_2EBC)Cl_2$ [Me_2EBC: 4,11-dimethyl-1,4,8,11-tetraazabicyclo[6.6.2]hexadecane] as the catalyst (Scheme 10.21). Using 1 mol% of catalyst, in combination with H_2O_2 resulted in the conversion of olefins such as cyclooctene, styrene, norbornene, and stilbene in moderate yields. Next to the desired epoxide, depending on substrate, side products such as those resulting from allylic oxidation were also observed. A stable $Mn^{IV}(OH)_2$ complex forms during the reaction, which was isolated and characterized by X-ray crystallography. The $Mn^{IV}(OH)_2$ complex was incapable of

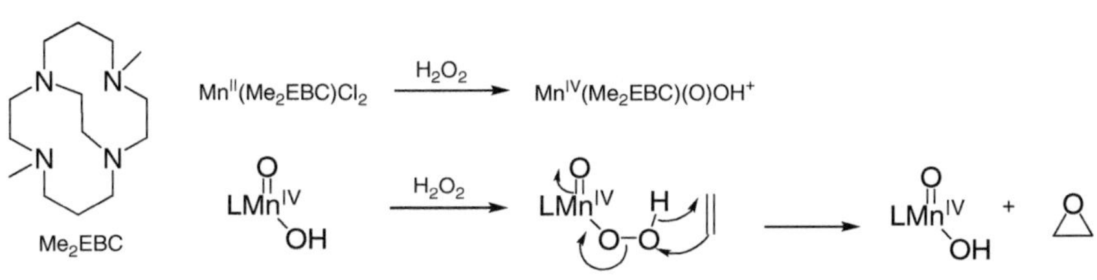

Scheme 10.21 Oxidation of $Mn^{II}(Me_2EBC)Cl_2$.

oxidizing alkenes despite that it formed upon oxidation of the $Mn^{II}(Me_2EBC)Cl_2$ with H_2O_2. Disproportionation to a Mn^{III} and a Mn^{V}–oxo species can be excluded even under basic conditions where the Mn^{IV} complex would be expected to exist as the oxo analogue. ^{18}O labeling with $H_2{}^{18}O$ was proposed to be followed by rapid exchange of the Mn^{IV}–oxo species with ^{18}O and hence if the Mn^{IV}–oxo species was directly involved in OAT, ^{18}O would be incorporated into the epoxide product. However only trace amounts of ^{18}O were observed in the product. In contrast with $H_2{}^{18}O_2$, O^{18} incorporation was observed. Taken together it is plausible that although the Mn^{IV}–oxo species itself does not react with alkenes, it may, as for the previous systems, activate H_2O_2 toward oxidation of substrates either as a Lewis acid or by formation of a permanganic acid species, which was observed tentatively by mass spectrometry, Scheme 10.21.

10.6 Enantioselective Manganese-Based Oxidation Catalysts

Due to the complex nature of the active oxidant in manganese catalysis in combination with a wide range of possible side products, asymmetric epoxidations are especially challenging to control and to optimize. Recently, epoxidations of electron-deficient olefins have received attention, because the resulting optically active epoxides are valuable precursors of many biologically active organic compounds. Especially biologically inspired chiral manganese-based catalysts, mimicking the naturally occurring metalloenzymes, have shown to be quite effective in the enantioselective epoxidation reaction. A number of reported chiral aminopyridines ligands are shown in Scheme 10.22. With these systems, high enantioselectivities and turnover numbers can be achieved in the enantioselective epoxidation of electron-deficient alkenes.

Ottenbacher et al. [49] reported the effect of steric and electronic properties of amino pyridine ligands on their reactivity as well as the nature of active species and the mechanism of oxygen transfer. With ligand L_4 (Scheme 10.22) as their starting point, the steric effects were increased around the pyridine moiety.

Catalyst loadings as low as 0.01–0.02 mol% were already sufficient to yield high conversion within four hours, although increasing the steric bulk from pyridine (**A**, Scheme 10.23) to isoquinolyl (**B**, Scheme 10.23) increased the enantioselectivity slightly, further increases had little effect. However, changing to a phenyl-substituted pyridine (**C**, Scheme 10.23) resulted in lower enantioselectivity.

The authors claimed this effect to be most likely due to the electron-accepting effect of the phenyl moiety prompting further investigation of electronic effects on the enantioselectivities. Four ligands were prepared based on the catalyst reported by Costas and coworkers [55], which showed that electron-donating moieties improved enantioselectivities. Especially the catalyst bearing the amino pyridine moiety (**D**, Scheme 10.24) showed high conversions with excellent enantioselectivities.

Based on their earlier studies, [56] involvement of a $LMn^{V}{=}O$ species was proposed, the formation of which is promoted by carboxylic acids. The presence of a (large) excess of a carboxylic acid can hamper ^{16}O–^{18}O exchange by blocking

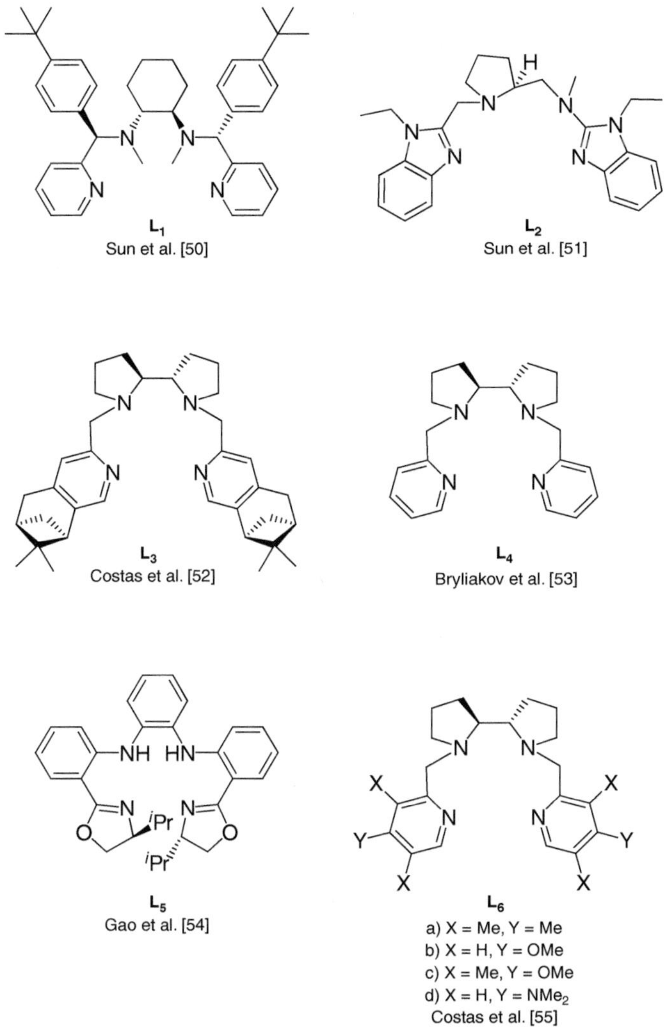

Scheme 10.22 Overview of chiral aminopyridine ligands. Source: Ottenbacher et al. [49]. L_1 [50], L_2 [51], L_3 [52], L_4 [53], L_5[54], and L_6 [55].

coordination; however, carboxylic acids are not essential to see reactivity and form active complexes. The oxidation of styrene showed ^{18}O incorporation in the epoxide and in the diol side product, supporting the initial hypothesis of formation of $LMn^{V}{=}O$, is still a viable option. Depending on the absence of carboxylic acid, the complex either forms a *cis*-HO-$Mn^{V}{=}O$ species or in the presence of carboxylic acid *cis*-RC(O)O-$Mn^{V}{=}O$, see Scheme 10.25.

The oxygen-transferring species was probed using para-substituted chalcones (Scheme 10.20, L_4). Their differences in reactivity and enantioselectivity were subjected to Hammett plot analysis with a $\rho = -1.51$ ($R = 0.995$) indicating that the active oxidant is electrophilic. Based on previously reported ρ values, the authors concluded that the olefin epoxidation proceeds via an electron transfer

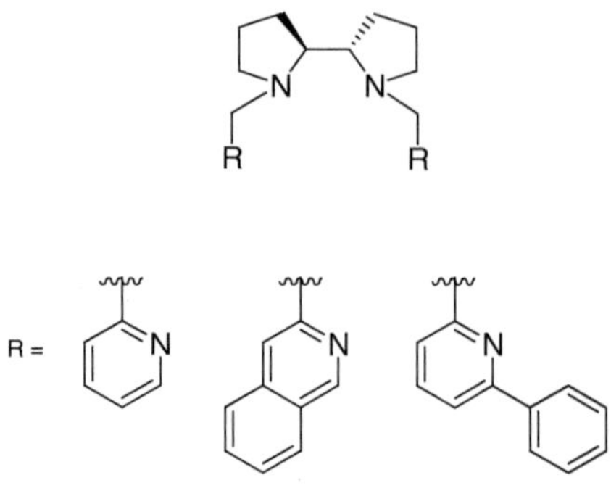

A B C

Scheme 10.23 Increased steric bulk on pyridine moiety.

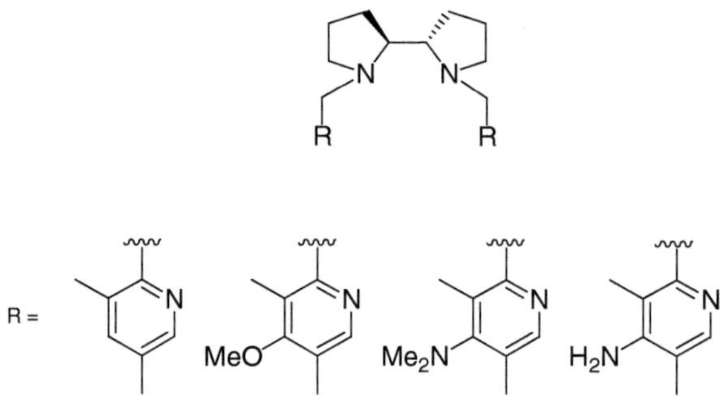

A B C

Scheme 10.24 Electronic-donating groups on the pyridine moiety.

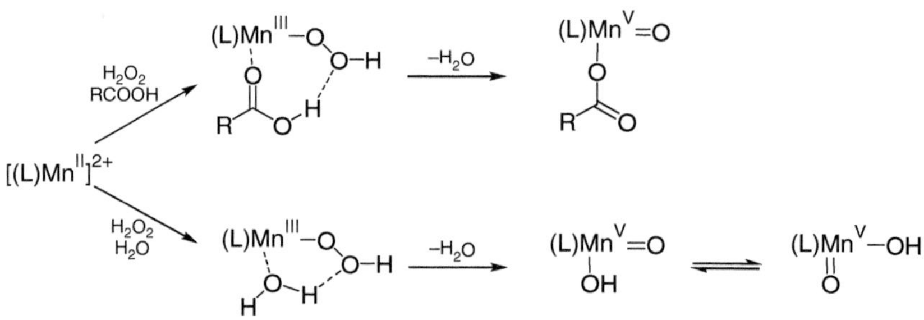

Scheme 10.25 Water and carboxylic acid-mediated mechanism toward Mn^{V}-oxo.

mechanism followed by formation of carbocationic intermediate. In the case of less reactive (electron-deficient) olefins, the transition state should be more product-like and therefore have higher stereocontrol because the resulting highly unstable cationic intermediate undergoes a fast ring closure, to form the corresponding epoxide at high ee. In contrast, for more reactive olefins, the transition state is more reagent-like, with relatively poorer stereocontrol, which leads to weaker enantioselectivity. The proposed OAT is depicted in Scheme 10.26.

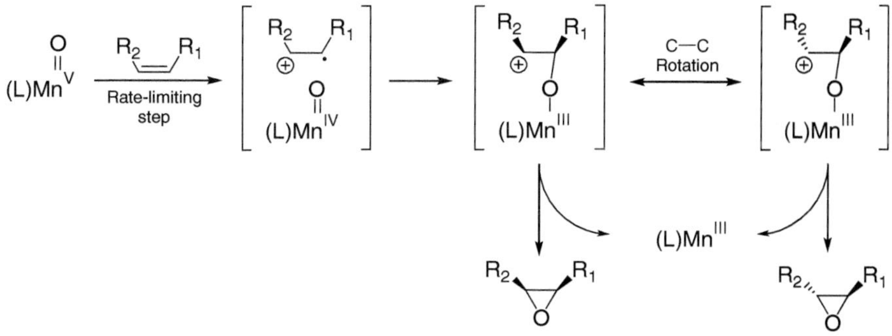

Scheme 10.26 Proposed mode of oxygen atom transfer.

10.7 Conclusions

In conclusion, multiple highly effective systems have been developed for the (enantioselective) epoxidation of various classes of alkenes. A general trend is that terminal alkenes are often more challenging with either no or lower conversion. From a mechanistic perspective, the challenging but rich spectroscopy of manganese complexes has allowed for extensive studies (EPR, O^{18} labeling) and formulation of possible reactive species. However, in many cases the reactivity of these catalysts, although highly desirable, has resulted in very limited success in definitive identification of the species responsible directly for OAT. Discussions of OAT are almost always based on observations after the transfer, cis/trans-isomerization, or product distribution when specific olefins are employed. Direct detection of the mode of oxygen transfer remains a challenge for now. In such studies, one should remain open to the possibility that minor changes to conditions can switch mechanistic pathways and that additives can have multiple effects and have chemistry themselves, for example, in the case of Sc^{3+}. Although the field of manganese-catalyzed epoxidations is well established, the key outstanding challenge is selective dihydroxylation at the level and utility seen for osmium catalysts. However, the scope of systems available and the high enantioselectivities achieved in epoxidation give considerable room for hope for the future development of this field.

References

1 Yudin, A.K. *Aziridines and Epoxides in Organic Synthesis*. Wiley.

2 Fallah-Mehrjardi, M., Kiasat, A.R., and Niknam, K. (2018). *J. Iran. Chem. Soc.* 15: 2033–2081.

3 Kolb, H.C., VanNieuwenhze, M.S., and Sharpless, K.B. (1994). *Chem. Rev.* 94: 2483–2547.

4 Lemieux, R.U. and Von Rudloff, E. (1955). *Can. J. Chem.* 33: 1710–1713.

5 Spannring, P., Yazerski, V.A., Chen, J. et al. (2015). *Eur. J. Inorg. Chem.* 2015: 3462–3466.
6 Talsi, E.P. and Bryliakov, K.P. (2012). *Coord. Chem. Rev.* 256: 1418–1434.
7 Krishnan, K.K., Thomas, A.M., Sindhu, K.S., and Anilkumar, G. (2016). *Tetrahedron* 72: 1–16.
8 de Ortiz, Montellano, P.R. (ed.). *Cytochrome P450*. Boston, MA: Springer US.
9 Collman, J.P., Brauman, J.I., Meunier, B. et al. (1984). *Proc. Natl. Acad. Sci. U.S.A* 81: 3245–3248.
10 Powell, M.F., Pai, E.F., and Bruice, T.C. (1984). *J. Am. Chem. Soc.* 106: 3277–3285.
11 Zhang, W., Loebach, J.L., Wilson, S.R., and Jacobsen, E.N. (1990). *J. Am. Chem. Soc.* 112: 2801–2803.
12 Jacobsen, E.N., Zhang, W., Muci, A.R. et al. (1991). *J. Am. Chem. Soc.* 113: 7063–7064.
13 Irie, R., Noda, K., Ito, Y. et al. (1991). *Tetrahedron: Asymmetry* 2: 481–494.
14 Méou, A., Garcia, M.-A., and Brun, P.J. (1999). *Mol. Catal. A Chem.* 138: 221–226.
15 Razenberg, J.A.S.J., Nolte, R.J.M., and Drenth, W. (1984). *Tetrahedron Lett.* 25: 789–792.
16 Ostovic, D. and Bruice, T.C. (1992). *Acc. Chem. Res.* 25: 314–320.
17 Adam, W., Roschmann, K.J., Saha-Möller, C.R., and Seebach, D. (2002). *J. Am. Chem. Soc.* 124: 5068–5073.
18 Saisaha, P., Pijper, D., van Summeren, R.P. et al. (2010). *Org. Biomol. Chem.* 8: 4444.
19 Yamada, T., Takai, T., Rhode, O., and Mukaiyama, T. (1991). *Bull. Chem. Soc. Jpn.* 64: 2109–2117.
20 Qi, J.Y., Li, Y.M., Zhou, Z.Y. et al. (2005). *Adv. Synth. Catal.* 347: 45–49.
21 Takahashi, A., Yamaki, D., Ikemura, K. et al. (2012). *Inorg. Chem.* 51: 7296–7305.
22 Teixeira, F., Mosquera, R.A., Melo, A. et al. (2014). *J. Phys. Chem. A* 118: 10788–10796.
23 Kurahashi, T., Hada, M., and Fujii, H. (2009). *J. Am. Chem. Soc.* 131: 12394–12405.
24 El-Bahraoui, J., Wiest, O., Feichtinger, D., and Plattner, D.A. (2001). *Angew. Chem. Int. Ed.* 40: 2073–2076.
25 Du, J., Miao, C., Xia, C. et al. (2018). *ACS Catal.* 8: 4528–4538.
26 Miao, C., Wang, B., Wang, Y. et al. (2016). *J. Am. Chem. Soc.* 138: 936–943.
27 Rivalta, I., Brudvig, G.W., and Batista, V.S. (2012). *Curr. Opin. Chem. Biol.* 16: 11–18.
28 Chen, Z., Yang, L., Choe, C. et al. (2015). *Chem. Commun.* 51: 1874–1877.
29 Hureau, C., Blondin, G., Charlot, M.-F. et al. (2005). *Inorg. Chem.* 44: 3669–3683.
30 Nodzewska, A. and Watkinson, M. (2018). *Chem. Commun.* 54: 1461–1464.
31 Steen, J.D., Stepanovic, S., Parvizian, M. et al. (2019). *Inorg. Chem.* 58: 14924–14930.
32 Qi, B., Lou, L.-L., Yu, K. et al. (2011). *Catal. Commun.* 15: 52–55.
33 Ho, K.-P., Wong, W.-L., Lam, K.-M. et al. (2008). *Chem. A Eur. J.* 14: 7988–7996.

34 Ho, K.-P., Chan, T.H., and Wong, K.-Y. (2006). *Green Chem.* 8: 900.

35 Lane, B.S., Vogt, M., DeRose, V.J., and Burgess, K. (2002). *J. Am. Chem. Soc.* 124: 11946–11954.

36 Richardson, D.E., Yao, H., Frank, K.M., and Bennett, D.A. (2000). *J. Am. Chem. Soc.* 122: 1729–1739.

37 Saisaha, P., Pijper, D., van Summeren, R.P.R.P. et al. (2010). *Org. Biomol. Chem.* 8: 4444–4450.

38 Dong, J.J.J.J., Saisaha, P., Meinds, T.G.T.G. et al. (2012). *ACS Catal.* 2: 1087–1096.

39 Saisaha, P., Dong, J.J.J., Meinds, T.G.G. et al. (2016). *ACS Catal.* 6: 3486–3495.

40 Brinksma, J., Feringa, B.L., Hage, R., and Kerschner, J. (2000). *Chem. Commun.*: 537–538.

41 Pijper, D., Saisaha, P., De Boer, J.W.W. et al. (2010). *Dalton Trans.* 39: 10375–10381.

42 Moretti, R.A., Du Bois, J., and Stack, T.D.P. (2016). *Org. Lett.* 18: 2528–2531.

43 Kwong, H.-K., Lo, P.-K., Lau, K.-C., and Lau, T.-C. (2011). *Chem. Commun.* 47: 4273–4275.

44 Leeladee, P. and Goldberg, D.P. (2010). *Inorg. Chem.* 49: 3083–3085.

45 Wang, S.H., Mandimutsira, B.S., Todd, R. et al. (2004). *J. Am. Chem. Soc.* 126: 18–19.

46 Bendix, J., Meyer, K., Weyhermüller, T. et al. (1998). *Inorg. Chem.* 37: 1767–1775.

47 de Boer, J.W., Brinksma, J., Browne, W.R. et al. (2005). *J. Am. Chem. Soc.* 127: 7990–7991.

48 Yin, G., Buchalova, M., Danby, A.M. et al. (2006). *Inorg. Chem.* 45: 3467–3474.

49 Ottenbacher, R.V., Samsonenko, D.G., Talsi, E.P., and Bryliakov, K.P. (2014). *ACS Catal.* 4: 1599–1606.

50 Wu, M., Wang, B., Wang, S. et al. (2009). *Org. Lett.* 11: 3622–3625.

51 Wang, B., Miao, C., Wang, S. et al. (2012). *Chem. Eur. J.* 18: 6750–6753.

52 Garcia-Bosch, I., Gómez, L., Polo, A. et al. (2012). *Adv. Synth. Catal.* 354: 65–70.

53 Ottenbacher, R.V., Bryliakov, K.P., and Talsi, E.P. (2011). *Adv. Synth. Catal.* 353: 885–889.

54 Dai, W., Li, J., Li, G. et al. (2013). *Org. Lett.* 15: 4138–4141.

55 Cussó, O., Garcia-Bosch, I., Font, D. et al. (2013). *Org. Lett.* 15: 6158–6161.

56 Lyakin, O.Y., Ottenbacher, R.V., Bryliakov, K.P., and Talsi, E.P. (2012). *ACS Catal.* 2: 1196–1202.

Index

Manganese Catalysis in Organic Synthesis, First Edition. Edited by Jean-Baptiste Sortais.

q

r

s